Electric Field and Potential

A Concept-Focused Text for JEE Main and Advanced, NEET, Olympiads, and Other Engineering & Medical Entrance Examinations

Each Problem is Solved with a Dedicated Problem-Solving Approach and Numerical Questions Range from Basic to JEE Advanced and Olympiad Level

Sanjay Kumar

M.Tech, PhD

Managing Director
Quanta Classes, Lucknow

Former Senior Faculty of Physics
Jain Classes, Jhansi
Bansal Classes, Madhya Pradesh
TATA Aarambh (Engineering & Medical Simplified), Lucknow

K 423 A Sector K Ashiyana Colony Lucknow

SK Education
Quanta Classes: K-423 A, Sector K, Ashiyana Colony, Lucknow (U.P.)
JEE (Main & Advanced) / NEET / Physics Olympiads / Foundation
📞 +91-9453763058
✉ spphysicsworld@gmail.com

Concepts and Problems in Physics
Title: Electric Field and Potential

Second Edition: 2025
ISBN: 9798411271546

© 2025, Sanjay Kumar
All rights reserved.

No part of this book may be reproduced, stored in a retrieval system, or transmitted in any form or by any means—electronic, mechanical, photocopying, recording, or otherwise—without the prior written permission of the author.

To the aspirants of JEE (Main & Advanced), NEET, and Physics Olympiads

With the hope that this work will inspire
a deeper interest in physics and serve as a reliable guide to its understanding.

This page intentionally left blank

Preface

This physics book is the product of more than twenty years of teaching and innovation in preparing students for JEE Main and Advanced, NEET, Physics Olympiads, and other competitive examinations. Our primary goals in writing this book are:

- To present the fundamental concepts and principles of physics that are essential for success in JEE Main, JEE Advanced, NEET, Physics Olympiads, and other engineering and medical entrance examinations.
- To maintain a balanced emphasis on both quantitative reasoning and conceptual understanding, with particular focus on areas where students commonly face difficulties.
- To systematically develop students' problem-solving skills and enhance their confidence.
- To engage students by incorporating real-world examples that connect physics to everyday experiences.

What's New?

This edition incorporates substantial improvements in structure, content, and teaching methodology. Many elements are newly developed and presented with a strong emphasis on learner-focused explanations. Below are five key highlights:

1. Each concept is explained in accessible, student-friendly language, followed by various types of solved problems. Every solution is structured around a problem-solving approach and includes explanatory discussion.
2. Checkpoint questions are strategically placed within sections, encouraging students to pause and test their understanding before moving forward. Thorough, fully worked-out solutions with a clear problem-solving approach to these checkpoint questions are provided at the end of each chapter, supporting effective self-assessment and independent learning.
3. Special care has been taken with tricky and conceptually dense topics to ensure they can be mastered with clarity and enjoyment.
4. After the theoretical discussions, each chapter concludes with a collection of miscellaneous solved problems that integrate multiple concepts.
5. To further enhance students' understanding, a comprehensive set of conceptual questions, practice problems, and multiple-choice questions—including problems from previous years' JEE Main and Advanced examinations—is provided at the end of each chapter. These exercises span various difficulty levels, denoted by the number of dots: straightforward (basic-level) problems by a single dot (•), intermediate problems by two dots (••), and challenging (advanced-level) problems by three dots (•••). Complete, step-by-step solutions for all end-of-chapter exercises, including hints where required and following a clear problem-solving approach, are provided at the end of each chapter to support independent practice and effective self-assessment.

I hope this book serves as a rigorous and insightful guide for students preparing for JEE Main, JEE Advanced, NEET, Physics Olympiads, and other equivalent engineering and medical entrance examinations. Your feedback, suggestions, and corrections are most welcome and will be gratefully acknowledged in future editions.

Sanjay Kumar
Lucknow, July 2025

Online Physics Classes
by
Dr. Sanjay Kumar

JEE (Main & Advanced) / NEET / Physics Olympiads / Foundation (IX - XII)
Quanta Classes: K 423 A Sector K Ashiyana Colony Kanpur Road Lucknow
Mo. +919453763058
Email: spphysicsworld@gmail.com

Contents

Preface   **vii**

1 Electric Charge and Field   1
- 1.1 Electric Charge- a Property of Matter 1
  - 1.1.1 Electric Fluid Model 1
  - 1.1.2 Atomic Model of Charge 1
- 1.2 Properties of Electric Charge 2
  - 1.2.1 Conservation of Charge 2
  - 1.2.2 Quantization of Charge 3
    - 1.2.2.1 Number of Electrons in 1C of Charge 3
  - 1.2.3 Charge is Always Associated with Mass 4
    - 1.2.3.1 Effect of Charge on the Mass of a Given Object 4
  - 1.2.4 Charge is Transferable 4
  - 1.2.5 An Accelerated Charge Always Radiates Energy 4
- 1.3 Quarks 5
- 1.4 Grounding a Conductor 5
  - 1.4.1 Check Point 1 5
- 1.5 Classification of Solids 6
  - 1.5.1 Conductors 6
  - 1.5.2 Insulators 6
  - 1.5.3 Semiconductors 6
  - 1.5.4 Superconductors 6
- 1.6 Methods of Charging 6
  - 1.6.1 Charging by Friction 6
    - 1.6.1.1 Friction is Used to Charge Insulators 6
    - 1.6.1.2 Charging Conductors by Friction is Very Difficult 6
    - 1.6.1.3 Triboelectric Series 6
  - 1.6.2 Charging by Direct Contact 7
  - 1.6.3 Charging by Induction 7
  - 1.6.4 Charging by Field Emission 8
  - 1.6.5 Charging by Thermionic Emission 8
  - 1.6.6 Charging by Photoelectric Effect 8
- 1.7 Coulomb's Law 8
  - 1.7.0.1 Value of k in CGS Units 9
  - 1.7.1 Experimental Verification of Coulomb's Law 9
  - 1.7.2 Vector Form of Coulomb's Law 10
- 1.8 Superposition Principle 11
- 1.9 Coulomb's Law in a Dielectric Medium 11
  - 1.9.0.1 Key Points Regarding Coulomb's Law 12
- 1.10 Equilibrium of Charged Particles 17
  - 1.10.1 Equilibrium of Three Point Charges 17
  - 1.10.2 Equilibrium of Symmetric Geometrical Point Charged Systems 17
  - 1.10.3 Equilibrium of Suspended Point Charge System 18
- 1.11 Check Point 2 24
- 1.12 The Electric Field 26
- 1.13 The Electric Field of Point Charges 27
  - 1.13.1 The Electric Field Due to a Single Point Charge 27
  - 1.13.2 Electric Field Due To a Group of Point Charges (Superposition of Electrostatic Fields) 27
- 1.14 Electric Dipole 32
  - 1.14.1 Dipole Moment 32
  - 1.14.2 Dipole Moment of a System of Discrete Charges 33
  - 1.14.3 Induced Dipole Moment 33
  - 1.14.4 Electric Field at any Point due to an Electric Dipole 33
    - 1.14.4.1 Electric field at an Axial Point, i.e., in End on Position 33
    - 1.14.4.2 At a Point on its Perpendicular Bisector, i.e. in Broadside on Position 33
    - 1.14.4.3 At any General Point 34
- 1.15 Check Point 3 35

- 1.16 Electric Field of a Continuous Charge Distribution 35
  - 1.16.1 Electric Field Due to a Uniformly Charged Rod of Finite Length at a Perpendicular Distance y 37
    - 1.16.1.1 Electric Field Due to an Infinite Line Charge 38
    - 1.16.1.2 Electric Field of a Finite Line Charge at a Point on its Extension 38
    - 1.16.1.3 Approximating Equations 1.62 and 1.63 on the Symmetry Plane 39
  - 1.16.2 Electric Field at the Centre of a Charged Circular Arc 41
  - 1.16.3 Electric Field at a Point on the Axis of a Thin Charged Ring 42
    - 1.16.3.1 Motion of a Negatively Charged Particle Close to the Centre of the Positively Charged Ring on it's Axis 43
  - 1.16.4 Electric Field on the Axis of a Charged Disc 43
- 1.17 The Shape of Lightning Rods 44
- 1.18 Check Point 4 47
- 1.19 Electric Field Lines 48
- 1.20 Theorem on Circulation of Vector $\vec{E}$ 48
  - 1.20.1 Deduction of Pattern of Field Lines 49
  - 1.20.2 Properties of Electric Field Lines 50
- 1.21 Check Point 5 50
- 1.22 Action of the Electric Field on Charges 51
  - 1.22.1 Motion of a charged Particle in a Uniform Electric Field 51
    - 1.22.1.1 Motion of a Charged Particle Along an Electric Field 51
    - 1.22.1.2 Motion of a Charged Particle Perpendicular to an Electric Field 52
- 1.23 Electric Dipole in an External Electric Field 53
  - 1.23.1 Electric force and torque on electric dipole in a uniform external electric field 53
  - 1.23.2 The Angular Acceleration and Time period of a Dipole in an External Uniform Electric Field 54
  - 1.23.3 A Dipole in a Non-uniform Field 54
- 1.24 Earnshaw's Theorem 57
- 1.25 Check Point 6 58
- 1.26 Conductors in Electrostatic Equilibrium 58
- 1.27 Limitations of Coulomb's Law for a Relativistically Moving Charge 59
  - 1.27.1 Retarded Field 60
- 1.28 Check Point 7 60
- 1.29 Solid Angle 61
- 1.30 Electric Flux 63
  - 1.30.0.1 The Electric Flux Through a Closed Surface 63
- 1.31 Check Point 8 65
- 1.32 Gauss's Law 66
  - 1.32.1 General Integral Form of Gauss's Law 67
  - 1.32.2 Differential Form of Gauss's Law (Optional) 70
- 1.33 Applications of Gauss's law 71
  - 1.33.1 Gauss's Law: Problem Solving Approach 71
  - 1.33.2 Cylindrical Symmetry 73
  - 1.33.3 Spherical Symmetry 76
- 1.34 Check Point 9 79
- 1.35 Conductors in Electrostatic Equilibrium 81
  - 1.35.1 Field in a Substance 81
    - 1.35.1.1 Micro- and Macroscopic Fields 81
    - 1.35.1.2 The Influence of a Substance on a Field 82
  - 1.35.2 Fields Inside and Outside a Conductor 82
    - 1.35.2.1 Inside a Conductor $E = 0$ 82
    - 1.35.2.2 The Field Near a Conductor Surface 83
  - 1.35.3 Mechanical Pressure (or Surface Density of Force) on the Surface of a Charged Conductor 84
  - 1.35.4 Faraday Cage Protection 85
- 1.36 Check Point 10 88
- 1.37 Questions and Exercises 89
  - 1.37.1 Conceptual Questions 89
  - 1.37.2 Problems 90
  - 1.37.3 Multiple Choice Questions 93
    - 1.37.3.1 Level 1 93
    - 1.37.3.2 Level 2 97
    - 1.37.3.3 Level 3 98
    - 1.37.3.4 Level 4 (Previous Years JEE Main & Advanced Questions) 99
- 1.38 Answer Keys and Solutions 109
  - 1.38.1 Check Point 1 109
  - 1.38.2 Check Point 2 110

		1.38.3 Check Point 3 113

- 1.38.3 Check Point 3 . . . 113
- 1.38.4 Check Point 4 . . . 114
- 1.38.5 Check Point 5 . . . 116
- 1.38.6 Check Point 6 . . . 116
- 1.38.7 Check Point 7 . . . 117
- 1.38.8 Check Point 8 . . . 120
- 1.38.9 Check Point 9 . . . 121
- 1.38.10 Check Point 10 . . . 124
- 1.38.11 Conceptual Questions . . . 125
- 1.38.12 Problems . . . 126
- 1.38.13 Multiple Choice Assignments . . . 140
  - 1.38.13.1 Level 1 . . . 140
  - 1.38.13.2 Level 2 . . . 140
  - 1.38.13.3 Level 3 . . . 140
  - 1.38.13.4 Level 4 . . . 140

2 Electric Potential   143

- 2.1 Electric Potential Energy and Potential Difference in Fields . . . 143
  - 2.1.1 Potential Difference in a Uniform Field . . . 146
  - 2.1.2 Electrostatic Potential Energy Difference and Potential Energy of a System of Two Point Charges . . . 147
- 2.2 Check Point 1 . . . 149
- 2.3 Electric Potential Energy of a System of Point Charges: Self Electric Energy . . . 149
  - 2.3.1 Potential Energy of a System of Two Charges in an External Field . . . 152
  - 2.3.2 Potential Energy of a Dipole in a Uniform Electric Field . . . 152
  - 2.3.3 The Electric Potential Due to a Point Charge . . . 153
- 2.4 Equipotential Surfaces . . . 154
  - 2.4.1 Various Equipotential Surfaces . . . 154
    - 2.4.1.1 Equipotential Surfaces For a Point Charge . . . 154
    - 2.4.1.2 Equipotential Surfaces Due to an Electric Dipole . . . 154
    - 2.4.1.3 Equipotential Surfaces Due to Two Identical Positive Charges: . . . 155
    - 2.4.1.4 Equipotential Surfaces of a Charged Wire of Infinite Length . . . 155
    - 2.4.1.5 Uniformly Charged Plane Surface of Infinite Dimensions . . . 155
  - 2.4.2 Properties of Equipotential surface . . . 156
- 2.5 Calculating the Field From the Potential: Potential Gradient . . . 156
- 2.6 Electric Potential For a System of Charges . . . 158
  - 2.6.1 Electric Potential at a Point $P(r, \theta)$ Due to an Electric Dipole . . . 161
- 2.7 Check Point 2 . . . 162
  - 2.7.1 Electric Potential due to a Line charge of Finite Length . . . 164
  - 2.7.2 Potential Due to an Infinite Charged Wire . . . 164
  - 2.7.3 Electric Potential due to a Charged Thin Plane Sheet . . . 164
  - 2.7.4 Electric Potential Due to a Charged Ring . . . 165
  - 2.7.5 Electric Potential Due to a Charged Disc at a Point on it's Geometric Axis . . . 166
  - 2.7.6 Electric Potential Due to a Charged Conducting Shell or Sphere . . . 168
  - 2.7.7 Electric Potential Due to a Non-Conducting Charged Sphere . . . 168
  - 2.7.8 Potential on the Edge of a Uniformly Charged Disc: . . . 169
- 2.8 Equipotentials and Conductors . . . 170
- 2.9 Connected Conducting Spheres . . . 173
- 2.10 Earthing . . . 174
- 2.11 Corona Discharge . . . 174
  - 2.11.1 Electric Potential Energy for Continuous Charge System . . . 175
    - 2.11.1.1 Self (or Internal) Energy of a Uniformly Charged Conducting Spherical Shell or Conducting Sphere 175
    - 2.11.1.2 Self Energy of a Uniformly Charged Non Conducting Solid Sphere . . . 175
- 2.12 The Van de Graaff Generator . . . 176
- 2.13 The Millikan Oil-Drop Experiment . . . 177
- 2.14 Check Point 3 . . . 178
- 2.15 Questions and Exercises . . . 179
  - 2.15.1 Conceptual Questions . . . 179
  - 2.15.2 Problems . . . 181
  - 2.15.3 Multiple Choice Questions . . . 186
    - 2.15.3.1 Level 2 . . . 188
    - 2.15.3.2 Level 3 . . . 189
    - 2.15.3.3 Level 4 (Previous Years JEE Main and Advanced Questions) . . . 190
- 2.16 Answer Keys and Solutions . . . 197
  - 2.16.1 Check Point 1 . . . 197

- 2.16.2 Check Point 2 198
- 2.16.3 Check Point 3 200
- 2.16.4 Conceptual Questions 201
- 2.16.5 Problems 205
- 2.16.6 Multiple Choice Assignments 212
  - 2.16.6.1 Level 1 212
  - 2.16.6.2 Level 2 212
  - 2.16.6.3 Level 3 212
  - 2.16.6.4 Level 4 212

Chapter 1
Electric Charge and Field

Electromagnetism governs virtually all that we see of the physical world. Electromagnetic forces control the structure of atoms and all materials. In this chapter, we introduce electric charge, a property of atomic constituents, and we discuss the fundamental law of the interaction of two charges at rest, Coulomb's law. This force law is as fundamental as the universal law of gravitation. The interaction between charges has the same space dependence as gravitation. To discuss the action at distance, we further introduce the concept of an electric field in terms of Faraday's electric field lines. Gauss' law, which relates electric fields to charges, is developed in terms of these electric field lines. Finally, we apply Gauss' law to a variety of physical situations.

1.1 Electric Charge- a Property of Matter

You have probably seen that on combing hair, on a dry day, hair is attracted to the comb. When we rub objects together such as wool against amber or silk against glass, we find that after rubbing, the objects acquire a property due to which they attract to each other; a silk cloth is attracted to a glass rod that it was rubbed against. The, rubbed objects can also attract other objects. For example, small bits of paper are attracted to a comb that had been rubbed through hair.

When objects behave in this way, they are said to be electrified (from the Greek word elektron meaning amber) or electrically charged.

You can easily electrify your body by vigorously rubbing your shoes on a wool carpet. Evidence of the electric charge on your body can be detected by touching a metal doorknob. Under the right conditions, you will feel a shock when you touch the metal doorknob. (Experiments such as these work best on a dry day because an excessive amount of moisture in the air can cause any charge you build up to "leak" from your body to the Earth.) There are two types of electric charge, as the following simple experiments show. Suppose a hard rubber rod that has been vigorously rubbed on fur is suspended by a string as shown in Figure 1.1. When a glass rod that has been rubbed on silk is brought near the rubber rod, the two attract each other (Fig. 1.1 a). On the other hand, if two charged rubber rods (or two charged glass rods) are brought near each other as shown in Figure 1.1b, the two repel each other. This observation shows that the rubber and glass have two different types of charge on them. Each type of charge repels the same type but attracts the opposite type. That is: **unlike charges attract; like charges repel**.

On the basis of above discussion, we can define electric charge as *a fundamental property of matter that gives rise to electric (more general term "electromagnetic") interactions*. Like mass, it is also a scalar quantity.

1.1.1 Electric Fluid Model

According to Benjamin Franklin all objects are full of an electric fluid. When you bring two objects (for example – a glass rod and a silk cloth) in contact, some electric fluid may transfer from one object to the other. As a result of which, one object gets a surplus of electric fluid and the other has a deficit of electric fluid. Franklin called the object with a surplus of electric fluid as positive or plus, and the object with a deficit of electric fluid as

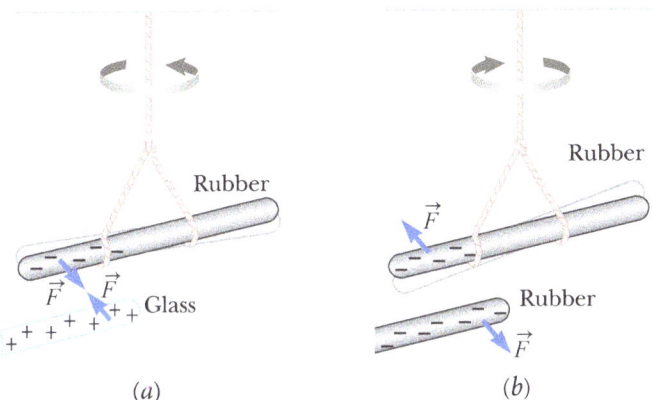

Figure 1.1: The electric force between (a) oppositely charged objects and (b) like-charged objects.

negative or minus. According to Franklin, initially, both of them (the glass rod and the silk cloth) have some amount of electric fluid, and after they are rubbed together, some of the electric fluid is transferred from the silk to the glass. The silk then has a deficit of electric fluid, so Franklin said the silk is negative. The glass has a surplus of electric fluid, so he said it is positive.

Franklin's choice was arbitrary in this model; he assumed the electric fluid was transferred from the silk to the glass. There was no way for Franklin to know whether that was true. Because there was no experimental evidence to determine which way the electric fluid flowed, Franklin could have imagined that electric fluid was transferred from the glass to the silk. Subsequent scientists have kept his arbitrary choice in their model of electric charge. It was only long afterward, with the discovery of electrons and protons, that electrons were found to be attracted to a positively charged glass rod while protons were repelled. Thus by convention electrons have a negative charge and protons have a positive charge. The magnitudes of charge on both the particles are equal.

1.1.2 Atomic Model of Charge

Today we have experimental evidence supporting the model that all objects are made up of atoms, and atoms are made up of three types of particles—neutrons, protons, and electrons. Neutrons are neutral, protons are positively charged and electrons are negatively charged particles. The neutrons and protons are tightly packed into the central region of the atom known as the nucleus. The electrons move rapidly outside the nucleus, forming a cloud. According to this model, when two objects are placed in contact, the electrons can move from atoms of the object in which these are loosely bound to atoms of the object in which electrons are relatively tightly bound. The protons and neutrons are tightly bound in the nucleus, and so they are not transferred.

Therefore, when glass is rubbed against silk, electrons are transferred from the glass to the silk. As a result of which, the glass has a surplus of protons and the silk has a deficit of protons (Fig.1.2). In terms of transferred electrons, we can also say that, the glass has a deficit of electrons and the silk has a surplus of electrons. As electrons and protons are particles of matter. Their motion is governed by Newton's laws. Electrons can move from one object to another when the objects are in contact, but neither electrons nor protons can jump through the air from one object to another. So. an object can not be charged by simply bringing it closer to a charged object.

Charge is represented by the symbol q (or sometimes Q). A macroscopic object, such as a plastic rod, has charge -
$$q = N_p e - N_e e = (N_p - N_e)e \qquad (1.1)$$
where N_p and N_e are the number of protons and electrons contained in the object. An object with an equal number of protons and electrons has no net charge (i.e., $q = 0$) and is said to be electrically neutral.

Thus, neutral does not mean "no charges" but, instead, no net charge.

If $N_e < N_p$, then from Eq.1.1, $q > 0$, i.e., an object is positively charged. Similarly, if $N_e > N_p$, then from Eq.1.1, $q < 0$, i.e., an object is negatively charged.

In practice, objects acquire a positive charge not by gaining protons, but by losing electrons. Protons are extremely tightly bound within the nucleus and cannot be added to or removed from atoms. Electrons, on the other hand, are bound rather loosely and can be removed without great difficulty. The process of removing an electron from the electron cloud of an atom is called ionization. An atom that is missing an electron is called a positive ion. Its net charge is $q = +e$.

Some atoms can accommodate an extra electron and thus become a negative ion with net charge $q = -e$. A saltwater solution is a good example. When table salt (the chemical sodium chloride, NaCl) dissolves, it separates into positive sodium ions Na^+ and negative chlorine ions Cl^-.

In ground state, Na atom has 11 protons and 11 electrons, i.e., $N_p = 11$ and $N_e = 11$. So, net charge on Na atom in ground state-
$$q_{Na} = (N_p - N_e)e = (11 - 11)e = 0$$
Similarly, in ground state of Cl atom, $N_p = 17$, $N_e = 17$, therefore net charge on it in ground state-
$$q_{Cl} = (N_p - N_e)e = (17 - 17)e = 0$$
When 1 electron is transferred from Na to Cl, Na ionizes to Na^+ and Cl ionizes to Cl^-.

In Na^+, we have $N_p = 11$, $N_e = 10$, therefore net charge on Na^+:
$$q_{Na^+} = (N_p - N_e)e = (11 - 10)e = +e$$
In Cl^-, we have $N_p = 17$, $N_e = 18$, therefore net charge on Cl^-:
$$q_{Cl^-} = (N_p - N_e)e = (17 - 18)e = -e$$

☞ Particles that exert electric forces are said to have an electric charge; particles that do not exert electric forces are said to have no electric charge. Thus, charge of a material body or particle is the property (acquired or natural) due to which it produces and experiences electric force. Some of naturally occurring charged particles are electrons, protons, α-particles etc.

So, Electric charge is thought of as the source of electric force, just as mass is the source of gravitational force.

Unit of Charge: Charge is a derived physical quantity and is measured in "coulomb (C)" in SI unit. 1 C is a very large amount of charge, in practice, we use smaller units milli coulomb (mC) (1 mC = 10^{-3}C), micro coulomb (μ C) $\left[1\ \mu C = 10^{-6}C\right]$, nano coulomb (nC) $\left[1\ nC = 10^{-9}C\right]$ and pico coulomb (pC) $\left[1\ pC = 10^{-12}C\right]$ etc.

CGS unit of charge is "electrostatic unit (esu)."
1C = 2997924579.9996 esu $\approx 3 \times 10^9$ esu of charge.
Dimensional formula of charge = $[M^0 L^0 T^1 A^1]$
Magnitude of the smallest known charge is $e = 1.6 \times 10^{-19} C$ (magnitude of charge of one electron or proton).

Note: Practically, mass of a body is always positive * whereas a charge can either be positive or negative. Particles are the sub-

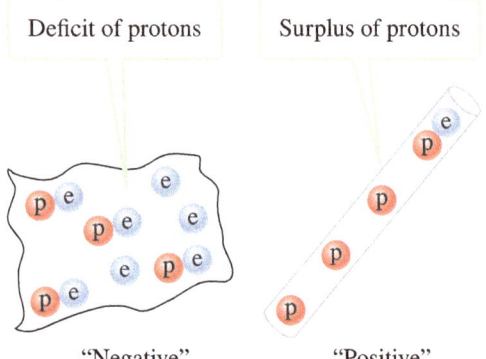

Figure 1.2: Electric charge model

stance and charge happens to be one of their intrinsic properties, just as mass is.

1.2 Properties of Electric Charge

Charging a body implies transfer of charge (electrons) from one body to another. Positively charged body means loss of electrons i.e. deficiency of electrons. Negatively charged body means excess of electrons. Since each electron has mass $m_e \approx 9.1 \times 10^{-31}$ kg. So, excess of electron means excess of mass. Therefore, the mass of a negatively charged body will always be greater than the original mass of the neutral body. A positively charged body means loss of electrons and hence loss of mass. If n electrons are transferred from a body A to body B, then the body A losses its mass by an amount of $\Delta m = nm_e$, whereas body B gains the same amount of mass. Here m_e is the mass of electron.

1.2.1 Conservation of Charge

Experiments show that- *"In an isolated system, total charge (sum of positive and negative charges) remains constant whatever change takes place in that system"*. This law is called **law of conservation of charge**. This law was first put forward by Benjamin Franklin which holds both for large-scale charged bodies and for atoms, nuclei, and elementary particles. No exceptions have ever been found. Thus, law of conservation of charge is a fundamental physical law like- law of conservation of energy, law of conservation of momentum (both linear and angular momentum).

Example of charge conservation occurs when an electron e^- (charge = $-e$) and its antiparticle, the positron e^+ (charge = $+e$), undergo an annihilation process, transforming into two gamma rays (high-energy light):
$$e^- + e^+ \rightarrow \gamma + \gamma \quad \text{(annihilation)} \qquad (1.2)$$
In pair production, the converse of annihilation, charge is also

---

*Theoretically, there is also a concept of $-ve$ mass. If we bring a negative mass near a positive mass, there is a gravitational repulsion between them.

1.2. PROPERTIES OF ELECTRIC CHARGE

conserved. In this process a gamma ray transforms into an electron and a positron:

$$2\gamma \to e^- + e^+ \quad \text{(pair production)} \quad (1.3)$$

In above Equations 1.2 and 1.3, we use two gamma photons in pair production and annihilation to ensure conservation of both energy and momentum. A single photon cannot satisfy both laws simultaneously in these processes.

Note: In above subsection we have seen that charge is conserved for an isolated system. So, unlike mass, the charge on a body does not change, whatever be its speed, however specific charge (q/m) depends on speed as mass depends on speed:

$$m = \frac{m_0}{\sqrt{1 - \frac{v^2}{c^2}}}$$

here m is called the dynamic mass and m_0 the rest mass.

EXAMPLE 1. Sixty four conducting drops each of radius 0.02 m and each carrying a charge of 5 μC are combined to form a bigger drop at constant temperature. Find the ratio of surface charge density of bigger drop to the smaller drop.

APPROACH Since smaller drops are combined at a constant temperature, there will not be any thermal expansion or contraction, and hence the principle of conservation of volume is applicable. By applying it, we can determine a relation between the radii of a smaller drop and the final bigger drop. Charge is also a conserved quantity; therefore, by applying the charge conservation principle, we can find a relation between the charge on a smaller drop and the final bigger drop. Now, the surface charge densities of smaller and bigger drops can be calculated by applying the relation:

$$\sigma = \frac{\text{total surface charge}}{\text{surface area}}$$

After finding the surface charge densities, we can calculate their ratio.

SOLUTION Let R denotes the radius of the combined drop, and r denotes the radius of the smaller drop. Conservation of volume gives:

$$\frac{4}{3}\pi R^3 = 64 \times \frac{4}{3}\pi r^3 \implies R = 4r$$

If q is the charge on each smaller drop, and Q is the charge on newly formed bigger drop, then by principle of conservation of charge, we can write:

$$Q = 64q;$$

If σ_{bigger} and σ_{smaller} denote the surface charge densities of bigger and smaller drops, then ratio of surface charge densities on the bigger and smaller drops

$$\frac{\sigma_{\text{bigger}}}{\sigma_{\text{smaller}}} = \frac{\frac{Q}{4\pi R^2}}{\frac{q}{4\pi r^2}} = \frac{Q}{q} \cdot \frac{r^2}{R^2} = 64\frac{r^2}{16r^2} = 4$$

$$\implies \sigma_{\text{bigger}} : \sigma_{\text{smaller}} = 4 : 1$$

1.2.2 Quantization of Charge

First, let's talk about what quantization means. Quantization is like putting things into specific groups or amounts instead of letting them be in between. It's when something can only be a certain amount, not anything in the middle. Imagine it's like having stairs instead of a smooth ramp.

Suppose, we have a quantity Q and let it is specified in terms of fundamental unit q, then according to quantization rule,

$$Q = Nq \quad (1.4)$$

where N is some integer which shows excess or deficiency of q. For example, if n is the number of excess of q, then $N = +n$ and if it is the deficiency of q, then $N = -n$. So, in terms of n Eq. 1.4 can also be written as:

$$Q = \pm q \quad (1.5)$$

Here, +ve sign shows the excess and −ve sign shows the deficiency of q.

Now, come to the charge. Quantization of charge refers to the observation that electric charge exists only in discrete or quantized amounts. In other words, charge is always found in specific multiples of a fundamental unit of charge. The fundamental unit of charge is the elementary charge, denoted as "e", which is approximately equal to 1.602×10^{-19} coulombs.

According to experimental observations, all charged particles in nature, such as electrons and protons, have charges that are integer multiples of the elementary charge (e). For example, an electron has a charge of "$-1e$", while a proton has a charge of "$+1e$". There are no particles found with charges that are fractions or arbitrary values of the elementary charge. This elementary charge can also be called "the quantum of charge".

Other quantized quantities are energy and angular momentum. The quantum of energy is $h\nu$ (i.e., photon) and that of angular momentum is $h/2\pi$. The quantum of mass not known till date.

The quantization of charge was first suggested by the experimental laws of electrolysis discovered by English experimentalist Faraday. In 1912, it was experimentally demonstrated by physicist Robert Millikan in his oil drop experiment. The experiment demonstrated that the charge on an object could only change in discrete steps, corresponding to the presence or absence of whole numbers of elementary charges. For example, if Q is the charge on object, then

$$Q = Ne \quad (1.6)$$

Here, N is the integer which represents excess or deficiency of fundamental unit of charge e.

If there is an excess of n protons on an object, then $N = n$. Therefore from Eq.1.4, the excess charge on object becomes:

$$Q = +ne \quad (1.7)$$

and for deficiency of n protons $N = -n$, therefore, Eq.1.4 takes form:

$$Q = -ne \quad (1.8)$$

According to atomic model, these are electrons not protons which move from one atom to other. So, as for as charge is concerned, the excess of electrons is equivalent to deficiency of protons and deficiency of electrons is equivalent to excess of protons. Therefore, above results can be generalized as:

$$Q = \pm ne \quad (1.9)$$

in which e, the elementary charge, has the approximate value $1.602 \times 10^{-19}C$ and $n = 0, 1, 2, ...$
+ve sign is used for deficiency and negative sign is used for excess of electrons.

☞ No unit of charge smaller than e has been detected on a free particle; current theories, however, propose the existence of particles called quarks (the constituent particles of protons and neutrons) having charges $\pm e/3$ and $\pm 2e/3$. Although there is considerable experimental evidence for such particles inside nuclear matter, free quarks have never been detected. For this reasons, we do not take their charges to be the elementary charge.

1.2.2.1 Number of Electrons in 1C of Charge

Here, we will find the number of deficient electrons in 1 C of positive charge. By quantization of charge (Eq.1.6), we have

$$Q = Ne \quad \text{or} \quad N = \frac{Q}{e}$$

Here, $Q = 1$ C, $e \approx 1.6 \times 10^{-19}C$, $N = ?$
Substituting these values in above equation and simplifying for N:

$$N = \frac{Q}{e} = \frac{1C}{1.6 \times 10^{-19}C} = 6.25 \times 10^{18}$$

Therefore, the number deficient electrons in 1C charge is

6.25×10^{18}, which is a huge number. Thus, the step size is very small as compared to the charges usually found on many cases. At macroscopic level, we deal with charges that are enormous compared to the magnitude of minimum charge i.e. e ($1.6 \times 10^{-19} C$). In this case, the increase or decrease in units of e is not very different from saying that charges are continuous. So, just for simplicity, at macroscopic level we can ignore the quantization of electric charge.

☞ As $1C$ charge contains 6.25×10^{18} electrons, which is a huge number. So "1 coulomb" is a huge amount of charge.

☞ We sometimes speak of the fundamental unit of charge e as the charge of an electron but e is a positive quantity. The charge of an electron, properly, is $-e$.

1.2.3 Charge is Always Associated with Mass

Every charged particle, such as electrons and protons, has mass. Mass is a property of matter that tells us how heavy or bulky something is. When particles have a charge, they also have a mass. charge cannot exist without mass, though mass can exist without charge. Particles such as neutrino or photon have no rest mass, so they cannot have any charge.

Reason: Particles interact with a special field called the Higgs field*, which is present throughout space. This interaction with the Higgs field gives particles their mass. The strength of this interaction depends on the particle's charge. So, in simple terms, charged particles have mass because they interact with the Higgs field, which gives them their weight or "heaviness".

1.2.3.1 Effect of Charge on the Mass of a Given Object

Let M_0 be the mass of a neutral object, m be the mass of an electron and $-e$ be the charge on an electron. If the object is given a positive charge Q by taking away $N = Q/e$ electrons from it, then the mass of the positively charged object will be-
$$M_{pos} = M_0 - mN = M_0 - mQ/e$$

Now, if another similar object is given a negative charge $-Q$ by putting $N = Q/e$ electrons on it, then, the mass of the negatively charged object-
$$M_{neg} = M_0 + mN = M_0 + mQ/e$$

☞ Here, you may wonder why the mass of an electron changes with velocity according to relativistic equation $m = m_0/\sqrt{1 - \frac{v^2}{c^2}}$, then why not it's charge?
The answer is that, the charge is a fundamental property of matter which is a quantum mechanical effect whereas the mass is a relativistic effect. These two effects are independent of each other and they do not effect each other.

1.2.4 Charge is Transferable

Charge is transferable from one charged body to another body which may be charged or uncharged, if they are put in contact. The process of charge transfer is called conduction. Whole of the charge cannot be transferred by conduction from one body to another except in case when a charged body is enclosed by a conducting body and connected to it (it will be discussed later in this chapter).

1.2.5 An Accelerated Charge Always Radiates Energy

Accelerated charge always radiates energy, which is a fundamental principle in classical electrodynamics. According to Maxwell's equations, when an electric charge undergoes acceleration, it emits electromagnetic radiation. This phenomenon is known as bremsstrahlung, which is a German term meaning "braking radiation".

When a charged particle accelerates or changes its velocity, it creates a changing electric field and a changing magnetic field around it. These changing fields propagate as electromagnetic waves, carrying energy away from the accelerating charge. This energy loss manifests as radiation.

The emitted radiation can have various forms, depending on the nature of the acceleration and the properties of the charged particle. For example, when an electron accelerates or decelerates in the vicinity of an atomic nucleus, it emits X-rays or gamma rays. In contrast, when a charged particle moves in a circular path, such as in a particle accelerator, it radiates synchrotron radiation.

EXAMPLE 2. Is it possible for a body to have an electric charge of $2.0 \times 10^{-19} C$? $3.2 \times 10^{-19} C$?
(A) Yes; yes (B) Yes; no
(C) No; yes (D) No; no

APPROACH According to quantization of charge, the net charge on any charged object is given by-
$$Q = \pm ne$$
Here, n is an integer value. Substitute given values of Q and $e = 1.6 \times 10^{-19} C$ and solve for n. Acceptable values of n are only the integer one, i.e., $n = 0, 1, 2, \ldots$

SOLUTION (C) In first case, $2.0 \times 10^{-19} C$, therefore-
$$n = \frac{2.0 \times 10^{-19} C}{1.6 \times 10^{-19} C} = \frac{2}{3}$$
Since, $n = 2/3$, is a fraction which is not acceptable.
In second case, $Q = 3.2 \times 10^{-19} C$, therefore-
$$n = \frac{Q}{e} = \frac{3.2 \times 10^{-19} C}{1.6 \times 10^{-19} C} = 2$$
Since, $n = 2$ is an integer value so it is acceptable.
Thus the first charge given is impossible, while the second charge given, twice the fundamental charge, is possible.

EXAMPLE 3. How much negative charge and how much positive charge are there on the electrons and the protons in a cup of water (0.25 kg)?

APPROACH To find total negative and positive charge in 250 g of water, we first calculate total number of electrons and protons in it. Then by using the expression, $Q = -ne$ we can find the total negative charge in it. $Q = +ne$ will give total positive charge in it.

SOLUTION The "molecular mass" of water is 18 g; therefore, number of moles in 250 g of water = 250/18 moles.
Since, each mole has N_A, i.e., 6.02×10^{23} molecules, therefore, number of molecules in 250/18 moles of water
$$= (250/18) \times 6.0 \times 10^{23}$$
Each molecule consists of two hydrogen atoms (one electron for each one) and one oxygen atom (eight electrons). Thus, there are 10 electrons in each molecule. Therefore, total number of electrons in 250 g of water-
$$n = (250/18) \times 6.0 \times 10^{23} \times 10$$
Total negative charge on 250 g of water,
$$Q = -ne = -(250/18) \times 6.02 \times 10^{23} \times 10 \, e$$
$$= -(250/18) \left(6.02 \times 10^{23}\right) \times 10 \times \left(1.60 \times 10^{-19} C\right)$$
$$= -1.3 \times 10^7 C$$

The positive charge on the protons is the opposite of this. You can also calculate it by using $Q = +ne$.

EXAMPLE 4. We have an iron piece of mass 56 milligram (Atomic number = 26, molar mass = 56 gram) if $10^{-6}\%$ times of

---
*The Higgs field is a special field that fills all of space. It's like an invisible substance that exists everywhere. When particles interact with this field, they gain mass. It's similar to how a ball moving through molasses (thick liquid) feels heavier and harder to push. The Higgs field slows down particles, giving them their mass or "heaviness". The discovery of the Higgs boson particle in 2012 confirmed the existence of this field and how it gives mass to other particles.

1.3. QUARKS

total electron are removed from this body then calculate charge appear on this body.
SOLUTION Number of moles
$$= \frac{\text{mass of substance}}{\text{molar mass}} = \frac{56 \times 10^{-3}}{56} = 10^{-3} \text{ moles}$$
Number of atom = No. of moles × Avogadro No.
$$= 10^{-3} \times 6.022 \times 10^{23} \approx 6 \times 10^{20}$$
Therefore, number of electrons in 56 milligram of iron,
$$= \text{Atomic number} \times \text{No. of atoms}$$
$$= 26 \times \left(6 \times 10^{20}\right)$$
So, number of electron removed $n = 26 \times 6 \times 10^{20} \times 10^{-8}$
Charge on body, $Q = ne$
$$= 156 \times 10^{12} \times 1.6 \times 10^{-19} \, C = 25 \, \mu C$$
Since, electrons are removed, therefore the nature of charge on the iron piece will be positive.

1.3 Quarks

Quarks are truly elementary particles that carry electric charges which are fractions of the elementary charge $\left(+\frac{2}{3}e \text{ or } -\frac{1}{3}e\right)$.

There are six types of quarks (referred to as six flavours of quarks) and these are: (i) up (u), (ii) charmed (c), (iii) top or truth (t), all having charge $+(2/3)e$, and (iv) down (d), (v) sideways or strange (s) (vi) bottom or beauty (b), all having charge $-(1/3)e$ (see Table 1.1).
The up, charmed, and top quarks have electric charges of $+2/3$ that of the proton, whereas the down, strange, and bottom quarks have charges of $-1/3$ that of the proton.

Table 1.1: Quarks and their charges

Quark	Symbol	Charge
Up	u	$+(2/3)e$
Down	d	$-(1/3)e$
Strange	s	$-(1/3)e$
Charmed	c	$+(2/3)e$
Bottom	b	$-(1/3)e$
Top	t	$+(2/3)e$

Two kinds (or flavors) of quarks, up (u) and down (d), are sufficient to describe normal matter. The electric charges are $+\frac{2}{3}e$ and $-\frac{1}{3}e$ for the u and d quarks, respectively. As shown in Fig.1.3, the proton consists of two u and one d valence quarks, and the neutron of one u and two d valence quarks:
$$p = +\frac{2e}{3} + \frac{2e}{3} - \frac{e}{3} = +e \quad \text{i.e., uud}$$

(a) Proton (uud), $q = +e$ (b) Neutron (udd), $q = 0$
Figure 1.3: Quark composition of proton and neutron

$$n = +\frac{2e}{3} - \frac{e}{3} - \frac{e}{3} = 0 \quad \text{i.e., udd}$$

There is firm experimental evidence of the existence of all six quarks and their six antiquarks within the nucleus, but free quarks have not been detected yet. Current theory implies that direct detection of quarks may, in principle, be impossible.

1.4 Grounding a Conductor

Earth is a conductor because of the presence of ions and moisture in it. It is also large enough that for many purposes it can be considered as a limitless reservoir of charge (i.e., the addition or subtraction of electrons has a negligible effect on it). So, the ground remains essentially neutral at all times. Grounding a conductor means to provide a conducting path between it and the Earth. The copper rod in Figure1.6 is connected to the ground through your body. When something is connected to the ground by a conductor, we say that it is grounded. Every building (with a contemporary electrical system) has a wire connected between the electrical system and a copper pipe. The copper pipe is connected through other copper pipes to the Earth. This ensures that the third connection on electrical sockets is grounded. The connection to the ground keeps electrical appliances from building up excess charge.

1.4.1 Check Point 1

1. •If you charge a pocket comb by rubbing it with a silk scarf, how can you determine if the comb is positively or negatively charged?
2. •Why does a shirt or blouse taken from a clothes dryer sometimes cling to your body?
3. •Explain why fog or rain droplets tend to form around ions or electrons in the air.
4. •A positively charged rod is brought close to a neutral piece of paper, which it attracts. Draw a diagram showing the separation of charge in the paper, and explain why attraction occurs.
5. •Why does a plastic ruler that has been rubbed with a cloth have the ability to pick up small pieces of paper? Why is this difficult to do on a humid day?
6. •Contrast the net charge on a conductor to the "free charges" in the conductor.
7. •Figures 1.4 (a) and (b) show how a charged rod placed near an uncharged metal object can attract (or repel) electrons. There are a great many electrons in the metal, yet only some of them move as shown. Why not all of them?

(b) Inducing a charge on an object connected to ground.

Figure 1.4: Comparison of Ben Franklin's model and our contemporary model of electric charge.

8. •The fact that the electron has a negative charge and the proton has a positive charge is due to a convention established by Benjamin Franklin. Would there have been any significant consequences if Franklin had chosen the opposite convention? Is there any advantage to naming charges plus and minus as opposed to, say, A and B?
9. •Small bits of paper are attracted to an electrically charged comb, but as soon as they touch the comb they are strongly

repelled. Explain this behaviour.
10. • Find the total electric charge of 2.5 kg of (a) electrons and (b) protons. Given: mass of an electron $\approx 9.1 \times 10^{-31}$ kg, mass of a proton $\approx 1.67 \times 10^{-27}$ kg.
11. •A container holds a gas consisting of 2.85 moles of oxygen molecules. One in a million of these molecules has lost a single electron. What is the net charge of the gas?
12. •A copper penny ($Z = 29$) has a mass of 3.10 grams. What is the total charge of all the electrons in the penny? Given that molar mass of Cu is 63.5 gram.

1.5 Classification of Solids

Solids can be classified in either of the following four categories:
(i) Conductors (ii) Insulators
(iii) Semiconductors (iv) Superconductors

1.5.1 Conductors
These are the materials that allow flow of charge through them. Examples include metals (such as copper in common lamp wire), tap water, and the human body (which is similar to tap water). This category generally comprises of metal but may sometimes contain non-metals too. (Ex. Carbon in form of graphite).

1.5.2 Insulators
These are the materials which do not allow movement of charge through them. Examples include rubber (such as the insulation on common lamp wire), plastic, glass, and chemically pure water.

1.5.3 Semiconductors
These are materials that are intermediate between conductors and insulators; Examples include silicon and germanium in computer chips.

1.5.4 Superconductors
These are materials that are perfect conductors, allowing charge to move without any hindrance.
In this chapter we discuss only conductors and insulators.

1.6 Methods of Charging

1.6.1 Charging by Friction
Charging by friction is the oldest form of charging. It was found that when an amber rod is rubbed with fur, the rod became negatively charged. The two bodies acquire opposite signs of electricity; one gets positively charged, while the other becomes negatively charged. When two bodies are charged by friction, they acquire the same magnitude of charge. Furthermore, the bodies retain these excess charges even when they are separated from each other.

Rubbing causes a transfer of electrons from one object to another. If some electrons are transferred from an object A to another object B, then A becomes positively charged and B negatively charged.

Note: Electrons are matter particles. Mass of electron is $m_e = 9.1 \times 10^{-31}$ kg. So, charging involves transformation of mass.

1.6.1.1 Friction is Used to Charge Insulators
The electrons in some materials—such as rubber—do not move freely. Such materials are known as insulators. When a surplus of charged particles (positive or negative) builds up on some part of an insulator, the excess remains there. So, if you hold one end of a rubber rod while the far end is being rubbed with silk (Fig.1.5a), the far end of the rod will acquire a surplus of electrons and those electrons will remain at that end, never flowing into your hand (Fig. 1.5b). So, insulators can be charged by friction.

Figure 1.5: (a) When a rubber rod is rubbed with a silk cloth, (b) the excess charge stays on the part of the rod that was in contact with the cloth.

1.6.1.2 Charging Conductors by Friction is Very Difficult
If you try to charge a copper rod in the same way, you will find that you cannot build up charge on the copper. Copper is an example of a conductor. A conductor is a material in which the charged particles (usually electrons) can flow freely. When you hold one end of the copper rod and rub the other end with silk, electrons are transferred from the silk to the copper rod, and those excess electrons are free to flow. Because like charges repel, the electrons move away from one another, which means they travel through the rod into your hand, as shown in Fig.1.6. The human body is also a conductor, so charged particles move freely through your body toward the Earth. If there are no insulators between you and the ground—such as when you are barefoot—the charge will continue to flow into the Earth. As a result, you and the copper rod remain neutral despite the rod being rubbed with silk.
• The human body consists largely of salt water. Pure water is not a very good conductor, but salt water, with its Na^+ and Cl^- ions, is. Consequently, humans are reasonably good conductors.

Now, considering a case, when two conductive materials are rubbed together. In this case, electrons can be transferred from one material to the other. This transfer occurs due to the different affinity for electrons that each material possesses. The material with a higher affinity for electrons tends to gain electrons and becomes negatively charged, while the other material loses electrons and becomes positively charged. This electron transfer leads to a charge separation between the two materials.

However, due to the conductive nature of the materials involved, the excess charges rapidly spread and distribute themselves evenly over the surface of the conductors. This redistribution occurs because the excess charges repel each other and move to the outer surface of the conductor, where they can be evenly distributed. As a result, any localized charge imbalance that occurs during the charging process quickly disappears, and the conductor as a whole remains electrically neutral. The excess charges are effectively cancelled out by their mutual repulsion and redistribution. From above discussion, we can say that charging by friction can be applied only if at least one material is insulator.

1.6.1.3 Triboelectric Series
In general, when two materials are rubbed together, the magnitude and sign of the charge that each material acquires depends on how strongly it holds onto its electrons. For example, if silk is rubbed against glass, the glass acquires a positive charge. It means electrons have moved from the glass to the silk, giving the

1.6. METHODS OF CHARGING

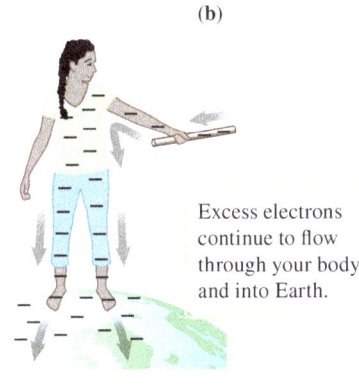

Figure 1.6: (a) When you rub a copper rod with a cloth, (b) the excess charge moves all over the rod and (c) through your body into the Earth.

silk a negative charge. If silk is rubbed against amber, however, the silk becomes positively charged, as electrons in this case pass from the silk to the amber.

Table 1.2: Triboelectric series

Material	Relative charging with rubbing
Air	Most positive
Asbestos	
Rabbit fur	
Glass	
Human hair, oily skin	
mica	
Nylon	
Wool (no charge)	
Lead	
Cat Fur	
Silk	
Aluminum	
Paper	Least positive
Cotton	Least negative
Wood	
Amber	
Rubber	
Vinyl (PVC)	
Teflon	
Ebonite	Most negative

Table 1.2, shows the relative charging due to rubbing for a variety of materials. This relative charging is known as triboelectric charging. The more plus signs associated with a material, the more readily it gives up electrons and becomes positively charged. Similarly, the more minus signs for a material, the more readily it acquires electrons. For example, we know that amber becomes negatively charged when rubbed against fur, but a greater negative charge is obtained if rubber, PVC, or Teflon is rubbed with fur instead. In general, when two materials in Table 1.2 are rubbed together, the one higher in the list becomes positively charged, and the one lower in the list becomes negatively charged. The greater the separation on the list, the greater the magnitude of the charge.

1.6.2 Charging by Direct Contact

Suppose, you have a conducting sphere as shown in Fig.1.7a. You can charge it by touching it with a charged object such as a negatively charged plastic rod (Fig.1.7b). When the rod is in contact with the sphere, some of the electrons are transferred from the rod to the sphere. Remember that charge cannot flow through the rod because it is an insulator, so you may need to roll the rod around the surface of the sphere in order to transfer charge from many parts of the rod. If we bring a negatively charged conducting rod in contact of an uncharged conducting sphere, then, off course, rolling is not required. In this case, electrons can easily be transferred from charged rod to uncharged sphere. Because electrons move freely throughout the conductor and because like charges repel, the electrons quickly redistribute themselves in the conductor, moving as far apart as possible. For a spherical conductor, "*as far apart as possible*" means that the *electrons are uniformly distributed on the outside surface of the sphere*. If you try to charge a conductor of another shape, the charge is again distributed on the outside surface, although for nonspherical shapes the charge distribution is not uniform.

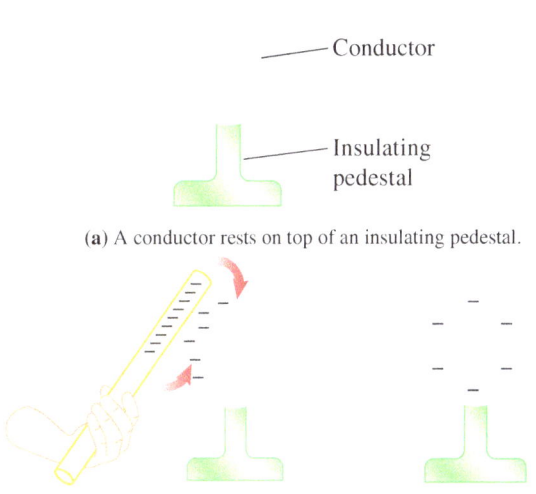

(a) A conductor rests on top of an insulating pedestal.

(b) When the conductor is touched by a charged rod, some charge is transferred from the rod to the conductor.

(c) The excess charge quickly distributes over the surface of the conductor.

Figure 1.7

- The uncharged body acquires the same sign of charge as the charged body. The total charge is distributed between the two bodies.

1.6.3 Charging by Induction

There is another way to build up charge on a conductor. Suppose you use the same equipment as in Figure1.7b. This time, however, you hold the negatively charged rod near but not touching the sphere as shown in Fig.1.8a. Because the electrons in the sphere are free to move, they flow to the side opposite the rod due to repulsion of negatively charged rod. If you ground the side of the conducting sphere where there is a surplus of electrons, those electrons will flow to the ground (Fig.1.8b). When you remove the connection to the ground, keeping the charged rod near the sphere, the net charge of the sphere is positive and the ground remains essentially neutral (Fig.1.8c). This excess positive charge on sphere is called called the induced positive charge. Now, if you remove the rod too, this charge get uniformly distributed over the surface of sphere due to repulsion.

Note that, no electrons are lost from the rod; it has the same charge throughout the process. In fact, you could reuse the rod to charge another conductor by same method without recharging the rod again.

"This process of charging, in which a charged plastic rod gives another body (metal sphere) a charge of opposite sign without losing any of its own charge, is called charging by induction. In this process the charged rod never touches the sphere."

Note: You can also charge two bodies by induction as follows-

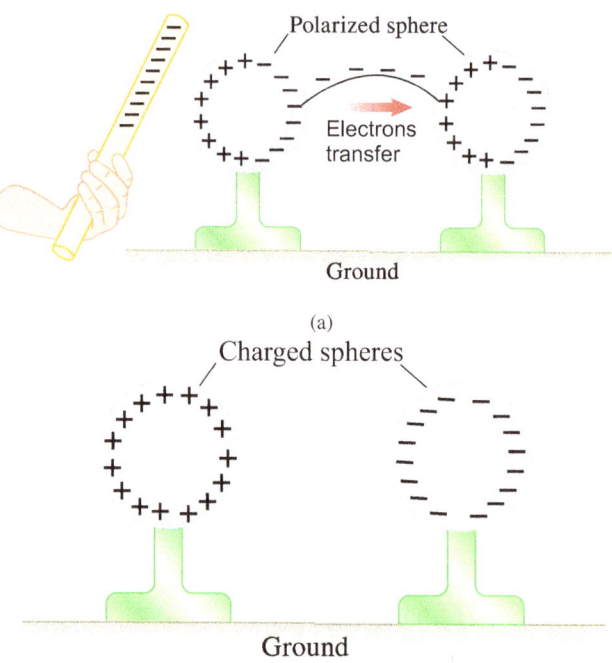

Figure 1.9: Charging by induction

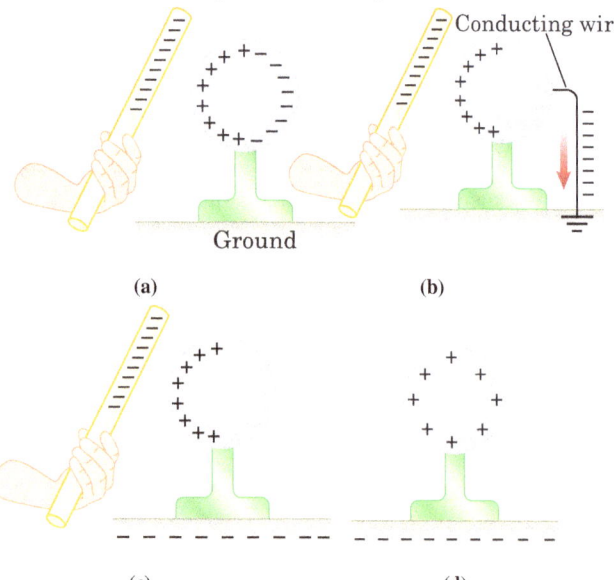

Figure 1.8: (a) When a charged rod is held near a neutral conductor, electrons in conductor move away from negatively charged rod and the conductor becomes polarized. (b) The far side of the conductor is connected to the ground, and some electrons flow out of the conductor into the Earth. (c) The connection is removed. (d) When the rod is removed, the electrons redistribute, leaving the surplus positive charge uniformly distributed over the surface of the conductor.

Take an isolated neutral conductor and then bring a charged rod near it. In Fig.1.9 the rod is negatively charged. Due to the charged rod, charges will induce on the conductor. Now, connect another neutral conductor with it. Due to repulsion of the negatively charged rod, some free electrons get transferred from the left conductor to the right conductor and due to deficiency of electrons positive charge appears on the left conductor and on the right conductor, there is an excess of electrons due to transfer from left conductor. Now disconnect the connecting wire and remove the rod. The first conductor will get negatively charged and the second conductor will get positively charged. ☞ Note that only conductors can be charged by induction.

1.6.4 Charging by Field Emission

When a large magnitude electric field $\left(\vec{E}\right)$ is applied near the metal surface, some electrons come out from the metal surface due to the electric force applied by the external electric field, and hence the metal becomes positively charged, as shown in Fig. 1.10.

Figure 1.10: Charging by field emission

1.6.5 Charging by Thermionic Emission

When the metal is heated at a high temperature then some electrons of metals are ejected and the metal becomes positively charged.

1.6.6 Charging by Photoelectric Effect

When light photons, having energy greater than work function* of material of a given metal sheet, are incident on it's surface then some electrons gain energy from these incident photons and come out of the metal surface. As a result, there becomes a deficiency of electrons in the metal surface. So, remaining metal becomes positively charged. These ejected electrons are called photo electrons. It is shown in Fig.1.11. In this figure, $h\nu$ is the energy of a photon having frequency ν. If ϕ represents work function of metal then for ejection of photoelectrons, the required condition is $h\nu \geq \phi$.

☞ Neutral does not mean "chargeless". A neutral body always has equal amount of positive and negative charge, i.e., there is a microscopic balance of –ve and +ve charge.

1.7 Coulomb's Law

Coulomb's law is a fundamental law in physics that describes the electrostatic interaction between point charges† quantitatively.

---

*The amount of energy that must be supplied to break the bond between a metal and one of it's electrons, is called the work function (ϕ). Each metal has its own characteristic work function.

†A point charge is a point like object with a non zero electric charge and non zero mass. Note that a point like object is small enough that its internal structure

1.7. COULOMB'S LAW

Figure 1.11: Charging by photoelectric effect

It is an empirical law, derived from observations rather than theoretical derivation.

It states that- the electrostatic force acting between two point charges is directly proportional to the of product of magnitude of charges and inversely proportional to the square of the distance between them.

Mathematically, the electrostatic force F that a particle of charge q_1 exerts on a particle of charge q_2 at a distance r, can be written as-

$$F \propto \frac{|q_1||q_2|}{r^2}$$

or
$$\boxed{F = k\frac{|q_1||q_2|}{r^2}} \qquad (1.10)$$

Here, k is a proportionality constant, its numerical value depends on the system of units used. its value is $8.99 \times 10^9 \, \text{N.m}^2/\text{C}^2$
In one significant figure, $k = 9 \times 10^9 \, \text{N.m}^2/\text{C}^2$
Eq.1.10 holds only for charged objects whose sizes are much

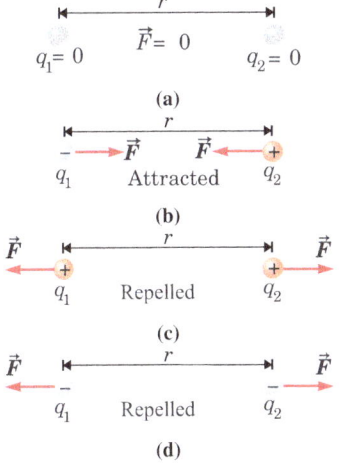

Figure 1.12: (a) There is no electrostatic force between neutral objects. (b) The electrostatic force between two oppositely charged objects is attractive. (c) The electrostatic force between two positively charged objects is repulsive. (d) The electrostatic force between two negatively charged objects is repulsive.

smaller than the distance between them. We often say that it holds only for point charges.

The directions of the forces the two charged particles exert on each other are always along the line joining them. If any one or both the charges are zero, then the electrostatic force between the particles will be zero (Fig.1.12a). When the charges q_1 and

q_2 have the same sign, either both positive or both negative, the forces are repulsive (see Fig.1.12c and 1.12d); when the charges have opposite signs, the forces are attractive (Fig.1.12b).

In the SI system, the constant k is expressed in the following form:

$$k = \frac{1}{4\pi\epsilon_0} \qquad (1.11)$$

Although the choice of this form for the constant k appears to make Coulomb's law needlessly complex, it ultimately results in a simplification of formulas of electromagnetism that are used more often than Coulomb's law.

Here, ϵ_0 ("epsilon nought") is called the electric permittivity* of free space. It's value, determined by the adopted value of the speed of light, is $\epsilon_0 = 8.854 \times 10^{-12} \, \text{C}^2/\text{N} \cdot \text{m}^2$

The constant k has the corresponding value (to three significant figures)

$$\boxed{k = \frac{1}{4\pi\epsilon_0} = 8.99 \times 10^9 \, \text{N} \cdot \text{m}^2/\text{C}^2}$$

For simplicity, we'll often use the approximate value

$$\boxed{k = \frac{1}{4\pi\epsilon_0} = 9.0 \times 10^9 \, \text{N} \cdot \text{m}^2/\text{C}^2}$$

With this choice of the constant k, Coulomb's law can be written as-

$$\boxed{F = \frac{1}{4\pi\epsilon_o}\frac{|q_1||q_2|}{r^2}} \qquad (1.12)$$

1.7.0.1 Value of k in CGS Units

In SI units, $k \approx 9.0 \times 10^9 \, \text{N.m}^2/\text{C}^2$
Here, we use the method of unit conversion. If n_1, u_1 are numeric value and unit respectively in SI, whereas n_2, u_2 in CGS, then

$n_1 u_1 = n_2 u_2$, gives

$$n_2 = n_1 \frac{u_1}{u_2} = 9.0 \times 10^9 \frac{\text{N.m}^2/\text{C}^2}{\text{dyne.cm}^2/\text{esu}^2}$$

$$= 9.0 \times 10^9 \frac{10^5 \text{dyne}.10^4 (\text{cm})^2/(3 \times 10^9 \text{esu})^2}{\text{dyne}.(\text{cm})^2/(\text{esu})^2} = 1$$

Therefore, in CGS, $k = 1 \, \text{dyne (cm)}^2/(\text{esu})^2$

1.7.1 Experimental Verification of Coulomb's Law

The experimental setup designed by Charles Augustin Coulomb (1736 - 1806) to measured electrical attractions and repulsions quantitatively, is shown in Fig.1.13. One small metal sphere is charged and fixed in place. A pair of small metal spheres is attached to a lightweight insulated rod. The rod is suspended from a torsion spring of known torsion spring constant.

Imagine that the pair of spheres is initially uncharged. The rod is twisted, bringing one of those spheres in contact with the charged fixed sphere and thus transferring charge. If the spheres are identical, half of the charge on the fixed sphere will be transferred to the movable sphere. So these spheres repel each other, causing the rod to twist. The amount of twist is measured. Because Coulomb knew the torsion spring constant, he could calculate the force exerted between the spheres. In many trials, Coulomb varied the amount of charge on the spheres as well as the torsion spring constant. From these trials, he found that-

"The electrostatic force acting between two charged spheres is

---

is of no importance. The electron can be treated as a point charge, since there is no experimental evidence for any internal structure. The proton does have internal structure—it contains three particles called quarks bound together—but, since its size is only about 10^{-15}m, it too can be treated as a point charge for most purposes. Any small charged bodies, provided that the sizes of these bodies are much smaller than the distances between them can also be treated like point charges. For example, a charged metal sphere of radius 10 cm can be treated as a point charge if it interacts with another such sphere 100 m away, but not if the two spheres are only a few centimetres apart.

*Electric permittivity is the measure of resistance that is encountered when forming an electric field (Section 1.12) in a particular medium. More electric permittivity means more resistance is offered by the medium in forming electric field inside the medium, i.e., the medium is more insulating.

Inside a conductor the electric field is always zero, therefore, its electric permittivity is infinite.

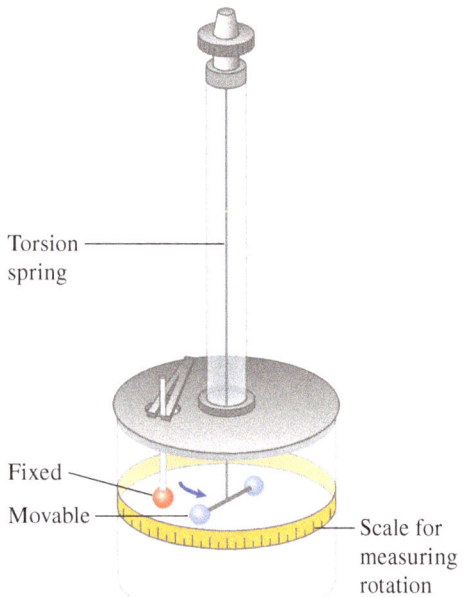

Figure 1.13: Torsion balance designed by Coulomb. Two (blue) spheres are attached to a rod suspended from a torsion spring. The fixed (red) sphere is positively charged, and then one of the movable spheres is brought in contact with the charged, fixed one. Because they are both conductors, charge is shared between the spheres, and the movable (blue) sphere becomes positive. Both spheres have charge of the same sign, so they are repelled, which causes the rod to rotate. The rod's rotation is measured using a scale (shown in yellow).

directly proportional to the of product of magnitude of charges and inversely proportional to the square of the distance between them."

☞ For given charges, if we plot a graph of the measured force versus the inverse square of the distance $(1/r^2)$, the graph should be a straight line and the slope of the line will give you the product of the charges divided by the electrostatic constant, (q_1q_2/k).

EXAMPLE 5. Find the electric force between two point charges of $1C$ each separated by 1 m apart in free space
SOLUTION By Coulomb's law (Eq.1.10), we have
$$F = k\frac{|q_1||q_2|}{r^2} \qquad \cdots (1)$$
Substituting, the values of k, q_1, q_2 and r in Eq.(1), we get-
$$F = (9.0 \times 10^9 \text{N.m}^2/\text{C}^2)\frac{|1\text{C}||1\text{C}|}{(1\text{m})^2} = 9.0 \times 10^9 \text{N}$$
which is a very large value and keeping these charges at 1 m apart, is practically very difficult. This is the reason, why we consider systems having smaller charges, such as- micro coulomb (1 μC = $10^{-6}C$), nano coulomb (1 nC = $10^{-9}C$) and pico coulomb (1 pC = $10^{-12}C$) etc.

1.7.2 Vector Form of Coulomb's Law

So far we have considered only the magnitude of the force between two point charges determined according to Coulomb's law. Force, being a vector, has directional properties as well. In the case of Coulomb's law, the direction of the force is determined by the relative sign of the two electric charges.
As illustrated in Fig.1.14, suppose we have two like point charges q_1 and q_2 at positions $\vec{r}_1$ and $\vec{r}_2$ respectively.
The position vector of charge q_1 with respect to q_2 is given by-
$$\vec{r}_{12} = \vec{r}_1 - \vec{r}_2$$
The unit vector in the direction of q_1 from q_2 is -
$$\hat{r}_{12} = \frac{\vec{r}_{12}}{r_{12}} = \frac{\vec{r}_1 - \vec{r}_2}{|\vec{r}_1 - \vec{r}_2|}$$

Here, r_{12} represents the magnitude of the vector $\vec{r}_{12}$, i.e., the separation between q_1 and q_2.
The repulsive electric force exerted by charge q_2 on q_1 is given by-

$$\boxed{\vec{F}_{12} = \frac{1}{4\pi\epsilon_0}\frac{q_1q_2}{r_{12}^2}\hat{r}_{12}} \qquad (1.13)$$

Similarly, electric force on charge q_2 due to charge q_1 is given by-
$$F_{21} = \frac{1}{4\pi\epsilon_0}\frac{q_1q_2}{r_{21}^2}$$
In vector form, above equation can be written as

$$\boxed{\vec{F}_{21} = \frac{1}{4\pi\epsilon_0}\frac{q_1q_2}{r_{21}^2}\hat{r}_{21}} \qquad (1.14)$$

Here $\hat{r}_{21}$ is a unit vector that points from charge q_1 to charge q_2; that is, it would be the unit vector in the direction of point charge q_2 if the origin of coordinates were at the location of charge q_1. Since, $\hat{r}_{21} = -\hat{r}_{12}$, therefore-

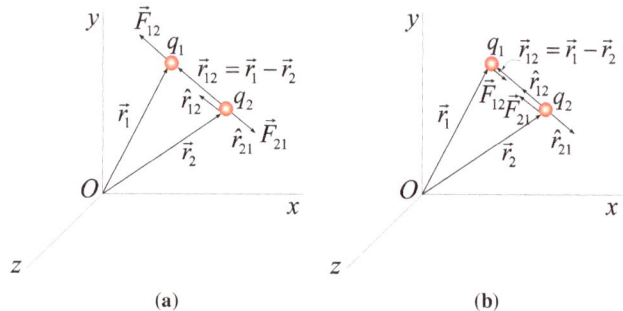

(a)   (b)

Figure 1.14: (a) Two point charges q_1 and q_2 of the same sign exert equal and opposite repulsive forces on one another. The vector $\vec{r}_{12}$ locates q_1 relative to q_2, and the unit vector $\hat{r}_{12}$ points in the direction of $\vec{r}_{12}$. Note that $\vec{F}_{12}$ is parallel to $\vec{r}_{12}$. (b) The two charges now have opposite signs, and the force is attractive. Note that $\vec{F}_{12}$ is antiparallel to $\vec{r}_{12}$

$$\boxed{\vec{F}_{21} = -\vec{F}_{12}} \qquad (1.15)$$

Thus, electric forces form an action-reaction pair.
The significance of Coulomb's law goes far beyond the description of the forces acting between charged particles. This law, when incorporated into the structure of quantum physics, correctly describes-
1. the electrical forces that bind the electrons of an atom to its nucleus.
2. the forces that bind atoms together to form molecules, and
3. the forces that bind atoms and molecules together to form solids or liquids.

Thus most of the forces of our daily experience that are not gravitational in nature are electrical. Moreover, unlike Newton's law of gravitation, which can be considered a useful everyday approximation of the more basic general theory of relativity, Coulomb's law is an exact result for stationary charges and not an approximation from some higher law. It holds not only for ordinary objects, but also for the most fundamental "point" particles such as electrons and quarks. Coulomb's law remains valid in the quantum limit (for example, in calculating the electrostatic force between the proton and the electron in an atom of hydrogen). When charged particles move at speeds close to the speed of light, such as in a high-energy accelerator, Coulomb's law does not give a complete description of their electromagnetic interactions; instead, a more complete analysis based on Maxwell's

1.8 Superposition Principle

The vector form of Coulomb's law is useful because it carries within it the directional information about $\vec{F}$ and whether the force is attractive or repulsive. Using the vector form is of critical importance when we consider the forces acting on an assembly of more than two charges. In this case, Eq.1.15 would hold for every pair of charges, and the total force on any one charge would be found by taking the vector sum of the forces due to each of the other charges. For example, the force on point charge q_1 in an assembly would be-

$$\boxed{\vec{F}_1 = \vec{F}_{12} + \vec{F}_{13} + \vec{F}_{14} + \cdots} \quad (1.16)$$

where $\vec{F}_{12}$ is the force on point charge q_1 due to point charge q_2, $\vec{F}_{13}$ is the force on point charge q_1 due to point charge q_3, and so on. Equation 1.16 is the mathematical representation of the **principle of superposition** applied to electric forces. It permits us to calculate the force due to any pair of charges as if the other charges were not present. For instance, the force $\vec{F}_{13}$ that point charge q_3 exerts on q_1 is completely unaffected by the presence of point charge q_2.

Note that - if the electromagnetic force were proportional to the square of the total source charge, then, the principle of superposition would not hold, since $(q_1 + q_2)^2 \neq q_1^2 + q_2^2$ (there would be an extra term "$2q_1q_2$" to consider). Superposition is not a logical necessity, but an experimental fact.

The electric force on test charge q_1, due to the point source charge q_i (say) ($i \neq 1$), not only depends on the position of q_i, it also depends on both velocities and on the acceleration.

Since, an electromagnetic signal or effect produced by the source charge q_i travels at the speed of light, so to find the effect of q_i on q_1, we need the position, velocity, and acceleration of q_i at some earlier time, when the signal left.

Hence, we can say that the principle of superposition does not hold in many situations when the source charge is moving. In this chapter, we will be considering only the stationary source charges, therefore, the principle of superposition is valid.

1.9 Coulomb's Law in a Dielectric Medium

A non conducting material (for example, air, glass, paper, or wood) is called a dielectric. If two point charges q_1 and q_2 are placed at separation r in a liquid dielectric medium of electric permittivity ϵ, then the electric force acting between them is given by-

$$\boxed{F = \frac{1}{4\pi\epsilon} \frac{|q_1|\,|q_2|}{r^2}} \quad (1.17)$$

If ϵ_r or κ denotes relative permittivity or dielectric constant of the medium, then-

$$\epsilon_r \text{ or } \kappa = \frac{\epsilon}{\epsilon_0}$$

or
$$\boxed{\epsilon = \epsilon_0 \kappa}$$

Substituting this value of ϵ in Eq.1.17, we get-

$$\boxed{F = \frac{1}{4\pi\epsilon_0 \kappa} \frac{q_1\,q_2}{r^2}}^\dagger \quad (1.18)$$

---

*Maxwell's equations will be discussed in chapter "Electromagnetic Waves"
†Eq.1.18 is applicable only for space filled with a liquid dielectric. According to Eq.1.18, the force depends on the fact that κ is a constant, which is only approximately true for most real materials. Finally, the microscopic electric field created by dielectric also applies electric force on the charged particles or

Eq.1.18 shows that effective electrostatic force between two point charges decreases to $1/\kappa$ times due to presence of dielectric medium.

> ☞ A dielectric medium consists of atoms which are neutral but not chargeless. When a dielectric or any other substance is present in vicinity of a charge, the positive and negative charges of atoms (nuclei and electrons) experience electric force, which in turn leads to a partial separation of these charges. These partially separated charges present on atoms apply additional electric force on charges which in combination with the force of interaction between q_1 and q_2 gives the resultant force. If we know the force of interaction between q_1 and q_2 in absence of dielectric medium and the additional electric force due to induced charges of the medium, we can forget about the presence of the substance itself while calculating the resultant force, since the role of the substance has already been taken into account with the help of induced charges. The effective force acting between charges q_1 and q_2 in presence of a dielectric is the result of superposition of actual force in absence of medium and the force applied by induced charges of medium due to it's presence.
>
> Thus, although the electrostatic force of q_1 on q_2 or vice versa remains unchanged, the effective force between the charges in a dielectric medium decreases to $1/\kappa$ times of it's value in free space.

For free space or air, $\epsilon = \epsilon_0$, therefore, $\kappa = \epsilon_0/\epsilon_0 = 1$
Now, Eq.1.18 can also be written as

$$F = \frac{1}{4\pi\epsilon_0} \frac{q_1 q_2}{(r\sqrt{\kappa})^2}$$

which is similar to an expression for electric force between two charged particles placed at a separation $r\sqrt{\kappa}$ in vacuum. So, the separation r of charged particles placed in a dielectric medium produces same electric force if these charged particles are placed at separation $r\sqrt{\kappa}$ in vacuum.

Now, we calculate the electric force between two point charges when the vacuum separation between the charges is partially filled with a dielectric medium of dielectric constant κ (Fig.1.15).

Suppose two charges q_1 and q_2 are separated by distance r in

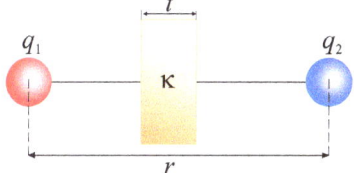

Figure 1.15

vacuum. If, we insert a liquid dielectric of dielectric constant κ and thickness $t(< r)$ between the charges, then, the width of vacuum part between the charges will become $(r-t)$ and effective vacuum width corresponding to the thickness t of dielectric slab is $t\sqrt{\kappa}$. Therefore, effective vacuum separation between the charges can be written as-

$$r_{\text{vacuum}} = (r - t + t\sqrt{\kappa})$$

So, in this case, Coulombs law can be written as -

$$\boxed{F = \frac{1}{4\pi\epsilon_0} \frac{q_1 q_2}{(r - t + t\sqrt{\kappa})^2}} \quad (1.19)$$

EXAMPLE 6. If two charges q_1 and q_2 are separated with distance 'r' and placed in a medium of dielectric constant κ. What will be the equivalent distance between charges in air for the same electrostatic force?

---

conductors in it (see subsection "?? ??" at page ?? of chapter "3. ??").

SOLUTION The electric force (F) between charges q_1 and q_2 separated by a distance r in a medium of dielectric constant κ is given by
$$F = \frac{1}{(4\pi\epsilon_0)}\frac{q_1 q_2}{\kappa r^2}$$
And electric force in air (F_{air}) between charges q_1 and q_2 separated by a distance r', is given by
$$F_{\text{air}} = \frac{1}{4\pi\epsilon_0}\frac{q_1 q_2}{r'^2}$$
If $F = F_{\text{air}}$, then
$$\frac{q_1 q_2}{4\pi\epsilon_0 \kappa r^2} = \frac{q_1 q_2}{4\pi\epsilon_0 r'^2} \implies r' = r\sqrt{\kappa}$$

1.9.0.1 Key Points Regarding Coulomb's Law

1. Electric forces always form action-reaction pairs, i.e., the force which first charge exerts on the second is equal and opposite to the force which the second charge exerts on the first.
2. The direction of force is always along the line joining the two charges.
3. Electrostatic force between two point charges is independent of presence or absence of other charges.
4. The electrostatic force is conservative, i.e., the work done in moving a point charge once round a closed path under the action of Coulomb's force, is zero.
5. The net Coulomb's force on two charged particles q_1 and q_2 separated by distance r in free space and in a uniform liquid dielectric medium, are-
   (i) $F = \dfrac{1}{4\pi\epsilon_0}\dfrac{q_1 q_2}{r^2}$ (in vacuum or free space)
   (ii) $F' = \dfrac{1}{4\pi\epsilon_0 \kappa}\dfrac{q_1 q_2}{r^2}$ (in the dielectric medium)
   Therefore, $\dfrac{F}{F'} = \dfrac{\epsilon}{\epsilon_0} = \kappa$

   So, the dielectric constant (κ) of a liquid dielectric medium, is numerically equal to the ratio of the force on two point charges in free space to that in the fully filled dielectric medium.
6. The law expresses the force between two point charges at rest. In applying it to the case of extended bodies of finite size care should be taken in assuming the whole charge of a body to be concentrated at its 'centre' as this is true only for spherical charged body, that too for external point.

   Although net electric force on both the charged particles changes in the presence of dielectric but force due to one charged particle on another charge particle does not depend on the medium between them.
7. Coulomb's law resembles Newton's inverse square law of gravitation, $F = Gm_1m_2/r^2$. Both are inverse square laws, and the charge q plays the same role in Coulomb's law that the mass m plays in Newton's law of gravitation. Main differences between the two forces are-
   - Practically, (not theoretically*) gravitational forces are always attractive, while electrostatic forces can be repulsive or attractive, depending on whether the two charges have the same or opposite signs.
   - Electric force depends on the nature of medium between the charges while gravitational force does not.
   - There is another important difference between the two laws. In using the law of gravitation, we were able to define mass from Newton's second law, $F = ma$, and then by applying the law of gravitation to known masses we could determine the constant G. In using Coulomb's law, we take the reverse approach—we define the constant k to have a particular value, and we then use Coulomb's law to determine the basic unit of electric charge as the quantity of charge that produces a standard unit of force.

     For example, consider the force between two equal charges of magnitude q. We could adjust q until the force has a particular value, say $1\,\text{N}$ for a separation of $r = 1\,\text{m}$, and define the resulting q as the basic unit of charge. *It is, however, more precise to measure the magnetic force between two wires carrying equal currents, and therefore the fundamental SI electrical unit is the unit of current, from which the unit of charge is derived.*

EXAMPLE 7. If a charged body is placed near a neutral conductor, will it attract the conductor or repel it?

ANSWER The charged body will attract the conductor.

EXPLANATION Suppose, a charged body A having charge $+q$ is brought near a neutral conductor (Fig.1.16) B, the charged body induces a redistribution of charges within the conductor. As here, the charged body is positively charged, it will attract the negative charges in the neutral conductor B towards it, causing an accumulation of the negative charge ($-q'$) on the side of conductor facing the charged body and an equal positive charge ($+q'$) accumulated on the side of the conductor facing away from the charged body A. Remember that the magnitude of induced charge produced on each side is always less than the magnitude of the source charge, i.e., $|q'| < |q|$ (See EXAMPLE 68). These induced charges can be assumed to be concentrated at their centre of charges.

Clearly, the separation between the centre of charge of positively charged body A from the centre of negative induced charge in conductor is smaller as compared to the separation between the centre of charge of positively charged body A and the centre of positive induced charge in neutral conductor B. Therefore, from $F \propto 1/r^2$, we can say that force of attraction between positively charged body A and negative induced charge of neutral body B is greater than the repulsion between positively charged body A and positive induced charge on the neutral conductor B. So, the resultant force is attractive. Thus, a positively charged body attracts a neutral body.

Similarly, we can show that a negatively charged body also attracts a neutral body.

From the above example, we can conclude that- "A charged body

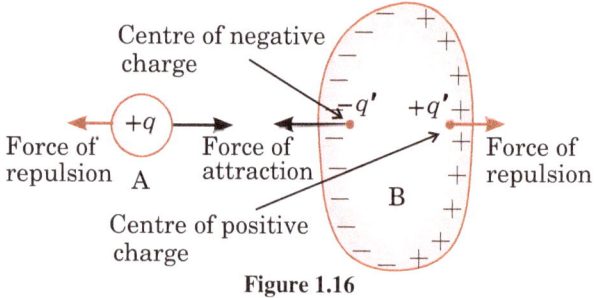

Figure 1.16

can attract a neutral body."

If there is attraction between two bodies then one of them may be neutral. But if there is repulsion between two bodies, both must be likely charged.

So, "repulsion is the sure test of electrification" whereas attraction is not.

EXAMPLE 8. Compare the magnitudes of the gravitational force of attraction and of the electric force of attraction between the electron and the proton in a hydrogen atom. According to Newtonian mechanics, what is the acceleration of the electron? Assume that the distance between these particles in a hydrogen atom is 5.3×10^{-11} m.

---
*Theoretically, there is a concept of negative mass. It is assumed that there is a gravitational repulsion between a positive and a negative mass.

1.9. COULOMB'S LAW IN A DIELECTRIC MEDIUM

APPROACH Gravitational force between two masses m_1 and m_2 separated by distance r, is given by-
$$F_g = G\frac{m_1 m_2}{r^2} \qquad (1)$$
with $G = 6.67 \times 10^{-11}$ N.m^2/kg^2
Electrostatic force between two point charges q_1 and q_2 separated by distance r, is given by-
$$F_e = k\frac{q_1 q_2}{r^2} \qquad (2)$$
with $k \approx 9.0 \times 10^9$ N.m^2/C^2
For electron-proton system- $m_1 = m_e = 9.11 \times 10^{-31}$ kg, $q_1 = -e = -1.6 \times 10^{-19}$ C, $m_2 = m_p = 1.67 \times 10^{-27}$ kg and $q_2 = +e = +1.6 \times 10^{-16}$ C
Substitute these values in Eq.(1) and (2) and then compare F_g and F_e.
SOLUTION From Eq.1, we get-
$$F_g = G\frac{m_e m_p}{r^2}$$
$$= \left(6.67 \times 10^{-11} \text{ N} \cdot \text{m}^2/\text{kg}^2\right)$$
$$\times \frac{(9.11 \times 10^{-31} \text{ kg})(1.67 \times 10^{-27} \text{ kg})}{(5.3 \times 10^{-11} \text{ m})^2}$$
$$= 3.6 \times 10^{-47} \text{ N}$$
and from Eq. 2, the magnitude of the electric force is-
$$F_e = \frac{1}{4\pi\epsilon_0}\frac{e \times e}{r^2}$$
$$= \left(8.99 \times 10^9 \text{ N} \cdot \text{m}^2/\text{C}^2\right) \times \frac{(1.60 \times 10^{-19} \text{C})^2}{(5.3 \times 10^{-11} \text{ m})^2}$$
$$= 8.2 \times 10^{-8} \text{ N}$$
The ratio of these forces is-
$$\frac{F_e}{F_g} = \frac{(8.2 \times 10^{-8} \text{ N})}{(3.6 \times 10^{-47} \text{ N})} = 2.3 \times 10^{39}$$
Thus the electric force is very large as compared to the gravitational force. Since the gravitational force is insignificant compared with the electric force, it can be neglected. The acceleration of the electron is then
$$a = \frac{F}{m} = \frac{8.2 \times 10^{-8} \text{ N}}{9.11 \times 10^{-31} \text{ kg}} = 9.0 \times 10^{22} \text{ m/s}^2$$
This is a very large acceleration. If it occurred along the electron's motion instead of centripetally, such an acceleration could boost the electron's velocity close to one-third of the speed of light in only a femtosecond $(10^{-15}$ s$)$.
Comment: For the ratio of the electric force and the gravitational force between the proton and electron, we would obtain the same immense value 2.3×10^{39} whatever the separation between the two particles, since both are inverse square forces. Also notice that for the given atomic-scale distance, the electric force has a measurable value, the same as weighing an 8 μg mass, whereas the gravitational force is far below the current limits of detection (the highest sensitivity attained by a measurement of force is near 10^{-20} N).

EXAMPLE 9. In EXAMPLE 3, what is the magnitude of the attractive force exerted by the electrons in a cup of water(0.25 kg water) on the protons in a second cup of water at a distance of 10 m?
APPROACH Substitute the values of charges, obtained in EXAMPLE 3, in Coulomb's force formula $F = k\frac{q_1 q_2}{r^2}$ and simplify for F_e.
SOLUTION According to the preceding example, the charge on the electrons in the cup is -1.3×10^7 C and the charge on the protons is $+1.3 \times 10^7$ C. If we treat both of these charges as point charges, the force on the protons is
$$F = \frac{1}{4\pi\epsilon_0}\frac{q_1 q_2}{r^2}$$
$$= \left(8.99 \times 10^9 \text{ N} \cdot \text{m}^2/\text{C}^2\right)\frac{(-1.3 \times 10^7 \text{C})(1.3 \times 10^7 \text{C})}{(10 \text{ m})^2}$$
$$= -1.5 \times 10^{22} \text{ N}$$
This is approximately the weight of a billion billion tons! This enormous attractive force on the protons is precisely canceled by an equally large repulsive force exerted by the protons in one cup on the protons in the other cup. Thus, the cups exert no net forces on each other.

EXAMPLE 10. The force of electrostatic repulsion between two fixed point charges separated by distance of 1 m is F. Now if we replace the two point charges by two metallic charged spheres each of radius 25 cm having the same charges as that of given point charges. then compare the force of repulsion in two cases.
APPROACH The electrostatic force acting between two point charges separated by distance r is given by
$$F = k\frac{q_1 q_2}{r^2}$$
So, if magnitudes of charges q_1 and q_2 remains fixed, then $F \propto 1/r^2$.

(a) Isolated charged metal sphere on an insulated stand.
(b) Centres of charges shift towards each other
(c) Centres of charges shift away from each other

Figure 1.17: Shifting of effective centers of charges when two charged metal spheres placed close to each other

In case of charged conducting spheres, due to mutual interaction between the charges, the effective distance (r) between centre of charges get altered, so the electrostatic force also get altered. If the charges of spherical conductors have opposite nature, then due to electrostatic attraction, the charge of each sphere moves towards the other and hence the effective distance between the centres of charge of the two spheres, decreases (Fig.1.17b). So, force of electrostatic attraction between the charges, increases. But, if the spheres have identical charges, then due to mutual repulsion between the charge of one sphere and the the other, the charge of of each sphere moves away from other sphere. So, in this case, the effective distance between the centres of charges of the two sphere, increases (Fig.1.17c) and hence force of electrostatic repulsion between the charged conducting sphere decreases.
SOLUTION Since, radii of both spheres (25 cm) cannot be

neglected as compared to separation between their centres (1 m), therefore, given spheres cannot be cosidered as point objects. Now, in 2nd case due to mutual repulsion, the effective distance between their centre of charges will be increased $(r_f > r_i)$ so force of repulsion decreases (as $F \propto \frac{1}{r^2}$)

EXAMPLE 11. Five Styrofoam balls A, B, C, D and E are used in an experiment. Several experiments are performed on the balls and the following observations are made:
(i) Ball A repels C and attracts B.
(ii) Ball D attracts B and has no effect on E.
(iii) A negatively charged rod attracts both A and E.
For your information, an electrically neutral Styrofoam ball is very sensitive to charge induction and gets attracted considerably, if placed nearby a charged body. What are the charges, if any, on each ball?
APPROACH Balls can be considered like extended bodies unlike just point particle. Also keep in mind that a neutral body is attracted towards both positive as well as negative charges. Now observe the given observations and decide the sign of the balls.
SOLUTION From (i), as A repels C, so both A and C must be likely charged. Either both are +ve or both are −ve. As A also attract B, so charge on B should be opposite of A or B may be uncharged conductor.
From (ii) as D has no effect on E, so both D and E should be uncharged and as B attracts uncharged D, so B must be charged and D must be an uncharged conductor.
From (iii), a negatively charged rod attracts the charged ball A, so A must be +ve and from experiment (i) C must also be +ve and B must be −ve.

ANSWER: The nature of charge on each ball is shown in following table:

A	B	C	D	E
+	−	+	0	0

EXAMPLE 12. A particle of mass m carrying charge '$+q_1$' is revolving around a fixed charge '$-q_2$' in a circular path of radius r. Calculate the period of revolution.
APPROACH To revolve the charged particle q_1 of mass m around the charge $-q_2$, with a constant angular speed ω in a circular orbit of radius r (Fig.1.18), the required centripetal force $(F_c = mr\omega^2)$ is provided by the electrostatic force of attraction of charge $-q_2$ on charge q_1 $\left(F_e = \frac{1}{4\pi\epsilon_0}\frac{q_1 q_2}{r^2}\right)$. Here, $\omega = 2\pi/T$. T is the time period of q_1 around the charge $-q_2$.

Figure 1.18

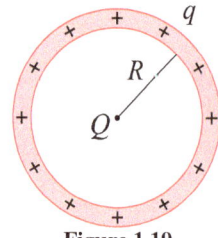

Figure 1.19

So, the condition for the charged particle q_1 to revolve around $-q_2$ in a circular orbit of radius r, is $F_e = F_c$, i.e.,

$$\frac{1}{4\pi\epsilon_0}\frac{q_1 q_2}{r^2} = mr\omega^2 \quad (1.20)$$

SOLUTION Substituting angular speed $\omega = \frac{2\pi}{T}$, in Eq.1.20, we get

$$\frac{1}{4\pi\epsilon_0}\frac{q_1 q_2}{r^2} = mr\left(\frac{2\pi}{T}\right)^2$$

or $\quad \dfrac{1}{4\pi\epsilon_0}\dfrac{q_1 q_2}{r^2} = \dfrac{4\pi^2 mr}{T^2}$

or $\quad T^2 = \dfrac{(4\pi\epsilon_0)\, r^2\, (4\pi^2 mr)}{q_1 q_2}\quad$ or $\quad T = 4\pi r\sqrt{\dfrac{\pi\epsilon_0 mr}{q_1 q_2}}$

here the vector $\vec{r}$ is drawn from source charge $(-q_2)$ to charge q_1.

EXAMPLE 13. A ring of radius R has uniformly distributed charge q. A point charge Q is placed at the centre of the ring (Fig.1.19).
(a) Find the increase in tension in the ring after the point charge is placed at its centre.
(b) Find the increase in force between the two semicircular parts of the ring after the point charge is placed at the centre.
(c) Using the result found in part (b) find the force that the point charge exerts on one half of the ring.
APPROACH If there is no charge at the centre, there is some tension in the ring due to repulsion of charge present on the ring.

(a) When Q is placed at the centre the tension increases by ΔT. The direction of ΔT is tangential to the ring at each point on it. Now, consider any infinitesimally small portion ACB(say) of the ring and resolve components of ΔT, at both ends, into two mutually perpendicular directions - one component towards centre of the ring whereas other perpendicular to it. In equilibrium, the magnitude of net component of ΔT towards centre of the ring will be equal to the magnitude of outward electrostatic force on the segment ACB.
SOLUTION Let the infinitesimally small segment ACB of the ring subtends a very small angle $d\theta$ at the centre of the ring (see Fig.1.20a). From Fig.1.20a, it is clear that horizontal compo-

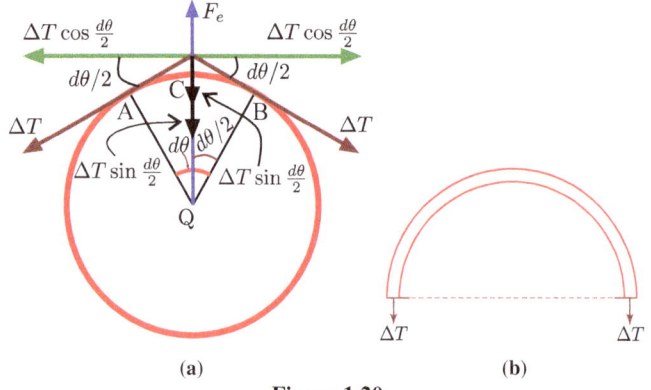

Figure 1.20

nents of ΔT at each end A and B is $\Delta T \cos\frac{d\theta}{2}$. These components are opposite in directions so get canceled.
Vertical components of ΔT at each end A and B is $\Delta T \sin\frac{d\theta}{2}$. These two components are directed towards the centre of the ring. Therefore, net inward force on the segment ACB due to tension of the ring is $2\Delta T \sin\frac{d\theta}{2}$. In mechanical equilibrium, this force will be equal to electrostatic repulsion F_e of Q at centre on the segment of the ring under consideration. Therefore,-

$$2\Delta T \sin\left(\frac{d\theta}{2}\right) = F_e \quad \text{(i)}$$

If λ is the linear charge density of the ring, then charge on the segment ACB will be -

$$dq = \text{length}ACB \times \lambda = (Rd\theta)\lambda$$

Therefore, $F_e = \dfrac{kQdq}{R^2} = \dfrac{kQR d\theta}{R^2}$, here R is the radius of the ring.

Since, $d\theta/2$ is a very small quantity, therefore, $\sin(d\theta/2) = d\theta/2$.

On substituting these values in Eq.(i), we get -

$$2\Delta T \cdot \frac{d\theta}{2} = \frac{kQ\lambda R d\theta}{R^2}$$

1.9. COULOMB'S LAW IN A DIELECTRIC MEDIUM

or $\Delta T = \dfrac{kQ\lambda}{R} = \dfrac{kQq}{2\pi R^2}$

(b) From Fig.1.20b it is clear that the answer is $2\Delta T = \dfrac{kQq}{\pi R^2}$

(c) Answer to (b) must be the answer to (c) also.

EXAMPLE 14. Three charges $q_1 = 1\ \mu C$, $q_2 = -2\ \mu C$ and $q_3 = 3\ \mu C$ are placed at the vertices of an equilateral triangle of side 1.0 m [Fig.1.21]. Find the net electric force acting on charge q_1.

APPROACH Charge q_2 will attract charge q_1 (along the line

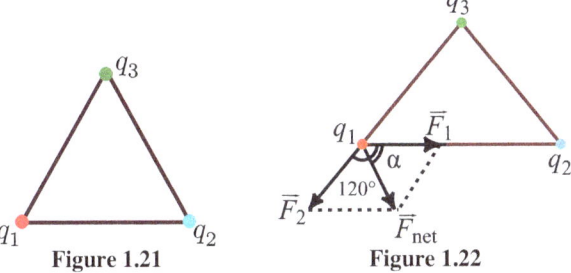

Figure 1.21   Figure 1.22

joining them) and charge q_3 will repel charge q_1 [Fig.1.22]. Therefore, two forces will act on q_1, one due to q_2 and another due to q_3. Since, the force is a vector quantity both of these forces (say $\vec{F}_1$ and $\vec{F}_2$) will be added by vector method. Now you can apply any one of the following methods of vector addition.

SOLUTION Method 1. In the Figure 1.22, the magnitude of force of q_2 on q_1

$\left|\vec{F}_1\right| = F_1 = \dfrac{1}{4\pi\epsilon_0} \cdot \dfrac{q_1 q_2}{r^2}$

$= \dfrac{(9.0 \times 10^9)(1.0 \times 10^{-6})(2.0 \times 10^{-6})}{(1.0)^2} = 1.8 \times 10^{-2}\ N$

Similarly, the magnitude of force of q_3 on q_1 -

$\left|\vec{F}_2\right| = F_2 = \dfrac{1}{4\pi\epsilon_0} \cdot \dfrac{q_1 q_3}{r^2}$

$= \dfrac{(9.0 \times 10^9)(1.0 \times 10^{-6})(3.0 \times 10^{-6})}{(1.0)^2} = 2.7 \times 10^{-2}\ N$

Now, $\left|\vec{F}_{net}\right| = \sqrt{F_1^2 + F_2^2 + 2F_1 F_2 \cos 120°}$

$= \left(\sqrt{(1.8)^2 + (2.7)^2 + 2(1.8)(2.7)\left(-\dfrac{1}{2}\right)}\right) \times 10^{-2}\ N$

$= 2.38 \times 10^{-2}\ N$

Angle made by F_{net} from the direction of F_1, is given by

$\tan\alpha = \dfrac{F_2 \sin 120°}{F_1 + F_2 \cos 120°}$

$= \dfrac{(2.7 \times 10^{-2})(0.87)}{(1.8 \times 10^{-2}) + (2.7 \times 10^{-2})\left(-\dfrac{1}{2}\right)}$

or $\alpha = 79.2°$

Thus, the net force on charge q_1 is $2.38 \times 10^{-2}\ N$ at an angle $\alpha = 79.2°$ with a line joining q_1 and q_2 as shown in the Fig. 1.22.

Method 2. In this method we assume a coordinate axes with q_1 at origin as shown in Fig.1.23. The coordinates of q_1, q_2 and q_3 in this coordinate system are $(0,0,0)$, $(1\ m, 0, 0)$ and $(0.5\ m, 0.87\ m, 0)$ respectively.

Now, force on charge q_1 due to charge q_2 is-

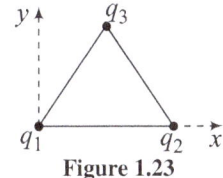

Figure 1.23

$\vec{F}_1 = \dfrac{1}{4\pi\epsilon_0} \cdot \dfrac{q_1 q_2}{|\vec{r}_1 - \vec{r}_2|^3}(\vec{r}_1 - \vec{r}_2)$

$= \dfrac{(9.0 \times 10^9)(1.0 \times 10^{-6})(-2.0 \times 10^{-6})}{(1.0)^3}$

$\times[(0-1)\hat{i} + (0-0)\hat{j} + (0-0)\hat{k}]$

$\Rightarrow \vec{F} = (1.8 \times 10^{-2}\hat{i})\ N$

and force on q_1 due to charge q_3 is given by-

$\vec{F}_2 = \dfrac{1}{4\pi\epsilon_0} \cdot \dfrac{q_1 q_3}{|\vec{r}_1 - \vec{r}_3|^3}(\vec{r}_1 - \vec{r}_3)$

$= \dfrac{(9.0 \times 10^9)(1.0 \times 10^{-6})(3.0 \times 10^{-6})}{(1.0)^3} \times$

$[(0-0.5)\hat{i} + (0-0.87)\hat{j} + (0-0)\hat{k}]$

$\Rightarrow \vec{F} = (-1.35\hat{i} - 2.349\hat{j}) \times 10^{-2}\ N$

Therefore, net force on q_1 is-

$\vec{F} = \vec{F}_1 + \vec{F}_2$

$= (0.45\hat{i} - 2.349\hat{j}) \times 10^{-2}\ N$

Note: Once you write a vector in terms of $\hat{i}$, $\hat{j}$ and $\hat{k}$, there is no need of writing the magnitude and direction of vector separately.

EXAMPLE 15. Four identical particles, each having charge $+q$, are fixed at the corners of a square of side L. A fifth point charge $-Q$ lies a distance z along the line perpendicular to the plane of the square and passing through the centre of the square (Fig.1.24). (a) Show that the force exerted by the other four charges on $-Q$ is

$$\vec{F} = -\dfrac{4kqQz}{\left[z^2 + (L^2/2)\right]^{3/2}}\hat{k}$$

Note that this force is directed toward the centre of the square whether z is positive ($-Q$ above the square) or negative ($-Q$ below the square). (b) If z is small compared with L, the above expression reduces to $\vec{F} \approx -(\text{constant})z\hat{k}$. Why does this imply that the motion of the charge $-Q$ is simple harmonic, and what is the period of this motion if the mass of $-Q$ is m?

APPROACH 1 In first approach we break the forces applied

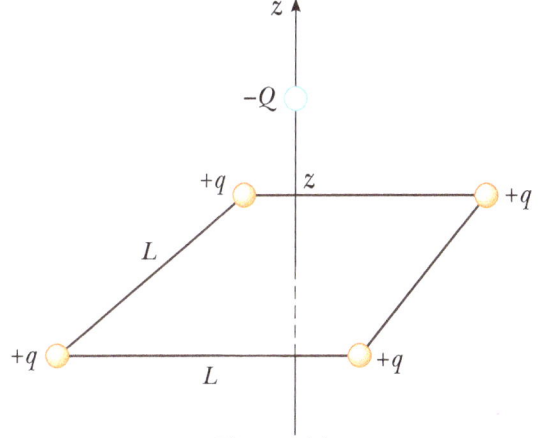

Figure 1.24

by all positive charges $+q$ on $-Q$ into horizontal and vertical components and then add these components to find the net force on Q.

SOLUTION (a) The distance from each corner to the centre of the square of side length L, is $L/\sqrt{2}$

From Fig.1.25, the distance from each positive charge to $-Q$ is-

$r = \sqrt{z^2 + \dfrac{L^2}{2}}$

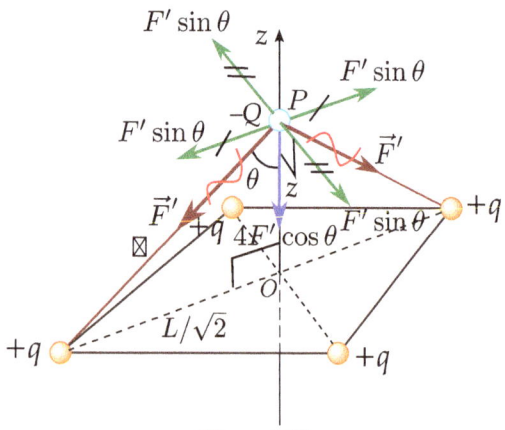

Figure 1.25

Each positive charge exerts a force directed along the line joining $+q$ and $-Q$, of magnitude

$$F' = k\frac{Qq}{r^2} = \frac{kQq}{z^2 + L^2/2}$$

The line of force makes an angle θ with the z-axis. From Fig.1.25

$$\cos\theta = \frac{z}{\sqrt{z^2 + L^2/2}}$$

The four charges together exert forces whose horizontal components $F'\sin\theta$ add to zero, while the vertical z-components $F'\cos\theta$ add to

$$\vec{F} = 4F'\cos\theta(-\hat{k})$$

$$\Rightarrow \quad \vec{F} = -\frac{4kQqz}{(z^2 + L^2/2)^{3/2}}\hat{k}$$

If m is the mass and $\frac{d^2z}{dt^2}$ is the acceleration of the charged particle corresponding to displacement z from position P, then above expression can also be written as

$$m\frac{d^2z}{dt^2} = -\frac{4kQqz}{(z^2 + L^2/2)^{3/2}}$$

$$\Rightarrow \quad \frac{d^2z}{dt^2} = -\frac{4kQqz}{m(z^2 + L^2/2)^{3/2}}$$

For $z \ll L$, above expression changes to-

$$\frac{d^2z}{dt^2} = -\left(\frac{8\sqrt{2}kQq}{mL^3}\right)z$$

Which is of the form of SHM, $\frac{d^2z}{dt^2} = -\omega^2 z$.

Here, $\omega = \sqrt{\left(\frac{8\sqrt{2}kQq}{mL^3}\right)}$ is the angular frequency of the SHM.

Time period of SHM, $T = \frac{2\pi}{\omega} = 2\pi\sqrt{\frac{mL^3}{8\sqrt{2}kQq}}$.

APPROACH 2 In second approach [Fig.1.26], we use Coulomb's law in vector form and then apply the superposition principle of electric forces

$$\vec{F} = \vec{F}_1 + \vec{F}_2 + \vec{E}_3 + \vec{F}_4$$

$$= \frac{q_1q_0(\vec{r}_p - \vec{r}_1)}{4\pi\epsilon_0|\vec{r}_p - \vec{r}_1|^3} + \frac{q_2q_0(\vec{r}_p - \vec{r}_2)}{4\pi\epsilon_0|\vec{r}_p - \vec{r}_2|^3} + \frac{q_3q_0(\vec{r}_p - \vec{r}_3)}{4\pi\epsilon_0|\vec{r}_p - \vec{r}_3|^3}$$

$$+ \frac{q_4q_0(\vec{r}_p - \vec{r}_4)}{4\pi\epsilon_0|\vec{r}_p - \vec{r}_4|^3}$$

SOLUTION The arrangement of charges is shown in Fig.1.26. According' to given problem,

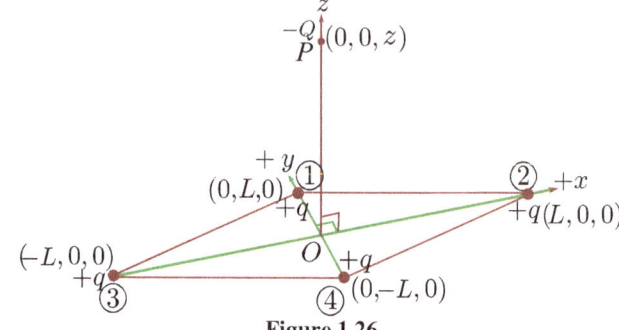

Figure 1.26

$q_1 = q_2 = q_3 = q_4 = +q$, $\vec{r}_1 = L\hat{j}$, $\vec{r}_2 = L\hat{i}$, $\vec{r}_3 = -L\hat{i}$; $\vec{r}_4 = -L\hat{j}$ and $\vec{r}_p = z\hat{k}$

$\therefore \quad \vec{r}_p - \vec{r}_1 = z\hat{k} - L\hat{i}$,

$$|\vec{r}_p - \vec{r}_1| = \sqrt{z^2 + L^2} = (z^2 + L^2)^{1/2}$$

$$\vec{r}_p - \vec{r}_2 = z\hat{k} - L\hat{j}, |\vec{r}_p - \vec{r}_2| = (z^2 + L^2)^{1/2}$$

$$\vec{r}_p - \vec{r}_3 = z\hat{k} + L\hat{i}, |\vec{r}_p - \vec{r}_3| = (z^2 + L^2)^{1/2}$$

and $\vec{r}_p - \vec{r}_4 = z\hat{k} + L\hat{j}, |\vec{r}_p - \vec{r}_4| = (z^2 + L^2)^{1/2}$

$$\vec{F} = \vec{F}_1 + \vec{F}_2 + \vec{F}_3 + \vec{F}_4$$

$$\Rightarrow \vec{F} = \frac{q_1q_0(\vec{r}_p - \vec{r}_1)}{4\pi\epsilon_0|\vec{r}_p - \vec{r}_1|^3} + \frac{q_2q_0(\vec{r}_p - \vec{r}_2)}{4\pi\epsilon_0|\vec{r}_p - \vec{r}_2|^3}$$

$$+ \frac{q_3q_0(\vec{r}_p - \vec{r}_3)}{4\pi\epsilon_0|\vec{r}_p - \vec{r}_3|^3} + \frac{q_4q_0(\vec{r}_p - \vec{r}_4)}{|\vec{r}_p - \vec{r}_4|^3}$$

Putting the values, we get

$$\vec{F} = -\frac{qQ}{4\pi\epsilon_0(z^2 + L^2)^{3/2}}\left[(z\hat{k} - L\hat{i}) + (z\hat{k} - L\hat{j}) + (z\hat{k} + L\hat{i})\right.$$

$$\left. + (z\hat{k} + L\hat{j})\right]$$

$$= -\frac{qQ}{4\pi\epsilon_0(z^2 + L^2)^{3/2}}(4z\hat{k})$$

Since, $\frac{1}{4\pi\epsilon_0} = k$, therefore, above expression can also be written as-

$$\vec{F} = -\frac{4kqQz}{(z^2 + L^2)^{3/2}}(\hat{k})$$

EXAMPLE 16. Three charges $q/2$, q and $q/2$ are placed at the corners A, B and C of a square of side 'a' as shown in Fig.1.27. What is the magnitude of electric field ($\vec{F}$) at the corner D of the square?

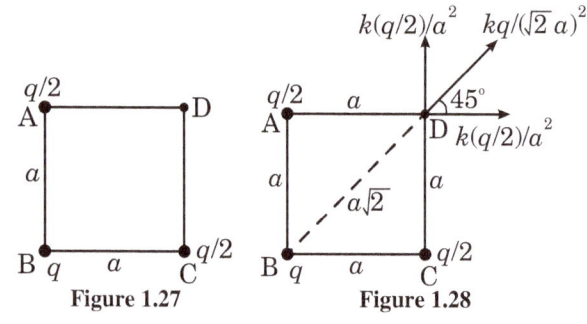

Figure 1.27   **Figure 1.28**

SOLUTION From Fig.1.28, the net electric field at D

$$(E_{net})_D = k\frac{q}{(a\sqrt{2})^2} + k\frac{q/2}{a^2}\cos 45° + k\frac{q/2}{a^2}\cos 45°$$

1.10. EQUILIBRIUM OF CHARGED PARTICLES

$$= \frac{kq}{2a^2} + \frac{\sqrt{2}kq}{2a^2} = \frac{q}{4\pi\epsilon_0 a^2}\left(\frac{1}{2} + \frac{1}{\sqrt{2}}\right)$$

1.10 Equilibrium of Charged Particles

If in a system of charged particles, the net electric force on every particle is zero, then they are said to be in equilibrium. In 1842, British mathematician Samuel Earnshaw proved that a collection of point charges cannot be maintained in a stable stationary equilibrium configuration solely by the electrostatic interaction of the charges. Discussion of this statement is given in "Earnshaw's theorem" in section 1.24.

1.10.1 Equilibrium of Three Point Charges

Suppose, three positive point charges q_1, q_2 and q are placed along a line as shown in Fig.1.29. The separation between charges q_1 and q_2 is r. Charge q is placed at distance x from charge q_1.

For the equilibrium of all the three charges, the net electrostatic

Figure 1.29

force on each of the above charges should be zero.
Now, for the equilibrium of charge q, the net electrostatic force on it should be zero, i.e., $F_q = 0$

$$\text{or} \quad k\frac{q_1 q}{x^2} - k\frac{q_2 q}{(r-x)^2} = 0 \quad (1.21)$$

Similarly, for the equilibrium of charge q_1, we have
$F_{q_1} = 0$, i.e., $-k\frac{q_1 q_2}{r^2} - k\frac{q_1 q}{x^2} = 0$

$$\text{or} \quad k\frac{q_1 q_2}{r^2} + k\frac{q_1 q}{x^2} = 0 \quad (1.22)$$

From Eq. (1.21), we have

$$\frac{q_1}{x^2} = \frac{q_2}{(r-x)^2} \Rightarrow \left(\frac{r-x}{x}\right)^2 = \frac{q_2}{q_1}$$

$$\Rightarrow \left(\frac{r-x}{x}\right) = \pm\sqrt{\frac{q_2}{q_1}} \Rightarrow \left(\frac{r}{x} - 1\right) = \pm\sqrt{\frac{q_2}{q_1}}$$

$$\Rightarrow \frac{r}{x} = 1 \pm \sqrt{\frac{q_2}{q_1}}$$

Solving for x, we get-

$$x = \frac{r}{1 \pm \sqrt{\frac{q_2}{q_1}}} \quad (1.23)$$

From Eq. (1.23), it is clear if q_1 and q_2 are unlike charges, then x will be imaginary. So, for a real value of x, both q_1 and q_2 must be like charges, i.e., either both of them must be positive or both of them must be negative.

Now, from Eq.(1.22), we have

$$q = -\frac{q_1 q_2}{r^2}\frac{x^2}{q_1} = -\frac{q_2}{r^2}x^2$$

on substituting the value of x, we get

$$q = -\frac{q_2}{r^2}\left(\frac{r}{1 \pm \sqrt{\frac{q_2}{q_1}}}\right)^2 \quad (1.24)$$

Negative sign in above Eq.(1.24) shows that, q and q_2 will be opposite in nature. But q_2 and q_1 are like charges, therefore, the nature of q is also opposite to q_1.

So, the result is that the nature of q_1 and q_2 will be similar whereas that of q is opposite to q_1 and q_2.

Eq. (1.23) and (1.24) can also be written as -

$$x = \frac{\sqrt{q_1}}{\sqrt{q_1} \pm \sqrt{q_2}}r \quad \text{and} \quad q = \frac{-q_1 q_2}{(\sqrt{q_1} \pm \sqrt{q_2})^2}$$

respectively.

EXAMPLE 17. Two particles A and B having charges 8.0×10^{-6}C and -2.0×10^{-6}C respectively are held fixed with a separation of 20 cm. Where should a third charged particle be placed so that it does not experience a net electric force?

APPROACH Since, the net electric force on charge C should be equal to zero, the force due to A and B must be equal and opposite in direction. Hence, the particle should be placed on the line AB. As A and B have charges of opposite signs, C cannot be between A and B. Also, A has larger magnitude of charge than B. So, we expect the charge C to be closer to B, the charge that has smaller magnitude, than A. The situation is shown in Figure 1.30

Figure 1.30

SOLUTION Suppose $BC = x$ and the charge on C is Q.

The force due to $A = \dfrac{(8\cdot 0\times 10^{-6}\text{C})Q}{4\pi\epsilon_0 (20\text{ cm}+x)^2}$.

The force due to $B = \dfrac{(2\cdot 0\times 10^{-6}\text{C})Q}{4\pi\epsilon_0 x^2}$.

They are oppositely directed and to have a zero resultant, they should be equal in magnitude. Thus,

$$\frac{8}{(20\text{ cm}+x)^2} = \frac{2}{x^2} \implies x = 20\text{ cm}$$

$$\implies \left(\frac{20\text{ cm}+x}{x}\right)^2 = \frac{8}{2} = 4 \implies \frac{20\text{ cm}+x}{x} = \pm 2$$

$$\implies \frac{20\text{ cm}}{x} = -1 \pm 2 \implies x = \frac{20}{-1 \pm 2}\text{cm}$$

From here, we get- $x = 20$ cm, $-\dfrac{20}{3}$ cm.

Since, x cannot be negative, therefore we take $x = 20$ cm.

EXAMPLE 18. A point charge $q_1 = 4q_0$ is placed at origin. Another point charge $q_2 = -q_0$ is placed at $x = 12$ cm. Charge of proton is q_0. The proton is placed on x-axis so that the electrostatic force on the proton in zero. In this situation, find the position of the proton from the origin.

SOLUTION Since the given charges have opposite natures, the zero electric field position will be on the line joining the given charges, outside the charges, and closer to the charge with the smaller magnitude.

This situation is shown in Fig.1.31. P is the position of the zero electric field point, which is located at a distance x_0 from point A, where a charge of $-q_0$ is placed. Now, if a proton is placed at P, it experiences zero electric force.

Since, net electric field is zero at point P, therefore-

Figure 1.31

$$\frac{q_0}{x_0^2} = \frac{4q_0}{(x_0 + 12)^2} \implies x_0 + 12 = 2x_0 \implies x_0 = 12\text{ cm}$$

Therefore, the position of proton from origin O is,

$$x = x_0 + 12 = 24\text{ cm}.$$

1.10.2 Equilibrium of Symmetric Geometrical Point Charged Systems

Figure 1.32 shows an equilateral triangle and a square of side length a. A charged particle of equal magnitude q is placed at each corner and a point charge Q at their circumcentre. If the system is in electrostatic equilibrium, then net electrostatic force on each point charge should be.

In equilibrium, to determine the value of charge Q placed at

18   Electric Charge and Field

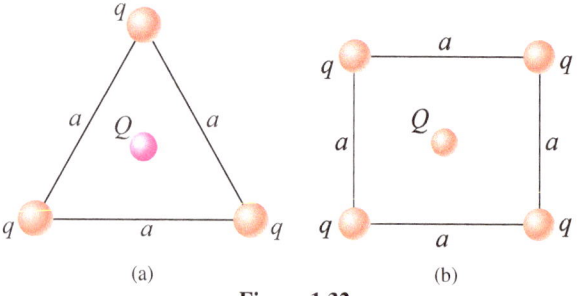

(a)   (b)
Figure 1.32

the centre, you have to apply the condition of equilibrium (i.e., $\Sigma \vec{F} = 0$) at any corner charge only not on the charge Q placed at circumcentre becase in that case each term get cancelled out and you wont find Q.

(i) Calculation of Q Placed at the Center of Triangle:
This situation with free body diagram of charge q at vertex A, is shown in Fig.1.33. From Fig.1.33, the distance $OA = (a/2) \sec 30° = a/\sqrt{3}$.
The electrostatic forces acting on charge q at A are-

Figure 1.33

1. The force $\vec{F}_{AB}$ (along BA) due to charge q at B.
2. The force $\vec{F}_{AC}$ (along CA) due to charge q at C.
3. The force $\vec{F}_{AO}$ (along OA) due to charge Q at centre O.

Note that, $|\vec{F}_{AB}| = |\vec{F}_{AC}| = k\frac{q^2}{a^2} = F$ (say)

The horizontal components of each of these two forces are $F \sin 30°$ and are oppositely directed, therefore they get cancelled with each other. However, the vertical components due to each force is $F \cos 30°$ and directed along OA, therefore they get added. So, the resultant of these forces will be $2F \cos 30°$ (along OA)

The electrostatic force applied by the charge Q at centre O on charge q at A is -

and $|\vec{F}_{AO}| = k\frac{Qq}{(a/\sqrt{3})^2} = k\frac{3Qq}{a^2}$

Forces F_{AO} and $2F \cos 30°$ both are directed along OA, therefore, net force on A-

$\Sigma F = F_{AO} + 2F \cos 30° = k\frac{3Qq}{a^2} + k\frac{q^2\sqrt{3}}{a^2}$

In equilibrium, $\Sigma F = 0$

$\Rightarrow \boxed{Q = -\frac{q}{\sqrt{3}}}$ (1.25)

Here, negative sign indicates that Q will be opposite in nature to q.

(ii) Calculation of Q Placed at the Center of Square: This situation with free body diagram of a charge q at C is shown in Fig.1.34. Considering AB along the positive direction of x axis and AD along the positive direction of y axis.
The electrostatic forces acting on charge q at C are-

1. Force $\vec{F}_{CA}$, due to charge q at A
2. Force $\vec{F}_{CB}$, due to charge q at B
2. Force $\vec{F}_{CD}$, due to charge q at D
2. Force $\vec{F}_{CO}$, due to charge Q at O

Net electrostatic force on charge q at C, is-

$$\Sigma \vec{F} = \vec{F}_{CA} + \vec{F}_{CB} + \vec{F}_{CD} + \vec{F}_{CO} \quad (1.26)$$

Here, $\vec{F}_{CA} = k\frac{q^2}{(a\sqrt{2})^2}\left(\frac{\hat{i}+\hat{j}}{\sqrt{2}}\right)$, $\vec{F}_{CB} = k\frac{q^2}{a^2}\hat{j}$, $\vec{F}_{CD} = k\frac{q^2}{a^2}\hat{i}$,

and $\vec{F}_{CO} = k\frac{qQ}{(a/\sqrt{2})^2}\left(\frac{\hat{i}+\hat{j}}{\sqrt{2}}\right) = k\frac{2qQ}{a^2}\left(\frac{\hat{i}+\hat{j}}{\sqrt{2}}\right)$

Substituting these values in Eq.1.26, we get-

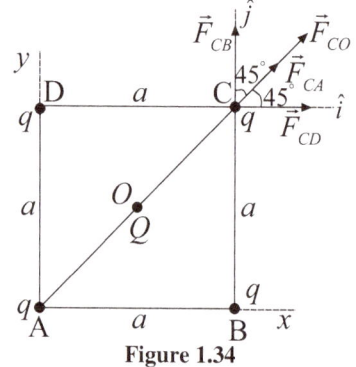

Figure 1.34

$\Sigma \vec{F} = k\frac{q}{a^2}\left[\left(\frac{q}{2\sqrt{2}} + q + Q\sqrt{2}\right)\hat{i} + \left(\frac{q}{2\sqrt{2}} + q + Q\sqrt{2}\right)\hat{j}\right]$

In equilibrium, $\Sigma \vec{F} \Rightarrow \frac{q}{2\sqrt{2}} + q + Q\sqrt{2} = 0$

$\Rightarrow \boxed{Q = \frac{-q(2\sqrt{2}+1)}{4}}$ (1.27)

Here, again negative sign shows that Q will be opposite in nature to q.

Comment: Students are advised to keep results 1.25 and 1.27 in their memory for simplifying such types of problems in a relatively quicker way.

1.10.3 Equilibrium of Suspended Point Charge System

If two charged point objects are suspended by threads in a given medium, then in equilibrium state, net force on each object must be zero. To solve such kind of problems, make FBD of of any one object (if necessary make FBDs of both objects). Resolve all forces acting on any one object in to two mutually perpendicular components along x and y direction (say). If ΣF_x and ΣF_y are the net forces along x and y directions, then, in equilibrium, $\Sigma F_x = 0$ and $\Sigma F_y = 0$. In other words, net force acting along positive direction of x axis must get balanced with net force acting in negative x direction and also net force acting along positive direction of y axis must get balanced with the net force acting in negative y direction. Don't forget to consider upthrust applied by medium if it's density is not negligible as compared to water. If V is the immersed volume of object in the medium, g is gravitational acceleration and ρ is the density of medium, then upthrust applied by medium on the object is $V\rho g$. The upthrust force of air can be neglected because average density of air at 1 atm and 20° is 1.204kg/m³ which is very small as compared to density of water which is 10^3kg/m³ at 1 atm and 20°.

To understand the working see following solved examples:

EXAMPLE 19. A simple electroscope for the detection and measurement of electric charge consists of two small foil-covered cork balls of 1.5×10^{-4}kg each suspended by threads 10 cm

1.10. EQUILIBRIUM OF CHARGED PARTICLES

long (see Fig.1.35). When equal electric charges are placed on the balls, the electric repulsive force pushes them apart, and the angle between the threads indicates the magnitude of the electric charge. If the equilibrium angle between the threads is 60°, what is the magnitude of the charge?

APPROACH Make FBD of any one sphere. Resolve all forces

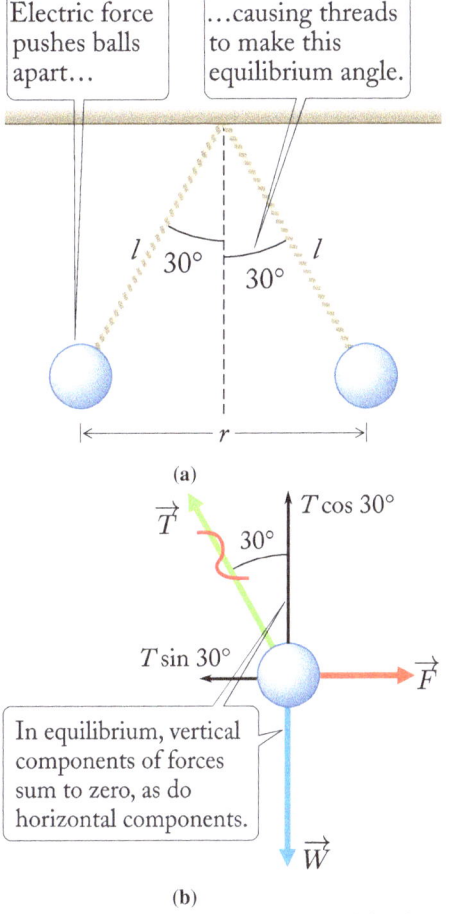

Figure 1.35: (a) Two equal charged balls suspended by threads. (b) "Free-body" diagram for the right ball.

acting on it in to two mutually perpendicular components along x and y direction (say). If ΣF_x and ΣF_y are the net forces along x and y directions, then, in equilibrium, $\Sigma F_x = 0$ and $\Sigma F_y = 0$.

SOLUTION Figure 1.35b shows a "free-body" diagram for right ball. The forces acting on the right ball are-
1. Electrostatic force F_e (along positive x direction)
2. Gravitational force mg (along negative y direction)
3. tension applied by string T (away from sphere along the string) x and y components of tension T are $T \sin 30°$ and $T \cos 30°$ respectively.

Net force on the ball, along x axis is-
$$\Sigma F_x = F - T \sin 30°$$
Net force on the sphere, along y axis is-
$$\Sigma F_y = T \cos 30° - W = T \cos 30° - mg$$
In equilibrium, $\sum F_x = 0 \Rightarrow F = T \sin 30°$ $\cdots$ (1)
and $\sum F_y = 0 \Rightarrow mg = T \cos 30°$ $\cdots$ (2)
We can eliminate the tension from above equations by taking the ratio of these equations, yielding
$$F = mg \tan 30° \quad \cdots (3)$$
From Fig.1.35a, we see that the distance between the balls is $r = 2l \sin 30°$, so, Coulomb's Law gives
$$F = \frac{1}{4\pi\epsilon_0} \frac{q^2}{(2l \sin 30°)^2} \quad \cdots (4)$$

Equating Eq.(3) and (4) for F, we get
$$mg \tan 30° = \frac{1}{4\pi\epsilon_0} \frac{q^2}{(2l \sin 30°)^2}$$
and $q = \sqrt{4\pi\epsilon_0 mg \tan 30°} \times 2l \sin 30°$
$= [(4\pi) \left(8.85 \times 10^{-12} \mathrm{C}^2/\mathrm{N \cdot m}^2\right) \left(1.5 \times 10^{-4} \mathrm{~kg}\right)$
$\times \left(9.81 \mathrm{~m/s}^2\right) (\tan 30°)]^{1/2} \times (2)(0.10 \mathrm{~m})(\sin 30°)$
$= 3.1 \times 10^{-8} \mathrm{C}$

EXAMPLE 20. Two identically charged spheres are suspended by strings of equal length L. The charge on each sphere is q and mass is m. Find the separation between them in equilibrium- (a) if the the system is located in gravitational field of earth, (b) if the system is located in gravity free space.

APPROACH Make FBD of either sphere. Resolve all forces

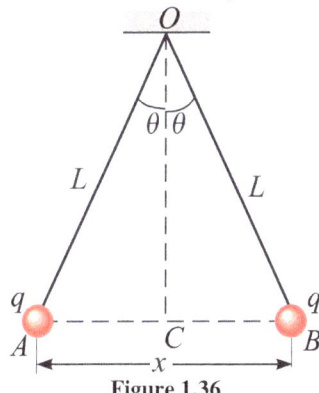

Figure 1.36

acting on it in to two mutually perpendicular components one along x and other along y direction (say). If ΣF_x and ΣF_y are the net forces along x and y directions, then, in equilibrium, $\Sigma F_x = 0$ and $\Sigma F_y = 0$.

If whole setup is placed in gravity free space, then there will be no downward gravitational force on any sphere. In this case, the angle between the strings becomes 180°.

SOLUTION FBD of left sphere is shown in Fig.1.37. The forces acting on the left sphere are-
1. Electrostatic force F_e (along negative x direction)
2. Gravitational force mg (along negative y direction)
3. tension applied by string T (away from sphere along the string) x and y components of tension T are $T \sin \theta$ and $T \cos \theta$ respectively. The separation between the spheres in equilibrium is x.
Net force on the sphere, along x axis is-
$$\Sigma F_x = T \sin \theta - F_e$$
Net force on the sphere, along y axis is-
$$\Sigma F_y = T \cos \theta - mg$$
In equilibrium,
$$\Sigma F_x = 0 \Rightarrow T \sin \theta = F_e \quad (i)$$
$$\text{and} \quad \Sigma F_y = 0 \Rightarrow T \cos \theta = mg \quad (ii)$$
On dividing Eq. (i) by (ii), we get-
$$\tan \theta = \frac{F_e}{mg}$$
Here, electrostatic force, $F_e = k \frac{q^2}{x^2}$, therefore,
$$\tan \theta = \frac{kq^2}{mgx^2}$$
If θ is small then-
$$\tan \theta \approx \sin \theta = \frac{x}{2L} \Rightarrow \frac{x}{2L} = \frac{kq^2}{x^2 mg}$$

Figure 1.37

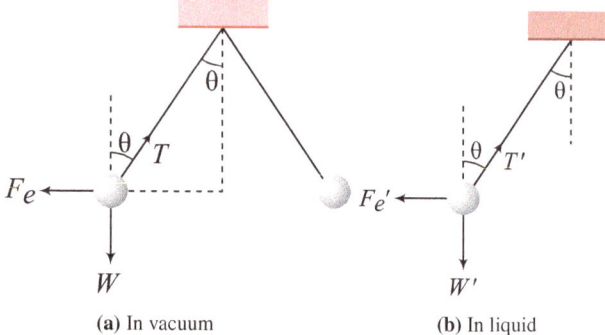

(a) In vacuum (b) In liquid
Figure 1.39

$$\Rightarrow \quad x^3 = \frac{2kq^2 L}{mg} \Rightarrow x = \left[\frac{q^2 L}{2\pi \epsilon_0 \, mg}\right]^{\frac{1}{3}}$$

(b) If whole set up is taken into a gravity free space or in an artificial satellite ($g_{\text{eff}} \simeq 0$), then from FBD shown in Fig.1.38

$$T = F_e = \frac{kq^2}{4L^2}$$

EXAMPLE 21. Two identical balls each having a density

Figure 1.38

ρ are suspended from a common point by two insulating strings of equal length. Both the balls have equal mass and charge. In equilibrium each string makes an angle θ with vertical. Now, both the balls are immersed in a liquid. As a result the angle θ does not change. The density of the liquid is σ. Find the dielectric constant of the liquid.

APPROACH Make FBD of either ball in air as well as in liquid. Write the force balance equations for equilibrium in air and in liquid and then solve for dielectric constant.

SOLUTION Each ball is in equilibrium under the following three forces:
(i) tension,
(ii) electric force and
(iii) weight, so, Lami's theorem* can be applied.

The weight of ball in air, $W = mg = V\rho g$
The upthrust applied by liquid is, $F_{th} = V\sigma g$
Therefore, the apparent weight of the ball in liquid,
$$W' = W - F_{th} = V\rho g - V\sigma g = V(\rho - \sigma)g$$
The electrostatic force between the balls in liquid-
$$F'_e = \frac{F_e}{\kappa}$$
where, F_e is the electrostatic force between the balls in air and κ = dielectric constant of liquid.

Applying Lami's theorem in vacuum (Fig.1.39a)
$$\frac{W}{\sin(90° + \theta)} = \frac{F_e}{\sin(180° - \theta)}$$
$$\Rightarrow \quad \frac{W}{\cos\theta} = \frac{F_e}{\sin\theta} \quad (1.28)$$

Similarly in liquid (Fig.1.39b),

---

*__Lami's Theorem:__ If $\vec{P}, \vec{Q}$ and $\vec{R}$ are three concurrent forces such that the angle between $\vec{Q}$ and $\vec{R}$ is α, between $\vec{R}$ and $\vec{P}$ it is β and between $\vec{P}$ and $\vec{Q}$ the angle is γ (Fig.1.49a), then in equilibrium-
$$\boxed{\frac{P}{\sin\alpha} = \frac{Q}{\sin\beta} = \frac{R}{\sin\gamma}}$$

$$\frac{W'}{\cos\theta} = \frac{F'_e}{\sin\theta} \quad (1.29)$$

Dividing Eq.(1.28) by Eq. (1.29), we get-
$$\frac{F}{F'} = \frac{W}{W'} = \frac{V\rho g}{V(\rho - \sigma)g} \quad (V = \text{volume of ball})$$
or $$\kappa = \frac{\rho}{\rho - \sigma}$$

Note: In the liquid F_e and W have changed. Therefore, T will also change.

EXAMPLE 22. For the system shown in Fig.1.40a, find Q for which resultant force on q is zero.

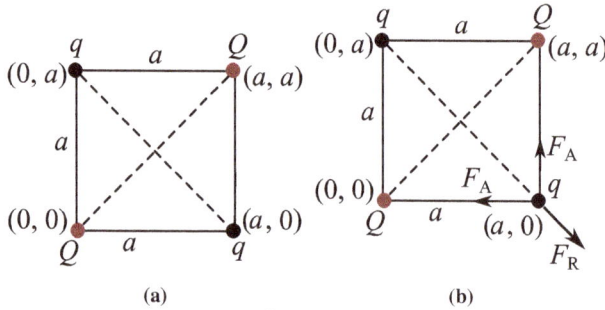

Figure 1.40

APPROACH Make FBD of charge q. Write force balance equations and simplify for required value of Q.

SOLUTION FBD of charge q at $(a, 0)$ is shown in Fig.1.40b. For force on q at $(a, 0)$ to be zero, charges q and Q must be of opposite nature.

Net force of attraction of both Q on q at $(a, 0)$ = force of repulsion of q at $(0, a)$ on q at $(a, 0)$.
$$\sqrt{F_A^2 + F_B^2 + 2F_A F_B \cos 90} = F_R$$

Here, F_A is the force between each Q and q whereas F_R is the repulsion between each q and q.

Simplifying above equation, we get
$$\sqrt{2} F_A = F_R \Rightarrow \sqrt{2}\frac{kQq}{a^2} = \frac{kq^2}{(\sqrt{2}a)^2}$$
$$\Rightarrow \quad q = 2\sqrt{2}Q.$$

Therefore, the required charge = $-2\sqrt{2}\, Q$

EXAMPLE 23. Two identically charged spheres are suspended by strings of equal length. The strings make an angle of 30° with each other in air (Fig.1.41). When suspended in a liquid of density 0.8 g/cc the angle remains same. What is the dielectric constant of liquid. Density of sphere = 1.6 g/cc.

APPROACH Make FBD of either ball in air and in liquid. Write translational equilibrium conditions in both case. Now simplify for dielectric constant κ of liquid.

SOLUTION In air: This situation is shown in Fig.1.42a. The forces acting on the left charged sphere are-

1.10. EQUILIBRIUM OF CHARGED PARTICLES

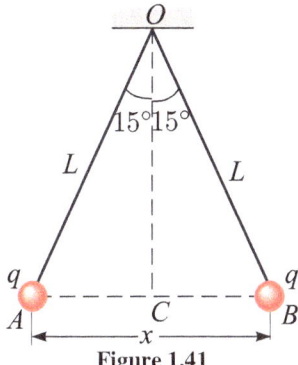

Figure 1.41

1. Gravitational force mg, acting in downward direction,
2. Electrostatic repulsive force F_e, acting towards left,
3. Tension in the string T, acting upward along the string.

In equilibrium, all forces and their components acting on left charged sphere, are shown in Fig.1.42
In horizontal direction we have-

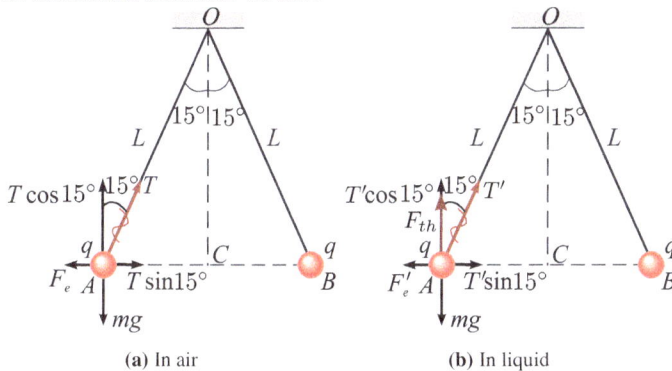

Figure 1.42

$$T \sin 15° = F_e \qquad (i)$$

In vertical direction we have-

$$T \cos 15° = mg \qquad (ii)$$

On dividing Eq. (i) by Eq. (ii), we get-

$$\boxed{\tan 15° = \frac{F_e}{mg}} \qquad (iii)$$

In liquid: This situation is shown in Fig.1.42a. When set up is immersed in the liquid medium as shown in Figure 1.42b, the forces on the sphere are -
1. Gravitational force mg, acting in downward direction,
2. Electrostatic repulsive force F_e', acting towards left,
3. Tension in the string T', acting upward along the string.
4. Upthrust of liquid F_{th}, acting in upward direction

In equilibrium, all forces and their components acting on left charged sphere, are shown in Fig.1.42b
In horizontal direction we have -

$$T' \sin 15° = F_e' \qquad (iv)$$

In vertical direction we have-

$$T' \cos 15° = mg - F_{th} \qquad (v)$$

On dividing Eq. (iv) by Eq. (v), we get-

$$\boxed{\tan 15° = \frac{F_e'}{mg - F_{th}}} \qquad (vi)$$

From Eq. (iii) and (vi), we can write-

$$\frac{F_e}{mg} = \frac{F_e'}{mg - F_{th}}$$

$$\Rightarrow \frac{F_e}{mg} = \frac{F_e/\kappa}{mg - F_{th}} \qquad [\because F_e' = F_e/\kappa]$$

$$\Rightarrow \kappa = \frac{mg}{mg - F_{th}}$$

$$\Rightarrow \kappa = \frac{mg}{mg - V\rho_l g} \qquad [\because F_{th} = V\rho_l g]$$

$$\Rightarrow \kappa = \frac{1}{1 - \frac{V\rho_l}{m}} = \frac{1}{1 - \frac{V\rho_l}{V\rho_s}} \qquad [\because m = V\rho_s]$$

$$\Rightarrow \kappa = \frac{1}{1 - \frac{0.8}{1.6}} = 2$$

EXAMPLE 24. Two small equally charged spheres, each of mass m, are suspended from the same point by silk threads of length l. In equilibrium, the distance between the spheres is x ($x \ll l$). Find the rate dq/dt with which the charge leaks off each sphere if their approach velocity varies as $v = a/\sqrt{x}$, where a is a constant.

APPROACH Initially, the charged spheres were in equilibrium when they were separated by distance x. Now consider only left sphere in equilibrium. The forces acting on this sphere are-

1. Gravitational force $m\vec{g}$ - acting downwards.
2. Tension applied by thread (T) - acting away along the thread.
3. Electrostatic repulsion F_e applied by second sphere- acting leftwards.

If ΣF_x and ΣF_y are the net electrostatic forces acting on the left sphere along x and y direction respectively, then, in equilibrium -
(i) $\Sigma F_x = T \sin\theta - F_e = 0 \quad \Rightarrow \quad T \sin\theta = F_e$, and
(ii) $\Sigma F_y = T \cos\theta - mg = 0 \quad \Rightarrow \quad T \cos\theta = mg$

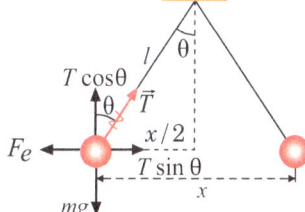

Figure 1.43

Dividing condition (i) by (ii), we get -

$$\tan\theta = \frac{F_e}{mg} \qquad (i)$$

But from the Figure 1.43,

$$\tan\theta = \frac{x}{2\sqrt{l^2 - \left(\frac{x}{2}\right)^2}} \cong \frac{x}{2l} \text{ as } x \ll l \qquad (ii)$$

and $$F_e = \frac{q^2}{4\pi\epsilon_0 x^2} \qquad (iii)$$

Now, substitute the values obtained in Eq.(ii) and (iii) in Eq.i and simplify for q. From here you will get a relation between q and x, i.e., $q = f(x)$ in equilibrium.

Differentiation of this relation with respect to time 't', will give you dq/dt, i.e., the rate of leakage of charge.

SOLUTION On substituting the values $\tan\theta$ and F_e from Eq.ii and iii in Eq. i, we get-

$$\frac{x}{2l} = \frac{q^2}{4\pi\epsilon_0 x^2 mg}$$

or $$q^2 = \frac{2\pi\epsilon_0 mg x^3}{l} \qquad (iii)$$

On differentiating Eq.(iii) with respect to time 't', we get-

$$2q\frac{dq}{dt} = \frac{2\pi\epsilon_0 mg}{l} 3x^2 \frac{dx}{dt}$$

According to the problem $\frac{dx}{dt} = v = \frac{a}{\sqrt{x}}$ (approach velocity is dx/dt) so,

$$\left(\frac{2\pi\epsilon_0 mg}{l} x^3\right)^{1/2} \frac{dq}{dt} = \frac{3\pi\epsilon_0 mg}{l} x^2 \frac{a}{\sqrt{x}}$$

Hence, $\frac{dq}{dt} = \frac{3}{2} a \sqrt{\frac{2\pi\epsilon_0 mg}{l}}$

EXAMPLE 25. Three identical charged balls each of charge 2 C are suspended from a common point P by silk threads of 2 m each (as shown in Fig.1.44). They form an equilateral triangle of side 1 m. Find the ratio of net electric force on a charged ball to the force between any two charged balls.

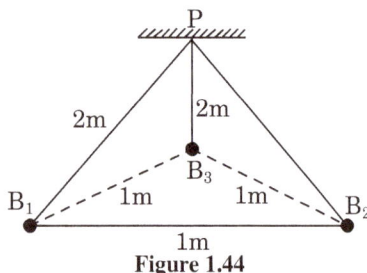

Figure 1.44

SOLUTION Fig.1.45 shows the FBD of ball B_2. Each ball has equal charge. Since, charged balls form equilateral triangle, therefore, the angle between forces applied by charged balls B_1 and B_3 on B_2 is $60°$.

If F is the magnitude of force applied by each charged particle

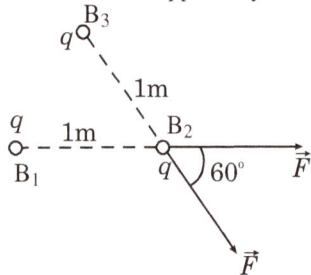

Figure 1.45

B_1 and B_3 on the particle B_2, then

$$F = k\frac{q_1 q_2}{r^2} = \frac{k(2)(2)}{(1)^2}$$

Therefore, net electric force on charged particle B_2 is

$$F_{B_2} = \sqrt{F^2 + F^2 + 2FF\cos 60°} = F\sqrt{3}$$

Therefore, $\frac{F_{net}}{F} = \frac{F\sqrt{3}}{F} = \sqrt{3}$

EXAMPLE 26. Given a cube with point charges q on each of its vertices. Calculate the force exerted on any of the charges due to rest of the 7 charges.

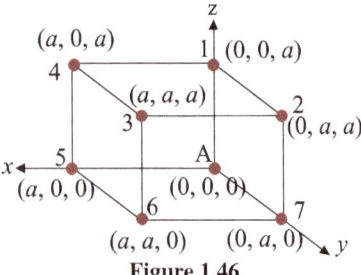

Figure 1.46

APPROACH All charges (1 to 7) are applying electrostatic forces simultaneously on charge at A. Therefore by applying superposition principle, you can find the net force on the charge at A.

SOLUTION The net force on particle A can be given by vector sum of force experienced by this particle due to all the other charges on vertices of the cube. For this we use vector form of coulomb's law $\vec{F} = \frac{kq_1 q_2}{|\vec{r}_1 - \vec{r}_2|^3}(\vec{r}_1 - \vec{r}_2)$

From the Fig.1.46 the forces acting on A due to charges at 1, 2, 3, ..., 7, are given as -

$\vec{F}_{A_1} = \frac{kq^2(-a\hat{k})}{a^3}$, $\vec{F}_{A_2} = \frac{kq^2(-a\hat{j}-a\hat{k})}{(\sqrt{2}a)^3}$,

$\vec{F}_{A_3} = \frac{kq^2(-a\hat{i}-a\hat{j}-a\hat{k})}{(\sqrt{3}a)^3}$, $\vec{F}_{A_4} = \frac{kq^2(-a\hat{i}-a\hat{k})}{(\sqrt{2}a)^3}$

$\vec{F}_{A_5} = \frac{kq^2(-a\hat{i})}{a^3}$, $\vec{F}_{A_6} = \frac{kq^2(-a\hat{i}-a\hat{j})}{(\sqrt{2}a)^3}$, $\vec{F}_{A_7} = \frac{kq^2(-a\hat{j})}{a^3}$

respectively

By superposition principle, the net force experienced by A can be given as -

$\vec{F}_{net} = \vec{F}_{A_1} + \vec{F}_{A_2} + \vec{F}_{A_3} + \vec{F}_{A_4} + \vec{F}_{A_5} + \vec{F}_{A_6} + \vec{F}_{A_7}$

$= \frac{-kq^2}{a^2}\left[\left(\frac{1}{3\sqrt{3}} + \frac{1}{\sqrt{2}} + 1\right)(\hat{i} + \hat{j} + \hat{k})\right]$

EXAMPLE 27. Five point charges, each of value $+q$ are placed at five vertices of a regular hexagon of side L m. What is the magnitude of the force on a point charge of value $-q$ coulomb placed at the centre of the hexagon?

APPROACH If there had been a sixth charge $+q$ at the remaining vertex of hexagon, the net force due to all the six charges on $-q$ at O would be zero (as the forces due to individual charges will balance each other).

Now if $\vec{f}$ is the force due to sixth charge and $\vec{F}$ due to remaining

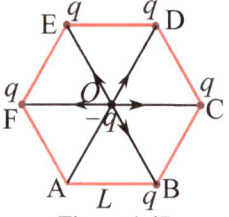

Figure 1.47

five charges. $\vec{F} + \vec{f} = 0 \Rightarrow \vec{F} = -\vec{f} \Rightarrow F = f$

Now, by using Coulomb's law, find f

SOLUTION By Coulomb's law, we have

$$f = k\frac{q_1 q_2}{r^2} = \frac{1}{4\pi\epsilon_0} \frac{q^2}{L^2}$$

$$= \frac{1}{4\pi\epsilon_0} \frac{q \times q}{L^2} = \frac{q^2}{4\pi\epsilon_0 L^2}$$

EXAMPLE 28. A particle A having a charge of 5.0×10^{-7}C is fixed in a vertical wall. A second particle B of mass 100 g and having equal charge is suspended by a silk thread of length 30 cm from the wall. The point of suspension is 30 cm above the particle A. Find the angle of the thread with the vertical when it stays in equilibrium.

APPROACH The situation is shown in Fig.1.48. Suppose the point of suspension is O and let θ be the angle between the thread and the vertical. Forces on the particle B are

(i) weight mg downwards
(ii) tension T along the thread and
(iii) electric force of repulsion F along AB.

For equilibrium, these forces should add to zero.
We will be solving the problem by using Lami's theorem. The distance between A and B can be obtained by sine rule.

Sine Rule: In any $\triangle ABC$ (Fig.1.49b), we have-

1.10. EQUILIBRIUM OF CHARGED PARTICLES

Figure 1.48

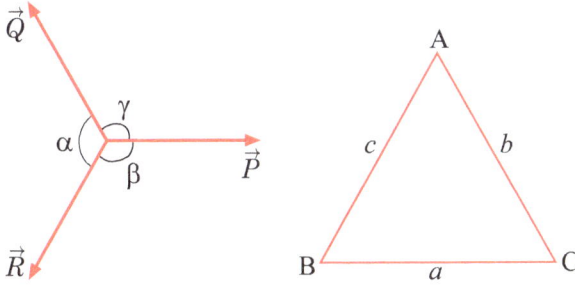

(a) Lami's Theorem (b) Sine Rule
Figure 1.49

$$\frac{a}{\sin A} = \frac{b}{\sin B} = \frac{c}{\sin C}$$

here a, b and c are sides against angles A, B and C respectively.

SOLUTION The FBD of problem is shown in Fig.1.48. Let B is the equilibrium position of ball and $\angle AOB = \theta$. Let us draw a perpendicular BL on side OA.
Note that $OA = OB$, so ΔOAB is an isosceles triangle.
Therefore, $\angle OAB = \angle OBA = \alpha$ (say)
So, in ΔOAB, $\angle AOB + \angle OAB + \angle OBA = 180°$
$\Rightarrow \theta + \alpha + \alpha = 180°$
$\Rightarrow 2\alpha = 180° - \theta \Rightarrow \alpha = 90° - \theta/2$
Again in right angle ΔBLA, $\alpha + 90° + \angle LBA = 180°$
$\Rightarrow \angle LBA = 90° - \alpha = 90° - (90° - \frac{\theta}{2})$
$= \frac{\theta}{2} = \angle CBX$ (opposite angles)

All angles between forces on the particle are shown in Fig.1.48.
By Lami's theorem at B, we have-

$$\frac{F_e}{\sin \angle OBE} = \frac{mg}{\sin \angle OBC}$$

$$\Rightarrow \frac{F_e}{\sin(180 - \theta)} = \frac{mg}{\sin(\theta + 90° - \frac{\theta}{2})}$$

$$\Rightarrow \frac{F_e}{\sin \theta} = \frac{mg}{\sin(90° + \frac{\theta}{2})} \Rightarrow \frac{F_e}{\sin \theta} = \frac{mg}{\cos \frac{\theta}{2}}$$

$$\Rightarrow \frac{F_e}{2 \sin \frac{\theta}{2} \cos \frac{\theta}{2}} = \frac{mg}{\cos \frac{\theta}{2}}$$

$$\Rightarrow F_e \cos \frac{\theta}{2} - 2mg \sin \frac{\theta}{2} \cos \frac{\theta}{2} = 0$$

$$\Rightarrow \cos \frac{\theta}{2} \left[F_e - 2mg \sin \frac{\theta}{2} \right] = 0$$

If $\cos \frac{\theta}{2} = 0$, then $\frac{\theta}{2} = \frac{\pi}{2}$ or $\theta = \pi$, which is not possible.

Now, $F_e - 2mg \sin \frac{\theta}{2} = 0$

$$\Rightarrow \sin \frac{\theta}{2} = \frac{F_e}{2mg} \qquad (i)$$

From sine rule, we have-

$$\frac{AB}{\sin \theta} = \frac{OA}{\sin \alpha} \Rightarrow AB = \frac{OA \sin \theta}{\sin \alpha}$$

$$\Rightarrow AB = \frac{OA (2 \sin \frac{\theta}{2} \cos \frac{\theta}{2})}{\sin(90° - \frac{\theta}{2})} = \frac{2 OA \sin \frac{\theta}{2} \cos \frac{\theta}{2}}{\cos \frac{\theta}{2}}$$

$$\Rightarrow AB = 2 OA \sin \frac{\theta}{2}$$

Electrostatic force between A and B is,

$$F_e = k \frac{q^2}{(AB)^2} = k \frac{q^2}{(2 OA \sin \frac{\theta}{2})^2} = k \frac{q^2}{4 OA^2 \sin^2 \frac{\theta}{2}}$$

On substituting this value in Eq.(i), we get -

$$\sin \frac{\theta}{2} = k \frac{q^2}{4(OA)^2 \sin^2 \frac{\theta}{2} (2mg)}$$

$$\Rightarrow \sin^3 \frac{\theta}{2} = k \frac{q^2}{8(OA)^2 mg}$$

On substituting given values in above expression, we get-

$$\sin^3 \frac{\theta}{2} = \frac{9 \times 10^9 \times 25 \times 10^{-14} \text{ N}}{4 \times 9 \times 10^{-2} \times 2 \times (100 \times 10^{-3} \text{ kg}) \times 9 \cdot 8 \text{ m s}^{-2}}$$

$$= 0 \cdot 0032$$

$$\Rightarrow \sin \frac{\theta}{2} = 0 \cdot 15 \Rightarrow \theta = 17°$$

EXAMPLE 29. A particle 'A' having a charge of 2×10^{-6} C and a mass of 100 g is fixed at the bottom of a smooth inclined plane of inclination 30°. Where should another particle B, having same charge and mass be placed on the incline so that it may remain in equilibrium?

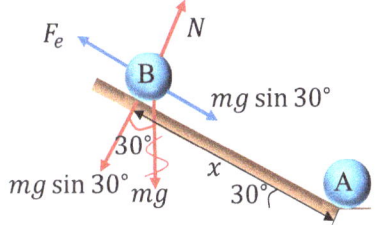

Figure 1.50

APPROACH Make FBD for B and apply the condition of translational static equilibrium along and perpendicular to the inclined plane of Fig.1.50

SOLUTION Suppose, the equilibrium distance of particle B from A is x. FBD of particle B, is shown in Fig.1.50.
The forces acting on the particle B are-
1. Gravitational force: mg (downwards)
2. Normal reaction of the plane: N (perpendicular to the plane in upward sense)
3. Electrostatic repulsion of A: F_e (along the plane in upward sense)

We resolve the gravitational force mg into components— one ($mg \sin 30°$) along the plane and other ($mg \cos 30°$) perpendicular to it.

In static equilibrium, we have -

Along the plane: $mg \sin 30° = F_e = \frac{kq^2}{x^2}$

$$\Rightarrow \frac{mg}{2} = \frac{kq^2}{x^2} \Rightarrow x = \sqrt{\frac{2kq^2}{mg}}$$

On substituting the given values- $q = 2 \times 10^{-6}$ C, $m = 0.1$ kg and $k = 9.0 \times 10^9$ N.m^2/C^2, we get -
$x = 0.27$ m $= 27$ cm

EXAMPLE 30. A rigid insulated wire frame in the form of a right angled triangle ABC, is set in a vertical plane as shown in the figure. Two beads of equal masses m each and carrying charges q_1 and q_2 are connected by a cord of length l and can slide without friction on the wires. Considering the case when

the beads are stationary, determine.
(a) (i) The angle α. (ii) The tension in the cord. (iii) The normal

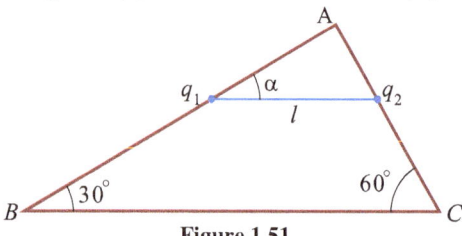

Figure 1.51

reaction on the beads.
(b) If the cord is now cut what are the values of the charges for which the beads continue to remain stationary?
APPROACH Make FBDs of both the beads. Write translational equilibrium conditions along and perpendicular to their respective planes and then simplify them for required variables.
SOLUTION The forces acting on the charge q_1 are its weight mg, normal reaction N_1, string tension T and Coulomb's force $F_e = q_1q_2/(4\pi\epsilon_0 l^2)$. Similarly, forces on the charge q_2 are its weight mg, normal reaction N_2, string tension T and Coulomb's force $F_e = q_1q_2/(4\pi\epsilon_0 l^2)$ The forces are shown in the Fig.1.52.

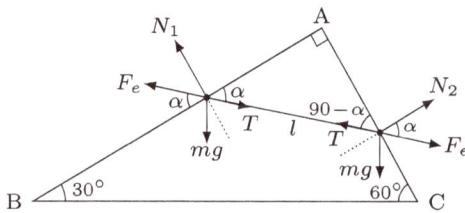

Figure 1.52

In triangle ABC, $\angle A = 90°$. Resolve the forces on q_1 and q_2 in the directions parallel and perpendicular to the sides of the frame. Balancing the forces on q_1, we get-

$$N_1 + F_e \sin\alpha = mg \cos 30° + T \sin\alpha \qquad (1.30)$$

$$F_e \cos\alpha + mg \sin 30° = T \cos\alpha \qquad (1.31)$$

and on q_2 to get

$$N_2 + F_e \cos\alpha = mg \cos 60° + T \cos\alpha, \qquad (1.32)$$

$$F_e \sin\alpha + mg \sin 60° = T \sin\alpha \qquad (1.33)$$

On simplifying equations (1.31) and (1.33), we get $\alpha = 60°$ and $T = mg + F_e$
Now, substitution of these values in equations (1.30) and (1.32) gives $N_1 = \sqrt{3}mg$ and $N_2 = mg$.
The tension becomes zero when the cord is cut. Substituting $T = 0$ in $T = mg + F_e$ gives $F_e = -mg$ i.e., $q_1q_2 = -4\pi\epsilon_0 l^2 mg$
Ans. (a) (i) $60°$ (ii) $\frac{1}{4\pi\epsilon_0}\frac{q_1q_2}{l^2} + mg$ (iii) $\sqrt{3}mg, mg$ (b) $q_1q_2 = -4\pi\epsilon_0 mgl^2$

EXAMPLE 31. Three particles, each of mass 1 g and carrying a charge q, are suspended from a common point by insulated massless strings, each 100 cm long. If the particles are in equilibrium and are located at the corners of an equilateral triangle of side length 3 cm, calculate the charge q on each particle. [Take $g = 10$ m/s^2.]
APPROACH Make FBD of either particle and apply the condition of translational equilibrium
SOLUTION Let the three charged particles, each of charge q, be located at the three corners of an equilateral triangle of side $a = 3$ cm in the horizontal plane. The distance of the charges from the centroid O is r. On applying sine rule in triangle OBC, we get -

$$\frac{r}{\sin \angle OBC} = \frac{a}{\sin \angle BOC}$$

$$\Rightarrow \frac{r}{\sin 30°} = \frac{a}{\sin 120°} \Rightarrow r = a/\sqrt{3} = \sqrt{3} \text{ cm}.$$

Consider the electrostatic force on the charge at A due to the

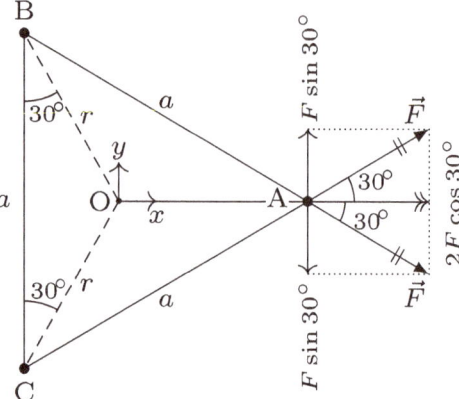

Figure 1.53

charges at B and C. The magnitude of forces due to each charge is

$$|\vec{F}| = \frac{q^2}{4\pi\epsilon_0 a^2}$$

and their directions are as shown in the Fig.1.53. Resolve $\vec{F}$ along and perpendicular to AO. The resultant electrostatic force on charge at A is given by

$$\vec{F}_e = 2F \cos 30° \hat{\imath} = \frac{\sqrt{3}q^2}{4\pi\epsilon_0 a^2}\hat{\imath}$$

The charges are connected with strings of length $l = 100$ cm. Let

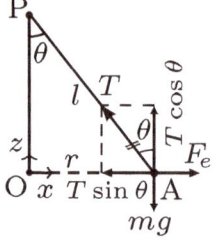

Figure 1.54

the other end of each string is hanging from the point P, which is vertically above O (Fig.1.54). In triangle OAP,

$$\tan\theta = \frac{r}{\sqrt{l^2 - r^2}} = \frac{\sqrt{3}}{\sqrt{100^2 - 3}} = 0.017$$

The forces on the charge at A are F_e, tension T, and weight mg. Resolve T in the horizontal and the vertical directions. The equilibrium condition on charge at A gives

$$T \sin\theta = F_e = \frac{\sqrt{3}q^2}{4\pi\epsilon_0 a^2} \qquad (1)$$

$$T \cos\theta = mg \qquad (2)$$

On dividing equation (1) by (2), we get

$$q = \left[\frac{4\pi\epsilon_0 a^2 mg \tan\theta}{\sqrt{3}}\right]^{1/2}$$

$$= \left[\frac{(0.03)^2 (10^{-3})(10)(0.017)}{9 \times 10^9 (1.73)}\right]^{1/2} = 3.17 \times 10^{-9} \text{C}$$

1.11 Check Point 2

1. •Suppose that the electric force between two point charges is attractive. What can you conclude about the signs of these point charges?
2. •Can two similarly charged bodies attract each other?
3. •Does in charging the mass of a body change?
4. •Why a third hole in a socket provided for grounding?

5. •Two balls, separated by some distance, carry equal electric charges and exert a repulsive electric force on each other. If we transfer a fraction of the electric charge of one ball to the other, will the electric force increase or decrease?

6. ••What would be the interaction force between two copper spheres, each of mass $1g$, separated by the distance 1 m, if the total electronic charge in them differed from the total charge of the nuclei by one percent? [Given: molar mass of copper = 63.54 gram, atomic number of copper = 29]

7. ••Six charges are kept at the vertices of a regular hexagon as shown in the Fig.1.55. If magnitude of force applied by $+Q$ on $+q$ charge is F, then net electric force on the $+Q$ is nF. Find the value of n.

Figure 1.55

Figure 1.56

8. ••Two stiff non conducting rods have length L each and have small balls connected to their ends. The rods are placed parallel to each other and the balls are connected by two identical springs as shown in Fig.1.56. When each ball is given a charge q, the system stays in equilibrium when it is in the shape of a square. If natural relaxed length of each spring is $L/2$ find the force constant (k) for them.

9. ••A charge Q is to be divided on two objects. What should be the values of the charges on the objects so that the force between the objects can be maximum?

10. ••Will a point charge have a position of stable equilibrium when it is at the mid-point between two other equal point charges, the signs of which are either the same or opposite to that of the first charge?

11. ••Three equal charges q of the same sign are located at the vertices of an equilateral triangle. What charge Q of opposite sign must be placed at the centre of the triangle in order for the resultant of the forces acting on each charge to be zero?

Multiple Choice Questions

12. ••Two point charges A and B, halving charges $+Q$ and $-Q$ respectively, are placed at certain distance apart and force acting between them is F. If 25% charge of A is transferred to B, then force between the charges becomes -
    (A) $\frac{4F}{3}$ (B) F (C) $\frac{9F}{16}$ (D) $\frac{16F}{9}$

13. ••Suppose the charge of a proton and an electron differ slightly. One of them is $-e$, the other is $(e + \Delta e)$. If the net of electrostatic force and gravitational force between two hydrogen atoms placed at a distance d (much greater than atomic size) apart is zero, then Δe is of the order of [Given: mass of hydrogen $m_h = 1.67 \times 10^{-27}$ kg]
    (A) 10^{-23}C (B) 10^{-37}C (C) 10^{-47}C (D) 10^{-20}C

14. ••Two identical charged spheres suspended from a common point by two massless strings of lengths l, are initially at a distance $d(d \ll l)$ apart because of their mutual repulsion. The charges begin to leak from both the spheres at a constant rate. As a result, the spheres approach each other with a velocity v. Then v varies as a function of the distance x between the spheres, as
    (A) $v \propto x^{-1/2}$ (B) $v \propto x^{-1}$ (C) $v \propto x^{1/2}$ (D) $v \propto x$

15. ••Three identically charged, small spheres each of mass m are suspended from a common point by insulated light strings each of length l. The spheres are always on vertices of an equilateral triangle of length of the sides $x(\ll l)$. The rate dq/dt with which charge on each sphere increases, if length of the sides of the equilateral triangle increases slowly according to law $\frac{dx}{dt} = \frac{a}{\sqrt{x}}$, is -
    (A) $\sqrt{\frac{3\pi\epsilon_0 mga^2}{l}}$ (B) $\sqrt{\frac{\pi\epsilon_0 mga^2}{l}}$
    (C) $\sqrt{\frac{\pi\epsilon_0 mga^2}{3}}$ (D) $\sqrt{\frac{2\pi\epsilon_0 mga^2}{l}}$

16. ••Two pith* balls carrying equal charges are suspended from a common point by strings of equal length, the equilibrium separation between them is r. Now the strings are rigidly clamped at half the height. The equilibrium separation between the balls now become [Fig.1.57]
    (A) $\left(\frac{2r}{\sqrt{3}}\right)$ (B) $\left(\frac{2r}{3}\right)$ (C) $\left(\frac{1}{\sqrt{2}}\right)^2$ (D) $\left(\frac{r}{\sqrt[3]{2}}\right)$

Figure 1.57

17. ••Two positive ions, each carrying a charge q, are separated by a distance d. If F is the force of repulsion between the ions, the number of electrons missing from each ion will be (e being the charge on an electron)
    (A) $\frac{4\pi\epsilon_0 Fd^2}{e^2}$ (B) $\sqrt{\frac{4\pi\epsilon_0 Fe^2}{d^2}}$
    (C) $\sqrt{\frac{4\pi\epsilon_0 Fd^2}{e^2}}$ (D) $\frac{4\pi\epsilon_0 Fd^2}{q^2}$

18. ••When air is replaced by a liquid dielectric medium of constant κ, the maximum force of attraction between two charges separated by a fixed distance
    (A) increases to κ times (B) remains unchanged
    (C) decreases to κ times (D) decreases to κ^{-1} times.

19. ••A charge q is placed at the centre of the line joining two equal charges Q. The system of the three charges will be in equilibrium if q is equal to
    (A) $-Q/4$ (B) $Q/4$ (C) $-Q/2$ (D) $Q/2$

20. ••Point charges $+4q, -q$ and $+4q$ are kept on the X-axis at point $x = 0, x = a$ and $x = 2a$ respectively. Then
    (A) only $-q$ is in stable equilibrium
    (B) all the charges are in stable equilibrium
    (C) all of the charges are in unstable equilibrium
    (D) none of the charges is in equilibrium.

21. ••A positively charged body 'A' attracts a body 'B' then charge on body 'B' may be:
    (A) positive (B) negative (C) zero (D) can't say

22. ••Three identical point charges are at the vertices of an equilateral triangle. A fourth, identical point charge is placed at the midpoint of one side of the triangle. As a result of the three electric force contributions from the vertex charges, the fourth charge is
    (A) in equilibrium and remains at rest
    (B) pushed toward the centre of the triangle
    (C) pushed outside the triangle
    (D) in unstable equilibrium

---

*A pith ball is a very small, lightweight styrofoam (or similar material) ball coated with conductive paint that picks up electric charge quite well.

1.12 The Electric Field

The concept of a field* was developed by Michael Faraday. According to him, an electric field is said to exist in the region of space around a charge, if another charge experiences an electric force (F_e) within it. Thus, the interaction between two charges is a two-step process—a charge produces an electric field around itself; this field then exerts an electric force on the other charge.

The electric field is a vector quantity generally denoted by $\vec{E}$.

In analogy with the gravitational field, we define the electric field intensity or strength associated with a certain collection of charges in terms of the force exerted on a positive test charge q_0 at a particular point, as

$$\boxed{E = \frac{F}{q_0}} \quad (1.34)$$

i.e., *the electric force per unit test charge is called the intensity of electric field at the position of test charge.* The direction of the vector $\vec{E}$ is the same as the direction of $\vec{F}$, because q_0 is a positive scalar.

The SI unit of electric field intensity is newton/coulomb (N/C), although it is more often given in the equivalent unit of volt/meter (V/m). The term volt will be discussed in next chapter, "Electric Potential".

Note the similarity with the gravitational field strength 'g', which can also be expressed as the force per unit mass in units of newton/kilogram. Both the gravitational and electric field strengths can be expressed as a force divided by a property (mass or charge) of the test body.

In Fig.1.58, the charge Q, which is producing the electric field, is

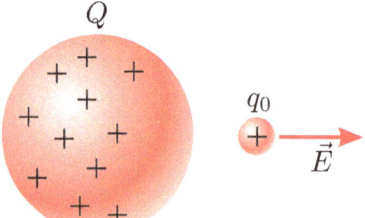

Figure 1.58

called a source charge and the charge q_0, which tests the effect of a source charge, is called a test charge. Note that the source charge Q must remain at its original location. However, if a charge q_0 is brought at any point around Q, Q itself is bound to experience an electrical force due to q_0 and will tend to move. To sort out this difficulty, take q_0 negligibly small. The force $\vec{F}$ is then negligibly small but the ratio $\vec{E} = \lim_{q_0 \to 0} \frac{F}{q_0}$ is finite and defines the electric field.

This limit in actuality cannot be taken to 0 because the test charge can never be smaller than the elementary charge e. If we are calculating (rather than measuring) the electric field due to a specified collection of charges at fixed positions, neither the magnitude nor the sign of q_0 affects the result.

Also, note that an electric field is a property of its source, the presence of the test charge is not necessary for the field to exist. The test charge is only used to detect the electric field.

When using $E = F/q_0$ we must assume that the test charge q_0 is small enough that it does not disturb the charge distribution

---
*Every position in a field is assigned a particular value of some quantity. For example, a temperature field is a scalar field and every position is associated with a temperature. If the field is a **vector field** (**for example gravitational or electric fields**), each position in the field is associated with a particular magnitude and direction.

responsible for the electric field. If vanishingly small test charge q_0 is placed near a uniformly charged metallic sphere, as in Fig. 1.59 (a) the charge on the metallic sphere, which produces the electric field, remains uniformly distributed. Now, if we consider a new test charge q'_0 which is great enough ($q'_0 \gg q_0$), as in Fig. 1.59b the charge on the metallic sphere is redistributed and the ratio of the force to the test charge is different. If in this case, the electric force on the test charge q'_0 is F', then-

$$\frac{F'}{q'_0} \neq \frac{F}{q_0}$$

Because of this redistribution of charge on the metallic sphere, the electric field it sets up is different from the field it sets up in the presence of the much smaller test charge q_0.

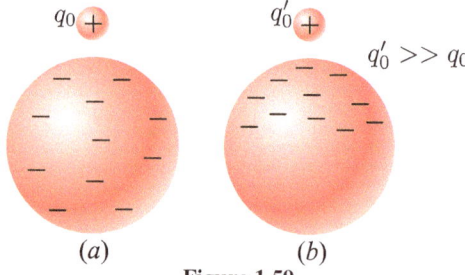

Figure 1.59

Important Points Regarding Electric Field

1. The static electric field produced by any point charge is always conservative. A conservative force or field must be a function of only spatial coordinates. The work done by such a force or field depends only on start and ending location in the field, and not on the actual path taken between those points. Now, suppose we place a point charge q near the surface of a conductor. It produces an induced charge over the surface of the conductor and this surface induced charge also produces an electric field in space which is also conservative. But if we move this point charge q, continuously, near the surface, the amount and distribution both of the induced surface charge changes continuously with time. So electric field produced by it also changes with time. Therefore, this induced electric field is time dependent and hence it is non-conservative.

2. The electric field of induced charge can be either conservative or non-conservative. In electrostatic situations (i.e., when charge remains in equilibrium under electric forces) without changing magnetic fields, it is conservative. In dynamic situations (i.e, when charge moves) or when there are changing magnetic fields, it can be non-conservative.

EXAMPLE 32. Suppose someone discovers that blue and yellow objects attract each other, that two blue objects repel each other, and that two yellow objects repel each other. The strength of this "chromatic interaction" is found to depend on color depth: The deeper the color, the greater the magnitude of the interaction. How would you define the magnitude and direction of the "chromatic field" of an object?

SOLUTION The gravitational field is defined as the gravitational force per unit of mass, with the field direction the same as the direction of the force. The electric field is defined as the electric force per unit of charge, with the field direction parallel to that of the force exerted on a positively charged particle. Therefore, the chromatic field can be defined as the chromatic force per unit of color, with the field direction parallel to that of the force exerted on a particle carrying some chosen color.

EXAMPLE 33. What happens when a high energy X-ray beam falls on a small metal ball suspended in a uniform electric field with the help of an insulated thread?

SOLUTION Since, the high energy X-rays cause ejection of the

1.13 The Electric Field of Point Charges

In this section we consider the electric field of point charges, first a single charge and then an assembly of individual charges. Later we generalize to continuous distributions of charge.

1.13.1 The Electric Field Due to a Single Point Charge

Let a positive test charge q_0 be placed a distance r from a point charge q. The magnitude of the force acting on q_0 is given by Coulomb's law.

$$\vec{F} = k\frac{qq_0}{r^2}\hat{r}$$

The magnitude of the electric field at the position of the test charge is, from Eq.1.34

$$\vec{E} = \frac{\vec{F}}{q_0} = k\frac{q}{r^2}\hat{r} = \frac{1}{4\pi\epsilon_0}\frac{q}{r^2}\hat{r} \quad (1.35)$$

where $\hat{r}$ is a unit vector directed from q towards q_0. The direction of $\vec{E}$ is the same as the direction of $\vec{F}$, along a radial line from q, pointing outward(i.e., along $\hat{r}$), if q is positive and inward (i.e., opposite to $\hat{r}$) if q is negative.

If q is negative, as in Fig.1.60c, the force on the test charge

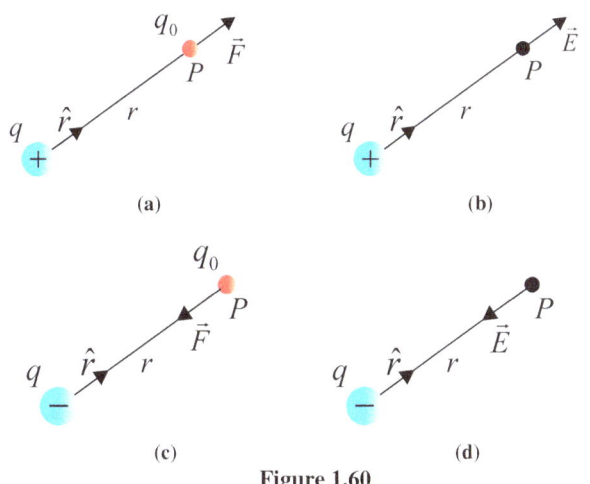

Figure 1.60

is toward the source charge, so the electric field at P is directed toward the source charge, as in Fig.1.60d.

Fig.1.60 shows the magnitude and direction of the electric field $\vec{E}$ at a point P due to positive and negative point charges.

EXAMPLE 34. (a) Does an electrically neutral particle that has mass interact with an electric field? (b) Does a charged particle interact with a gravitational field?

SOLUTION (a) No. A particle has no extent and therefore cannot be polarized. So, an uncharged particles* don't interact with electric fields.
(b) Yes, because any particle which is charged would certainly have mass. so it will be, interacted with a gravitational field.

EXAMPLE 35. A water droplet of mass 3.00×10^{-12} kg is located in the air near the ground during a stormy day. An atmospheric electric field of magnitude 6.00×10^3 N/C points vertically downward in the vicinity of the water droplet. The droplet remains suspended at rest in the air. What is the electric charge on the droplet?

APPROACH Since, the droplet is hovering at rest in the air in a downward atmospheric electric field and gravitational field of earth, so there will be a downward gravitational force of earth on the droplet and to keep it in equilibrium, there must be an equal upward force on it. This upward force is provided by downward electric field $\vec{E}$.

SOLUTION Suppose, the mass of droplet is m and it's charge

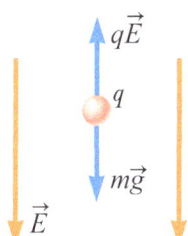

Figure 1.61: FBD of water droplet

is q. If the downward electric field is $\vec{E}$, then The forces acting on the droplet are-
1. Downward gravitational force $m\vec{g}$
2. Upward electrostatic force $q\vec{E}$

The free body diagram (FBD) of the droplet is shown in Fig.1.61. In equilibrium, we have

$$qE = mg \quad \Rightarrow \quad q = \frac{mg}{E}$$

On substituting the given values, we get

$$q = \frac{(3.00 \times 10^{-12} \text{ kg})(9.80 \text{ m/s}^2)}{6.00 \times 10^3 \text{ N/C}} = 4.90 \times 10^{-15}\text{C}$$

Since, the direction of electrostatic force is opposite to the electric field $\vec{E}$, therefore the nature of charge on the drop is negative. So, the charge on the droplet, is -4.90×10^{-15}C

EXAMPLE 36. A vertical electric field of magnitude 4.9×10^5 N/C just prevents a water droplet of a mass 0.1 g from falling. Find the magnitude of charge on the droplet. (Given $g = 9.8$ m/s²)

SOLUTION Let m is the mass of droplet and q is the charge on it. In equilibrium under gravitational field $\vec{g}$ and electric field $\vec{E}$, the forces acting on the charged droplet, are-
1. Downward Gravitational force $m\vec{g}$
2. Upward electric force $q\vec{E}$.

If q is positive, then electric force on q will be in the direction of $\vec{E}$ and if q is negative then it will be opposite to $\vec{E}$.

Since, the droplet is in equilibrium under these two forces only, therefore, net force on it must be zero.

Taking upward direction positive, we have

$$qE - mg = 0 \quad \Rightarrow \quad q = \frac{mg}{E}$$

Clearly, if E is positive, i.e., in upward direction, then q will be positive and if E is negative, i.e., in downward direction, then q will also be negative.

Substituting the given values in above expression, we get

$$q = \frac{(0.1 \times 10^{-3}) \times 9.8}{4.9 \times 10^5} = 2 \times 10^{-9} \text{ C}$$

1.13.2 Electric Field Due To a Group of Point Charges (Superposition of Electrostatic Fields)

To find $\vec{E}$ for a group of N point charges, the procedure is as follows: (1) Calculate E_i due to each charge i at the given point as if it were the only charge present. (2) Add these separately calculated fields vectorially to find the resultant field $\vec{E}$ at the

---

*An uncharged body is a different case. It is a collection of particles. It has equal amount of positive and negative charge. So it is always attracted towards a charged body or charged particle.

point. In equation form,
$$\vec{E} = \vec{E}_1 + \vec{E}_2 + \vec{E}_3 + \cdots$$
$$= \sum \vec{E}_i \quad (i = 1, 2, 3, \ldots, N) \quad (1.36)$$

The sum is a vector sum, taken over all the charges. *Eq. 1.36 is an example of the application of the principle of superposition, which states that at a given point the electric fields due to separate charge distributions simply add up (vectorially) or superimpose independently.* This principle may fail when the magnitudes of the fields are extremely large*, but it will be valid in all situations we discuss in this text.

Above equation can also be written as-
$$\vec{E} = k \sum_i \frac{q_i}{r_i^2} \hat{r}_i \quad (1.37)$$

where r_i is the distance of point P from i^{th} source charge q_i. The unit vector $\hat{r}$ is directed from q_i toward P.

If some more charges are added, more terms are added to the summation. However, there is no change to the terms that were already there, provided that the original charges do not move.

EXAMPLE 37. Two point charges are separated by a distance of 10.0 cm. One has a charge of $-25\mu C$ and the other $+50\mu C$. (a) Determine the direction and magnitude of the electric field at a point P between the two charges that is 2.0 cm from the negative charge (Fig.(1.62)a). (b) If an electron (mass = 9.11×10^{-31} kg) is placed at rest at P and then released, what will be its initial acceleration (direction and magnitude)?

APPROACH (a) The electric field at P will be the vector sum

Figure 1.62

of the fields created separately by Q_1 and Q_2. The field due to the negative charge Q_1 points toward Q_1, and the field due to the positive charge Q_2 points away from Q_2. Thus both fields point to the left as shown in Fig. 1.62b, and we can add the magnitudes of the two fields together algebraically, ignoring the signs of the charges.

(b) In this part, we use Newton's second law $\left(\sum \vec{F} = m\vec{a}\right)$ to find the acceleration, where $\sum \vec{F} = q\Sigma\vec{E}$.

SOLUTION (a) Each field is due to a point charge as given by Eq. 1.35, $E = kq/r^2$. The total field points to the left and has magnitude

$$E = k\frac{Q_1}{r_1^2} + k\frac{Q_2}{r_2^2} = k\left(\frac{Q_1}{r_1^2} + \frac{Q_2}{r_2^2}\right)$$
$$= (9.0 \times 10^9 \text{ N} \cdot \text{m}^2/\text{C}^2)\left(\frac{25 \times 10^{-6}\text{C}}{(2.0 \times 10^{-2}\text{ m})^2} + \frac{50 \times 10^{-6}\text{C}}{(8.0 \times 10^{-2}\text{ m})^2}\right)$$
$$= 6.3 \times 10^8 \text{ N/C}$$

(b) The electric field points to the left, so the electron will feel a force to the right since it is negatively charged. Therefore the acceleration $a = F/m$ (Newton's second law) will be to the right. The force on a charge q in an electric field E is $F = qE$. Hence the magnitude of the electron's initial acceleration is
$$a = \frac{F}{m} = \frac{qE}{m} = \frac{(1.60 \times 10^{-19}\text{C})(6.3 \times 10^8 \text{ N/C})}{9.11 \times 10^{-31} \text{ kg}} = 1.1 \times 10^{20} \text{ m/s}^2$$

*Large electric field can cause the material to become polarized, which can interact with the external electric field.

EXAMPLE 38. Two point charges A and B of magnitude $+8 \times 10^{-6}$ C and -8×10^{-6} C respectively are placed at a distance d apart. The electric field at the middle point O between the charges is 6.4×10^4 NC^{-1}. Calculate the distance 'd' between the point charges A and B.

APPROACH Electric fields at point O, due to both charges, are directed along the same direction (here it is directed from O to B). Therefore, by superposition principle, the magnitude of net electric field at point O will be equal to the sum of their magnitudes, i.e.

SOLUTION $\quad E_O = 2 \times \frac{kq}{(d/2)^2} \implies E_O = 8\frac{kq}{d^2}$

Figure 1.63

$$\implies d^2 = \frac{8 \times 9 \times 10^9 \times 8 \times 10^{-6}}{6.4 \times 10^4} \implies d = 3 \text{ m}$$

EXAMPLE 39. Charges q_1 and q_2 are located on the x axis, at distances a and b, respectively, from the origin as shown in Fig.1.64.

(a) Find the components of the net electric field at the point P, which is at position $(0, y)$.

(b) Evaluate the electric field at point P in the special case that $|q_1| = |q_2|$ and $a = b$.

(c) Find the electric field, in part (b), when point P is at a distance $y \gg a$ from the origin. $|q_1| = |q_2|$ and $a = b$.

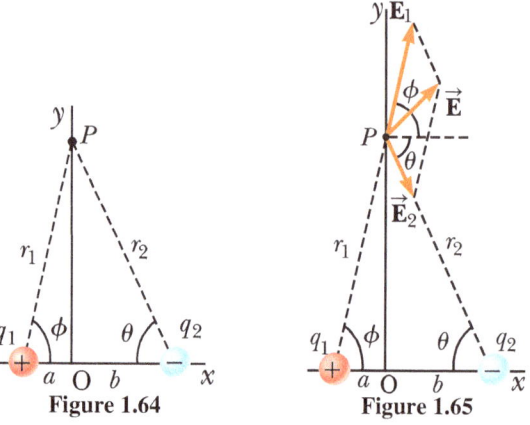

Figure 1.64   Figure 1.65

(a) APPROACH First of all calculate the intensities of electric fields at point P due to charges q_1 and q_2, then add these electric field vectors to find the net electric field at point P.

SOLUTION Find the magnitude of the electric field at P due to charge q_1: $E_1 = k\frac{|q_1|}{r_1^2} = k\frac{|q_1|}{a^2+y^2}$

Find the magnitude of the electric field at P due to charge q_2: $E_2 = k\frac{|q_2|}{r_2^2} = k\frac{|q_2|}{b^2+y^2}$

Write the electric field vector for each charge in unit-vector form:
$$\vec{E}_1 = k_e\frac{|q_1|}{a^2+y^2}\cos\phi \hat{i} + k_e\frac{|q_1|}{a^2+y^2}\sin\phi \hat{j},$$
$$\vec{E}_2 = k_e\frac{|q_2|}{b^2+y^2}\cos\theta \hat{i} - k_e\frac{|q_2|}{b^2+y^2}\sin\theta \hat{j}$$

The X and Y components of the net electric field vector:
$$E_x = E_{1x} + E_{2x} = k_e\frac{|q_1|}{a^2+y^2}\cos\phi + k_e\frac{|q_2|}{b^2+y^2}\cos\theta,$$
$$E_y = E_{1y} + E_{2y} = k_e\frac{|q_1|}{a^2+y^2}\sin\phi - k_e\frac{|q_2|}{b^2+y^2}\sin\theta$$

(b) APPROACH Fig.1.66 shows the situation when $|q_1| = |q_2|$ and $a = b$. Such symmetrical charge distribution is called an electric dipole.

Because Fig.1.66 is a special case of the general case shown in

1.13. THE ELECTRIC FIELD OF POINT CHARGES

Fig.1.65, so, we can take the result of part (a) and substitute the appropriate values of the variables.

SOLUTION Evaluate the values of E_x and E_y obtained in from

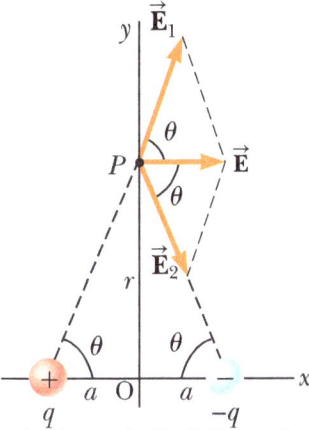

Figure 1.66: When the charges in Fig 1.65 are of equal magnitude and equidistant from the origin, the situation becomes symmetric as shown here.

part (a) with $a = b$, $|q_1| = |q_2| = q$, and $\phi = \theta$:

$$E_x = k\frac{q}{a^2+y^2}\cos\theta + k\frac{q}{a^2+y^2}\cos\theta = 2k\frac{q}{a^2+y^2}\cos\theta$$

$$E_y = k\frac{q}{a^2+y^2}\sin\theta - k\frac{q}{a^2+y^2}\sin\theta = 0$$

From the geometry in Fig.1.66, evaluate $\cos\theta$:

$$\cos\theta = \frac{a}{r} = \frac{a}{(a^2+y^2)^{1/2}}$$

Now substitute this value of $\cos\theta$ in the above expression for E_x:

$$E_x = 2k\frac{q}{a^2+y^2}\left[\frac{a}{(a^2+y^2)^{1/2}}\right] = k\frac{2aq}{(a^2+y^2)^{3/2}}$$

(c) In the solution to part (b), because $y \gg a$, neglect a^2 compared with y^2 and write the expression for E in this case:

$$E \approx k\frac{2aq}{y^3}$$

EXAMPLE 40. Two identical positive point charges, each of magnitude q, are placed on the axis at $x = -a$ and $x = +a$, as shown in Fig.1.67.
(a) Find an expression of net electric field on x axis and plot the variation of electric field E along the x-axis.
(b) Find an expression of net electric field on y axis and plot the variation of E along the y-axis.

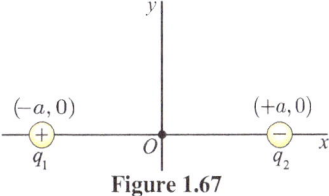

Figure 1.67

(a) APPROACH Let $\vec{E}_1$ and $\vec{E}_2$ be the electric fields due to q_1 and $q_{2'}$ respectively. Because q_1 is positive, $\vec{E}_1$ points away from q_1 everywhere, and because q_2 is also positive, so, $\vec{E}_2$ also points away from q_2 everywhere (here, $q_1 = q_2 = q$).

By principle of superposition, the net electric field at any position is: $\vec{E} = \vec{E}_1 + \vec{E}_2$.

SOLUTION (a) 1. Draw the charge configuration and place the field point A on the x axis at the appropriate place. Draw vectors representing the electric field at A due to each point charge. Repeat this procedure for field point B (Fig.1.68):

Figure 1.68: Because q_1 is a positive charge, $\vec{E}_1$ points away from q_1 at both point A and point B. Because q_2 is a positive charge, $\vec{E}_2$ points away from q_2 at both point A and point B.

2. Calculate $\vec{E}$ at point A using, $r_{1A} = |x_A + a|$ and $r_{2A} = |x_A - a|$:

$$\vec{E} = \vec{E}_1 + \vec{E}_2 = \frac{kq_1}{r_{1A}^2}\hat{r}_{1A} + \frac{kq_2}{r_{2A}^2}\hat{r}_{2A}$$

$$= \frac{kq_1}{(x_A+a)^2}\hat{i} + \frac{kq_2}{(x_A-a)^2}\hat{i}$$

$$= kq\left[\frac{1}{(x_A+a)^2}\hat{i} + \frac{1}{(x_A-a)^2}\hat{i}\right] \quad (\because q_1 = q_2 = q)$$

Special Positions: (i) When $x \to a$, then $\vec{E} = +\infty\,\hat{i}$
(ii) When $x \to +\infty$, then $\vec{E} = 0\,\hat{i}$

(b) Calculate $\vec{E}$ at point B, where $r_{1B} = |x_B + a|$ and $r_{2B} = |x_B - a|$:

$$\vec{E} = \vec{E}_1 + \vec{E}_2 = \frac{kq_1}{r_{1B}^2}\hat{r}_{1B} + \frac{kq_2}{r_{2B}^2}\hat{r}_{2B}$$

$$= \frac{kq_1}{(x_B+a)^2}\hat{i} + \frac{kq_2}{(x_B-a)^2}(-\hat{i})$$

$$= kq\left[\frac{1}{(x_B+a)^2}\hat{i} + \frac{1}{(x_B-a)^2}(-\hat{i})\right] \quad (\because q_1 = q_2 = q)$$

Special Positions: (i) When $x \to a$, then $\vec{E} = -\infty\,\hat{i}$
(ii) When $x = 0$, then $\vec{E} = 0\,\hat{i}$
(iii) When $x \to -a$, then $\vec{E} = +\infty\,\hat{i}$

(c) If we consider another point C in the left of charge q_1 (not shown in the diagram,), the calculation will be similar to solution step 2(a) except that the net field is directed towards the negative direction of x-axis (students are advised to think this situation and draw diagram themselves). In this case-

$$\vec{E} = \vec{E}_1 + \vec{E}_2 = \frac{kq_1}{r_{1C}^2}\hat{r}_{1C} + \frac{kq_2}{r_{2C}^2}\hat{r}_{2C}$$

$$= \frac{kq_1}{(x_C-a)^2}(-\hat{i}) + \frac{kq_2}{(x_C+a)^2}(-\hat{i})$$

$$= kq\left[\frac{1}{(x_C-a)^2} + \frac{1}{(x_C+a)^2}\right](-\hat{i}) \quad (\because q_1 = q_2 = q)$$

Special Positions: (i) When $x \to a$, then, $\vec{E} = -\infty\hat{i}$
(ii) When $x \to -\infty$, then, $\vec{E} = 0\,\hat{i}$

E vs x graph: The resultant electric field at source points close to q_1 is dominated by the field $\vec{E}_1$ due to q_1. There is one point between q_1 and q_2 where the resultant electric field is zero. Since, $q_1 = q_2 = q$, therefore it is the origin, where net electric field due to both the charges is zero. A test charge placed at this point would experience no electric force.

A sketch of E versus x for this charge configuration is shown in Fig.1.69.

(b) APPROACH We'll assume that q is positive when drawing pictures, but the solution should allow for the possibility that q is negative. The question does not ask about any specific point, so we will be looking for a symbolic expression in terms of the unspecified position y. Fig.1.70 shows the charges, the coordinate system, and the two electric field vectors $\vec{E}_1$ and $\vec{E}_2$. Each of these field points away from its source charge because of the assumption that q is positive. We need to find the vector sum

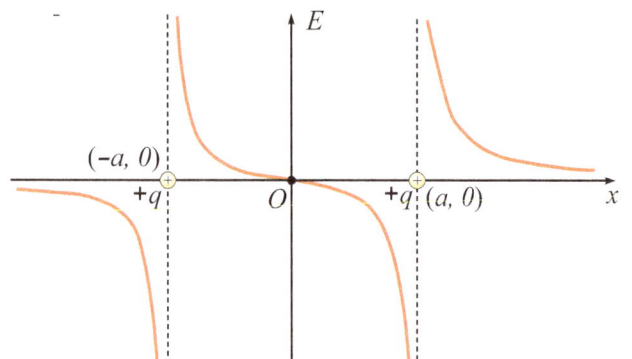

Figure 1.69: The variation of electric field E along the x axis. Electric field directed along $+ve$ x-axis is taken as poisitive

$$\vec{E}_{net} = \vec{E}_1 + \vec{E}_2$$

Before rushing into a calculation, we can make our task much easier by first thinking qualitatively about the situation. For example, the fields $\vec{E}_1$ and $\vec{E}_2$ both lie in the xy-plane, hence we can conclude without any calculations that $(E_{net})_z = 0$. Next, look at the x-components of the fields. The fields $\vec{E}_1$ and $\vec{E}_2$ have equal magnitudes and are tilted away from the y-axis by the same angle θ. Consequently, the x-components of $\vec{E}_1$ and $\vec{E}_2$ will cancel when added. The only component we need to calculate is $(E_{net})_y$.

SOLUTION The y-component of the field is

$$E_1 \sin\theta + E_2 \sin\theta = 2E_1 \sin\theta$$

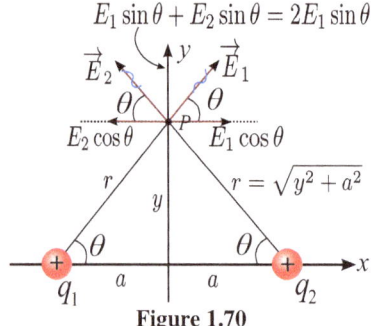

Figure 1.70

$$(E_{net})_y = (E_1)_y + (E_2)_y = 2(E_1)_y$$

where we used the fact that fields $\vec{E}_1$ and $\vec{E}_2$ have equal y-components.
Vector $\vec{E}_1$ is at angle θ from the x-axis, so its y-component is

$$(E_1)_y = E_1 \sin\theta = k\frac{q_1}{r^2}\sin\theta$$

where r is the distance from q_1. This expression for $(E_1)_y$ is correct, but it is not yet sufficient. Both the distance r and the angle θ vary with the position y and need to be expressed as functions of y. From the Pythagorean theorem, $r = (y^2 + a^2)^{1/2}$. Thus,

$$\sin\theta = \frac{y}{r} = \frac{y}{(y^2+a^2)^{1/2}}$$

By combining these pieces, we see that $(E_1)_y$ is

$$(E_1)_y = k\frac{q}{y^2+a^2}\frac{y}{(y^2+a^2)^{1/2}}$$

$$\Rightarrow \quad (E_1)_y = k\frac{qy}{(y^2+a^2)^{3/2}}$$

$$\therefore \quad (E_{net})_y = 2(E_1)_y = k\frac{2qy}{(y^2+a^2)^{3/2}}$$

The other two components of $\vec{E}_{net}$ are zero, hence the electric field of the two charges at a point on the y-axis is

$$\vec{E}_{net} = k\frac{2qy}{(y^2+a^2)^{3/2}}\hat{j} = \frac{1}{4\pi\epsilon_0}\frac{2qy}{(y^2+a^2)^{3/2}}\hat{j}$$

$$\left[\because \quad k = \frac{1}{4\pi\epsilon_0}\right]$$

This is the electric field only at points on the y-axis. Furthermore, this expression is valid only for $y > 0$. The electric field to the downward of the charges points in the opposite direction. Our expression is valid for both positive and negative q. A negative value of q makes $(E_{net})_y$ negative, which would be an electric field pointing in negative direction of y-axis.

Tracing E vs y Graph

1. **Maximum and Minimum values of Electric Field:** For maximum value of E, on y axis, we have

$$\frac{dE}{dy} = 0$$

We have already obtained the expression for $\vec{E}$, on y axis as -

$$\vec{E}_{net} = k\frac{2qy}{(y^2+a^2)^{3/2}}\hat{j}$$

Therefore,

$$\frac{dE}{dy} = 2kq\frac{d}{dy}\frac{y}{(y^2+a^2)^{3/2}}$$

$$= 2kq\left[\frac{(y^2+a^2)^{3/2}1 - y\frac{3}{2}(y^2+a^2)^{1/2}2y}{(y^2+a^2)^3}\right]$$

$$= 2kq(y^2+a^2)^{1/2}\left[\frac{a^2-2y^2}{(y^2+a^2)^3}\right]$$

Therefore, $\frac{dE}{dy} = 0 \Rightarrow 2kq(y^2+a^2)^{1/2}\frac{a^2-2y^2}{(y^2+a^2)^3} = 0$

or $\quad y = \pm\frac{a}{\sqrt{2}}$

So, the value of electric field will be maximum at $y = \pm\frac{a}{\sqrt{2}}$

Substituting this value of y in the expression for E, we get

$$\vec{E}_{max} = \pm 2kq\frac{a/\sqrt{2}}{\left(\frac{a^2}{2}+a^2\right)^{3/2}}\hat{j} = \pm\sqrt{\frac{16}{27}}\frac{kq}{a^2}\hat{j}$$

☞ If a point charge is placed at a distance $y \pm a/\sqrt{2}$, it experiences the maximum Coulomb's force

2. From the equation for $\vec{E}$, at $y = 0$, the electric field $\vec{E} = 0$.
3. When $y \to \pm\infty$, then, again $\vec{E} \to 0$

Keeping these points in mind, the variation of E with y is given in Fig.1.71.

EXAMPLE 41. A charge $+q$ is at $x = a$ and a second charge

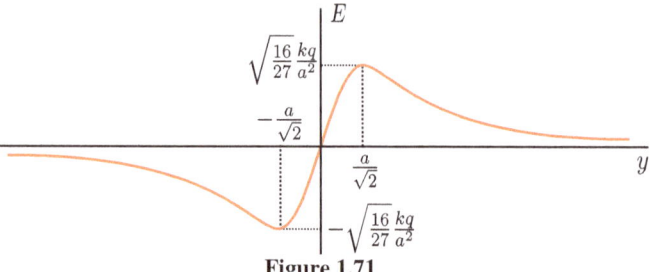

Figure 1.71

$-q$ is at $x = -a$ (Figure 1.73). (a) Find the electric field on the

1.13. THE ELECTRIC FIELD OF POINT CHARGES

x axis at an arbitrary point $x > a$. (b) Find the limiting form of the electric field for $x \gg a$ (c) Plot the variation of E along the x-axis.

APPROACH We calculate the electric field at point P using

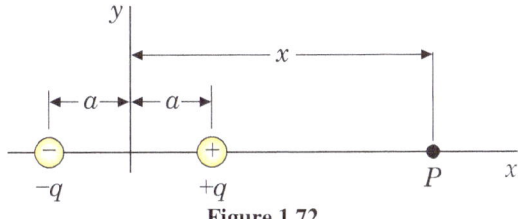

Figure 1.72

the principle of superposition, $\vec{E}_P = \vec{E}_{1P} + \vec{E}_{2P}$. For $x > a$, the electric field $\vec{E}_+$ due to the positive charge is in the $+x$ direction and the electric field $\vec{E}_-$ due to the negative charge is in the $-x$ direction. The distances are $x - a$ to the positive charge and $x - (-a) = x + a$ to the negative charge.

SOLUTION (a) 1. Draw the charge configuration on a coordinate axis and label the distances from each charge to the field point (Fig.1.73):

2. Calculate $\vec{E}$ due to the two charges for $x > a$: (Note that the

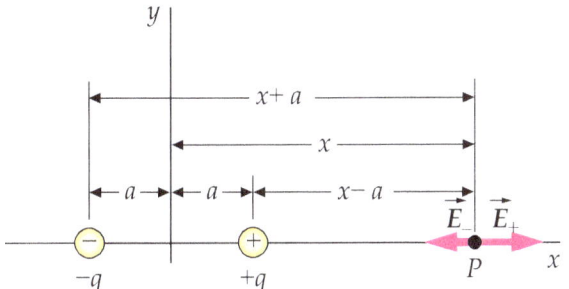

Figure 1.73

equation given below holds only for $x > a$.)

$$\vec{E} = \vec{E}_+ + \vec{E}_- = \frac{kq}{[x-a]^2}\hat{i} + \frac{kq}{[x-(-a)]^2}(-\hat{i})$$

$$= kq\left[\frac{1}{(x-a)^2} - \frac{1}{(x+a)^2}\right]\hat{i}$$

On simplifying, we get-

$$\vec{E} = kq\left[\frac{(x+a)^2 - (x-a)^2}{(x+a)^2(x-a)^2}\right]\hat{i} = kq\frac{4ax}{(x^2-a^2)^2}\hat{i} \quad x > a$$

(b) In the limit $x \gg a$, we can neglect a^2 compared with x^2 in the denominator:

$$\vec{E} = kq\frac{4ax}{(x^2-a^2)^2}\hat{i} \approx kq\frac{4ax}{x^4}\hat{i} = \frac{4kqa}{x^3}\hat{i} \quad x \gg a$$

The answer approaches zero as x approaches infinity, which is as expected.

(c) Fig.1.74 shows E_x versus x for all x, for $q = 1.0$ nC and $a = 1.0$ m. For $|x| \gg a$ (far from the charges), the field is given by

$$\vec{E} = \frac{4kqa}{|x|^3}\hat{i} \quad |x| \gg a$$

Between the charges, the contribution from each charge is in the negative direction. An expression for $\vec{E}$ at distance $x(<a)$, is

$$\vec{E} = -\frac{kq}{(x-a)^2}\hat{i} - \frac{kq}{(x+a)^2}\hat{i} \quad -a < x < a$$

In the range, $-a \leq x \leq +a$: corresponding to $x \to \pm a$, above equation gives-

$$\vec{E} \to -\infty \hat{i}$$

At origin, i.e., at $x = 0$, above equation gives-

$$\vec{E} = -\frac{kq}{(-a)^2}\hat{i} - \frac{kq}{(a)^2}\hat{i} = -\frac{2kq}{a^2}\hat{i}$$

In right side of charge $+q$, if the point is very close to it, i.e, $x \to a$, then point (3) of part (a), gives

$$\vec{E} = kq\lim_{x \to +a}\frac{4ax}{(x^2-a^2)^2}\hat{i} = +\infty\,\hat{i}$$

In right side of charge $+q$, if the point is very far from it, i.e, $x \to \infty$, then point (3) of part (a), gives

$$\vec{E} = kq\lim_{x \to \infty}\frac{4ax}{(x^2-a^2)^2}\hat{i} = 0\,\hat{i}$$

In left of charge $-q$, the net electric field at distance x from the origin is given by

$$\vec{E} = \vec{E}_+ + \vec{E}_- = \frac{kq}{[|x|+a]^2}\hat{i} + \frac{kq}{[|x|-a]^2}\hat{i}$$

$$= kq\left[\frac{1}{(|x|+a)^2} + \frac{1}{(|x|-a)^2}\right]\hat{i}$$

or $\quad \vec{E} = kq\dfrac{4a|x|}{(|x|^2-a^2)^2}\hat{i}$

when, $x \to a$, then

or $\quad \vec{E} = kq\lim_{x \to -a}\dfrac{4a|x|}{(|x|^2-a^2)^2}\hat{i} = +\infty\hat{i}$

and when $x \to -\infty$, then

$$\vec{E} = kq\lim_{x \to -\infty}\frac{4a|x|}{(|x|^2-a^2)^2}\hat{i} = 0\hat{i}$$

Keeping these points in mind, the variation of E with x is given in Fig.1.74

EXAMPLE 42. Three point charges lie along the x axis

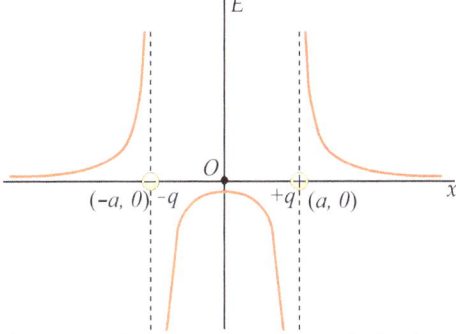

Figure 1.74: A plot of E versus x on the x axis for the given charge distribution

as shown in Fig.1.75. The positive charge $q_1 = 24.0$ μC is at $x = 3.00$ m, the positive charge $q_2 = 6.00$ μC is at the origin, and the resultant force acting on q_3 is zero. What is the x coordinate of q_3?

APPROACH Since the charges q_1 and q_2 are of same nature

Figure 1.75

(here both are positive), therefore the null point lies between them near the smaller charge.

If x is the distance of null point from smaller charge q_2, then net electric force on any charge q_3, placed at this point will be zero,

i.e.,
$$|\vec{F}_{32}| = |\vec{F}_{31}| \qquad (i)$$

here $\vec{F}_{32}$ is the electrostatic force of charge q_2 on the charge q_3 and $\vec{F}_{31}$ is the electrostatic force of charge q_1 on the charge q_3. Now, find the forces $\vec{F}_{32}$ and $\vec{F}_{31}$ by using Coulomb's law and substitute these values in Eq.(i) to get the required value of x.

☞ Be careful that Eq.(i) is the relation between magnitudes only. So, for each value of x obtained from Eq.(i), the magnitudes of forces on charge q_3 will be equal but it does not tell anything about directions of forces on q_3. Since the position of null point should be between q_1 and q_2, therefore, only $+ve$ values of x will be acceptable.

SOLUTION $k\dfrac{|q_2||q_3|}{x^2} = k\dfrac{|q_1||q_3|}{(3.00-x)^2}$

$\Rightarrow \quad (3.00-x)^2 |q_2| = x^2 |q_1|$

On simplifying this quadratic equation for x, we find that the positive root is $x = 1$ m. There is also a second root, $x = -3$ m. This is another location at which the magnitudes of the forces on q_3 are equal, but both forces are in the same direction at this location.

EXAMPLE 43. Point charges q and $-q$ are located at the vertices of a square with diagonals $2l$ as shown in Fig.1.76. Find the magnitude of the electric field strength at a point located symmetrically with respect to the vertices of the square at a distance x from its centre.

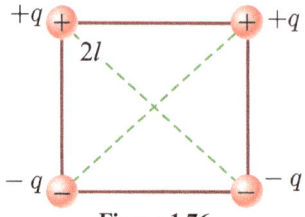

Figure 1.76

APPROACH Apply superposition principle of electric field:
$$\vec{E} = \vec{E}_1 + \vec{E}_2 + \vec{E}_3 + \vec{E}_4$$
$$= \dfrac{q_1(\vec{r}_p - \vec{r}_1)}{4\pi\epsilon_0 |\vec{r}_p - \vec{r}_1|^3} + \dfrac{q_2(\vec{r}_p - \vec{r}_2)}{4\pi\epsilon_0 |\vec{r}_p - \vec{r}_2|^3} + \dfrac{q_3(\vec{r}_p - \vec{r}_3)}{4\pi\epsilon_0 |\vec{r}_p - \vec{r}_3|^3} + \dfrac{q_4(\vec{r}_p - \vec{r}_4)}{4\pi\epsilon_0 |\vec{r}_p - \vec{r}_4|^3}$$

SOLUTION The arrangement of charges is shown in Fig.1.77.

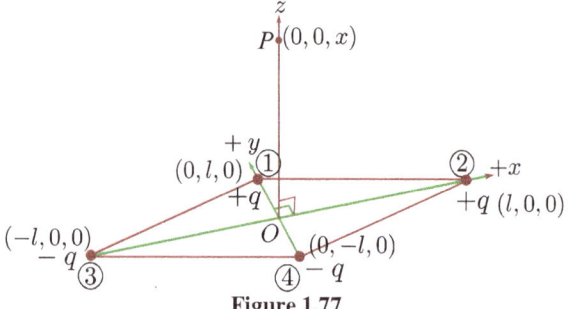

Figure 1.77

According to given problem,
$q_1 = +q, \vec{r}_1 = l\hat{j}$, $q_2 = +q$, $\vec{r}_2 = l\hat{i}$
$q_3 = -q$, $\vec{r}_3 = -l\hat{i}$; $q_4 = -q$, $\vec{r}_4 = -l\hat{j}$,
$\vec{r}_p = x\hat{k}$

$\therefore \quad \vec{r}_p - \vec{r}_1 = x\hat{k} - l\hat{j}$,

$|\vec{r}_p - \vec{r}_1| = \sqrt{x^2 + l^2} = \left(x^2 + l^2\right)^{1/2}$

$\vec{r}_p - \vec{r}_2 = x\hat{k} - l\hat{i}$, $\quad |\vec{r}_p - \vec{r}_2| = \left(x^2 + l^2\right)^{1/2}$

$\vec{r}_p - \vec{r}_3 = x\hat{k} + l\hat{i}$, $\quad |\vec{r}_p - \vec{r}_3| = \left(x^2 + l^2\right)^{1/2}$

and $\vec{r}_p - \vec{r}_4 = x\hat{k} + l\hat{j}$, $|\vec{r}_p - \vec{r}_4| = \left(x^2 + l^2\right)^{1/2}$

$\vec{E} = \vec{E}_1 + \vec{E}_2 + \vec{E}_3 + \vec{E}_4$

$\Rightarrow \vec{E} = \dfrac{q_1(\vec{r}_p - \vec{r}_1)}{4\pi\epsilon_0 |\vec{r}_p - \vec{r}_1|^3} + \dfrac{q_2(\vec{r}_p - \vec{r}_2)}{4\pi\epsilon_0 |\vec{r}_p - \vec{r}_2|^3}$

$\qquad + \dfrac{q_3(\vec{r}_p - \vec{r}_3)}{4\pi\epsilon_0 |\vec{r}_p - r_3|^3} + \dfrac{q_4(\vec{r}_p - \vec{r}_4)}{|\vec{r}_p - \vec{r}_4|^3}$

Putting the values, we get

$\vec{E} = \dfrac{q}{4\pi\epsilon_0 (x^2+l^2)^{3/2}} \left[(x\hat{k} - l\hat{j}) + (x\hat{k} - l\hat{i}) - (x\hat{k} + l\hat{i}) \right.$

$\qquad \left. -(x\hat{k} + l\hat{j})\right]$

$\quad = \dfrac{q}{4\pi\epsilon_0 (x^2+l^2)^{3/2}} (-2l\hat{j} - 2l\hat{i})$

$\therefore \quad E = \dfrac{q}{4\pi\epsilon_0 (x^2+l^2)^{3/2}} \sqrt{(-2l)^2 + (-2l)^2}$

$\Rightarrow \quad E = \dfrac{2ql\sqrt{2}}{4\pi\epsilon_0 (x^2+l^2)^{3/2}} = \dfrac{ql}{\sqrt{2}\pi\epsilon_0 (x^2+l^2)^{3/2}}$

EXAMPLE 44. A point charge of 10 C is placed at the origin. At what location on the X–axis should a point charge of 40 μC be placed so that the net electric field is zero at $x = 2$ cm on the X–axis?

SOLUTION Suppose point A is located at $x = x_0$ from the origin O, where the second charge is placed, which is 40 μC. As the given charges have the same nature, the zero-field point will be between the charges and closer to the smaller charge, i.e., 10 μC. In Fig.1.78, P is the zero field point at distance 2 cm from origin O where 10 μC charge is placed.

Now, $E_P = \dfrac{k \times 10}{2^2} - \dfrac{k \times 40}{(x_0 - 2)^2} = 0 \implies \dfrac{1}{2} = \dfrac{2}{x_0 - 2}$

```
    10μC         40μC
     O      P     A
     |      |     |————— X
     0    2 cm   x₀
     |←2 cm→|←(x₀−2) cm→|
```
Figure 1.78

$\Rightarrow \qquad x_0 = 6$ cm

1.14 Electric Dipole

Two equal and opposite point charges $+q$ and $-q$ placed a short distance ($2l$, say) apart, form an electric dipole. The particles are attracted to each other, but their separation is maintained so that the distance between them remains constant as shown in Fig.1.79.

```
  −q       O   p→      +q
   ⊖————————————————————⊕
   |←————— 2l ——————→|
```
Figure 1.79

1.14.1 Dipole Moment

The product of the magnitude of either charge and the distance between the charges is called the electric dipole moment. The electric dipole moment–also called the dipole moment–is a vector quantity whose direction is along the axis of dipole pointing from negative towards the positive charge in a dipole. It is denoted by $\vec{p}$.

Magnitude of dipole moment is given as–

$$\boxed{p = q.(2l) = 2ql} \qquad (1.38)$$

1.14. ELECTRIC DIPOLE

The S.I. unit of dipole moment is "coulomb-metre".
The practical unit of dipole moment is debye.
1 debye (D) = 3.3×10^{-30} coulomb-meter.
The direction of $\vec{p}$ identifies the orientation of the dipole, and the dipole-moment magnitude $p = 2ql$ determines the electric field strength.

1.14.2 Dipole Moment of a System of Discrete Charges

Since dipole moment is a vector quantity, hence in case of two or more than two dipoles, the resultant dipole moment will be the vector sum of the dipole moments of individual dipoles, i.e.,
$$\vec{p} = \vec{p}_1 + \vec{p}_2 + \vec{p}_3 + \dots$$
The real dipoles are the chemical dipoles such as H_2O and CO_2. H_2O has some dipole moment but CO_2 being linear has zero dipole moment.

1.14.3 Induced Dipole Moment

Many molecules, such as water (H_2O), have *permanent* electric dipole moments. In other molecules like: CO_2 and CH_4 (called *nonpolar molecules*) and in every isolated atom, the centres of the positive and negative charges coincide (Fig.1.80a) and thus no dipole moment is set up. However, if we place an atom or a nonpolar molecule in an external electric field, the field distorts the electron orbits and separates the centres of positive and negative charge (Fig.1.80b). Because the electrons are negatively charged, they tend to be shifted in a direction opposite the field. This shift sets up a dipole moment $\vec{p}$ that points in the direction of the field. This dipole moment is said to be *induced* by the field, and the atom or molecule is then said to be *polarized* by the field (that is, it has a positive side and a negative side). When the field is removed, the induced dipole moment and the polarization disappear.

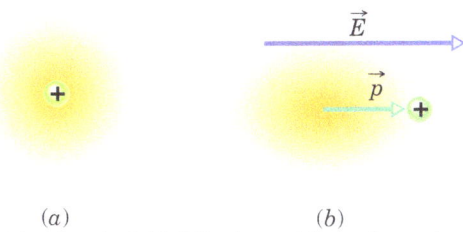

(a)   (b)

Figure 1.80: The electric field shifts the positive and negative charges, creating a dipole.

☞ In non-polar molecules, centre of the positive charge coincides with centre of negative charge, i.e., separation between the centres of positive and negative charges is zero, therefore net dipole moment of non-polar molecules is zero.

☞ When non polar material is placed in an external field, centre of negative charge shifts opposite to applied electric field. So, in presence of external field, the centres of opposite charges do not coincide. Therefore, a net dipole moment induces in it from centre of negative charge to centre of positive charge, i.e., in the direction of applied electric field.

1.14.4 Electric Field at any Point due to an Electric Dipole

1.14.4.1 Electric field at an Axial Point, i.e., in End on Position

Let us consider that there is no external field, and we want to determine the electric field produced by the dipole at a point P on the axis of the dipole AB [Fig.1.81]. If O is the centre of the dipole, then $OP = r > l$.

Let E_A and E_B are the magnitudes of electric fields at the axial

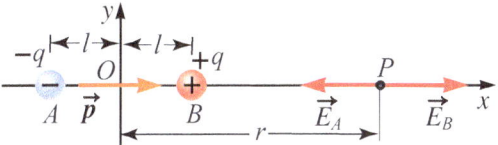

Figure 1.81

position P due to charges $-Q$ at A and $+Q$ at B respectively, then
$$\vec{E}_A = k\frac{q}{(r+l)^2} \text{ (Along } \overrightarrow{PO}) \tag{1.39}$$

$$\vec{E}_B = k\frac{q}{(r-l)^2} \text{ (Along } \overrightarrow{OP}) \tag{1.40}$$

Net intensity of electric field at P
$$\vec{E} = E_B - E_A \text{ (Along } \overrightarrow{OP})$$
$$= kq\left(\frac{1}{(r-l)^2} - \frac{1}{(r+l)^2}\right) \text{ (Along } \overrightarrow{OP})$$
$$= kq\left(\frac{(r+l)^2 - (r-l)^2}{(r^2-l^2)^2}\right) \text{(Along } \overrightarrow{OP})$$
$$= kq\left(\frac{4rl}{(r^2-l^2)^2}\right) \text{(Along } \overrightarrow{OP})$$
$$= k\frac{2(2ql)r}{(r^2-l^2)^2} \text{ (Along } \overrightarrow{OP})$$
$$= k\frac{2pr}{(r^2-l^2)^2} \text{(Along } \overrightarrow{OP})$$

$$\Rightarrow \boxed{E = k\frac{2pr}{(r^2-l^2)^2} \text{(Along } \overrightarrow{OP})} \tag{1.41}$$

In vector form,

$$\Rightarrow \boxed{\vec{E} = k\frac{2\vec{p}r}{(r^2-l^2)^2} \text{ (Along } \overrightarrow{OP})} \tag{1.42}$$

As $r \gg l$, therefore

$$\boxed{E = k\frac{2p}{r^3} \text{ (Along } \overrightarrow{OP}} \tag{1.43}$$

In vector form

$$\boxed{\vec{E} = k\frac{2\vec{p}}{r^3} \text{ (Along } \overrightarrow{OP}} \tag{1.44}$$

In this case, $\vec{E}$ is in the direction of $\vec{p}$.

Note: E vs r graph in axial position will be similar to graph shown in Fig.1.74. The only difference is that letter a is used for l and x is used for r.

1.14.4.2 At a Point on its Perpendicular Bisector, i.e. in Broadside on Position

Let us consider mid point O of dipole as the origin. Let P be any point at distance $OP = r$ on the perpendicular bisector of the electric dipole. $\vec{E}_1$ and $\vec{E}_2$ are the intensities at P due to charges $+q$ and $-q$ respectively.

$$\vec{E}_1 = k\frac{q}{r^2+l^2} \quad \text{(along } \overrightarrow{BP} \text{ direction)} \tag{1.45}$$

$$\vec{E}_2 = k\frac{q}{r^2+l^2} \quad \text{(along } \overrightarrow{PA} \text{ direction)} \tag{1.46}$$

From Eq.1.45 and 1.46, it is clear that $E_1 = E_2$
Now to get the resultant of $\vec{E}_1$ and $\vec{E}_2$ we resolve $\vec{E}_1$ and $\vec{E}_2$ into components:
(i) Vertical components $E_1\sin\theta$, and $E_2\sin\theta$ are equal in magnitude and oppositely directed to each other, therefore they get cancelled.
(ii) Horizontal components $E_1\cos\theta$ and $E_2\cos\theta$ are directed along same direction, so they get added

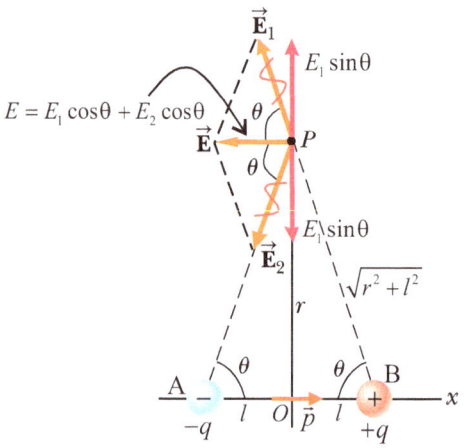

Figure 1.82

Therefore, net electric field
$$E = E_1\cos\theta + E_2\cos\theta = 2E_1\cos\theta$$
Now, from Fig.1.82,
$$\cos\theta = \frac{l}{\sqrt{r^2+l^2}}$$
Therefore, above equation gives
$$E = 2k\left(\frac{q}{r^2+l^2}\right)\left(\frac{l}{\sqrt{r^2+l^2}}\right)$$
$$\Rightarrow \quad E = k\frac{2ql}{(r^2+l^2)^{3/2}} \qquad (1.47)$$

Since, $2ql = p$ (dipole moment), therefore Eq.(1.47) gives-
$$\boxed{E = k\frac{p}{(r^2+l^2)^{3/2}}} \qquad (1.48)$$

In vector form-
$$\boxed{\vec{E} = -k\frac{\vec{p}}{(r^2+l^2)^{3/2}}} \qquad (1.49)$$

In this case, the direction of $\vec{E}$ is opposite to that of $\vec{p}$. If $r^2 \gg l^2$, then,
$$\boxed{E = k\frac{p}{r^3}} \qquad (1.50)$$

In vector form-
$$\boxed{\vec{E} = -k\frac{\vec{p}}{r^3}} \qquad (1.51)$$

Thus, we see that for $r \gg l$, the intensity of electric field in axial position of dipole is double in magnitude to that of intensity at the same distance on perpendicular bisector.

Special Cases: (i) At $r = 0$, the electric field $E = k\frac{p}{l^3}$. It is the maximum value of electric field in broad side on position
(ii) At $r \to \pm\infty$, the electric field $E \to 0$.
Keeping these point in mind, the variation of E with r is plotted in Fig.1.83.

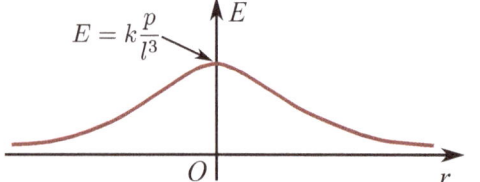

Figure 1.83: Variation of E with r in broad side on position

1.14.4.3 At any General Point

Let $P(r, \theta)$ be any general point where intensity of electric field due to electric dipole is to be determined. The direction of dipole moment is along axis A to B.
Now, resolving vector $\vec{p}$ into two components
(i) $p_1 = p\cos\theta$ (along the direction $\overrightarrow{OP}$),
(ii) $p_2 = p\sin\theta$ (perpendicular to $\overrightarrow{OP}$)
Point P is now at end on position of a dipole of dipole moment

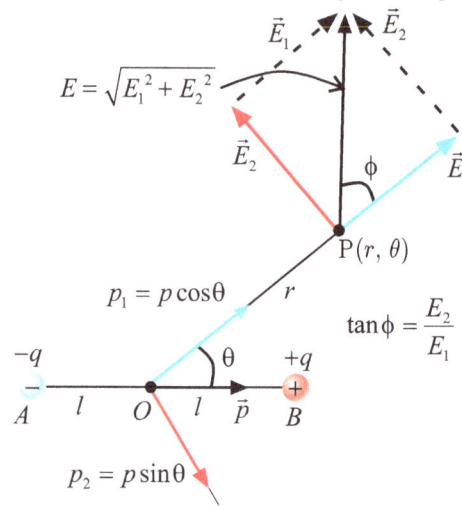

Figure 1.84

p_1 and it is on perpendicular bisector position of dipole of dipole moment p_2.
Electric field $\vec{E}_1$ at P due to dipole of dipole moment p_1 will be along $\overrightarrow{OP}$ vector. From Eq.1.43, the magnitude of $\vec{E}_1$ is given by
$$E_1 = k\frac{2p_1}{r^3} = k\frac{2p\cos\theta}{r^3}$$
Electric field $\vec{E}_2$ at P due to dipole of dipole moment p_2 will be perpendicular to $\overrightarrow{OP}$ and opposite to the direction of $\vec{p}_2$ vector. From Eq.1.50, the magnitude of $\vec{E}_2$ is given by
$$E_2 = k\frac{2p_2}{r^2} = k\frac{p\sin\theta}{r^3}$$
Resultant electric field at P,
$$E = \sqrt{E_1^2 + E_2^2} = k\frac{p}{r^3}\sqrt{4\cos^2\theta + \sin^2\theta}$$
$$\Rightarrow \quad E = k\frac{p}{r^3}\sqrt{4\cos^2\theta + (1-\cos^2\theta)}$$
$$\Rightarrow \quad \boxed{E = k\frac{p}{r^3}\sqrt{3\cos^2\theta + 1}} \qquad (1.52)$$

The angle between the resultant field and the radial direction $\overrightarrow{OP}$ (O is the centre point of the dipole axis and P is that at which electric field is being calculated) is ϕ. From Fig.1.84, we can write
$$\tan\phi = \frac{E_2}{E_1}$$
$$\Rightarrow \quad \boxed{\phi = \tan^{-1}\left(\frac{\tan\theta}{2}\right)} \qquad (1.53)$$

Special cases:
1. $\theta = 0°$: In this case P is on the axis of the dipole. This position is called an end-on position.
In this case,
$$E_1 = k\frac{2p\cos 0°}{r^3} = k\frac{2p}{r^3},$$
$$E_2 = k\frac{p\sin 0°}{r^3} = 0$$
and $E = k\frac{p}{r^3}\sqrt{3\cos^2 0 + 1} = k\frac{2p}{r^3}$

$$\phi = \tan^{-1}\left(\frac{\tan\theta}{2}\right) = \tan^{-1} 0 = 0$$

2. $\theta = 90°$: In this case P is on the perpendicular bisector of the dipole axis.

$$E_1 = k\frac{2p\cos 90°}{r^3} = 0,$$

$$E_2 = k\frac{p\sin 90°}{r^3} = k\frac{p}{r^3}$$

and $E = k\frac{p}{r^3}\sqrt{3\cos^2 90° + 1} = k\frac{p}{r^3}$,

$$\phi = \tan^{-1}\left(\frac{\tan 90}{2}\right) = \tan^{-1}\infty = \frac{\pi}{2}$$

1.15 Check Point 3

1. ••Shown in Figure 1.85 is an array of six charges. The top charges are all positive and of the same magnitude, $+Q$, and there is a distance a between the charges. At a distance $2a$ below this line of charges are three other charges, with the outer two being $+Q$ and the central one $-Q$. What is the net electric field at a point P at the centre of the charge array?

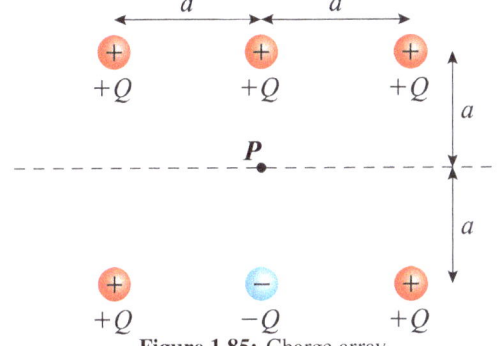

Figure 1.85: Charge array

2. ••A charge $q = 1\mu C$ is placed at point $(1\text{ m}, 2\text{ m}, 4\text{ m})$. Find the electric field at point $P(0, -4\text{m}, 3\text{ m})$.

3. ••A positive point charge $50\ \mu C$ is located in the plane xy at the point with radius vector $\vec{r}_0 = 2\hat{i} + 3\hat{j}$, where $\hat{i}$ and $\hat{j}$ are the unit vector of the x and y axis. Find the vector of the electric field strength $\vec{E}$ and its magnitude at the point with radius vector $\hat{r} = 8\hat{i} - 5\hat{j}$. Here r_0 and r are expressed in metres.

4. ••Four identical positive charges (each of magnitude q) are fixed at the corners of a square of side l. Find electric field at point P which is at a distance $z\ (\ll l)$ lying on the line perpendicular to the plane of the square passing through the centre of square. Now, if we place a test charge q_0 at point P, find the net electric force on it.

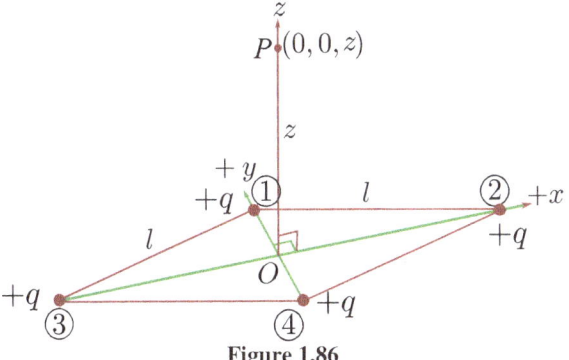

Figure 1.86

5. ••Charges Q_1 and Q_2 are at point A and B of a right angle triangle OAB [Fig.1.87]. The resultant electric field at point O is perpendicular to the hypotenuse, then find the value of Q_1/Q_2.

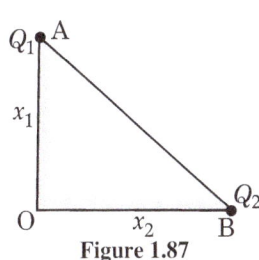

Figure 1.87

6. ••A positively charged ball hangs from a long silk thread. We wish to measure E at a point in the same horizontal plane as that of the hanging charge. To do so, we put a positive test charge q_0 at the point and measure F/q_0 [Fig.1.88]. Will F/q_0 be less than, equal to, or greater than E at the point in the question?

Figure 1.88

7. ••In Fig.1.89, find distance from $4Q$, where total electric field is zero.

Figure 1.89

8. ••Infinite charges of equal magnitude $20\mu c$ each are placed at $x = 1$ cm, 2 cm, 4 cm, 8 cm, 16 cm, $\ldots \infty$. Calculate intensity of electric field at origin.

9. ••Infinite charges are placed on x-axis such that $+q$ at $x = a, -q$ at $x = \sqrt{2}a, +q$ at $x = \sqrt{3}a, -q$ at $x = 2a, +q$ at $x = \sqrt{5}a$ etc. Calculate electric field intensity at origin.

Multiple Choice Questions

10. ••A small electric dipole is placed at origin with its dipole moment directed along positive x-axis. The direction of electric field at point $(2, 2\sqrt{2}, 0)$ is along
    (A) z-axis (B) y-axis
    (C) negative y-axis (D) negative z-axis

11. ••What is the angle between electric dipole moment and the electric field strength due to dipole on the broadside on position.
    (A) $0°$ (B) $90°$
    (C) $180°$ (D) none of these

1.16 Electric Field of a Continuous Charge Distribution

So far we have dealt with only charged particles because Coulomb's law applies only to charged particles. However, most charged objects— like charged comb, charged sphere, charged sheet, line charge etc. — are not particles. Instead, they are extended bodies. Although every macroscopic object consists of very large numbers of charged particles-protons and electrons-it is not practical to calculate the individual field of each of these particles and then add them vectorially. Instead, we shall treat any macroscopic charged object as having a continuous charge distribution (Fig.1.90). The field set up by a continuous charge distribution can be computed by dividing the distribution into infinitesimal elements dq. Each element of charge establishes a field $d\vec{E}$ at a point P, and the resultant field at P is then found from the superposition principle by adding (that is, integrating) the field contributions due to all the charge elements, i.e.,

$$\vec{E} = \int d\vec{E} \qquad (1.54)$$

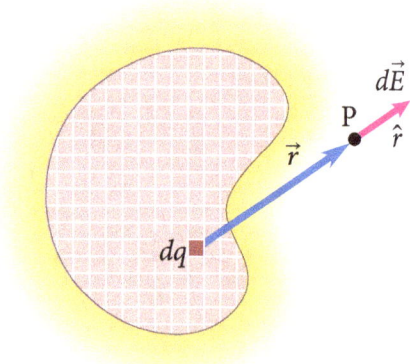

Figure 1.90

In Fig.1.90, the electric field produced by small charge element dq at point P is,

$$d\vec{E} = k\frac{dq}{r^2}\hat{r} \qquad (1.55)$$

Therefore, net electric field at point P, will be given as-

$$\vec{E} = \int d\vec{E} = k\int \frac{dq}{r^2}\hat{r} \qquad (1.56)$$

In order to evaluate this integral, we must express dq, $1/r^2$, and $\hat{r}$ in terms of the same coordinate(s). To do so, it is necessary to express the charge on the object in terms of a charge density. There are three types of charge densities:

1. Volume charge density ρ (lowercase Greek letter rho) is the amount of charge per unit volume (Fig. 1.91). For a uniform charge distribution,

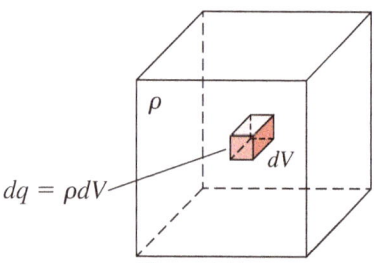

Figure 1.91

$$\rho \equiv \frac{Q}{V} \qquad (1.57)$$

Volume charge density has the dimensions charge per volume, with SI units C/m^3. The amount of charge dq in a small volume dV can be written in terms of the volume charge density:

$$\begin{pmatrix} \text{charge contained} \\ \text{in small volume} \end{pmatrix} = \begin{pmatrix} \text{charge per} \\ \text{unit volume} \end{pmatrix} \times (\text{volume})$$

$$\boxed{dq = \rho \, dV} \qquad (1.58)$$

2. Surface charge density σ (lowercase Greek letter sigma) is the amount of charge per unit area (Fig. 1.92). For a uniform charge distribution,

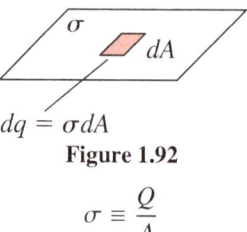

Figure 1.92

$$\sigma \equiv \frac{Q}{A}$$

The dimensions of σ are charge per area, with SI units C/m^2. The amount of charge dq in a small area dA can be written in terms of the surface charge density:

$$\begin{pmatrix} \text{charge contained} \\ \text{in small area} \end{pmatrix} = \begin{pmatrix} \text{charge per} \\ \text{unit area} \end{pmatrix} \times (\text{area})$$

$$\boxed{dq = \sigma \, dA} \qquad (1.59)$$

3. Linear charge density λ (lowercase Greek letter lambda) is the amount of charge per unit length (Fig.1.93). For a uniform charge distribution,

Figure 1.93

$$\lambda \equiv \frac{Q}{L} \qquad (1.60)$$

The dimensions of λ are charge per length, with SI units C/m. The amount of charge dq in a small length element dL can be written in terms of the linear charge density:

$$\begin{pmatrix} \text{charge contained in} \\ \text{small length element} \end{pmatrix} = \begin{pmatrix} \text{charge per} \\ \text{unit length} \end{pmatrix} \times (\text{length})$$

$$\boxed{dq = \lambda \, dL} \qquad (1.61)$$

Problem solving tactics for calculating the electric field from continuous charge distributions:

1. Identify the type of charge distribution and compute the charge density λ, σ or ρ.
2. Divide the charge distribution into infinitesimal charges dq, each of which will act as a tiny point charge.
3. The amount of charge dq, i.e., within a small element dl, dA or dV is
   $dq = \lambda dl$ (charge distributed along length)
   $dq = \sigma dA$ (charge distributed over a surface)
   $dq = \rho dV$ (charge distributed throughout a volume)
4. Draw at point P the dE vector produced by the charge dq. The magnitude of dE is

$$dE = \frac{1}{4\pi\epsilon_0}\frac{dq}{r^2}$$

Vector $d\vec{E}$ is along radial line joining dq to P, $d\vec{E}$ is directed away for positive charge dq while directed towards dq for negative dq.

5. Resolve the $d\vec{E}$ vector into its components. Identify any special symmetry features to show whether any component(s) of the field that are not get cancelled by other components.
6. Write the distance r and any trigonometric factors in terms of given coordinates and parameters.
7. The electric field is obtained by summing over all the infinitesimal contributions.

$$\vec{E} = \int d\vec{E} = \int \frac{dq}{4\pi\epsilon_0 r^2}$$

1.16.1 Electric Field Due to a Uniformly Charged Rod of Finite Length at a Perpendicular Distance y

Let us consider a rod of length L as shown in Fig.1.94. suppose a positive charge Q is uniformly distributed over the surface of the rod along it's length. P is an arbitrarily positioned point, where the electric field $\vec{E}$ is to be calculated.

The x axis is along the rod, and the y axis passes through the field point P. Let y be the perpendicular distance of P from the x axis. To calculate the electric field $\vec{E}$ at P, we separately calculate E_x and E_y. Using Equations 1.55 and 1.56, first find the electric field increment $d\vec{E}$ at P due to an arbitrary increment dq of the charge distribution. Then integrate each component of $d\vec{E}$ over the entire charge distribution. (Because Q is distributed uniformly, the linear charge density λ equals Q/L.)

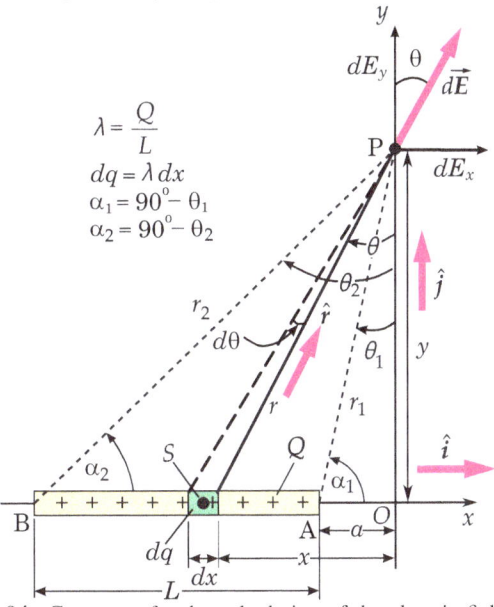

Figure 1.94: Geometry for the calculation of the electric field at field point P due to a uniformly charged rod.

Stepwise Calculation

1. The situation is shown in Fig.1.94. In addition, figures also shows an arbitrary increment of the line charge at point S (at OS = x) that has a length dx and a charge dq, and the electric field $d\vec{E}$ at P due to dq. The electric field vector $d\vec{E}$ is drawn by assuming if dq is positive.

2. $\vec{E} = E_x \hat{i} + E_y \hat{j}$. Now, we find the expressions for dE_x and dE_y in terms of dE_r and θ, where dE_r, is the component of $d\vec{E}$ in the direction away from S toward P:
$$d\vec{E} = dE_r \hat{r}$$
so $\quad dE_x = dE_r \sin\theta \quad$ and $\quad dE_y = dE_r \cos\theta$

3. First we solve for E_x. Express dE_r using Equation 1.55, where r is the distance from the source point S to the field point P. From Fig.1.94 $\cos\theta = y/r$. In addition, use $dq = \lambda dx$:
$$dE_r = \frac{k\, dq}{r^2} \quad \text{and} \quad \sin\theta = \frac{x}{r}$$
so $\quad dE_x = \frac{k\, dq}{r^2}\sin\theta = \frac{k\sin\theta \lambda dx}{r^2}$

4. Integrate the step-3 result:

$$dE_x = \int_a^{a+L} \frac{k\sin\theta \lambda dx}{r^2} = k\lambda \int_a^{a+L} \frac{\sin\theta dx}{r^2}$$

5. Now, change the integration variable from x to θ. From Figure 1.94, find the relation between x and θ and between r and θ.
$$\tan\theta = \frac{x}{y}, \quad \text{so} \quad x = \frac{y}{\tan\theta}$$
$$\cos\theta = \frac{y}{r}, \quad \text{so} \quad r = \frac{y}{\cos\theta} = y\sec\theta$$

6. Differentiate the step 5 result to obtain an expression for dx (the field point P remains fixed, so y is constant):
$$dx = y\sec^2\theta\, d\theta$$

7. Substitute $y\sec^2\theta d\theta$ for dx and $y\sec\theta$ for r in the integral in step 4 and simplify:
$$\int_a^{a+L} \frac{\sin\theta\, dx}{r^2} = \int_{\theta_1}^{\theta_2} \frac{\sin\theta\, y\, \sec^2\theta\, d\theta}{y^2 \sec^2\theta}$$
$$= \frac{1}{y}\int_{\theta_1}^{\theta_2} \sin\theta\, d\theta \quad (y \neq 0)$$

8. Evaluate the integral and solve for E_x:
$$E_x = k\lambda \frac{1}{y}\int_{\theta_1}^{\theta_2} \sin\theta\, d\theta = -\frac{k\lambda}{y}(\cos\theta_2 - \cos\theta_1)$$
Again from Fig.1.94, $\theta_1 = 90° - \alpha_1$ and $\theta_2 = 90° - \alpha_2$, therefore in terms of angles $\alpha_1 =$ and α_2, above expression can also be written as
$$E_x = -\frac{k\lambda}{y}(\sin\alpha_2 - \sin\alpha_1)$$
Now, substituting $\cos\theta_2 = y/r_2$ and $\cos\theta_1 = y/r_1$ in first expression, we get
$$E_x = -\frac{k\lambda}{y}\left(\frac{y}{r_2} - \frac{y}{r_1}\right)$$
$$= k\lambda\left(\frac{1}{r_1} - \frac{1}{r_2}\right) \quad (r_1 > 0 \text{ and } r_2 > 0)$$
From this expression, it is clear that $E_x > 0$ at all points on the x axis in the region $x > a$. Above expression is not defined at end points of the rod, because at end point A, $r_1 = 0$ and at end point B, $r_2 = 0$.

9. Similar to E_x, use steps steps 3-7 to get E_y:
$$E_y = k\lambda\frac{1}{y}\int_{\theta_1}^{\theta_2}\cos\theta\, d\theta = \frac{k\lambda}{y}(\sin\theta_2 - \sin\theta_1)$$
In terms of α_1 and α_2,
$$E_y = \frac{k\lambda}{y}(\cos\alpha_2 - \cos\alpha_1)$$
From step 5, $\cos\theta_2 = y/r_2$ or $y = r_2\cos\theta_2$, and $\cos\theta_1 = y/r_1$ or $y = r_1\cos\theta_1$ therefore,
$$E_y = k\lambda\left(\frac{\sin\theta_2}{y} - \frac{\sin\theta_1}{y}\right) = k\lambda\left(\frac{\sin\theta_2}{r_2\cos\theta_2} - \frac{\sin\theta_1}{r_1\cos\theta_1}\right)$$
or $\quad E_y = k\lambda\left(\frac{\tan\theta_2}{r_2} - \frac{\tan\theta_1}{r_1}\right) \quad (y \neq 0)$

And in terms of α_1 and α_2 above expression can be written as
$$E_y = k\lambda\left(\frac{\cot\alpha_2}{r_2} - \frac{\cot\alpha_1}{r_1}\right) \quad (y \neq 0)$$

Note that if point P lies on the x-axis, then $y = 0$. In this case, you should not write, $\theta_1 = \theta_2 = 180°$, or $\alpha_1 = \alpha_2 = 0°$. It is because the first expression for E_y: $E_y = \frac{k\lambda}{y}(\sin\theta_2 - \sin\theta_1)$ is valid everywhere in the xy plane but not on the x axis where ($y = 0$). The two tangent functions in the expression for E_y are given by

$\tan\theta_1 = \dfrac{a}{y}$ and $\tan\theta_2 = \dfrac{a+L}{y}$

and neither of these functions is defined on the x axis (where $y = 0$). Therefore, to get E_y on x axis, we have to use Eq.1.55. By recognizing that on the x axis $\hat{r} = \pm\hat{i}$, we can see that Equation 1.55 tells us that $d\vec{E} = \pm dE\hat{i}$, which implies $E_y = 0$.

10. Combine steps 8 and 9 to obtain an expression for the electric field at P:
$$\vec{E} = E_x\hat{i} + E_y\hat{j}$$

Making Sense of the Result: Consider the plane that is perpendicular to and bisecting the rod. At points on this plane, symmetry dictates that $\vec{E}$ points directly away from the centre of the rod. That is, we expect that $E_x = 0$ throughout this plane. At all points on this plane $r_1 = r_2$. The step- 8 result gives $E_x = 0$ if $r_1 = r_2$, as expected.

Generalization of Result

Now, consider Fig.1.95. The electric field at point P due to a thin uniformly charged rod located on the z axis is given by $\vec{E} = E_z\hat{k} + E_R\hat{R}$, where

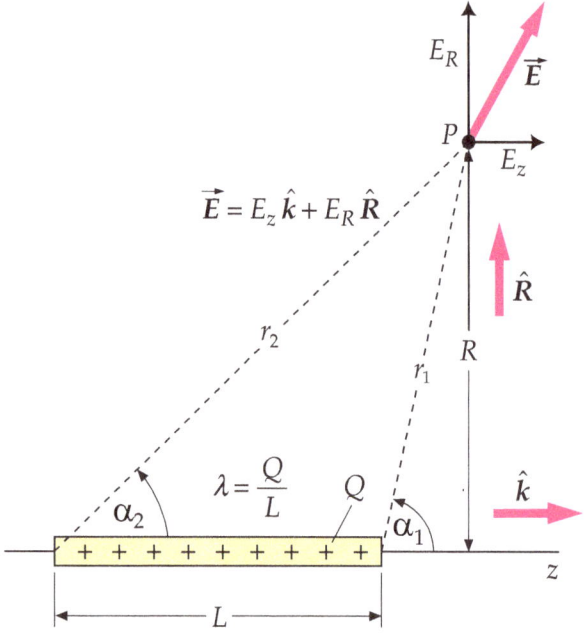

Figure 1.95: The electric field due to a uniformly charged thin rod.

$$E_z = -\dfrac{k\lambda}{R}(\sin\alpha_2 - \sin\alpha_1)$$

or $E_z = k\lambda\left(\dfrac{1}{r_1} - \dfrac{1}{r_2}\right)$ $(r_1 \neq 0)$ and $(r_2 \neq 0)$ (1.62)

and $E_R = \dfrac{k\lambda}{R}(\cos\alpha_2 - \cos\alpha_1)$

or $E_R = k\lambda\left(\dfrac{\cot\alpha_2}{r_2} - \dfrac{\cot\alpha_1}{r_1}\right)$ $(R \neq 0)$ (1.63)

The expressions for E_z (Equation (1.62)) are undefined at the end points of the thin charged rod and the expressions for E_R (Equation (1.63)) are undefined at all points on the z axis (where $R = 0$). However, $E_R = 0$ at all points where $R = 0$.

Here, it is to be noted that α_1 and α_2 should be used with proper sign.

1.16.1.1 Electric Field Due to an Infinite Line Charge

If $L \to \infty$, then from the Figure 1.95, we see that $\alpha_1 \to \pi$ and $\alpha_2 \to 0$. Now, put these values in Equations (1.62) and (1.63)

and solve for E_z and E_R. The resultant electric field at point P can be obtained by using the expression-
$$\vec{E} = E_z\hat{k} + E_R\hat{R}$$

Step Wise Calculations

1. Choose the first expression for the electric field in each of Equations (1.62) and (1.63):
$$E_z = -\dfrac{k\lambda}{R}(\sin\alpha_2 - \sin\alpha_1)$$
$$E_R = \dfrac{k\lambda}{R}(\cos\alpha_2 - \cos\alpha_1)$$

2. Take the limit as both $\alpha_1 \to \pi$ and as $\alpha_2 \to 0$
$$E_z = -\dfrac{k\lambda}{R}(\sin 0 - \sin\pi) = -\dfrac{k\lambda}{R}(0 - 0) = 0$$
$$E_R = \dfrac{k\lambda}{R}(\cos 0 - \cos\pi) = \dfrac{k\lambda}{R}(1 - (-1)) = 2\dfrac{k\lambda}{R}$$

3. Express the electric field in vector form:
$$\vec{E} = E_z\hat{k} + E_R\hat{R} = 0\hat{k} + \dfrac{2k\lambda}{R}\hat{R} = \dfrac{2k\lambda}{R}\hat{R}$$
$$\implies \boxed{\vec{E} = \dfrac{2k\lambda}{R}\hat{R}} \quad (1.64)$$

Thus, the magnitude of electric field decreases inversely with the radial distance from the line charge.

Special Results: Semi Infinite Wire

For a semi-infinite wire, as shown in Fig.1.96, we have-
$$\alpha_1 = \pi, \quad \alpha_2 = \pi/2$$

Figure 1.96

Therefore, $E_z = -\dfrac{k\lambda}{R}(\sin\alpha_2 - \sin\alpha_1) = -\dfrac{k\lambda}{R}\left(\sin\dfrac{\pi}{2} - \sin\pi\right)$
$$= -\dfrac{k\lambda}{R}(1 - 0) = -\dfrac{k\lambda}{R}$$

and $E_R = \dfrac{k\lambda}{R}(\cos\alpha_2 - \cos\alpha_1) = \dfrac{k\lambda}{R}\left(\cos\dfrac{\pi}{2} - \cos\pi\right)$
$$= \dfrac{k\lambda}{R}(0 - (-1)) = \dfrac{k\lambda}{R}$$

In this case, net electric field,
$$E = \sqrt{E_z^2 + E_R^2} = \sqrt{(-k\lambda/R)^2 + (k\lambda/R)^2} = \dfrac{k\lambda}{R}\sqrt{2}$$

The angle between net electric field vector $\vec{E}$ and length of the rod, i.e., z-axis, is given by-
$$\theta = \tan^{-1}\dfrac{E_R}{E_z} = \tan^{-1}1 = \dfrac{\pi}{4}$$

1.16.1.2 Electric Field of a Finite Line Charge at a Point on its Extension

EXAMPLE 45. A charge Q is uniformly distributed along the z axis, from $z = -\tfrac{1}{2}L$ to $z = +\tfrac{1}{2}L$. Show that for large values of z the expression for the electric field of the line charge on the z axis approaches the expression for the electric field of a point charge Q at the origin.

APPROACH To find $\vec{E}$, at an axial point P, use Eq. (1.62).

Step wise Calculation

1. The electric field on the z axis has only a z component, given by Equation (1.62):
$$E_z = k\lambda\left(\dfrac{1}{r_2} - \dfrac{1}{r_1}\right)$$

2. The line charge along the z axis, the field point P, r_1 and r_2 are shown in Figure 1.97:

3. Substitute with $r_1 = z - \tfrac{1}{2}L$ and $r_2 = z + \tfrac{1}{2}L$ into the step 1

1.16. ELECTRIC FIELD OF A CONTINUOUS CHARGE DISTRIBUTION

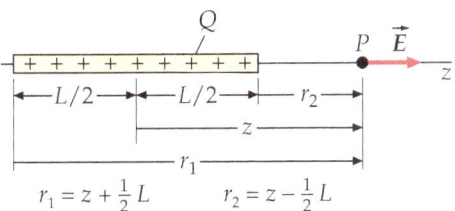

Figure 1.97: Geometry for the calculation of the electric field on the axis of a uniformly charged rod.

result and simplify:

$$E_z = k\lambda \left(\frac{1}{z - \frac{1}{2}L} - \frac{1}{z + \frac{1}{2}L} \right)$$

$$= \frac{kQ}{L} \frac{L}{z^2 - \left(\frac{1}{2}L\right)^2}$$

$$= \frac{kQ}{z^2 - \left(\frac{1}{2}L\right)^2} \quad \left(z > \frac{1}{2}L\right)$$

This is the required expression for electric field at an axial position of a uniformly charged rod.
Note that the result is valid for the region $L/2 < z < \infty$ only. The result is not valid in the region $-L/2 < z < +L/2$?
For $z \gg L$, the above result gives-

$$E_z \approx \frac{kQ}{z^2} \quad (z \gg L)$$

This expression is the same as the expression for the electric field of a point charge Q located at the origin.

EXAMPLE 46. What is the electric field at any point on the axis of a uniformly charged rod of length 'L' having linear charge density 'λ'? The point is separated from the nearer end by distance 'a' [Fig.1.98].
Note: Although this type of problem has been already discussed

Figure 1.98

in above article, here we solve it by using the method of integration.
SOLUTION Consider an elementary length dx of the rod at distance x, from point P, where electric field E is to be determined. The elemental charge, $dq = \lambda dx$
The small electric field at P due to this element,

$$dE = k \cdot \frac{\lambda dx}{x^2}$$

or $\quad E = k\lambda \int_a^{a+L} \frac{1}{x^2} dx = k\lambda \left[-\frac{1}{x} \right]_a^{a+L} = k\lambda \left[\frac{-1}{a+L} + \frac{1}{a} \right]$

Thus, $\quad E = \frac{\lambda}{4\pi\epsilon_o} \left[\frac{1}{a} - \frac{1}{L+a} \right] \quad \left[\because k = \frac{1}{4\pi\epsilon_0} \right]$

1.16.1.3 Approximating Equations 1.62 and 1.63 on the Symmetry Plane

EXAMPLE 47. A charge Q is uniformly distributed along the z axis, from $z = -\frac{1}{2}L$ to $z = +\frac{1}{2}L$. (a) Find an expression for the electric field on the $z = 0$ plane as a function of R, the radial distance of the field point from the z axis. (b) Show that for $R \gg L$, the expression found in Part (a) approaches that of a point charge at the origin of charge Q. (c) Show that for $R \ll L$, the expression found in Part (a) approaches that of an infinitely long line charge on the z axis with a uniform linear charge density $\lambda = Q/L$.

APPROACH The charge distribution is uniform and the linear charge density is $\lambda = Q/L$. The line charge is along the z axis and the field point is in the $z = 0$ plane. We use Equations (1.62) and (1.63) to find the electric field expression for part (a). The electric field due to a point charge decreases inversely with the square of the distance from the charge. Examine the part (a) result to see how it approaches that of a point charge at the origin for $R \gg L$. The electric field due to a uniform line charge of infinite length decreases inversely with the radial distance from the line (Eq.1.64). Examine the Part (a) result to see how it approaches the expression for the electric field of a line charge of infinite length for $R \ll L$.

SOLUTION (a) Step wise solution is given below-
1. From Eq.(1.62) and (1.62), we have-

$$E_z = -\frac{k\lambda}{R}(\sin\alpha_2 - \sin\alpha_1), \quad E_R = \frac{k\lambda}{R}(\cos\alpha_2 - \cos\alpha_1)$$

2. The charge configuration with the line charge on the z axis from $z = -\frac{1}{2}L$ to $z = +\frac{1}{2}L$ is shown in Fig.1.99. The field point P is in the $z = 0$ plane at a distance R from the origin.

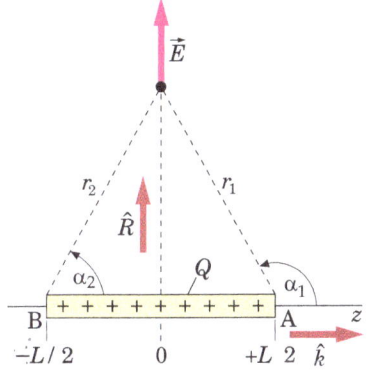

Figure 1.99

3. From the Figure 1.99, $r_1 = r_2$, therefore, the interior angle at the right end B of the rod will also be α_2. Therefore, at point B, we can write $\alpha_1 + \alpha_2 = \pi$, so $\sin\alpha_1 = \sin(\pi - \alpha_2) = \sin\alpha_2$ and $\cos\alpha_1 = \cos(\pi - \alpha_2) = -\cos\alpha_2$. Substitution of these values in equations of step 1 gives:

$$E_z = -\frac{k\lambda}{R}(\sin\alpha_2 - \sin\alpha_2) = 0,$$

$$E_R = \frac{k\lambda}{R}(\cos\alpha_2 + \cos\alpha_2) = \frac{2k\lambda}{R}\cos\alpha_2$$

4. Express $\cos\alpha_2$ in terms of R and L and substitute into the step-3 result:

$$\cos\alpha_2 = \frac{\frac{1}{2}L}{\sqrt{R^2 + \left(\frac{1}{2}L\right)^2}}$$

so, $E_R = \frac{2k\lambda}{R} \frac{\frac{1}{2}L}{\sqrt{R^2 + \left(\frac{1}{2}L\right)^2}} = \frac{k\lambda L}{R\sqrt{R^2 + \left(\frac{1}{2}L\right)^2}}$

5. Express the electric field in vector form, and substitute Q for λL:

$$\vec{E} = E_z\hat{k} + E_R\hat{R} = 0\hat{k} + E_R\hat{R}$$

so $\vec{E} = E_R\hat{R} = \frac{kQ}{R\sqrt{R^2 + \left(\frac{1}{2}L\right)^2}}\hat{R}$

(b) The step wise solution is given below-
1. Examine the step-5 result of part (a). If $R \gg L$ then $R^2 + \left(\frac{1}{2}L\right)^2 \approx R^2$. Substitute R^2 for $R^2 + \left(\frac{1}{2}L\right)^2$:

$$\vec{E} \approx \frac{kQ}{R\sqrt{R^2}}\hat{R} = \frac{kQ}{R^2}\hat{R} \quad (R \gg L)$$

2. This (approximate) expression for the electric field decreases inversely with the square of the distance from the origin, just as it would for a point charge Q at the origin.

$$\vec{E} \approx \frac{kQ}{R^2}\hat{R} \quad (R \gg L)$$

(c) Examine the Part (a), step - 5 result. If $R \ll L$ then $R^2 + \left(\frac{1}{2}L\right)^2 \approx \left(\frac{1}{2}L\right)^2$. Substitute $\left(\frac{1}{2}L\right)^2$ for $R^2 + \left(\frac{1}{2}L\right)^2$. This (approximate) expression for the electric field falls off inversely with the radial distance from the line charge, just as the exact expression for an infinite line charge (1.64) would.

$$\vec{E} \approx \frac{k\lambda L}{R\sqrt{\left(\frac{1}{2}L\right)^2}}\hat{R} = \frac{2k\lambda}{R}\hat{R} \quad (R \ll L)$$

EXAMPLE 48. Figure 1.100 shows a non-conducting rod that has a uniform positive charge density $+\lambda$ and a total charge Q along its right half, and a uniform negative charge density $-\lambda$ and a total charge $-Q$ along its left half. What is the direction and magnitude of the net electric field at point P that shown in Fig. 1.100?

SOLUTION When we consider a segment dx at distance x from

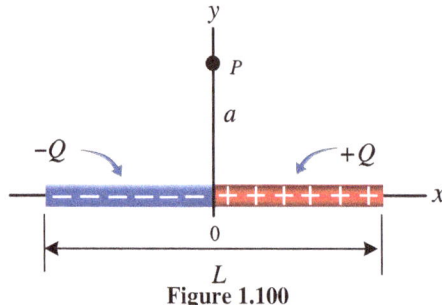

Figure 1.100

O on the right side of the rod, the charge on this segment will be $dq = \lambda dx$, as shown in Fig. 1.101.

The electric field $d\vec{E}_+$ at P due to this segment is directed outwards and away from the positive charge dq and has a magnitude:

$$dE_+ = k\frac{dq}{r^2} = k\frac{\lambda dx}{r^2}$$

A symmetric segment on the opposite side of the rod, but with a negative charge, creates an electric field $d\vec{E}_-$ that is directed inwards and toward this segment and has the same magnitude as $d\vec{E}_+$, i.e. $dE_+ = dE_-$. The resultant electric field $d\vec{E}$ from both symmetric segments will be a vector to the left (Fig. 1.101), and its magnitude will be given by:

$$dE = dE_+ \cos\theta + dE_+ \cos\theta = 2dE_+ \cos\theta$$

From Fig.1.101, $\cos\theta = x/r$ and $r = \sqrt{x^2 + a^2}$, therefore

$$dE = 2k\frac{\lambda dx}{r^2}\frac{x}{r} = k\lambda(x^2 + a^2)^{-3/2}(2x)dx$$

The total electric field at P due to all segments of the rod is found by integrating dE from $x = 0$ to only $x = L/2$, since the negative charge of the rod is considered in evaluating dE. Thus:

$$E = \int dE = k\lambda\int_{x=0}^{x=L/2}(x^2 + a^2)^{-3/2}(2xdx) \quad \cdots (1)$$

We can simplify this integration by using substitution method. Let $x^2 + a^2 = u$, then $2xdx = du$
When $x = 0$, then $u = x^2 + a^2$ gives $u = 0^2 + a^2 = a^2$ and when $x = L/2$, then $u = (L/2)^2 + a^2$.
Substituting these values and limits in Eq.(1), we get

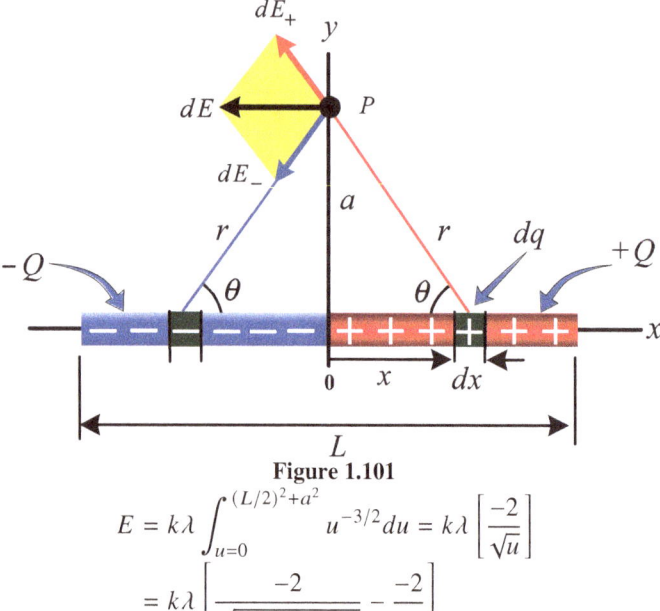

Figure 1.101

$$E = k\lambda\int_{u=0}^{(L/2)^2+a^2} u^{-3/2}du = k\lambda\left[\frac{-2}{\sqrt{u}}\right]$$

$$= k\lambda\left[\frac{-2}{\sqrt{(L/2)^2 + a^2}} - \frac{-2}{a}\right]$$

$$= 2k\lambda\left[\frac{1}{a} - \frac{1}{\sqrt{(L/2)^2 + a^2}}\right]$$

When we use the fact that the magnitude of the charge Q is given by $Q = \lambda L/2$, we get:

$$E = \frac{4kQ}{L}\left[\frac{1}{a} - \frac{1}{\sqrt{(L/2)^2 + a^2}}\right]$$

When P is very far away from the rod, i.e. $a \gg L$, we can neglect $(L/2)^2$ in the denominator of this equation and hence get $E \approx 0$. In this situation, the two oppositely charged halves of the rod would appear to point P as if they were two coinciding point charges and hence have a zero net charge.

EXAMPLE 49. An infinite sheet of charge is lying on the xy-plane as shown in Fig. 1.102. A positive charge is distributed uniformly over the plane of the sheet with a charge per unit area σ. Calculate the electric field at a point P located a distance a from the plane.

SOLUTION Let us divide the plane into narrow strips parallel to the y-axis. A strip of width dx can be considered as an infinitely long wire of charge per unit length λ given as:

$$\lambda = \lim_{L\to\infty}\frac{\text{charge on the length } L \text{ of strip of thickness } dx}{L}$$

$$= \lim_{L\to\infty}\frac{\sigma(L(dx))}{L} = \sigma dx$$

From Eq. (1.64), at point P, the strip sets up an electric field $d\vec{E}$ lying in the xz-plane of magnitude:

$$dE = 2k\frac{\lambda}{r} = 2k\frac{\sigma dx}{r}$$

This electric field vector can be resolved into two components $d\vec{E}_x$ and $d\vec{E}_z$. When we consider the entire sheet of charge, we find a symmetrically opposite component corresponding to each component of $d\vec{E}_x$. So, it will sum to zero. Therefore, the resultant electric field at point P will be in the z-direction, perpendicular to the sheet. From Fig. 1.102, we find the following:

$$dE_z = dE\sin\theta$$

and hence:

$$E = \int dE_z = 2k\sigma\int_{-\infty}^{+\infty}\frac{\sin\theta dx}{r}$$

To perform the integration of this expression, we must first relate

1.16. ELECTRIC FIELD OF A CONTINUOUS CHARGE DISTRIBUTION

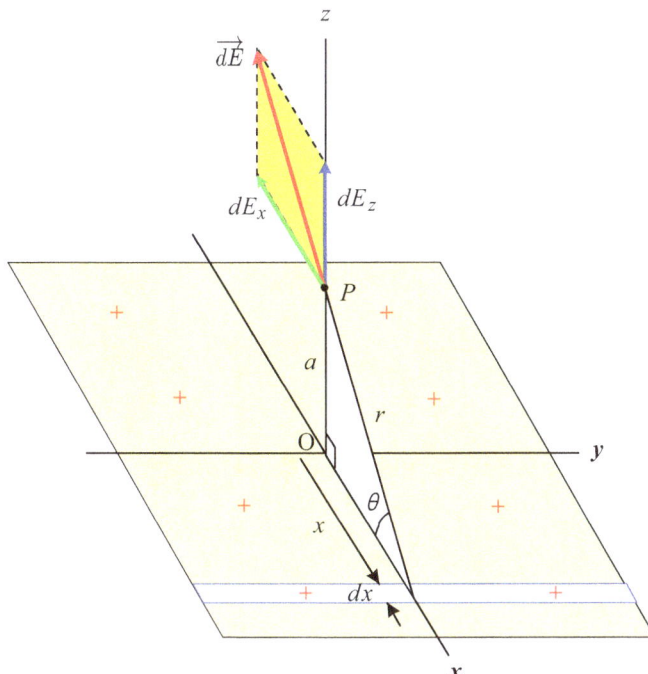

Figure 1.102

the variables θ, x, and r. One approach is to express θ and r in terms of x. From the geometry of Fig.1.102, we have: $r = \sqrt{x^2 + a^2}$ and $\sin\theta = \dfrac{a}{r} = \dfrac{a}{\sqrt{x^2+a^2}}$

Therefore, $E = 2k\sigma a \displaystyle\int_{-\infty}^{+\infty} \dfrac{dx}{x^2+a^2} = 2k\sigma a \left[\dfrac{1}{a}\tan^{-1}\dfrac{x}{a}\right]_{-\infty}^{+\infty}$

$= 2k\sigma \left[\tan^{-1}(\infty) - \tan^{-1}(-\infty)\right]$

$= 2k\sigma \left[\dfrac{\pi}{2} + \dfrac{\pi}{2}\right] = 2\pi k\sigma = 2\pi \dfrac{1}{4\pi\epsilon_0}\sigma = \dfrac{\sigma}{2\epsilon_0}$

i.e., $\boxed{E = \dfrac{\sigma}{2\epsilon_o}}$ (1.65)

We note that the distance a from the plane to the point P does not appear in the final result of E. This means that the electric field set up at any point by an infinite plane sheet of charge is independent of how far the point is from the plane. In other words, the electric field is *uniform* and *normal* to the plane.

Also, the same result is obtained if the point P lies below the xy-plane. That is, the field below the plane has the same magnitude as that above the plane but as a vector it points in the opposite direction.

EXAMPLE 50. Two infinite plane sheets with uniform surface charge densities $+\sigma$ and $-\sigma$ are placed parallel to each other with separation d (Fig.1.103). Find the electric field between the sheets, above the upper sheet, and below the lower sheet.

APPROACH Equation (1.65) gives the electric field due to a single infinite plane sheet of charge. To find the field due to two such sheets, we combine the fields by using the principle of superposition (Fig.1.103).

SOLUTION From Eq.(1.65), both $\vec{E}_1$ and $\vec{E}_2$ have the same magnitude at all points, independent of distance from either sheet:

$$E_1 = E_2 = \dfrac{\sigma}{2\epsilon_0}$$

$\vec{E}_1$ is everywhere directed away from sheet 1, and $\vec{E}_2$ is everywhere directed toward sheet 2. Between the sheets, $\vec{E}_1$ and $\vec{E}_2$ reinforce each other; above the upper sheet and below the lower

Figure 1.103: Finding the electric field due to two oppositely charged infinite sheets. The sheets are seen edge-on; only a portion of the infinite sheets can be shown!

sheet, they cancel each other. Thus the total field is

$$\vec{E} = \vec{E}_1 + \vec{E}_2 = \begin{cases} 0 & \text{above the upper sheet} \\ \dfrac{\sigma}{\epsilon_0}\hat{j} & \text{between the sheets} \\ 0 & \text{below the lower sheet} \end{cases} \quad (1.66)$$

Discussion: Because we considered the sheets to be infinite, our result does not depend on the separation d. Our result shows that the field between oppositely charged plates is essentially uniform if the plate separation is much smaller than the dimensions of the plates.

important Point Electric fields are not some kind of physical substance that "flows". So, it is not correct to say that the field $\vec{E}_1$ of sheet 1 would be unable to "penetrate" sheet 2, and that field $\vec{E}_2$ caused by sheet 2 would be unable to "penetrate" sheet 1. Infact, the electric fields $\vec{E}_1$ and $\vec{E}_2$ depend on only the individual charge distributions that create them. The total field at every point is just the vector sum of $\vec{E}_1$ and $\vec{E}_2$.

EXAMPLE 51. A force of 10 N acts on a charged particle placed between two thin large charged metal plates. Now, one plate is removed, find the force acting on that particle.

SOLUTION If σ is the magnitude of surface charge density on the metal plates, then from Eq.1.66, the electric field between the plates:

$$E = \dfrac{\sigma}{\epsilon_0}$$

Now, according to given problem, 10 N force is acting on a charged particle placed between the plates. So, if Q is the charge on charged particle placed between the plates, then

$$F = QE = \dfrac{Q\sigma}{\epsilon_0} \implies \dfrac{Q\sigma}{\epsilon_0} = 10 \text{ N} \quad \cdots (1)$$

If one plate is removed then from Eq.1.65, electric field produced due to single plate will be

$$E' = \dfrac{\sigma}{2\epsilon_0}$$

Therefore, the force on the charged particle due to single plate

$$F' = QE' = \dfrac{Q\sigma}{2\epsilon_0} = \dfrac{1}{2}\left(\dfrac{Q\sigma}{\epsilon_0}\right)$$

From Eq.(1), $Q\sigma/\epsilon_0 = 10$ N, therefore above equation gives

$$F' = \dfrac{1}{2}(10 \text{ N}) = 5 \text{ N}$$

1.16.2 Electric Field at the Centre of a Charged Circular Arc

Let us consider a thin circular arc of radius R. Suppose, total positive charge Q is uniformly distributed over it. This arc forms central angle of α rad (Fig.1.104a). To find the electric field at the centre P of this arc, we place coordinate axes such that the axis of symmetry of the arc lies along the x axis and the origin is at the arc's center. If λ represent the linear charge density of this arc which has a length $R\alpha$, then:

$$\lambda = \dfrac{Q}{R\alpha}$$

For an arc element ds subtending an angle $d\theta$ at P, we have:

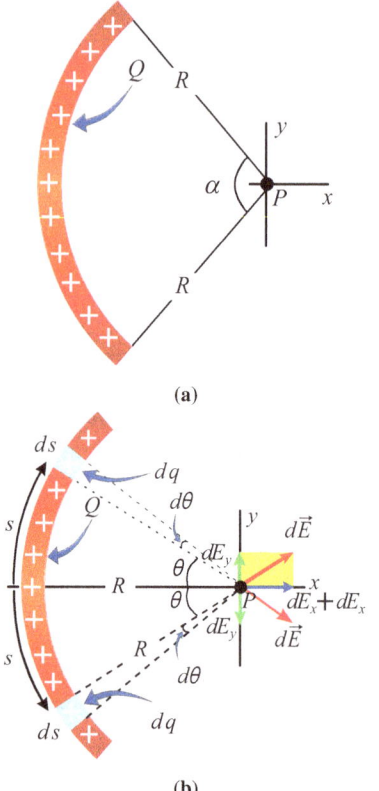

(a)

(b)

Figure 1.104: (a) A circular arc of radius R, central angle α, and centre P has a uniformly distributed positive charge Q. (b) The figure shows the electric fields $d\vec{E}$ at P due to lower and upper arc elements (each of length ds and charge dq). From symmetry, the vertical components of all elements cancel out and the total field is along the x-axis

$ds = R d\theta$

Therefore, the charge dq on this arc element will be given by:

$$dq = \lambda ds = \frac{Q}{R\alpha} R d\theta = \frac{Q}{\alpha} d\theta$$

To find the electric field at point P, we first calculate the magnitude of the electric field dE at P due to this element of charge dq, [Fig.1.104b], as follows:

$$dE = k \frac{dq}{R^2} = \frac{kQ}{R^2 \alpha} d\theta \quad (1.67)$$

This field has a vertical component $dE_y = dE \sin\theta$ along the positive y-axis and a horizontal component $dE_x = dE \cos\theta$ along the positive x-axis, as shown in Fig.1.104b. The y-component created at P by any charge element dq is canceled by a symmetric charge element on the opposite side of the arc. Thus, the perpendicular components of all of the charge elements sum to zero. The horizontal component will take the form:

$$dE_x = dE \cos\theta = \frac{kQ}{R^2 \alpha} \cos\theta d\theta \quad (1.68)$$

Consequently, the total electric field at P due to all elements of the arc is given by the integration of the x-component as follows:

$$E = \int dE_x = \frac{kQ}{R^2 \alpha} \int_{-\alpha/2}^{+\alpha/2} \cos\theta d\theta = \frac{kQ}{R^2 \alpha} [\sin\theta]_{-\alpha/2}^{+\alpha/2}$$

$$= \frac{kQ}{R^2 \alpha} \left[\sin\frac{\alpha}{2} - \sin\left(-\frac{\alpha}{2}\right) \right]$$

$$\Rightarrow \quad E = \frac{kQ}{R^2 \alpha} \left[\sin\frac{\alpha}{2} + \sin\frac{\alpha}{2} \right]$$

or $$\boxed{E = \frac{kQ}{R^2} \frac{\sin\alpha/2}{\alpha/2}} \quad (1.69)$$

So, the total electric field at P will be along the x-axis and it's magnitude given by Eq.(1.69).

There are three special cases to Eq. (1.69):

1. $\alpha = 0$ (Point charge)
   When we apply the limiting case
   $$\lim_{\alpha \to 0} [\sin(\alpha/2)/(\alpha/2)] = 1, \text{ we get: } E = \frac{kQ}{R^2}$$

2. $\alpha = \pi$ (A Semicircular arc)
   When we substitute with $\sin(\pi/2)/(\pi/2) = 2/\pi$, we get:
   $$E = \frac{2kQ}{\pi R^2}$$

3. $\alpha = 2\pi$ (A ring of radius R)
   When we substitute with $\sin(2\pi/2) = 0$, we get: $E = 0$
   This is an expected result, since we shall see that Eq. (1.76) gives $E = 0$ when P is at the centre of the ring, i.e. when $a = 0$.

1.16.3 Electric Field at a Point on the Axis of a Thin Charged Ring

Fig.1.105 shows a ring-shaped conductor with radius a carries a total charge Q uniformly distributed around it. Axis of the ring is along x axis. P is an arbitrary point that lies on the axis of the ring at a distance x from its centre. Our target variable is the electric field at such a point as a function of the coordinate x.

☞This is a clear case of the superposition of electric fields. Note that the charge is distributed continuously around the ring rather than in a number of point charges.

Suppose, the ring divided into infinitesimal segments of length ds, Each segment has charge dQ and acts as a point-charge source of electric field as shown in Fig.1.105. Let $d\vec{E}$ be the electric field from one such segment; the net electric field at P is then the sum of all contributions $d\vec{E}$ from all the segments that make up the ring.

Now, Consider two segments at the top and bottom of the ring: The contributions $d\vec{E}$ to the field at P from these segments have the same x-component but opposite y-components. Hence the total y-component of field due to this pair of segments is zero. When we add up the contributions from all such pairs of segments, the total field $\vec{E}$ will have only a component along the ring's symmetry axis (the x-axis), with no component perpendicular to that axis (that is, no component in yz-plane). So the field at P is described completely by its x-component E_x.

CALCULATION OF $\vec{E}$ AT POINT P: If r be the distance of

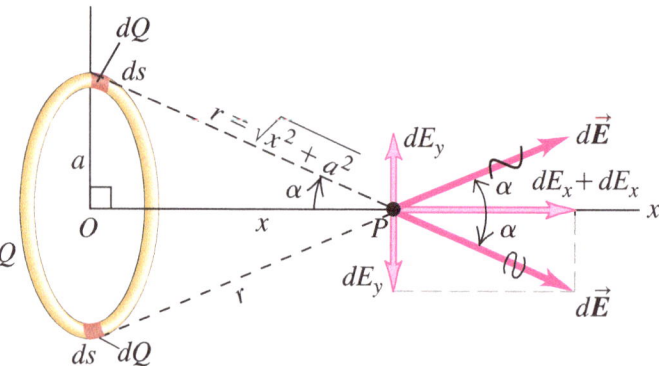

Figure 1.105: The total electric field at P is along the x axis. The perpendicular component of the field at P due to upper charge element segment dQ is cancelled by the perpendicular component due to lower charge element segment dQ.

point P from the upper charge segment dQ, then from Fig.1.105, we have

$$r = \sqrt{x^2 + a^2} \quad (1.70)$$

1.16. ELECTRIC FIELD OF A CONTINUOUS CHARGE DISTRIBUTION

and, $\cos \alpha = \dfrac{x}{r}$ (1.71)

The magnitude of the electric field at P due to the upper segment ds of charge dQ is

$$dE = k\dfrac{dQ}{r^2} \quad (1.72)$$

The x-component of this field-

$$dE_x = dE \cos \alpha \quad (1.73)$$

Substituting, the values of dE, $\cos \alpha$ and r from Eq.1.72, 1.71 and 1.70 respectively in Eq.1.73, we get

$$dE_x = \left(k\dfrac{dQ}{r^2}\right)\dfrac{x}{r} = \dfrac{kx}{(x^2+a^2)^{3/2}}dQ \quad (1.74)$$

All elements of the ring make the same contribution to the field at P because they are all equidistant from this point. Thus, we can integrate to obtain the total field at P

$$E_x = \int \dfrac{kx}{(x^2+a^2)^{3/2}}dQ$$

Since x does not vary as we move from point to point around the ring, all the factors on the right side except dQ are constant and can be taken outside the integral. The integral of dQ is just the total charge Q, and we finally get

$$E_x = \dfrac{kx}{(x^2+a^2)^{3/2}} \int dQ$$

or $\boxed{E_x = \dfrac{kQx}{(x^2+a^2)^{3/2}}}$ (1.75)

As Q is positive, therefore, field is directed away from the centre of the ring. In vector form Eq.(1.75) can be written as-

$$\boxed{\vec{E} = E_x \hat{i} = k\dfrac{Qx}{(x^2+a^2)^{3/2}}\hat{i}} \quad (1.76)$$

Here, $k = 1/4\pi\epsilon_0$

Special Cases
1. When $x = 0$ i.e., the field point P is at the centre of the ring, the Eq.1.76 gives: $\vec{E} = 0$.
2. When $x \gg a$, i.e., the field point P is much farther from the ring than its size, the denominator in Eq.1.76 becomes approximately equal to x^3, and the expression becomes approximately

$$\vec{E} = \dfrac{1}{4\pi\epsilon_0}\dfrac{Q}{x^2}\hat{i}$$

In other words, when we are so far from the ring that its size a is negligible in comparison to the distance x, its field is the same as that of a point charge.

Tracing E vs x Graph
1. **Maximum and Minimum values of Electric Field:** For maximum value of E, on x axis, we have

$$\dfrac{dE}{dx} = 0$$

From Eq.1.76, the expression for E, on x axis as-

$$E = k\dfrac{Qx}{(x^2+a^2)^{3/2}}$$

Therefore,

$$\dfrac{dE}{dx} = kQ\dfrac{d}{dx}\dfrac{x}{(x^2+a^2)^{3/2}}$$

$$= kQ\left[\dfrac{(x^2+a^2)^{3/2}1 - x\frac{3}{2}(x^2+a^2)^{1/2}2x}{(x^2+a^2)^3}\right]$$

$$= kQ(x^2+a^2)^{1/2}\left[\dfrac{a^2-2x^2}{(x^2+a^2)^3}\right]$$

Therefore, $\dfrac{dE}{dx} = 0 \Rightarrow kQ(x^2+a^2)^{1/2}\dfrac{a^2-2x^2}{(x^2+a^2)^3} = 0$

or $x = \pm\dfrac{a}{\sqrt{2}}$

So, the value of electric field will be maximum at $x = \pm\dfrac{a}{\sqrt{2}}$.

Substituting this value of x in the expression for E, we get

$$E_{\max} = \pm kQ\dfrac{a/\sqrt{2}}{\left(\dfrac{a^2}{2}+a^2\right)^{3/2}} = \pm\dfrac{2}{\sqrt{27}}\dfrac{kQ}{a^2}$$

2. We have already shown that, at $x = 0$, the electric field $E = 0$.
3. When $x \to \pm\infty$, then, again $\vec{E} \to 0$

Keeping these points in mind, the variation of E with x is given in Fig.1.106

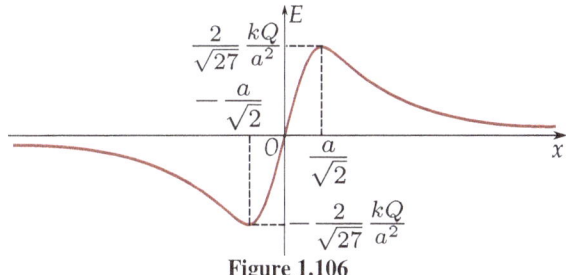

Figure 1.106

1.16.3.1 Motion of a Negatively Charged Particle Close to the Centre of the Positively Charged Ring on it's Axis

In the expression for the field (Eq.1.76) due to a ring of charge Q, we let $x \ll a$, which results in

$$E_x = \dfrac{kQ}{a^3}x$$

Thus, from $F_x = Q_0 E_x$ the force on a charge $-Q_0$ placed near the centre of the ring is

$$F_x = -k\dfrac{QQ_0}{a^3}x$$

Negative sign shows that the force is restoring in nature. Because this force has the form of SHM, the motion will be simple harmonic.

If mass of the charged particle is m and it's acceleration towards the centre of the ring is d^2x/dt^2, then above equation can also be written as-

$$m\dfrac{d^2x}{dt^2} = -k\dfrac{QQ_0}{a^3}x \implies \dfrac{d^2x}{dt^2} = -\dfrac{kQQ_0}{ma^3}x$$

$$\implies \dfrac{d^2x}{dt^2} = -\omega^2 x$$

here, $\omega = \sqrt{\dfrac{kQQ_0}{ma^3}}$ is the angular frequency of SHM.

Time period of oscillation of the particle about the centre of the ring-

$$T = \dfrac{2\pi}{\omega} = 2\pi\sqrt{\dfrac{ma^3}{kQQ_0}}$$

Note: Always look for the condition of SHM as $F \propto -x$

1.16.4 Electric Field on the Axis of a Charged Disc

Let us consider a point P on the axis of a uniformly charged disc of radius R at a distance x from the centre of it. Suppose, the surface charge density of it is σ.

We have already calculated the electric field at an axial point of a uniformly charged ring. Now, the disc can be considered as made of number of concentric very thin continuous rings with continuously varying radii from $r = 0$ to $r = R$. To find the electric field on the axis of the disc, we first consider an elementary concentric

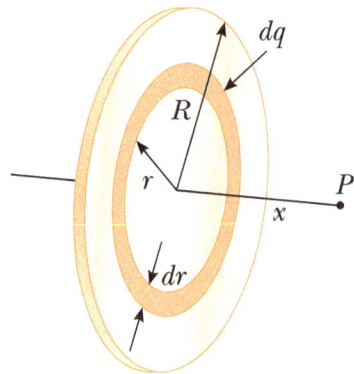

Figure 1.107: A uniformly charged disc of radius R. The electric field at an axial point P is directed along the central axis, perpendicular to the plane of the disc.

ring of radius r and thickness dr as shown in Fig.1.107. The surface area of this ring is $2\pi r dr$. The charge dq on this ring is $2\pi r dr \sigma$. We can easily find the electric field at point P due to this elementary charged ring and then integration from $r = 0$ to $r = R$ gives the electric field at P due to the complete uniformly charged disc of radius R.

Now, from Eq.1.75, the electric field at point P due to this elementary charged ring

$$dE_x = \frac{kx}{(x^2 + r^2)^{3/2}} (2\pi\sigma r dr)$$

To obtain the total field at P, we integrate this expression over the limits $r = 0$ to $r = R$. Here, x is a constant.

$$E_x = \pi k \sigma x \int_0^R \frac{2r dr}{(x^2 + r^2)^{3/2}} = 2\pi k \sigma \left(1 - \frac{x}{(x^2 + R^2)^{1/2}}\right)$$

Since, $k = 1/4\pi\epsilon_0$, therefore

$$E_x = \frac{\sigma}{2\epsilon_0}\left(1 - \frac{x}{(x^2 + R^2)^{1/2}}\right) \quad (1.77)$$

This result is valid for all values of $x > 0$ and $x < 0$, but for $x = 0$ the answer is not valid as there is discontinuity at $x = 0$. We can calculate the field close to the disc along the axis by assuming that $R \gg x$; thus, the expression in bracket reduces to unity to give us the near-field approximation.

In this case, just right side of the centre of disc, the electric field

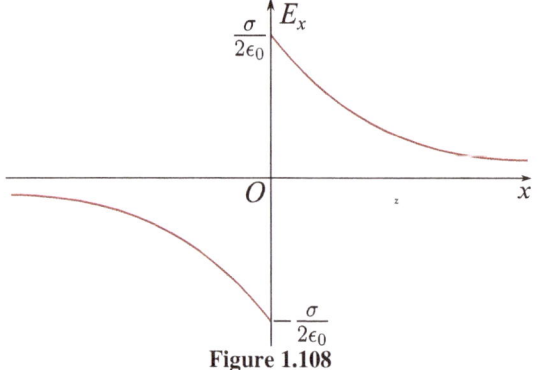

Figure 1.108

is directed towards the positive direction of x-axis and is given by-

$$E_x = 2\pi k \sigma = \frac{\sigma}{2\epsilon_0} \quad (\text{if} \quad x \to 0^+)$$

Just left side of the centre of disc, the electric field is directed towards the negative direction of x-axis and is given by-

$$E_x = -2\pi k \sigma = -\frac{\sigma}{2\epsilon_0} \quad (\text{if} \quad x \to 0^-)$$

In this case, the disc can be regarded as a plane sheet of charge.

If Q is the the total charge on the disc, then $\sigma = Q/\pi R^2$. Substituting this value of σ in Eq.1.77, we get

$$E_x = \frac{Q}{2\pi\epsilon_0 R^2}\left(1 - \frac{x}{(x^2 + R^2)^{1/2}}\right) \quad (1.78)$$

This is an expression of electric field at any point on the axis of the charged disc in terms of it's charge Q.

1.17 The Shape of Lightning Rods

According to Benjamin Franklin, the lightning was a giant electrical spark. To support his theory, he suggested that a metal rod be placed on top of a tall structure to capture "electric fluid"— what we call charged particles.

After conducting his experiments that supported his theory of lightning as a giant spark, Franklin came up with a plan to avoid lightning strikes. He knew that objects could be charged by rubbing. He also knew that you cannot charge a metal object—a conductor—by rubbing it if you connect the conductor to ground. Today we know why: The Earth acts as a source and sink of electrons. If you build up a few excess electrons on a grounded conductor, these electrons quickly travel through the conductor and along the pathway to ground (Fig.1.109).

Franklin reasoned that during a storm, the atmosphere builds up

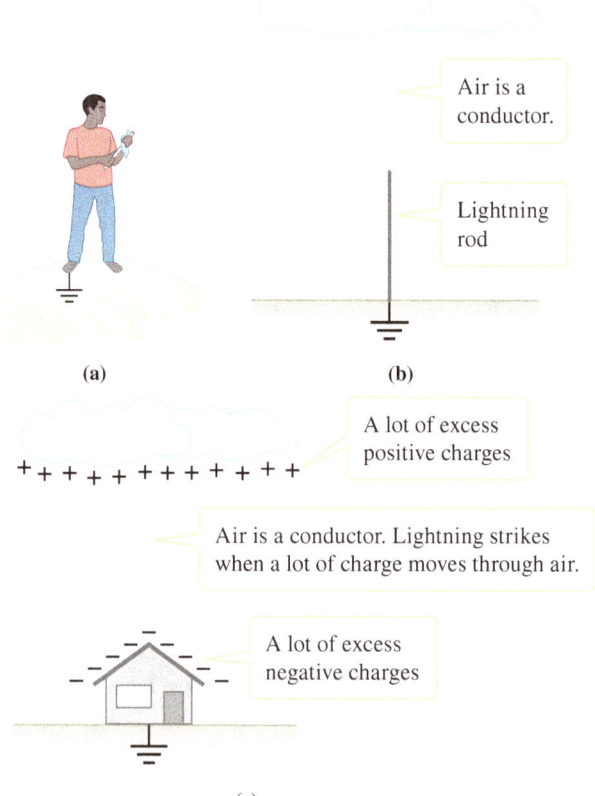

Figure 1.109: (a) Franklin knew that a grounded person cannot charge a conducting rod by rubbing. (b) According to Franklin, lightning strikes could be prevented if a town put up a lot of lightning rods. Then the atmosphere would be connected to ground, so it could not build up charge. (c) Without a lightning rod, Franklin reasoned, the atmosphere discharges in a violent strike.

excess charge much as a glass rod builds up excess charge when it is rubbed with silk. He knew from his experience with charged objects that a spark can make its way through the air. His indoor experiments demonstrated that if he used a pointed object (like

1.17. THE SHAPE OF LIGHTNING RODS

a knitting needle) to draw charge from an object through the air, only a small spark occurred compared to the larger spark drawn to a blunt object (like his thumb). Franklin reasoned that when a small charge builds up in the atmosphere, a pointed lightning rod will draw the small charge through the air and into the Earth continuously (Fig. 1.109b). If only a small charge travels through the air, it will not be visible and there will be no giant spark of lightning.

Without a lightning rod, charge still builds up in the atmosphere. Because buildings are connected to the Earth, they attract charge of the opposite sign (Fig. 1.109c). When sufficient charge builds up, the air acts as a conductor and a large charge is transferred in a giant lightning spark. Such a violent spark causes great damage to buildings and can be very dangerous. Franklin published the following recommendations for lightning rods:

1. Just outside each building an iron rod **should be planted 3 feet to 4 feet into the moist ground**.
2. The rod should extend 6 feet to 8 feet **above the tallest part of the structure**.
3. On top of the rod should be a foot of brass wire **sharpened to a fine point.**

Benjamin Wilson—a contemporary of Franklin's—believed that lightning rods should be blunt. He argued that a pointed rod would draw down lightning that might have just passed harmlessly overhead. Wilson argued that pointed lightning rods were more dangerous than having no rods at all.

In this section, we explore both types of rods (Fig.1.110) in following two solved examples. In order for air to breakdown[*] and become a conductor, the electric field in the air must be 3×10^6 N/C. Let's assume that in order for a lightning rod to work, the electric field at its surface must equal to breakdown electric field. We will calculate the amount of charge on the surface of each conductor. The one with the least amount of charge is the better design because a smaller amount of charge on the surface of the conductor means a smaller amount of charge travels through the air. Calculating the charge on each rod gives us a way to compare their effectiveness. A rod that does not require much charge to have a strong electric field on its surface will not generate a big spark to discharge the atmosphere. Because both designs use conductors, we will assume the excess charge is uniformly spread over the surface of the lightning rod.

EXAMPLE 52. At the end of Wilson's rod was a cannonball

Figure 1.110: If there is a ball at the end of the lightning rod, a lot of charge builds up before the atmosphere discharges. If there is no ball at the end of the lightning rod, little charge builds up because the atmosphere continually discharges through the rod

---
[*]If the magnitude of an electric field in air is so great, that the air becomes ionized and begins to conduct electricity, then the electric field is called the breakdown field for air and the phenomenon is called **dielectric breakdown** of air

(Fig.1.111) with a radius of $R = 0.1 m$. If the electric field on the surface of the ball is $E = 3 \times 10^6$ N/C (breakdown field strength for air), what is the charge on the ball?

APPROACH Because the ball is so much larger than the thickness of the supporting rod, we will ignore that rod and model Wilson's device as a suspended sphere connected to ground. The ball is a sphere, so we can use the magnitude of $\vec{E}(r) = (kQ/r^2)\hat{r}$ to find the charge on the ball.

SOLUTION For the charge Q on the surface of the sphere, where $r = R$, we have-

$$E(R) = k\frac{Q}{R^2}$$

$$\Rightarrow \quad Q = \frac{ER^2}{k} = \frac{(3 \times 10^6 \text{N/C})(0.1 \text{ m})^2}{(8.99 \times 10^9 \text{N} \cdot \text{m}^2/\text{C}^2)}$$

$$\Rightarrow \quad Q = 3.3 \times 10^{-6} C = 3.3 \mu C$$

Result This is not very much charge, especially compared to the amount of charge involved in a lightning strike, which is of the order of hundreds of coulombs.

EXAMPLE 53. Let's model the end of a Franklin rod as a tiny ball of radius $R = 2$ mm - something like the end of a knitting needle. If the electric field on the surface of the ball is $E = 3 \times 10^6$ N/C, what is the charge on the ball?

APPROACH Repeat the calculation from above example for the Wilson rod, this time for a much smaller ball. Solve for the charge q on the surface of the sphere, where $r = R$.

SOLUTION $E(R) = k\frac{q}{R^2}$

$$\Rightarrow \quad q = \frac{ER^2}{k} = \frac{(3 \times 10^6 \text{N/C})(0.002 \text{ m})^2}{(8.99 \times 10^9 \text{N} \cdot \text{m}^2/\text{C}^2)}$$

$$\Rightarrow \quad q = 1.3 \times 10^{-9} C = 1.3 \text{ nC}$$

So, the charge on Franklin's rod is about 2500 times lower than on Wilson's.

From above two examples, we can say that an ideal Franklin rod (one that is infinitely thin) would be better than the blunt Wilson rod.

EXAMPLE 54. Find an expression for the electric field at the tip of Franklin's pointed lightning rod. Evaluate Franklin's design by considering a point very close to the end of the rod.

APPROACH Fig. 1.111 shows a vertical rod of length L and P is a point at a distance d from the upper end of the rod. For a limiting case, when P is just above the upper tip of the rod, we have $d \to 0$. As the charge will get distributed uniformly on the metal rod, therefore, to find electric field at point P, we have to consider a differential charge element $dq(=\lambda \, dl)$ of length dl of the rod at distance l from the point P. Here λ is the linear charge density of the rod. Since, each elementary piece of the rod produces electric field, at point P, in the same direction, therefore, the integration from $l = d$ to $l = d + L$ will give the net electric field at P due to complete length of the charged rod.

Electric field at P due to elementary charge dq, is given by-

$$d\vec{E} = k\frac{dq}{r^2}\hat{j}, \text{ with } r = l$$

To derive an expression for $d\vec{E}$, integrate this expression from $l = d$ to $l = d + L$

For the electric field just above the tip of the rod, we have $d \to 0$.

SOLUTION Electric field at P, due to the infinitesimal charge $dq = \lambda dl$, is-

$$d\vec{E} = k\frac{dq}{l^2}\hat{j} \quad \Rightarrow \quad d\vec{E} = k\frac{\lambda dl}{l^2}\hat{j}$$

Now, on integrating over the entire length of the rod from $l = d$ to $l = d + L$, we get

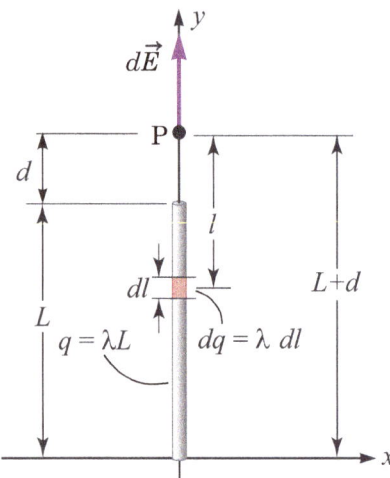

Figure 1.111

$$\int_0^E d\vec{E} = \int_0^l k\frac{\lambda dl}{l^2}\hat{j}$$

$$\Rightarrow \vec{E} = k\lambda\left[-\frac{1}{l}\right]_d^{d+L}\hat{j} = -k\lambda\left[\frac{1}{d+L} - \frac{1}{d}\right]\hat{j}$$

$$\Rightarrow \vec{E} = -k\lambda\left[\frac{-L}{d(d+L)}\right]\hat{j} = k\left[\frac{\lambda L}{d(d+L)}\right]\hat{j}$$

The total charge on the rod is $q = \lambda L$. Therefore, on substituting $\lambda L = q$ in above expression, we get-

$$\vec{E} = \frac{kq}{d(d+L)}\hat{j} \quad (1.79)$$

If point P is far from the rod ($d \gg L$), then $d + L \approx d$. In this case, the electric field at P is given by

$$\vec{E} = \frac{kq}{d^2}\hat{j}$$

So, $\vec{E}$ approaches the electric field due to a charged particle. For the point, just above the rod, $d \to 0$, therefore

$$\vec{E} = \lim_{d \to 0}\left[\frac{kq}{d(d+L)}\right] = \infty$$

So, it is clear that, as point P gets closer to the end of the rod, the electric field increases. At the tip of the rod, the electric field approaches infinity. So, in principle, even a very tiny charge on the rod would lead to a very high electric field at its end. Franklin's rod would continually draw a small amount of charge from the atmosphere and prevent a large lightning strike.

EXAMPLE 55. A thin non-conducting ring of radius R has a linear charge $\lambda = \lambda_0 \cos\theta$, where λ_0 is the value of λ at $\theta = 0$ [see Fig.1.112]. Find the net electric dipole moment for this charge distribution.

APPROACH The charge density of the given ring is $\lambda = $

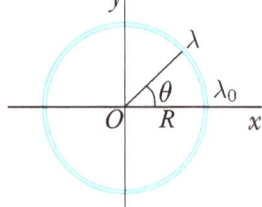

Figure 1.112

$\lambda_0 \cos\theta$. The value of $\cos\theta$ is positive for $-\pi/2 < \theta < \pi/2$ and negative for $\pi/2 < \theta < 3\pi/2$. So, the charge of the ring is positive in first and fourth quadrants whereas, it is negative in second and third quadrants. Since, $|\cos\theta| = |\cos(2\pi \pm \theta)| = |\cos(\pi \pm \theta)|$. Therefore, corresponding to each positive elementary charge in first and fourth quadrant, there will be an equal and opposite charge in third and fourth quadrant. So, each elementary charge is able to form an electric dipole.

Let us consider a small part of length dl of the ring as shown in Fig.1.113a. If this part subtends angle $d\theta$ at the centre of the ring, then

$$dl = Rd\theta$$

The small charge on this part of ring is given as-

$$dq = \lambda\, dl = (\lambda_0 \cos\theta)(Rd\theta)$$

The symmetrically opposite part to this part has charge $-dq$. So, these parts form a dipole. The location of charge $-dq$ may be considered in quadrant fourth or in quadrant second (Fig. 1.113). For, Fig.1.113a, dipole length is $2R$, whereas for dipole shown in Fig.1.113b, its length is $2R\cos\theta$.

Now, for the dipole considered in Fig.1.113(a), the magnitude of dipole moment of this elementary dipole is given by-

$$dp = 2Rdq = 2R\lambda_0 \cos\theta\, Rd\theta$$

The direction of this dipole moment is from $-dq$ to $+dq$.
x and y components of this dipole moment are,

$$dp_x = 2R\lambda_0 \cos\theta R d\theta \cos\theta \quad (1.80a)$$

$$dp_y = 2R\lambda_0 \cos\theta R d\theta \sin\theta \quad (1.80b)$$

To get, total electric dipole moment of the ring, we have to integrate these expressions from $\theta = -\pi/2$ to $+\pi/2$. For the

(a) (b)

Figure 1.113: Charged ring

dipole shown in Fig.1.113b, the dipole length is $2R\cos\theta$. It's dipole moment is given by-

$$dp = 2R\cos\theta\, dq = 2R\cos\theta\,(\lambda_0 \cos\theta)\, Rd\theta$$

$$\Rightarrow dp = 2R^2\lambda_0\cos^2\theta\, d\theta \quad (1.81)$$

The direction of this dipole moment is from $-dq$ to $+dq$ (Fig1.113b), i.e., along +ve x- direction. In this case, y- component of dipole moment is zero.

To get net dipole moment of the ring, we have to integrate it from $\theta = -\pi/2$ to $+\pi/2$.

Now, we find the net dipole moment of the ring by using both above methods, i.e., by considering both the dipoles of Fig.1.113.

SOLUTION Method 1. Considering, the dipole shown in Fig.1.113, From, Eq.(1.80) the $x\ y$ components of dipole moment are -

$$dp_x = 2R^2\lambda_0\cos^2\theta d\theta$$

and $$dp_y = 2R\lambda_0\cos\theta\sin\theta R\, d\theta$$

respectively.
On integrating both of the above expressions, from $\theta = -\pi/2$ to $+\pi/2$, we get-

$$p_x = \int_{-\pi/2}^{+\pi/2} 2R^2\lambda_0 \cos^2\theta\, d\theta$$

and $$p_y = \int_{-\pi/2}^{+\pi/2} 2R^2\lambda_0 \cos\theta \sin\theta\, d\theta$$

$$\Rightarrow p_x = \int_{-\pi/2}^{+\pi/2} 2R^2\lambda_0\cos^2\theta\, d\theta$$

$$= R^2\lambda_0 \int_{-\pi/2}^{+\pi/2} (1+\cos 2\theta)\, d\theta$$

$$= R^2\lambda_0 \left[[\theta]_{-\pi/2}^{+\pi/2} + \left[\frac{\sin 2\theta}{2}\right]_{-\pi/2}^{+\pi/2}\right] = \pi R^2 \lambda_0$$

$$\implies p_x = \pi R^2 \lambda_0$$

$$\&\ p_y = R^2\lambda_0 \int_{-\pi/2}^{+\pi/2} \sin 2\theta\, d\theta = R^2\lambda_0 \left[-\frac{\cos 2\theta}{2}\right]_{-\pi/2}^{+\pi/2} = 0$$

Method 2. In this method, we integrate Eq.(1.81) from limit $\theta = -\pi/2$ to $\theta = +\pi/2$
Integrating, we get-

$$p = 2R^2\lambda_0 \int_{-\pi/2}^{\pi/2} \cos^2\theta R d\theta$$

Simplifying as like we have done for p_x, we get
$$p = \pi R^2 \lambda_0$$
Here, $\vec{p}$ is directed along the +ve direction of x axis.

1.18 Check Point 4

1. ●●In Fig.1.114, a thin glass rod forms a semicircle of radius $R = 5.00$ cm. Charge is uniformly distributed along the rod, with $+q = 4.50$ pC in the upper half and $-q = -4.50$ pC in the lower half. What are the (a) magnitude and (b) direction (relative to the positive direction of the x axis) of the electric field $\vec{E}$ at the centre (P) of the semicircle?

2. ●●In Fig.1.115, two curved plastic rods, one of charge $+q$ and the other of charge $-q$, form a circle of radius $R = 8.50$ cm in an xy plane. The x axis passes through both of the connecting points, and the charge is distributed uniformly on both rods. If $q = 15.0$ pC, what are the (a) magnitude and (b) direction (relative to the positive direction of the x axis) of the electric field $\vec{E}$ produced at P, the centre of the circle?

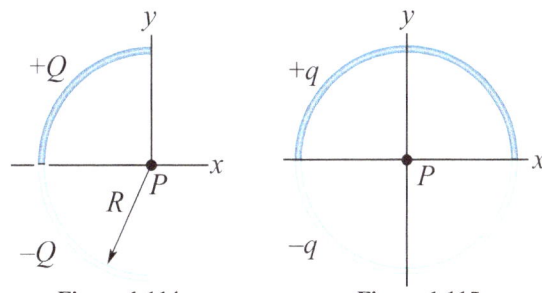

Figure 1.114   Figure 1.115

3. ●●Find the electric dipole moment of a non-conducting ring of radius a, made of two semicircular rings having linear charge densities $-\lambda$ and $+\lambda$ as shown in Fig. 1.116.

4. ●●Find the electric dipole moment of a non-conducting ring of radius a, having linear charge densities $+\lambda$ and $-\lambda$ and arranged as shown in Fig.1.117.

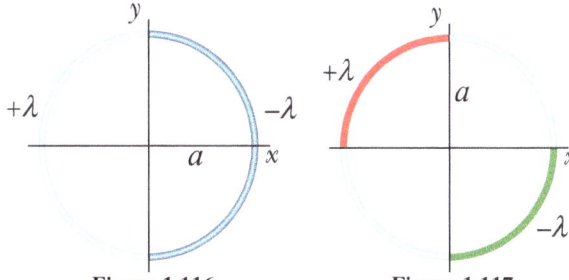

Figure 1.116   Figure 1.117

5. ●●Fig.1.118 shows two discs and a flat ring, each with the same uniform charge Q. Rank the objects according to the magnitude of the electric field they create at points P (which are at the same vertical heights) in decreasing order.

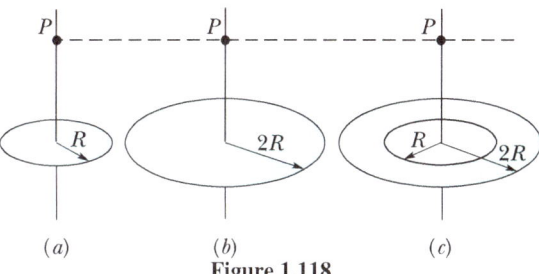

Figure 1.118

6. ●●A positive point charge, Q, is located at a distance h directly above the centre of a charged thin non-conducting circular plate of radius R. The plate carries a total positive charge, Q, spread uniformly over its surface area. What will be the electrical force on the point charge?

7. ●●●A thin nonconducting ring of radius R has a linear charge density $\lambda = \lambda_0 \cos\phi$, where λ_0 is a constant, and azimuthal angle ϕ is shown in Fig.1.119. Find the magnitude of the electric field strength
   (a) at the centre of the ring;

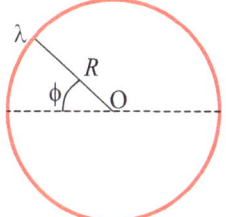

Figure 1.119

(b) on the axis of the ring as a function of the distance z from its centre. Investigate the obtained function at $z \gg R$.

8. ●●A thread carrying a uniform charge λ per unit length has the configurations shown in Fig. 1.120a and 1.120b. Assuming a curvature radius R to be considerably less than the length of the thread, find the magnitude of the electric field strength at the point O.

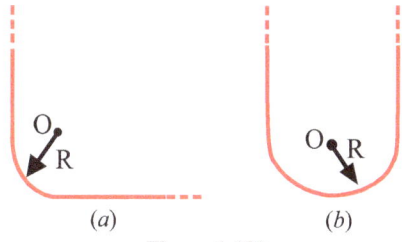

Figure 1.120

Multiple Choice Questions

9. ●●Supposing that the earth has a surface charge density of 1 electron/m^2; calculate electric field just outside earth's surface. The electronic charge is -1.6×10^{-19}C and earth's radius is 6.4×10^6m. $\left(\epsilon_0 = 8.9 \times 10^{12} \text{C}^2/\text{N.m}^2\right)$

   (A) 1.8×10^{-8} N/C   (B) $+1.8 \times 10^{-8}$N/C
   (C) 1.8×10^{-9} N/C   (D) $+1.8 \times 10^{-9}$N/C

10. ●●A circular wire of radius R carries a total charge Q distributed uniformly over its circumference. A small length of the wire subtending angle θ at the centre is cut off. Find the electric field at the centre due to the remaining portion.

    (A) $Q/(4\pi^2\epsilon_0 R^2)\sin\theta$ (B) $Q/(4\pi^2\epsilon_0 R^2)\sin\theta/2$
    (C) $Q/(4\pi^2\epsilon_0 R^2)\sin\theta$ (D) $Q/(4\pi^2\epsilon_0 R^2)\sin\theta/8$

1.19 Electric Field Lines

The picture of field lines (sometimes called lines of force) was invented by Faraday to develop an intuitive non-mathematical way of visualizing electric fields around charged configurations. An electric field line is, in general, a curve drawn in such a way that the tangent to it at each point is in the direction of the net field at that point (Fig.1.121). An arrow on the curve is obviously necessary to specify the direction of electric field from the two possible directions indicated by a tangent to the curve. A field line is a space curve, i.e., a curve in three dimensions.

To understand the dependence of the field lines on the area, or

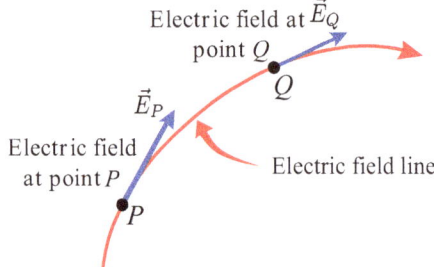

Figure 1.121: The direction of the electric field at any point is the tangent to the electric field line at this point

rather the solid angle subtended by an area element, let us try to relate the area with the solid angle, a generalization of angle to three dimensions. Recall how a (plane) angle is defined in two-dimensions. Let a small transverse line element Δl be placed at a distance r from a point O. Then the angle subtended by Δl at O can be approximated as $\Delta \theta = \Delta l / r$. Similarly, in three-dimensions the solid angle subtended by a small perpendicular plane area ΔS, at a distance r, can be written as $\Delta \Omega = \Delta S / r^2$. In Fig.1.122, for two points P_1 and P_2 at distances r_1 and r_2 from the charge, the element of area subtending the solid angle $\Delta \Omega$ is $r_1^2 \Delta \Omega$ at P_1 and an element of area $r_2^2 \Delta \Omega$ at P_2, respectively. The number of lines (say n) cutting these area elements are the same. The number of field lines, cutting unit area element is therefore $n/(r_1^2 \Delta \Omega)$ at P_1 and $n/(r_2^2 \Delta \Omega)$ at P_2, respectively. Since n and $\Delta \Omega$ are common, the strength of the field clearly has a $1/r^2$ dependence.

The electric field is strong where field lines are close together

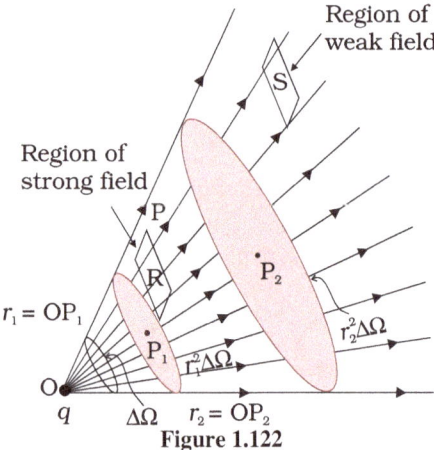

Figure 1.122

and weak where they are far apart (Fig.1.123). (More specifically, if you imagine a small surface perpendicular to the field lines, the magnitude of the field is proportional to the number of lines that cross the surface divided by the area.)

☞ Electric field lines represent the net electric field at any point. This means that they take into account the contribution of all the charges in the region.

Rules for Drawing Electric Field Lines: The rules for drawing

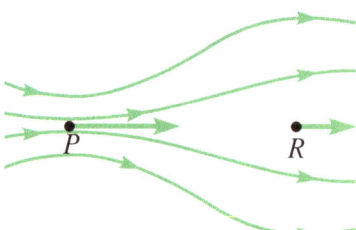

Figure 1.123: The magnitude of the electric field at point P is larger than the magnitude at R.

electric field lines are as follows:
- The lines must begin on a positive charge and terminate on a negative charge. In the case of an excess of one type of charge, some lines will begin or end infinitely far away.
- The number of field lines emerging from a positive charge or ending at a negative charge is proportional to the magnitude of the charge.
- At large distances from a system of charges that has a non-zero net charge, the field lines are equally spaced and radial, as if they emanated from (or terminated on) a single point charge equal to the total charge of the system.
- Electric field lines cannot cross each other. The electric field at any point has a unique direction; if two field lines crossed, the field would have two directions at the same point (Fig.1.124)

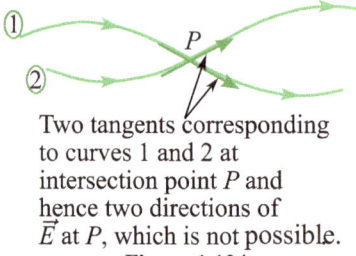

Figure 1.124

We choose the number of field lines starting from any positively charged object to be Cq and the number of lines ending on any negatively charged object to be $C|q|$, where C is an arbitrary proportionality constant and q is charge on the object. Once C is chosen, the number of lines is fixed. For example, if object 1 has charge Q_1 and object 2 has charge Q_2, then the ratio of number of lines is $N_2/N_1 = Q_2/Q_1$. The electric field lines for two isolated point charges of magnitude $+q$ and $|-2q|$ are shown in Fig.1.125. Because the magnitudes of charges are in the ratio of $1:2$, the number of field lines that originated from $+q$ and terminated on $-2q$ are also in the ratio of $1:2$. For example, 8 field line are originated from $+q$ and 16 lines are terminated on $-2q$, so $N_1 : N_2 = 8 : 16$, i.e., $1:2$.

1.20 Theorem on Circulation of Vector $\vec{E}$

In mechanics we have studied that any stationary field of central forces is conservative, i.e. the work done by the forces of this field is independent of the path and depends only on the position of the initial and final points. This property is inherent in the electrostatic field, viz. the field created by a system of fixed charges. If we take a unit positive charge for the test charge and carry it from initial position point i of a given field $\vec{E}$ to final position point f, the elementary work of the forces of the field done over the distance $d\vec{l}$ is-
$$dW = \vec{F_e} \cdot d\vec{l} = \vec{E} \cdot d\vec{l}$$
[because, for unit positive test charge, $\vec{F_e} = q_0 \vec{E} = 1\vec{E} = \vec{E}$]
and the total work of the field forces over the distance between points i and f is defined as

1.20. THEOREM ON CIRCULATION OF VECTOR $\vec{E}$

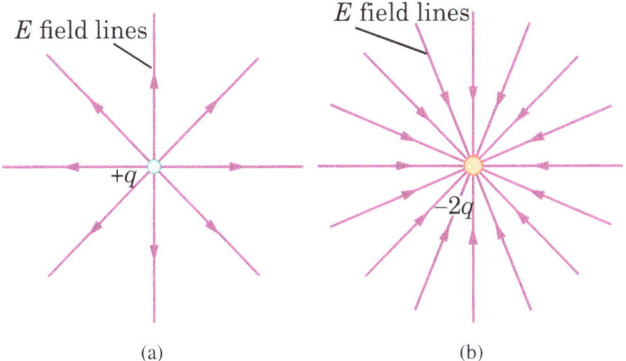

Figure 1.125: Electric field lines for a point charge (a) near a positive charge the field lines point radially away from the charge. The lines start on the positive charge and end at infinity. (b) near a negative charge the field lines point radially inward. They start at infinity and end on a negative charge and are more dense where the field is more intense. notice that the number of lines drawn for part (b) is twice the number drawn for part (a), a reflection of the relative magnitudes of the charges

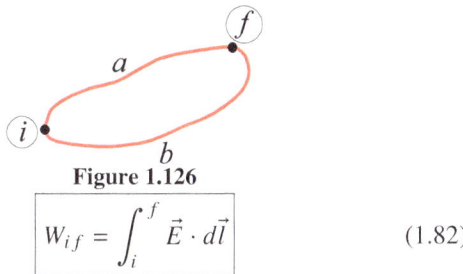

Figure 1.126

$$W_{if} = \int_i^f \vec{E} \cdot d\vec{l} \quad (1.82)$$

This integral is taken along a certain line (path) and is therefore called the line integral.

We shall now show that from the independence of line integral (1.82) of the path between two points it follows that when taken along an arbitrary closed path, this integral is equal to zero. Integral (1.82) along a closed path is called the circulation of vector $\vec{E}$ and is denoted by $\oint$.

Thus, we state that circulation of vector $\vec{E}$ in any electrostatic field is equal to zero, i.e.,

$$\oint \vec{E} \cdot d\vec{l} = 0 \quad (1.83)$$

This statement is called the *theorem on circulation of vector $\vec{E}$*. In order to prove this theorem, we break an arbitrary closed path into two parts iaf and fbi (Fig.1.126). Since line integral (1.82), we denote it by $\left(\int_{12}\right)$, does not depend on the path between points i and f, we have $\int_{if}^{(a)} = \int_{if}^{(b)}$. On the other hand, it is clear that $\int_{if}^{(b)} = -\int_{fi}^{(b)}$, where $\int_{fi}^{(b)}$ is the integral over the same segment b but taken in the opposite direction. Therefore,

$$\oint \vec{E} \cdot d\vec{l} = \int_{if}^{(a)} \vec{E} \cdot d\vec{l} + \int_{fi}^{(b)} \vec{E} \cdot d\vec{l}$$
$$= \int_{if}^{(a)} \vec{E} \cdot d\vec{l} - \int_{if}^{(b)} \vec{E} \cdot d\vec{l} = 0,$$

A field having property (1.83) is called the potential field. Hence, any electrostatic field is a potential field. The term 'potential' will be discussed in next chapter.

☞ In next chapter, we will see that the term potential is defined only for conservative force field, therefore, the field for which Eq.1.83 holds, is also known as a conservative field.

The theorem on circulation of vector $\vec{E}$ makes it possible to draw a number of important conclusions without resorting to calculations. Let us consider two examples.

EXAMPLE 56. The field lines of an electrostatic field $\vec{E}$ cannot be closed. Why?

SOLUTION Suppose, the opposite is true and some lines of field $\vec{E}$ are closed like Fig.1.127. The arrows on the contour indicate the direction of circumvention as well as the direction of electric field.

Taking the circulation of vector $\vec{E}$ along the line shown in this figure, we get-

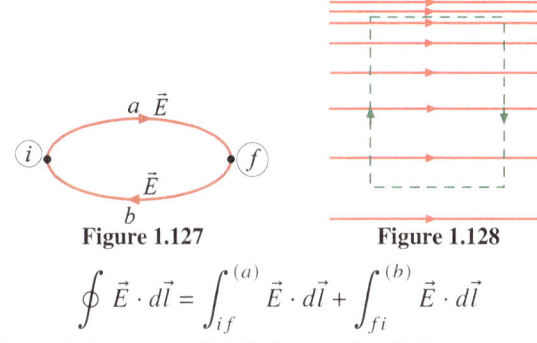

Figure 1.127   Figure 1.128

$$\oint \vec{E} \cdot d\vec{l} = \int_{if}^{(a)} \vec{E} \cdot d\vec{l} + \int_{fi}^{(b)} \vec{E} \cdot d\vec{l}$$

For both the terms of right hand side of above expression, $\vec{E}$ and $d\vec{l}$ are directed in the same direction. Therefore, the scalar product $\vec{E} \cdot d\vec{l} > 0$. So,

$$\oint \vec{E} \cdot d\vec{l} > 0$$

It contradicts the theorem 1.20. This means that electric field lines cannot form any closed curve in an electrostatic field: *the lines emerge from positive charges and terminate on negative ones (or go to infinity).*

EXAMPLE 57. Is the configuration of an electrostatic field shown in Fig.1.128 possible?

SOLUTION No, it is not. It can be easily shown by applying the theorem on circulation of vector $\vec{E}$ to the closed contour shown in the Fig.1.128 by the dashed line. The arrows on the contour indicate the direction of circumvention. With such a special choice of the contour, the contribution to the circulation from its vertical parts is equal to zero, since in this case $\vec{E} \perp d\vec{l}$ and $\vec{E} \cdot d\vec{l} = 0$. It remains for us to consider the two horizontal segments of equal lengths. The Fig.1.128 shows that the contributions to the circulation from these regions are opposite in sign, and unequal in magnitude (the contribution from the upper segment is larger since the field lines are denser, and hence the value of $\vec{E}$ is larger). Therefore, the circulation of $\vec{E}$ differs from zero, which contradicts to (1.83).

1.20.1 Deduction of Pattern of Field Lines

Fig.1.129 shows the sketch of field lines for two point charges $2q$ and $-q$.

The pattern of field lines can be deduced by considering the following points:

1. **Symmetry:** For every point above the line joining the two charges there is an equivalent point below it. Therefore, the pattern must be symmetrical about the line joining the two charges.

2. **Near field:** Very close to a charge, its own field predominates. Therefore, the lines are radial and spherically symmetric. The high density of lines near the charges indicates a region of

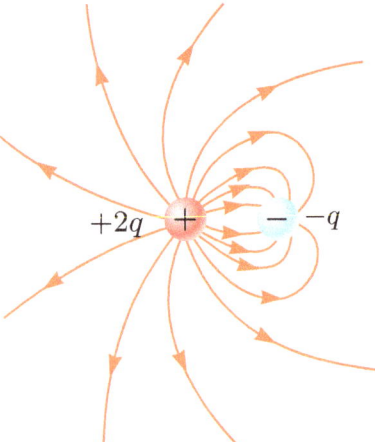

Figure 1.129: Two field lines leave $+2q$ for every one that terminates on $-q$.

strong electric field.
3. **Far field:** Far from the system of charges, the pattern should look like that of a single point charge of value $(2Q - Q) = +Q$, i.e., the lines should be radially outward.
4. **Null point:** The point at which $E = 0$, is called null point. No lines should pass through this point.
5. **Number of lines:** Number of lines is directly proportional to magnitude of charge. So, number of field lines that leave $+2Q$ is twice that of terminating on charge $-Q$.

The field lines around a dipole [Fig.1.130a], show clearly a vivid pictorial description of the mutual attraction between the two charges. The total charge of the dipole is zero, but because the charges are separated, the electric field does not vanish. Instead, the field lines start from positive charge and terminate at negative charge.

The field lines around a system of two positive charges (q, q) [Fig.1.130b] give a vivid pictorial description of their mutual repulsion.

Fig.1.131 shows the electric field lines for a configuration of two positive and two negative charges, all of equal magnitude.

1.20.2 Properties of Electric Field Lines

The field lines follow some important general properties:
1. Field lines start from positive charges and end at negative charges. If there is a single charge, they may start or end at infinity.
2. In a charge-free region, electric field lines can be taken to be continuous curves without any breaks.
3. Tangent to the lines of force at any point gives the direction of the electric field.
4. Two field lines can never cross each other. (If they did, the field at the point of intersection will not have a unique direction, which is absurd.)
5. Electrostatic field lines do not form any closed loops. This follows from the conservative nature of electric field.
6. Lines of force originate or terminate perpendicular to the metal surface. Reason is given in section 1.26
7. The density of the lines at any point (the number of lines per unit area through a surface element normal to the lines) is proportional to the magnitude of the field there.

1.21 Check Point 5

1. •The electric field lines for a system of two charges are shown in Figure 1.132 (a) What is the sign of charge 1? (b) What is the sign of charge 2? (c) Is the magnitude of charge 1 greater than, less than, or equal to the magnitude of charge 2? Explain.

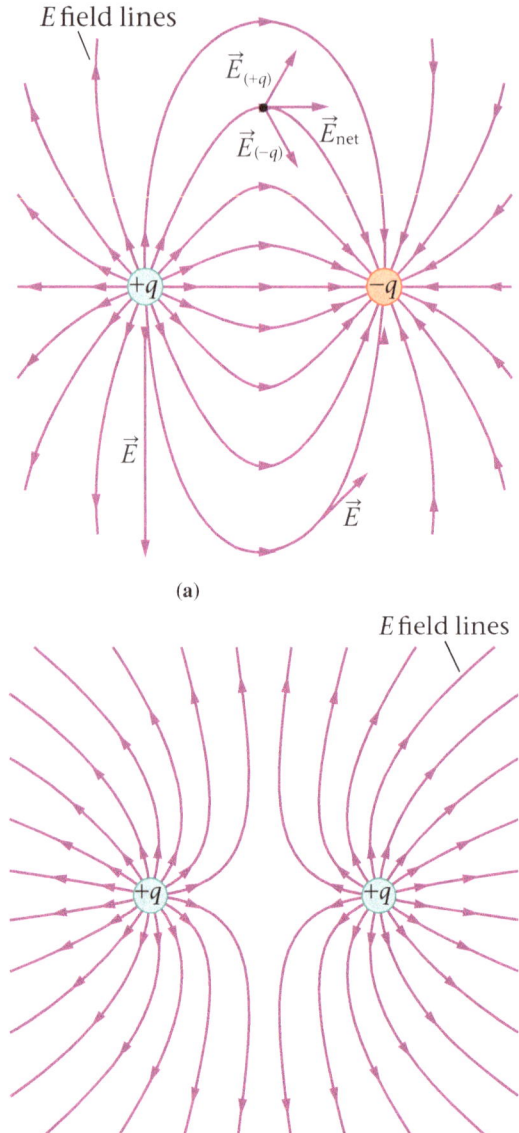

Figure 1.130: Electric field lines for systems of charges (a) The electric field lines for a dipole form closed loops that become more widely spaced with distance from the charges. Note that at each point in space, the electric field vector $\vec{E}$ is tangent to the field lines. (b) All of the field lines in a system with charges of the same sign extend to infinity

2. •Rank the magnitudes of the electric field at points A, B, and C shown in Figure 1.133 (greatest magnitude first).
3. •For the drawing shown in Figures 1.134, write the ratio of charges q_1 and q_2.
4. •The electric field lines surrounding three charges are shown in Figure 1.135 (a) Which of the charges q_A, q_B, and q_C are positively charged, and which are negatively charged? (b) Rank the charges in order of increasing magnitude.
5. •Sketch the electric field lines for the system of charges shown in Fig.1.136.
6. •Sketch the electric field lines for the system of charges described in Figure 1.137.

Multiple Choice Questions

7. •Which lines in Figure 1.138 cannot represent an electric field? Explain.
8. •In the Figure 1.139, rank points 1 – 4 in order of increasing field strength.

1.22. ACTION OF THE ELECTRIC FIELD ON CHARGES

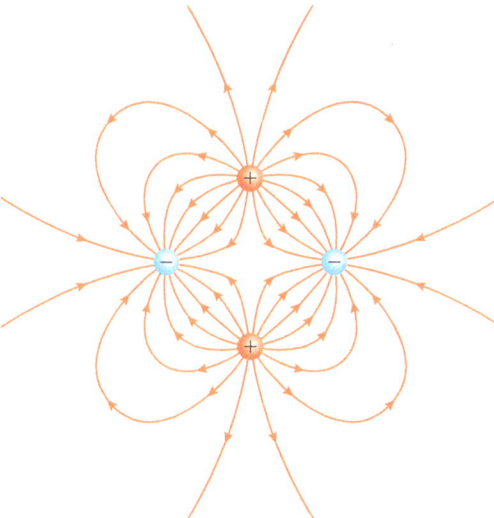

Figure 1.131: Electric field lines for a configuration of two positive and two negative charges, all of equal magnitude. Note that the field lines always start on a positive charge and end on a negative charge when there are equal numbers of both charges.

Figure 1.134

Figure 1.135

Figure 1.136

Figure 1.137

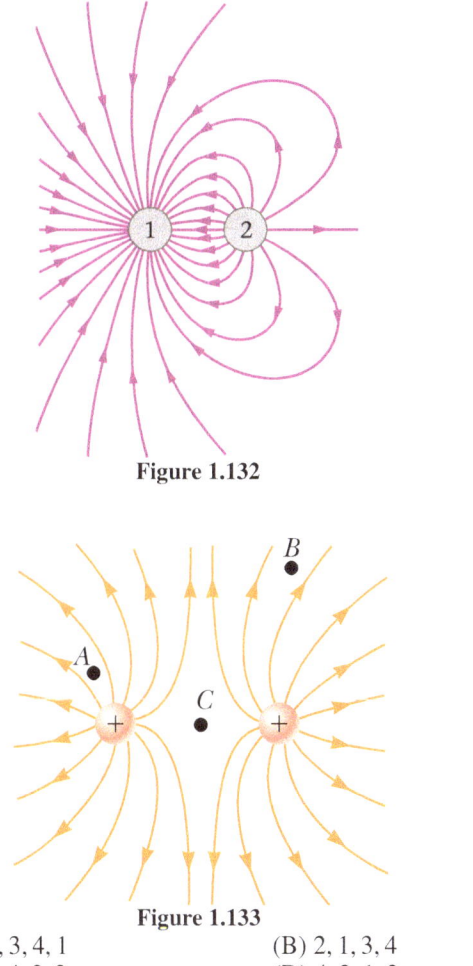

Figure 1.132

Figure 1.133

(A) 2, 3, 4, 1 (B) 2, 1, 3, 4
(C) 1, 4, 3, 2 (D) 4, 3, 1, 2

1.22 Action of the Electric Field on Charges

A uniform electric field can exert a force on a single charged particle and can exert both a torque and a net force on an electric dipole.

1.22.1 Motion of a charged Particle in a Uniform Electric Field

When a particle of charge q and mass m is in an external electric field of strength $\vec{E}$, a force $q\vec{E}$ will be exerted on this particle. If $q\vec{E}$ is the only acting force on the particle, then according to Newton's second law, $\Sigma \vec{F} = m\vec{a}$, the acceleration of the particle will be given by:

$$\vec{a} = q\vec{E}/m \quad (1.84)$$

1.22.1.1 Motion of a Charged Particle Along an Electric Field

Consider a particle of positive charge q and mass m in a uniform horizontal electric field $\vec{E}$ produced by two charged plates that

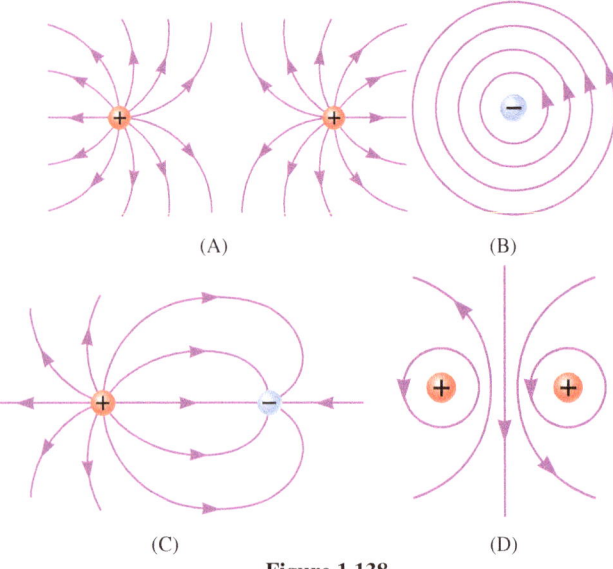

(A) (B)

(C) (D)

Figure 1.138

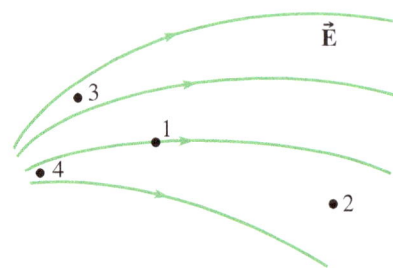

Figure 1.139

are separated by a distance d as shown in Fig.1.140.

If the particle is released from rest at the positive plate and $q\vec{E}$ is

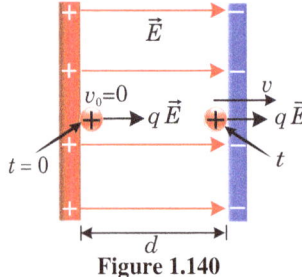

Figure 1.140

the only force that acts on the particle, then the particle will move horizontally along the x - axis with an acceleration $\vec{a} = q\vec{E}/m$. In such a case, we can apply the kinematics equations on the initial and final motion as follows:

- The particle's time of flight t:

$$x = v_0 t + \frac{1}{2} a t^2 \implies d = 0 + \frac{1}{2}\frac{qE}{m}t^2$$

$$\implies t = \sqrt{\frac{2md}{qE}} \qquad (1.85)$$

- The speed of the particle v:

$$v = v_0 + at \implies v = 0 + \frac{qE}{m}\sqrt{\frac{2md}{qE}}$$

$$\implies v = \sqrt{\frac{2qEd}{m}} \qquad (1.86)$$

- The kinetic energy of the particle K:

$$K = \frac{1}{2}mv^2$$

$$\implies K = \frac{1}{2}m\left(\sqrt{\frac{2qEd}{m}}\right)^2 = qEd \qquad (1.87)$$

The last result can also be obtained from the application of the work-energy theorem $W = \Delta K$ because $W = (qE)d$ and $\Delta K = K_f - K_i = K$.

EXAMPLE 58. In Fig.1.140, assume that the charged particle is a proton of charge $q = +e$. The proton is released from rest at the positive plate. In this case, each of the two oppositely charged plates which are $d = 2$ cm apart has a charge per unit area of $\sigma = 5\mu C/m^2$. (a) What is the magnitude of the electric field between the two plates? (b) What is the speed of the proton as it strikes the second plate?

SOLUTION (a) Since, the electric field arises from two infinite thin plates, therefore,

$$E = \frac{\sigma}{2\epsilon_0} + \frac{\sigma}{2\epsilon_0} = \frac{\sigma}{\epsilon_0}$$

$$= \frac{5 \times 10^{-6} C/m^2}{8.85 \times 10^{-12} C/N \cdot m^2} = 5.65 \times 10^5 \text{ N/C}$$

(b) We first find the proton's acceleration from Newton's second law:

$$a = \frac{F}{m} = \frac{eE}{m} = \frac{(1.6 \times 10^{-19} C)(5.65 \times 10^5 \text{ N/C})}{1.67 \times 10^{-27} \text{ kg}}$$

$$= 5.41 \times 10^{13} \text{ m/s}^2$$

Then, using $x = v_0 t + \frac{1}{2}at^2$, we find that $d = \frac{1}{2}at^2$. Thus:

$$t = \sqrt{\frac{2d}{a}} = \sqrt{\frac{2(0.02 \text{ m})}{5.41 \times 10^{13} \text{ m/s}^2}} = 2.72 \times 10^{-8} \text{ s}$$

Finally, we use $v = v_0 + at$ to find the speed of the proton as follows:

$$v = at = \left(5.41 \times 10^{13} \text{ m/s}^2\right)\left(2.72 \times 10^{-8} \text{ s}\right)$$

$$= 1.47 \times 10^6 m/s$$

1.22.1.2 Motion of a Charged Particle Perpendicular to an Electric Field

Let a uniform electric field $\vec{E}$ is created between two parallel, charged plates as shown in Fig.1.141. A charged particle of charge q enters at O (0, 0) in the field symmetrically between the plates with an initial speed v_0. The length of each plate is l.

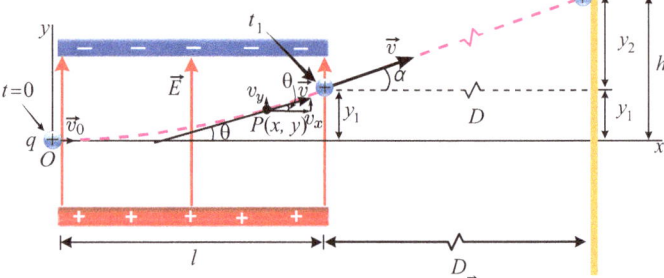

Figure 1.141: The effect of an upward force $q\vec{E}$ exerted on charged particle projected horizontally with speed v_0 into an upward uniform electric field $\vec{E}$

The acceleration of the charged particle is $a = \frac{qE}{m}$ in upward direction. The horizontal velocity remains v_0 as there is no acceleration in horizontal direction. Thus, the time taken by charged particle in reaching to a general point $P(x, y)$ in electric field $\vec{E}$ is given by-

$$t = \frac{x}{v_0} \qquad (1.88)$$

The upward component of the velocity of the charged particle at point $P(x, y)$ in the field region is

$$v_y = v_{0y} + a_y t = 0 + \frac{qEx}{mv_0}$$

or $\quad v_y = \frac{qEx}{mv_0} \quad (1.89)$

At point P, the horizontal component of velocity remains $v_x = v_0$. The angle θ made by the resultant velocity with the original direction is given by

$$\tan\theta = \frac{v_y}{v_x} = \frac{qEx}{mv_0^2}$$

Thus, the charged particle deviates by an angle,

$$\theta = \tan^{-1}\frac{qEx}{mv_0^2} \quad (1.90)$$

Vertical displacement y can be given by-

$$y = v_{0y}t + \frac{1}{2}a_y t^2 = 0 + \frac{1}{2}\frac{qEx^2}{mv_0^2}$$

$$\Rightarrow \boxed{y = \frac{1}{2}\frac{qEx^2}{mv_0^2}} \quad (1.91)$$

This is an expression for path of a charged particle in a uniform electric field. Since Eq.(1.91) is of the form of a parabola, therefore the path of charged particle will be parabolic in uniform electric field.

When, charged particle emerges from the field, $x = l$, from Eq.1.91, the vertical displacement can be written as-

$$y_1 = \frac{1}{2}\frac{qEl^2}{mv_0^2} \quad (1.92)$$

The deviation of charged particle at $x = l$, is given by-

$$\alpha = \tan^{-1}\frac{qEl}{mv_0^2} \quad (1.93)$$

The extra vertical distance y_2 that the charged particle will move before hitting the screen, which is located at a horizontal distance D from the plates, is given by:

$$y_2 = D\tan\alpha = D\frac{qEl}{mv_0^2}$$

Finally, the total vertical distance h that the charged particle will move is:

$$h = y_1 + y_2 = \frac{qEl}{mv_0^2}\left(\frac{l}{2} + D\right)$$

EXAMPLE 59. An electron enters the region of a uniform electric field as shown in Fig.1.141, with $v_0 = 3.00 \times 10^6$ m/s and $E = 200$ N/C. The horizontal length of the plates is $l = 0.100$ m.
(A) Find the acceleration of the electron while it is in the electric field.
(B) If the electron enters the field at time $t = 0$, find the time at which it leaves the field.
(C) If the vertical position of the electron as it enters the field is $y_0 = 0$, what is its vertical position when it leaves the field?

SOLUTION (A) The charge on the electron has an absolute value of $q = 1.60 \times 10^{-19}$ C, and $m_e = 9.11 \times 10^{-31}$ kg. Therefore, acceleration of electron-

$$\vec{a} = -\frac{eE}{m_e}\hat{j} = -\frac{(1.60 \times 10^{-19}\text{C})(200 \text{ N/C})}{9.11 \times 10^{-31} \text{ kg}}\hat{j}$$

$$= -3.51 \times 10^{13}\hat{j}\text{ m/s}^2$$

(B) The horizontal distance across the field is $l = 0.100$ m. Using Equation (1.88) with $x = l$, we find that the time at which the electron exits the electric field is

$$t = \frac{l}{v_0} = \frac{0.100 \text{ m}}{3.00 \times 10^6 \text{ m/s}} = 3.33 \times 10^{-8} \text{ s}$$

(C) Vertical displacement of electron is given by-

$$y = 0 + \frac{1}{2}at^2 = -\frac{1}{2}\left(3.51 \times 10^{13} \text{ m/s}^2\right)\left(3.33 \times 10^{-8} \text{ s}\right)^2$$

$$= -0.0195 \text{ m} = -1.95 \text{ cm}$$

If the electron enters just below the negative plate in Figure 1.141 and the separation between the plates is less than the value we have just calculated, the electron will strike the positive plate.

EXAMPLE 60. A block of mass m having a charge q is placed on a smooth horizontal table and is connected to a wall through an unstressed spring of spring constant k as shown in Fig.1.142. A horizontal electric field E parallel to the spring is switched on. Find the extension in the spring. Now, if we turn off the field, find the amplitude of the resulting motion of the block.

Figure 1.142

SOLUTION Suppose, the initial position of the block is at $x = 0$. At this instant, the spring has it's natural length. Now, if we turn on the horizontal electric field $\vec{E}$ as shown in Fig.1.143, it applies an electric force $F_e = qE$ on the block in the direction of electric field that causes the spring to stretch to an equilibrium length $x = x_0$. In equilibrium, the forces acting on the block are-
1. Gravitational force $m\vec{g}$: acting downwards.

Figure 1.143

2. Normal reaction $\vec{N}$: acting upwards.
3. Electric force $\vec{F}_e$ ($F_e = qE$): acting towards right
4. Spring force $\vec{F}_s$ ($F_s = kx$): acting towards left.
For vertical equilibrium of the block, $N = mg$.
For horizontal equilibrium, we have-

$$F_s = F_e \implies kx_0 = qE \implies x_0 = \frac{qE}{k}$$

Now, if we turn-off the electric field, the block begins to oscillate about mean position $x = 0$ with amplitude of $x_0 = qE/k$.

1.23 Electric Dipole in an External Electric Field

1.23.1 Electric force and torque on electric dipole in a uniform external electric field

Fig.1.144 shows an electric dipole AB of length $2l$ in a uniform external electric field $\vec{E}$. The charges at A and B are $-q$ and $+q$ respectively. The direction of dipole moment vector $\vec{p}$, is from $-q$ to $+q$ i.e., from point A to B. The angle between dipole moment vector $\vec{p}$ (which is along $\overrightarrow{AB}$) and electric field vector $\vec{E}$ is θ. The electric forces on $-q$ and $+q$ have same magnitude ($= qE$) but their directions are opposite to each other (Fig.1.144a). The electric force on $-q$ is opposite to $\vec{E}$ and on $+q$, it is parallel to $\vec{E}$.

The net force on the dipole is $F_{net} = -qE + qE = 0$ Thus, net force on an electric dipole in a uniform external electric field is zero, so the dipole will have no translational acceleration.

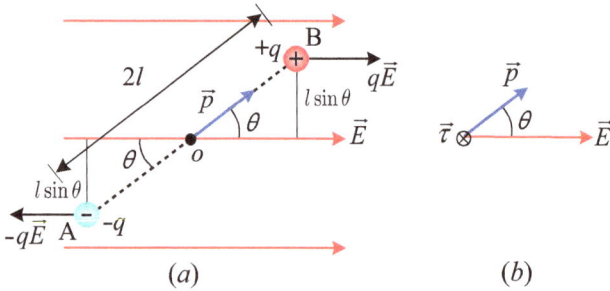

(a)   (b)

Figure 1.144: (a) An electric dipole has an electric dipole moment $\vec{p}$ in an external uniform electric field $\vec{E}$. The angle between $\vec{p}$ and $\vec{E}$ is θ. The line connecting the two charges represents their rigid connection and their centre of mass is assumed to be midway between them. (b) Representing the electric dipole by a vector $\vec{p}$ in the external electric field $\vec{E}$ and showing the direction of the torque $\vec{\tau}$ into the page by the symbol $\otimes$

Since, the lines of action of these forces, are not same, so they produce torque about the centre of mass of the dipole. This torque tends to rotate the dipole to bring $\vec{p}$ into alignment with $\vec{E}$.
If $\vec{\tau}_1$ and $\vec{\tau}_2$ are the respective torques produced by forces on $+q$ and $-q$ about the centre of mass of the dipole, then from Fig.1.144, the magnitude of net torque on the dipole-
$$\tau = \tau_1 + \tau_2 = qE(l\sin\theta) + qE(l\sin\theta)$$
$$= 2qlE\sin\theta = pE\sin\theta$$
$$\Rightarrow \quad \tau = pE\sin\theta \quad (1.94)$$
In vector form, torque on dipole due to external uniform electric field $\vec{E}$ can be written as-
$$\vec{\tau} = \vec{p} \times \vec{E} \quad (1.95)$$
which is consistent with the directional relationships for the cross product, as shown by the three vectors in 1.144b
Unit: Unit of torque is "newton-meter".
Maximum and Minimum Values of Torque:
(i) Maximum torque applied by electric field on the dipole is given by-
$$\tau_{max} = pE(\sin\theta)_{max}$$
Since, $(\sin\theta)_{max} = 1$ and corresponding to it $\theta = \pi/2$, therefore-
$$\tau_{max} = pE$$
(ii) Minimum torque applied by electric field on the dipole is given by-
$$\tau_{max} = pE\,|(\sin\theta)_{min}|$$
Since, $|(\sin\theta)_{min}| = 0$ and corresponding to it $\theta = 0$, therefore-
$$\tau_{min} = 0$$
The torque is greatest when $\vec{p}$ is perpendicular to $\vec{E}$, zero when $\vec{p}$ is aligned with or opposite to $\vec{E}$.

EXAMPLE 61. Two charges each of magnitude 0.01 C and separated by a distance of 0.4 mm constitute an electric dipole. If the dipole is placed in a uniform electric field $|\vec{E}|$ of 10 dyne/C making $30°$ angle with $\vec{E}$, find the magnitude of torque acting on it.
APPROACH To determine the magnitude of torque of electric field on the dipole, apply Eq.1.94:
$$\tau = pE\sin\theta \quad (1)$$
here, $p = q(2l)$ (see Eq.1.38) is the dipole moment.
SOLUTION Given: $q = 0.01$ C, $2l = 0.4$ mm $= 4.0 \times 10^{-4}$ m and $E = 10$ dyne/C $= 10^{-4}$ N/C
Dipole moment of dipole,
$$p = q(2l) = 0.01 \times 4.0 \times 10^{-4} = 4.0 \times 10^{-6} \text{ Cm}$$
Substituting given values and the value of dipole moment in Eq.(1), we get
$$\tau = pE\,\sin\theta = 4 \times 10^{-6} \times 10 \times 10^{-5} \times \sin 30°$$

$$= 4 \times 10^{-6-5+1} \times \frac{1}{2} = 2.0 \times 10^{-10} \text{ Nm}$$

1.23.2 The Angular Acceleration and Time period of a Dipole in an External Uniform Electric Field

Since, the torque applied by electric field on the dipole is restoring, therefore we can write equation (1.94) as follows-
$$\tau = -pE\sin\theta$$
If I is the moment of inertia of dipole about an axis passing through its centre and perpendicular to the plane of the Fig.1.144a, then torque can also be written as $\tau = I\frac{d^2\theta}{dt^2}$, here $d^2\theta/dt^2$ is the angular acceleration of the dipole. For small angular displacement, $\sin\theta \approx \theta$, therefore, we can write above equation as-
$$I\frac{d^2\theta}{dt^2} = -pE\theta$$
$$\Rightarrow \quad \frac{d^2\theta}{dt^2} = -\frac{pE}{I}\theta \quad (1.96)$$
which is of the form of angular SHM: $\frac{d^2\theta}{dt^2} = -\omega^2\theta$, with $\omega = \sqrt{pE/I}$
Thus, if in uniform electric field the dipole is rotated by a small angular displacement θ and released, the electric dipole performs angular SHM with time period-
$$\boxed{T = \frac{2\pi}{\omega} = 2\pi\sqrt{\frac{I}{pE}}} \quad (1.97)$$

1.23.3 A Dipole in a Non-uniform Field

Suppose that a dipole is placed in a non-uniform electric field in which the electric field is changing along x-axis. Let E and $E + dE$ are magnitudes of electric fields at positions A and B respectively. The first response of the dipole is to rotate until it is aligned with the field, with the dipole's positive end pointing in the same direction as the field. Now, however, there is a slight difference between the forces acting on the two ends of the dipole. This difference occurs because the electric field is not same at positions A, and B. The net force exerted by external non-uniform electric field on the dipole is given by-
$$F = \underbrace{q\,(E+dE)}_{\text{Electric force on charge }+q\text{ at B}} - \underbrace{qE}_{\text{Electric force on charge }-q\text{ at A}}$$

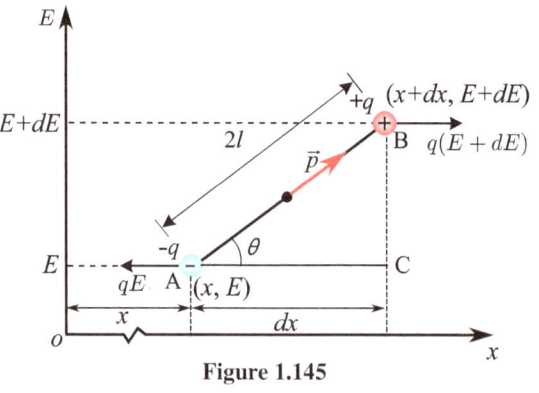

Figure 1.145

$$\Rightarrow F = q\frac{dE}{dx}dx = q\,dx\frac{dE}{dx} \quad (1.98)$$

1.23. ELECTRIC DIPOLE IN AN EXTERNAL ELECTRIC FIELD

From Fig.1.145, $dx = AC = 2l\cos\theta$, therefore, on substituting this value of dx in above equation, we get-

$$F = q(2l\cos\theta)\frac{dE}{dx} = p\cos\theta\frac{dE}{dx} \quad (1.99)$$

Eq.(1.99) gives the electric force on a dipole inclined at an angle θ with the direction of non uniform electric field $\vec{E}$.

When the dipole is aligned in the direction of field, $\theta = 0$, therefore, $\cos\theta = 1$. In this case, Eq.1.98 becomes-

$$F = (2ql)\frac{dE}{dx} = p\frac{dE}{dx} \quad (1.100)$$

So, once the dipole is aligned, the rightward force on its positive end is slightly stronger than the leftward force on its positive end. This causes a net force towards the stronger region of the field. For any non-uniform electric field (except at maximum or minimum field points*), the net force on a dipole is toward the direction of the strongest field. Because any finite-size charged object, such as a charged rod or a charged disc, has a field strength that increases as you get closer to the object, we can conclude that a dipole will experience a net force toward any charged object.

EXAMPLE 62. A dipole consists of charges $+2\,\mu C$ and $-2\,\mu C$ separated by a distance of $4\,cm$. It is placed in a non-uniform electric field given by

$$E(x) = 200\,x^2\ \text{N/C},$$

where x is in metres. The axis of the dipole makes an angle of $60°$ with the x-axis.

Calculate the magnitude of the net force acting on the dipole when its centre is at $x = 0.50\,m$.

SOLUTION Given: $q = 2\times 10^{-6}$ C, $2l = 0.04$ m, $\theta = 60°$

The electric field is

$$E(x) = 200\,x^2 \quad \Rightarrow \quad \frac{dE}{dx} = 400x.$$

At $x = 0.50$ m:

$$\frac{dE}{dx} = 400\times 0.50 = 200\ \text{N/C·m}.$$

The force on the dipole is

$$F = q(2l\cos\theta)\frac{dE}{dx}.$$

Substituting the value in above expression, we get

$$F = (2\times 10^{-6})(0.04\times\cos 60°)(200)$$
$$= (2\times 10^{-6})(0.04\times 0.5)(200) = 8.0\times 10^{-6}\ \text{N}$$
$$\Rightarrow F = 8.0\times 10^{-6}\ \text{N} = 8.0\ \mu\text{N}$$

EXAMPLE 63. A wheel having mass m, charge $+q$ and $-q$ on diametrically opposite points, is in equilibrium on a rough inclined plane in the presence of uniform vertical electric field $\vec{E}$ as shown in Fig.1.146. Find the magnitude of $\vec{E}$.

APPROACH A wheel can be considered as a rigid body. The system containing wheel, attached with charges $+q$ and $-q$ at it's diametrically opposite ends, is shown in Fig.1.146.

The forces acting on the wheel are-
1. Gravitational force $m\vec{g}$: acting in downward direction
2. Normal reaction $\vec{N}$: acting perpendicular to the inclined plane
3. Force of static friction $\vec{f}_s$: acting along the plane in upward sense.
4. Electrostatic force of upward electric field $\vec{E}$: On charge $-q$,

---
*For maximum or minimum value of electric field, $\frac{dE}{dx} = 0$, therefore from Eq.1.100, $F = 0$.
So, if the dipole is kept at the point where the electric field is at its maximum or minimum, then the force on it will be zero.

Figure 1.146

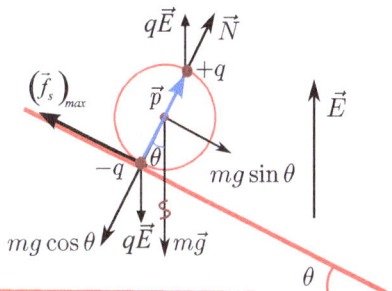

Figure 1.147

it is $-q\vec{E}$ and directed opposite to $\vec{E}$ i.e., in downward direction and on charge $+q$, it is $+q\vec{E}$ in the direction of $\vec{E}$, i.e., in upward direction (Fig.1.147). These two forces are equal in magnitude (qE) but opposite in direction. So, the net electrostatic force on the wheel due to dipole attached to it, is zero. However, the lines of action of these electrostatic forces are not passing through the centre of mass (CM) of the wheel, so it will definitely provide a mechanical torque about CM.

Conditions for Static Equilibrium for the Wheel-Dipole System

Let positive direction of x axis is along the inclined plane in downward direction and positive direction of y axis perpendicular to it in upward sense. In static equilibrium:

1. Conditions for Translational Equilibrium: For translational equilibrium along x and y direction-
$$\Sigma F_x = 0, \Sigma F_y = 0$$

2. Condition for Rotational Equilibrium: In rotational equilibrium, the net torque on the dipole - wheel system about centre of mass (CM),
$$\Sigma \tau_{CM} = 0$$

Now, apply these conditions and simplify for minimum required value of electric field.

SOLUTION If μ is the coefficient of static friction between the wheel and the inclined plane, then for translational equilibrium of the wheel, we have-

1. $\Sigma F_y = 0 \Rightarrow \quad N - mg\cos\theta = 0$
$$\Rightarrow N = mg\cos\theta \quad (i)$$

2. $\Sigma F_x = 0 \Rightarrow \quad (f_s)_{max} - mg\sin\theta = 0$
$$\Rightarrow (f_s)_{max} = mg\sin\theta \quad (ii)$$

In limiting translational equilibrium, we have-
$$(f_s)_{max} = \mu_s N$$

Substituting this value of $(f_s)_{max}$ in Eq.(ii), we get-
$$\mu_s N = mg\sin\theta \quad (iii)$$

From Eq.(i) and (iii), we get-
$$\mu_s = \tan\theta \quad (iv)$$

And for rotational equilibrium, the net torque on the wheel about it's centre of mass should be zero, i.e.,

$$\Sigma\tau = \tau_{\text{electric field}} + \tau_{mg} + \tau_{\text{friction}} = 0 \quad (v)$$

Considering, perpendicularly outward to the plane of figure as a positive direction of torque, the torque of electric field about the centre of mass of the wheel

$$\tau_{\text{electric field}} = pE \sin\theta = +2qrE \sin\theta \qquad \text{(vi)}$$

The direction of this torque is normally outward to the plane of the figure.

Torque of weight mg about the centre of mass of the wheel is zero, because it passes through the centre of mass of the wheel.

$$\tau_{mg} = 0 \qquad \text{(vii)}$$

Torque of force of friction about the centre of mass of the wheel-

$$\tau_{\text{friction}} = -r \times \mu_s N \qquad \text{(viii)}$$

The direction of this torque is perpendicularly inward to the plane of the Fig.1.147

Substituting the value of $\mu_s N$, from Eq. (iii) in Eq.(viii), we get-

$$\tau_{\text{friction}} = -rmg \sin\theta \qquad \text{(ix)}$$

Using, Eq.(vi), (vii) and (ix) in Eq. (v), we get -

$$2qrE \sin\theta - rmg \sin\theta = 0$$

$$\implies E = \frac{mg}{2q} \qquad \text{(x)}$$

This is the required electric field needed for equilibrium of the wheel.

EXAMPLE 64. A system consists of a thin uniformly charged wire ring of radius R and a very long uniformly charged thin rod oriented along the axis of the ring, with one of its ends coinciding with the centre of the ring. The total charge of the ring is equal to Q. The linear charge density of the rod is λ. Find the interaction force between the ring and the rod.

Note: To determine the total electric force between the rod and ring, we can use following two approaches-

APPROACH 1. Fig.1.148 shows a thin charged wire ring of radius R carries a total charge Q uniformly distributed around it. A very long uniformly charged wire having uniform charge density λ is placed along it's axis in such a way that it's one end coincides with the centre of the ring.

Now, consider an elementary length dx of the rod at distance

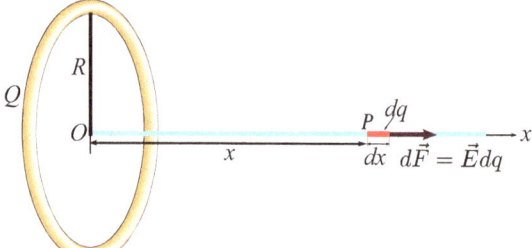

Figure 1.148

x from the centre of the ring and find the electric field E of the ring at the position of the elementary charge of rod. If the elementary charge of the rod is $dq (= \lambda dx)$ then electric force on this elementary charge due to electric field of ring is-

$$dF = E dq \implies dF = E\lambda dx$$

The integration of this expression from $x = 0$ to $x = \infty$, will give the net force F on the rod exerted by charged ring, i.e.,

$$F = \int_0^\infty E\lambda dx \qquad (1.101)$$

SOLUTION The electric field at an axial position of charged ring is given by-

$$E = \frac{kQx}{(R^2 + x^2)^{3/2}} \qquad (1.102)$$

Substituting this value in Eq.(1.101), we get-

$$F = \int_0^\infty \left(\frac{kQ\lambda x}{(R^2 + x^2)^{3/2}} \right) \lambda dx$$

or

$$F = kQ\lambda \int_0^\infty \left(\frac{x}{(R^2 + x^2)^{3/2}} \right) \lambda dx \qquad (1.103)$$

To solve right hand integration, we let-

$$R^2 + x^2 = t \implies x dx = \frac{1}{2} dt$$

when $x \to 0$, then $t \to R^2$
and when $x \to \infty$, then $t \to \infty$

Now, substituting these values in Eq.(1.103), we get

$$F = \frac{kQ\lambda}{2} \int_{R^2}^\infty \frac{dt}{t^{3/2}}$$

$$= \frac{kQ\lambda}{2} \int_{R^2}^\infty t^{-3/2} dt = kQ\lambda \left[-\frac{1}{\sqrt{t}} \right]_{R^2}^\infty = \frac{kQ\lambda}{R}$$

So, the force on the rod is-

$$\boxed{F = \frac{kQ\lambda}{R}} \qquad (1.104)$$

By, Newton's third law, a force of equal in magnitude but opposite in direction, will act on the ring.

APPROACH 2. In second approach, we first calculate the electric field of the rod at any point on the circumference of ring. Since the ring is symmetrical about the rod, therefore, the magnitude of electric field of the rod will be equal for each position on the circumference of the ring.

Now, calculate the force on an elementary segment of ring due to this field. Break it in component form and use symmetrically opposite segment to know which component will get cancelled and which will sum to provide the net force on the ring. Now integration of this component for the total circumference gives us the net force on ring due to charged rod.

SOLUTION Electric field of the charged thin rod, at any point

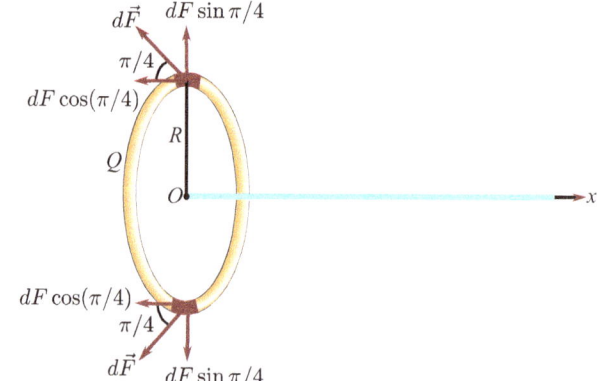

Figure 1.149

on the circumference of the ring is given by -

$$E_{\text{rod}} = \frac{k\lambda\sqrt{2}}{R}$$

This field is directed at angle $\pi/4$ from the negative direction of x-axis (Fig.1.149)

Now consider an elementary part of ring having charge dQ. The electric force on this part of the ring is given by-

$$dF = E_{\text{rod}} dQ = \frac{k\lambda\sqrt{2}}{R} dQ$$

This force has two components-

1. $dF_x = dF \cos(\pi/4)$: acting along the negative direction of the axis of the rod.
2. $dF_y = dF \sin(\pi/4)$: acting perpendicular to the axis of the

ring, i.e., acting along the outward radius vector of the ring. Now, Consider two segments at the top and bottom of the ring: The perpendicular force component on the top segment $dF \sin(\pi/4)$ get cancelled by equal and opposite perpendicular force component of bottom segment. Hence the total perpendicular component of force due to this pair of segments is zero. When we add up the components of all such pairs of segments, the total force $\vec{F}$ will have only a component along the negative direction of ring's symmetry axis (the $-x$ -axis), with no component perpendicular to that axis (that is, no component in yz-plane). So, net force on the ring is given by-

$$F = \int dF_x = \int dF \sin\frac{\pi}{4} = \int_0^Q \frac{k\lambda\sqrt{2}}{R} dQ \sin\frac{\pi}{4} = \int_0^Q \frac{k\lambda}{R} dQ$$

or $\boxed{F = \frac{kQ\lambda}{R}}$ (1.105)

As expected, Eq.1.104 and 1.105 are same. If the charge density on the ring is also same as that of rod, then-
$$Q = \lambda(2\pi R)$$
Substituting this value and $k = 1/4\pi\epsilon_0$ in Eq.1.104 or 1.105 gives us the result-

$$\boxed{F = \frac{\lambda^2}{2\epsilon_0}} \quad (1.106)$$

1.24 Earnshaw's Theorem

According to this theorem, *A charged particle cannot be held in a stable equilibrium by electrostatic forces alone.* As an example, imagine three negative charges at the corners of an equilateral triangle in a horizontal plane. Now, put a positive charge at the centre of the triangle and ignore gravity for the moment (although including it would not change the results). The net electrostatic force on the positive charge is zero at the center. So the charged particle is in equilibrium. Now, suppose we displace it from its equilibrium position and observe the direction of net force at new position, we find that it is not restoring, i.e., not towards the centre of the triangle, so the equilibrium will be unstable.

This theorem also holds for a complicated arrangement of charges held together in fixed relative positions—with rods. For example, consider two equal charges fixed on a rod. This combination too cannot be in a stable equilibrium in some electrostatic field, because, the total force on the rod cannot be restoring for displacements in every direction.

The extension of the argument shows that no rigid combination of any number of charges can have a position of stable equilibrium in an electrostatic field in free space.

If want to make equilibrium stable, we have to use pivots or other mechanical constraints. For example, consider a hollow tube in which a charge can move back and forth freely, but not sideways. Now, produce inward electric fields at both the ends by fixing two identical positive charges at the ends of tube. Near the centre of the tube, electric fields also directed towards the sidewall of the tube[Fig.1.150]. This time, the equilibrium is a stable one. The sidewall restricts its lateral motion by applying non-electrical normal force to it whereas inward electric fields provide restoring force to it. A theoretical proof of the theorem, by applying Gauss's law, is given in 'EXAMPLE 82'

EXAMPLE 65. Earnshaw's theorem says that no particle can be in stable equilibrium under the action of electrostatic forces alone. Consider, however, point P at the centre of a square of four equal positive charges, as in Figure 1.151. If you put a positive test charge there it might seem to be in stable equilibrium. Every one of the four external charges pushes it toward P. Yet Earnshaw's theorem holds. Can you explain how?

Figure 1.150: A charge can be in equilibrium if there are mechanical constraints.

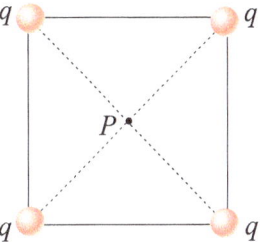

Figure 1.151

SOLUTION The equilibrium of positive test charge at the point P is unstable. Because, on shifting it from mean position (including along diagonal) there is no net restoring force on it towards the position P. So, the equilibrium is unstable. For a stable equilibrium, the total force on the test charge should always be restoring for displacements in every direction.

EXAMPLE 66. Earnshaw's theorem states that a point charge cannot be in stable equilibrium while purely electrostatic forces act on the point charge.

Consider a ring that is uniformly positively charged, with a positive charge at the centre. It appears that the centre charge suffers an identical repulsive force from every direction. How can the theorem be true?

SOLUTION If we shift the positive charge along the symmetric axis of the ring, there will be a repulsion of ring charge on it. As a result of it, the charge moves away from the center. So, the equilibrium at the centre of the ring is unstable. For a stable equilibrium, there must be a restoring force on the charge.

EXAMPLE 67. If a dielectric in the form of a sphere is introduced into a homogeneous electric field. A, B and C are three points as shown in Fig.1.152, then find the points where the intensity of electric field increases and where it decreases.

ANSWER The intensity of electric field at points A, C will in-

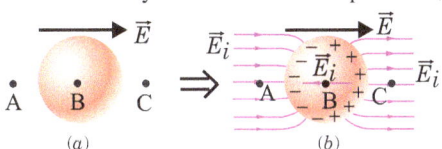

Figure 1.152

crease and at B it will decrease.

Reason: The dielectric polarises when placed in an external electric field. $-ve$ charge produces towards A and an equal positive charge towards B (Fig.1.152). These induced charges produce their own electric fields. In Fig.??b induced fields are represented by $\vec{E}_i$. Clearly at points A the induced field $\vec{E}_i$ due to negative charge is towards the sphere which enhances $\vec{E}$. Similarly, the induced field $\vec{E}_i$ at point C due to positive induced charge is away from the sphere which also enhances external field $\vec{E}$. So, at A and C, the net electric field will be greater than E. At point B, the induced electric field is from right to left which opposes the external electric field, so net electric field at B will be less than E.

EXAMPLE 68. Show that the charges of each sign induced on a conductor B, when it is approached by a point charge charge A of magnitude $+q$ (Fig.1.153), is always less than q.

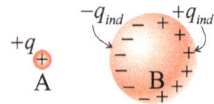

Figure 1.153

SOLUTION Since the sum of the charges in the system as a whole is not zero, and the conductor B does not enclose the charges $+q$, part of the lines of forces from $+q$ must go to infinity (or terminate on other conductors), and only a fraction of the lines of force from $+q$ terminate in the induced charge of B. The induced charge (q_{ind}) is thus less than q.

EXAMPLE 69. If a positive point charge is brought in an electric field, how the field near the point charge get effected.

Answer Suppose the electric field's direction is to the right at a given observation point, and a positive point charge is placed to the left of this point. The additional electric field produced by the given point charge at the observation point is directed away from the positive point charge, that is, towards the right. In this case, both electric fields are directed in the same direction at the observation point, and hence the net electric field increases.

Now, let's consider a scenario where the positive point charge is placed to the right of the observation point. In this case, at the observation point, the electric field produced by the positive point charge will be directed to the left, i.e., opposite to the given electric field. As a result, the field at the observation point decreases.

1.25 Check Point 6

Multiple Choice Questions

1. •A charged drop of mass m floats in air in electric field E. The magnitude of charge on the drop is-
   (A) mg/E (B) E/mg
   (C) m/Eg (D) none of these

2. ••A point charge q moves from point P to S along the path PQRS in a uniform electric field $\vec{E}$ pointing parallel to the positive direction of the x-axis. The coordinates of the point P, Q, R and S are $(a, b, 0)$, $(2a, 0, 0)$, $(a, -b, 0)$ and $(0, 0, 0)$ respectively. The work done by the field in the above process is given by the expression-
   (A) qaE (B) $-qaE$
   (C) $q\sqrt{(a^2+b^2)}E$ (D) $3qE\sqrt{(a^2+b^2)}$

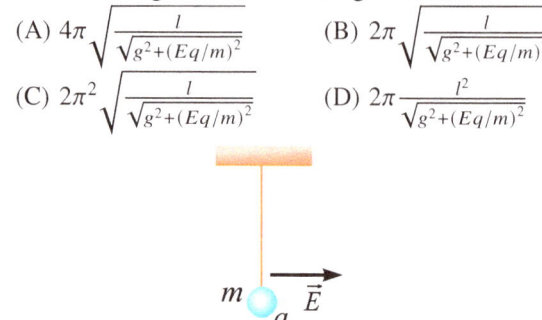

Figure 1.154

3. ••A simple pendulum consists of a small sphere of mass m suspended by a thread of length l. The sphere carries a positive charge q. The pendulum is placed in a uniform electric field of strength E directed vertically upwards. With what period will the pendulum oscillate if the electrostatic force acting on the sphere is less that the gravitational force? Assume the oscillations to be small

   (A) $T = 2\pi\left(\frac{l}{g}\right)^{1/2}$ (B) $T = 2\pi\left(\frac{ml}{qE}\right)^{1/2}$
   (C) $T = 2\pi\left[\frac{l}{\left(g-\frac{qE}{m}\right)}\right]^{1/2}$ (D) $T = 2\pi\left[\frac{l}{\left(g+\frac{qE}{m}\right)}\right]^{1/2}$

4. •••A simple pendulum of mass m and length l carries a charge q. It's time period when it is suspended in a uniform electric field region as shown in Fig.1.155, is-
   (A) $4\pi\sqrt{\frac{l}{\sqrt{g^2+(Eq/m)^2}}}$ (B) $2\pi\sqrt{\frac{l}{\sqrt{g^2+(Eq/m)^2}}}$
   (C) $2\pi^2\sqrt{\frac{l}{\sqrt{g^2+(Eq/m)^2}}}$ (D) $2\pi\frac{l^2}{\sqrt{g^2+(Eq/m)^2}}$

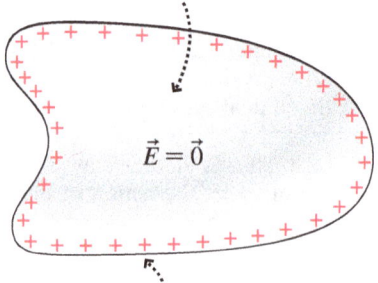

Figure 1.155

5. ••An electric dipole, made up of a positive and a negative charge, each of magnitude 1 μC and placed at a distance 2 cm apart, is placed in an electric field 10^5 N/C. Compute the maximum torque which the field can exert on the dipole.
   (A) 6×10^{-3} N.m   (B) 3×10^{-3} N.m
   (C) 4×10^{-3} N.m   (D) 2×10^{-3} N.m

6. ••An electric dipole of dipole moment p is placed in a uniform electric field E in stable equilibrium position. Its moment of inertia about the centroidal axis is I. If it is rotated slightly from its mean position find the period of small oscillation.
   (A) $T = 2\pi\sqrt{\frac{PE}{I}}$ (B) $T = 2\pi\sqrt{\frac{I}{PE}}$
   (C) $T = 4\pi\sqrt{\frac{I}{PE}}$ (D) $T = 4\pi^2\sqrt{\frac{I}{PE}}$

1.26 Conductors in Electrostatic Equilibrium

A good electric conductor like copper, although electrically neutral, contains charges (electrons) that aren't bound to any atom and are free to move about within the material. When no net motion of charge occurs within a conductor, the conductor is said to be in **electrostatic equilibrium**. An isolated conductor (one that is insulated from ground) has the following properties:

1. *In electrostatic equilibrium, the electric field is zero everywhere inside the conducting material. In terms of field lines, we can say that, there are no field lines within the conducting material*[Fig.1.156].

   Reason: If the electric field within a conducting material is The electric field inside the conductor is zero.

All excess charge is on the surface.
Figure 1.156: The electric field inside e a charged conductor.

nonzero, it exerts a force on each of the mobile charges (usually electrons) and makes them move preferentially in a certain direction. With mobile charge in motion, the conductor cannot be in electrostatic equilibrium.

2. *Any excess charge on an isolated conductor resides entirely on its surface.*
   Reason: This Property is a direct result of the $1/r^2$ repulsion between like charges described by Coulomb's law. If by some means an excess of charge is placed inside a conductor, the repulsive forces between the like charges push them as far apart as possible, causing them to quickly migrate to the surface. So, the electric field is zero within the conducting material, but is not necessarily zero outside.

3. *The electric field just outside a charged conductor is perpendicular to the conductor's surface (Fig.1.157). In terms of field lines, we can say that- the field lines that start or stop on the surface of a conductor are perpendicular to the surface where they intersect it.*
   Proof: Let us start with a contradictory statement, that elec-

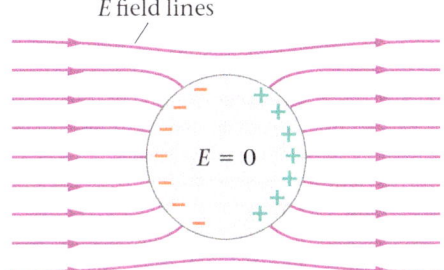

Figure 1.157: Electric field lines meet a conducting surface at right angles and the electric field $\vec{E}$ vanishes inside the conductor

tric field lines are not perpendicular to the conductor surface and the angle between $\vec{E}$ and conductor surface is θ ($\neq \pi/2$). The component of field along the surface (Fig.1.158)
$$E_x = E\cos\theta$$
The electric force applied by this field on the electronic charge $q(=-e)$ on the surface
$$F_x = qE_x = qE\cos\theta$$
As in electrostatic equilibrium, charges on the surface remains at rest, i.e., no net force acting along the surface. Therefore,
$$F_x = 0 \Rightarrow qE\cos\theta = 0$$
$$\Rightarrow \cos\theta = 0 \Rightarrow \theta = \pi/2$$
which contradicts our initial assumption that $\vec{E}$ is not perpendicular to the surface. Therefore, it is proved that just outside a conductor, the electric field $\vec{E}$ is perpendicular to it's surface.

Figure 1.158: This situation is impossible if the conductor is in electrostatic equilibrium. If the electric field $\vec{E}$ had component parallel to the surface, an electric force would be exerted on the charges along the surface and they would move to the left.

4. *On an irregularly shaped conductor, the charge accumulates at sharp points, where the radius of curvature of the surface is smallest. In terms of field lines this can be stated that the field lines are much denser near the more sharply curved surface than the less sharply curved surface.*[Fig.1.159b]

Figure 1.159: (a) Repulsive forces on a charge constrained to move along a curved surface due to two of its neighbors. The parallel components of the forces $(F_\parallel)$ determine the spacing between the charges. (b) For a conductor in electrostatic equilibrium, the surface charge density is largest where the radius of curvature of the surface is smallest and the electric field just outside the surface is strongest there.

along the surfaces of the conductor. On flat surfaces, repulsive forces between neighboring charges push parallel to the surface, making the charges spread apart evenly. On a curved surface, only the components of the repulsive forces parallel to the surface, $F_\parallel$, are effective at making the charges spread apart (Fig.1.159a). If charges were spread evenly over an irregular surface, the parallel components of the repulsive forces would be smaller for charges on the more sharply curved regions (since, at more sharply curved surface, the angle between $\vec{E}$ and $F_\parallel$ is larger that that of less sharply curved surface) and charge would tend to move toward these regions.

The electric field lines just outside a conductor are densely packed at sharp points because each line starts or ends on a surface charge. Since the density of field lines reflects the magnitude of the electric field, the electric field outside the conductor is largest near the sharpest points of the conducting surface. The electric field near very sharp points may be strong enough to ionize the air around it.

EXAMPLE 70. Fig.1.160 shows a positively charged metal sphere above a conducting plate with a negative charge. Sketch the electric field lines.

APPROACH Field lines start on positive charges and end on

Figure 1.160

negative charges. Thus we draw the field lines from the positive sphere to the negative plate, perpendicular to both surfaces, as shown in Fig.1.161. The single field line that goes upward tells us that there is a field above the sphere, but that it is weak.

1.27 Limitations of Coulomb's Law for a Relativistically Moving Charge

Coulomb's law:
$$\vec{F} = \frac{1}{4\pi\varepsilon_0}\frac{q_1q_2}{r^2}\hat{r}$$

gives the electrostatic force between two stationary charges. In practice, it is also applied when charges move very slowly compared to the speed of light, so that the finite time taken for interaction can be ignored. The law assumes that the force depends only on the instantaneous separation between the charges, as if

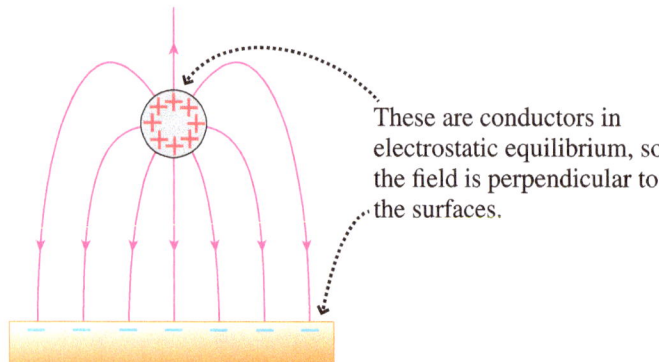

Figure 1.161: Drawing field lines from sphere to plate.

any change in position were transmitted instantly.

In reality, the force on a charge is determined by the electric field at its location:

$$\vec{F} = q\vec{E}$$

The electric field itself is generated by other charges, but any change in their state does not influence the field everywhere instantly. Instead, information about the change propagates outward at the finite speed of light c.

This fact becomes significant in two distinct cases:

- **Charges at very large separations:** If the charges are at rest and have always been at rest, Coulomb's law still holds, even if they are separated by vast distances (e.g., light-years). The electric field is static, so there is no new information that needs to travel. However, if one charge suddenly moves or changes, the other will continue to experience the field due to its *previous* state for a time interval equal to the signal travel time, r/c. In such cases, the assumption of instantaneous interaction breaks down.
- **Charges moving at relativistic speeds:** Even if the distance is not large, if a charge moves at a speed comparable to the speed of light, it generates not just an electric field but also a magnetic field. The force it exerts on another charge depends not only on their separation but also on velocity, acceleration, and the geometry of motion. Coulomb's law, which accounts only for static electric interactions, is therefore invalid in such scenarios.

Because of these limitations, applying Coulomb's law blindly to moving charges or across astronomical distances leads to incorrect predictions. The correct description requires the concept of *retarded fields* (see subsection **1.27.1** "Retarded Field,"' page "60"), which take into account both the finite speed of propagation and the fact that moving charges influence the field in more complicated ways. However, a detailed treatment of retarded fields belongs to advanced electrodynamics and is **beyond the scope of JEE, NEET, or Olympiad syllabi**. Here, we will only introduce the idea and definition for awareness.

> **Note for Exams**
>
> In JEE, NEET, and Olympiads, Coulomb's law is always applied in the *static or quasi-static limit*. You will not be required to calculate using retarded fields. Only the basic idea and definition are included here for completeness.

1.27.1 Retarded Field

A **retarded field** is the electric (or magnetic) field observed at a point in space, not based on the charge's current position, but on where the charge was at an earlier time. This is because changes in the field travel at the speed of light, not instantly.

Example: Imagine a point charge placed exactly 1 light-year away from you in vacuum. Suppose the charge suddenly changes its position or starts moving. Will you feel the change in the electric force immediately? The answer is **no**. Electromagnetic effects (such as changes in electric or magnetic fields) do not propagate instantaneously—they travel at the finite speed of light, $c = 3 \times 10^8$ m/s.

The distance between you and the charge is:
$$\text{Distance} = 1 \text{ light-year} = c \times 1 \text{ year}$$

So the time taken for any change in the charge's motion or position to affect you is:

$$\text{Time taken} = \frac{\text{Distance}}{\text{Speed}} = \frac{1 \text{ light-year}}{c} = \frac{c \times 1 \text{ year}}{c} = 1 \text{ year}$$

Therefore, for the next 1 year, you will continue to experience the electric field *as if the charge were still at its old position*.

This delayed influence is due to the finite speed at which electromagnetic information travels and is known as the **retarded field**. It reflects the state of the charge (position, velocity, etc.) from 1 year ago—not the present moment.

1.28 Check Point 7

1. ••A simple pendulum has a bob of mass m which carries a charge q on it. Length of the pendulum is L. There is a uniform electric field E in the region. Calculate the time period of small oscillations for the pendulum about its equilibrium position in following cases:
   (a) E is vertically down having magnitude $E = mg/q$
   (b) E is vertically up having magnitude $E = 2mg/q$
   (c) E is horizontal having magnitude $E = mg/q$
   (d) $\vec{E}$ has magnitude of $E = \sqrt{2}mg/q$ and is directed upward making an angle of 45° with the horizontal.

2. ••In a region of space an electric field line is in the shape of a semicircle of radius R. Magnitude of the field at all point is E. A particle of mass m having charge q is constrained to move along this field line. The particle is released from rest at A [Fig.1.162].

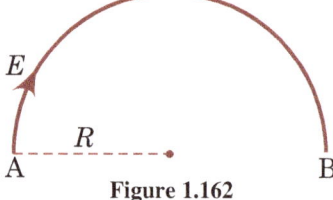

Figure 1.162

   (a) Find its kinetic energy when it reaches point B.
   (b) Find the acceleration of the particle when it is at midpoint of the path from A to B.

3. ••Electric field $E = -bx + a$ exists in a region parallel to the x direction (a and b are positive constants). A charge particle having charge q and mass m is released from the origin $x = 0$. Find the acceleration of the particle at the instant its speed becomes zero for the first time after release.

4. ••Two identical blocks resting on a frictionless, horizontal surface are connected by a light spring having a spring constant $k = 100$ N/m and an unstretched length $L_i = 0.400$ m as shown in Figure 1.163a. A charge Q is slowly placed on each block, causing the spring to stretch to an equilibrium length $L = 0.500$ m as shown in Figure 1.163b. Determine the value of Q, modelling the blocks as charged particles.

Multiple Choice Questions

5. ••In absence of gravity, an electron falls from rest through a vertical distance h in a uniform and vertically upward directed electric field E. The direction of electric field is now reversed, keeping its magnitude the same. A proton

1.29. SOLID ANGLE

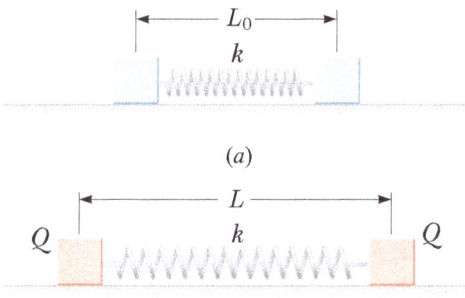

(a)

(b)

Figure 1.163

is allowed to fall from rest in it through the same vertical distance h. The time of fall of the electron, in comparison to the time of fall of the proton is
(A) smaller   (B) 5 times greater
(C) 10 times greater (D) equal

6. ••A toy car with charge q moves on a frictionless horizontal plane surface under the influence of a uniform electric field $\vec{E}$. Due to the force $q\vec{E}$, its velocity increases from 0 to 6 m s^{-1} in one second duration. At that instant the direction of the field is reversed. The car continues to move for two more seconds under the influence of this field. The average velocity and the average speed of the toy car between 0 to 3 seconds are respectively-
(A) 2 m s^{-1}, 4 m s^{-1} (B) 1 m s^{-1}, 3 m s^{-1}
(C) 1 m s^{-1}, 3.5 m s^{-1} (D) 1.5 m s^{-1}, 3 m s^{-1}

7. ••A particle of mass m and charge q is placed at rest in a uniform electric field E and then released. The kinetic energy attained by the particle after moving a distance y is
(A) qEy (B) qE^2y
(C) qEy^2 (D) q^2Ey

8. ••A charged particle (mass m and charge q) moves along X axis with velocity v_0. When it passes through the origin it enters a region having uniform electric field $\vec{E} = -E\hat{j}$ which extends upto $x = d$ [Fig.1.164]. Equation of path of electron in the region $x > d$ is:
(A) $y = \frac{qEd}{m\, v_0^2}x$ (B) $y = \frac{qEd}{m\, v_0^2}(x-d)$
(B) $y = \frac{qEd}{m\, v_0^2}\left(\frac{d}{2}-x\right)$ (D) $y = \frac{qEd^2}{m\, v_0^2}x$

9. ••A small point mass carrying some positive charge on it, is released from the edge of a table. There is a uniform electric field in this region in the horizontal direction [Fig.1.165]. Which of the following options then correctly describe the trajectory of the mass [Fig.1.166]? (Curves are drawn schematically and are not to scale)

Figure 1.164   Figure 1.165

10. ••A ring has charge Q and radius R. If a charge q is placed at its centre then the increase in tension in the ring is [Fig.1.167]
(A) $\frac{Qq}{4\pi\epsilon_0 R^2}$ (B) zero
(C) $\frac{Qq}{4\pi^2\epsilon_0 R^2}$ (D) $\frac{Qq}{8\pi^2\epsilon_0 R^2}$

11. •A point Charge q is placed inside the cavity of a metallic shell. Which one of the diagram [Fig.1.168], correctly

Figure 1.166

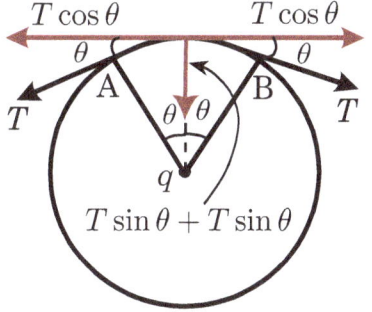

Figure 1.167

represents the electric lines of force.

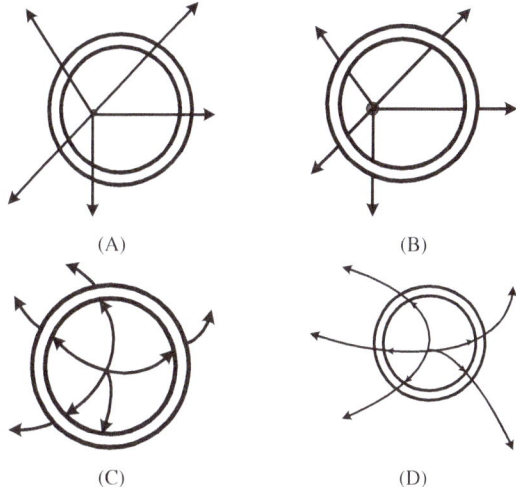

Figure 1.168

1.29 Solid Angle

The solid angle is the extension of the concept of angle from two to three dimension. So let's start with a two dimensional geometry- a circle (Fig.1.169) and pick two rays OA and OC starting from the centre O. They will divide the circumference in two parts ABC and ADC, called arcs. The length of each arc divided by the length of the radius will be the measure of the angle subtended by the arc itself, i.e.,

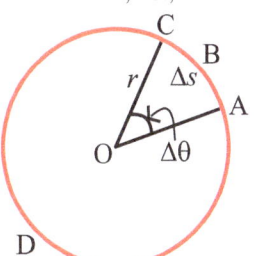

Figure 1.169: Plane angle

From Fig.1.169, we have

$$\text{angle} = \frac{\text{arc}}{\text{radius}}$$

or
$$\Delta\theta = \frac{\text{arcABC}}{OA} = \frac{\Delta s}{r} \quad (1.107)$$

Here, Δs is the arc length and r is the radius of the circle. The SI unit of plane angle is "radian (rad)".

Now, extend this idea to three dimensions. Instead of a circle take a sphere, and instead of picking two rays pick a cone centered in the centre of the sphere [Fig.1.170]. The cone will cross the surface of the sphere: and now to define the *solid angle* measure the area (ΔA) of the surface bounded by the cone, divided by the square of the length of the radius (R^2) (so that we have an area divided by an area).

i.e., solid angle,

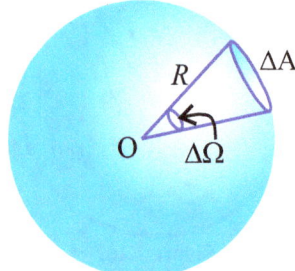

Figure 1.170: A solid angle measured on a sphere.

$$\Delta\Omega = \frac{\Delta A}{R^2} \quad (1.108)$$

The SI unit of solid angle is steradian (sr).

Here, it is important to note that the line joining centre of sphere O to ΔA is normal to ΔA.

Since, plane angles and the solid angles are the ratios of same physical quantities, therefore they are dimensionless quantities. Note that a small surface area as seen from a short distance can cover the same solid angle as a large area as seen from a long distance. For example, in Fig.1.171 different areas A and $A'(A' > A)$, at positions R, R' covers same solid angle at O.

Definition: *The solid angle is a three dimensional angle subtended by an object (two dimensional or three dimensional) at a certain point in the space. It merely depends on the relative distance of the object and its configuration with respect to the given point in the space. Solid angle subtended by a straight line or a point is always zero.*

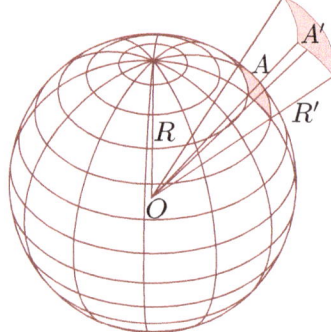

Figure 1.171

In Fig.1.172, we see that the projection of the area element perpendicular to the radius vector is $\Delta A \cos\theta$. Thus, the quantity $\Delta A \cos\theta/R^2$ is equal to the solid angle $\Delta\Omega$ that the surface element ΔA subtends at the *origin O*. We also see that $\Delta\Omega$ is equal to the solid angle subtended by the area element of a spherical surface of radius R.

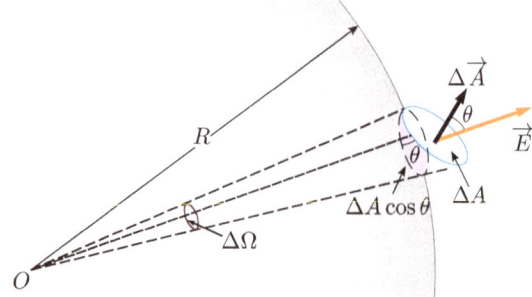

Figure 1.172

☞ If the line joining O to ΔA makes an angle θ with the normal to ΔA (Fig.1.172), we can write:

$$\Delta\Omega = \frac{\Delta A \cos\theta}{R^2}$$

A complete circle subtends an angle -

$$\theta = \frac{\Delta s}{r} = \frac{2\pi r}{r} = 2\pi \text{ rad}$$

at the centre. In fact, any closed curve subtends an angle 2π at any of the internal points. Similarly, a complete sphere subtends a solid angle-

$$\Omega = \frac{A}{R^2} = \frac{4\pi R^2}{R^2} = 4\pi \text{ sr}$$

at the centre. Also, any closed surface subtends a solid angle 4π sr at any internal point.

So for, we have defined the solid angles at an internal point. Now, we find the plane and solid angles at an external point.

From Fig.1.173, it is clear that on gradually closing the curve,

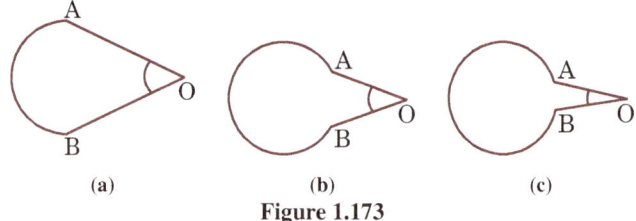

Figure 1.173

the angle subtended by arc AB at an external point O, finally diminishes to zero. So, we can say that a closed curve subtends zero angle at an external point. Similarly, a closed surface also subtends zero solid angle at an external point.

Calculation of Solid Angle at the Centre of a Sphere:

Mathematically, the solid angle can be defined as-

$$\Omega = \frac{A}{R^2}$$

The definition of a solid angle 'Ω,' is analogous to the definition of a plane angle. Just as the arc length s is everywhere perpendicular to the radius r, the area A must be everywhere perpendicular to the radius.

Fig.1.174 shows a sphere of radius R. Let us consider a conical

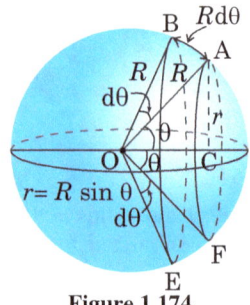

Figure 1.174

section AOF having vertex at O. Suppose semi vertex angle of this cone is $\angle AOC = \theta$ and radius of its base is r. Now, consider

a strip $ABEF$ of radius r on this sphere. If $\angle AOB = d\theta$, then thickness of the strip $AB = Rd\theta$ and radius of strip, $r = R\sin\theta$. Curved area of this strip is given by -
$$dA = 2\pi r \times \text{(thickness } AB\text{)}$$
∵ $r = R\sin\theta$, therefore -
$$dA = 2\pi R \sin\theta \, R \, d\theta$$
$$\Longrightarrow \quad A = 2\pi R^2 \int_0^\theta \sin\theta d\theta \quad \Longrightarrow \quad A = 2\pi R^2 \, (1 - \cos\theta)$$
∴ $$\Omega = \frac{A}{R^2} = \frac{2\pi R^2 \, (1 - \cos\theta)}{R^2} = 2\pi \, (1 - \cos\theta)$$
$$\boxed{\Omega = 2\pi \, (1 - \cos\theta) = 4\pi \sin^2\frac{\theta}{2}} \quad (1.109)$$

Eq.(1.109) gives the relation between semi vertex plane angle and the solid angle formed at the vertex of cone having its vertex at the centre of the sphere of radius R.
At $\theta = 0°$, $\Omega = 2\pi(1 - \cos 0°) = 0$ sr
For a hemispherical surface, at it's centre we have-
$\theta = 90°$, $\Omega = 2\pi(1 - \cos 90°) = 2\pi$ sr
For a complete spherical surface, at it's centre, we have-
$\theta = 180°$, $\Omega = 2\pi(1 - \cos 180°) = 4\pi$ sr
Not only a spherical surface, all closed surfaces subtend a solid angle of 4π sr at their centres.

1.30 Electric Flux

Latin: flux = "to flow"

Analogy With Flow of Water and Concept of Flux: Imagine holding a ring with inside area A in a stream of water flowing with velocity $\vec{v}$, as shown in Fig.1.175. The area vector, $\vec{A}$, of the ring is defined as a vector with magnitude A pointing in a direction perpendicular to the plane of the ring. Of course, for a ring, disc or other "open" surface that is not part of a three-dimensional volume, there are two possible directions perpendicular to the surface. Choose the direction of $\vec{A}$ so that it makes the smallest possible angle with respect to the velocity vector. In Fig.1.175a, the area vector of the ring is parallel to the flow velocity, and the flow velocity is perpendicular to the plane of the ring. The product Av gives the amount of water passing through the ring per unit time, where v is the magnitude of the flow velocity. If the plane of the ring is tilted with respect to the direction of the flowing water (Fig.1.175b), the amount of water flowing through the ring per second, is given by $Av\cos\theta$, where θ is the angle between the area vector of the ring and the direction of the velocity of the flowing water. The amount of water flowing through the ring is called the flux, $\Phi = Av\cos\theta = \vec{A} \cdot \vec{v}$. Since flux is a measure of volume per unit time, its units are cubic meters per second (m^3/s). For the case of the electric field, we define an analogous

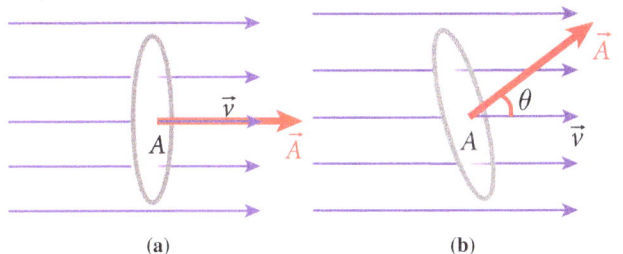

Figure 1.175: Water flowing with velocity of magnitude v through a ring of area A. (a) The area vector is parallel to the flow velocity. (b) The area vector is at an angle θ to the flow velocity.

quantity and call it *electric flux*. We should however note that there is no *flow* of a physically observable quantity unlike the case of water flow.
Consider a uniform electric field of magnitude E passing through a given area A (Fig.1.176). Again, the area vector is $\vec{A}$, with a direction normal to the surface of the area and a magnitude A. The angle θ is the angle between the vector electric field and the area vector, as shown in Fig.1.176. The electric field passing through a given area A is called the electric flux and is given by-
$$\Phi_E = \vec{E} \cdot \vec{A} = EA\cos\theta \quad (1.110)$$
In simple words, the electric flux is proportional to the number of electric field lines passing through the area.
The flux through a surface of area A has a maximum value EA when the surface is perpendicular to the field (i.e. when $\theta = 0°$), and is zero when the surface is parallel to the field (i.e. when $\theta = 90°$).
When we apply Eq.(1.110), it is often best to sketch $\vec{E}$ and $\vec{A}$ with their tails touching and θ is the angle between the two vectors From Eq.(1.110), we can also say that the electric flux Φ_E is the

Figure 1.176: Field lines for a uniform electric field through an area A that is at an angle of $(90° - \theta)$ to the field.

magnitude of the electric field $|\vec{E}|$ times the magnitude of the component of the area vector $\vec{A'}$ that is parallel to the electric field (Fig.1.176):
$$\Phi_E = EA' \quad (1.111)$$
where $A' = A\cos\theta$. The unit of electric flux is $NC^{-1}m^2$.
If you were to shine a light parallel to the electric field, the shadow cast by a surface of area A would have an area equal to the magnitude of $\vec{A'}$ (Fig.1.176).

A More General Expression for Electric Flux
In Eq.1.110 we have assumed that the electric field $\vec{E}$ was constant over the surface. What if a surface is curved and/or the field varies with position? In such a case, we divide the surface into many infinitesimally small pieces, each piece is so small that we can consider it as essentially flat and the field is uniform over it (Fig.1.177). If $d\vec{A}$ is the area vector of one such a small piece and the electric field at this piece is $\vec{E}$, then the flux through one such small piece is $d\Phi_E = \vec{E} \cdot d\vec{A}$, and the flux through the entire surface comes from integrating the fluxes through each small pieces, i.e.,
$$\Phi_E = \int_{\text{surface}} \vec{E} \cdot d\vec{A} \quad (1.112)$$
Equation 1.177 is a surface integral, which means it must be evaluated over the surface in question. Generally, the value of Φ_E depends both on the field pattern and on the surface.

1.30.0.1 The Electric Flux Through a Closed Surface
Our final step, to calculate the electric flux through a closed surface such as a box, a cylinder, or a sphere, requires nothing new. We've already learned how to calculate the electric flux

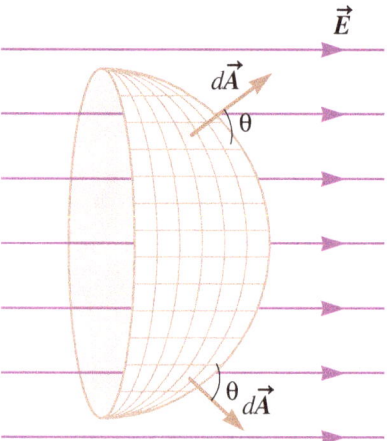

Figure 1.177

through flat and curved surfaces, and a closed surface is nothing more than a surface that happens to be closed.

However, the mathematical notation for the surface integral over a closed surface differs slightly from what we've been using. It is customary to use a little circle on the integral sign to indicate that the surface integral is to be performed over a closed surface. With this notation, the electric flux through a closed surface is

$$\Phi_E = \oint \vec{E} \cdot d\vec{A} = \oint E_n dA \qquad (1.113)$$

where E_n represents the component of the electric field normal to the surface.

Note that only the notation has changed. The electric flux is still the summation of the fluxes through a vast number of tiny pieces, that now cover a closed surface.

Direction of Area Vector: A closed surface has a distinct inside and outside. The direction of the area vector $d\vec{A}$ is chosen so that the vector points outward from the surface. If the area element is not part of a closed surface, the direction of the area vector is chosen so that the angle between the area vector and the electric field vector is less than or equal to 90°.

In case of a closed surface, the sign of the flux depends on the angle between $\vec{E}$ and $d\vec{A}$ as follows:

1. If $\theta < 90°$, then $\vec{E}$ crosses the surface from the inside to the outside and hence $d\Phi_E = \vec{E} \cdot d\vec{A}$ is positive.
2. If $\theta = 90°$, then $\vec{E}$ grazes the surface and hence $d\Phi_E = \vec{E} \cdot d\vec{A}$ is zero.
3. If $90° < \theta < 180°$, then $\vec{E}$ crosses the surface from the outside to the inside and hence $d\Phi_E = \vec{E} \cdot d\vec{A}$ is negative.

Consider the closed surface in Fig.1.178. The vectors $d\vec{A}_i$ point in different directions for the various surface elements, but for each element they are normal to the surface and point outward. At the element labeled ①, the field lines are crossing the surface from the inside to the outside and $\theta < 90°$; hence, the flux $d\Phi_1 = \vec{E} \cdot d\vec{A}_1$ through this element is positive. For element ②, the field lines graze the surface (perpendicular to $d\vec{A}_2$); therefore, $\theta = 90°$ and the flux is zero. For elements such as ③, where the field lines are crossing the surface from outside to inside, $90° < \theta < 180°$ and the flux is negative because $\cos\theta$ is negative. The *net* flux through the surface is proportional to the net number of lines leaving the surface, where the net number means the number of lines leaving the surface minus the number of lines entering the surface. If more lines are leaving than entering, the net flux is positive. If more lines are entering than leaving, the net flux is negative.

Now we're ready to calculate the flux through a closed surface.

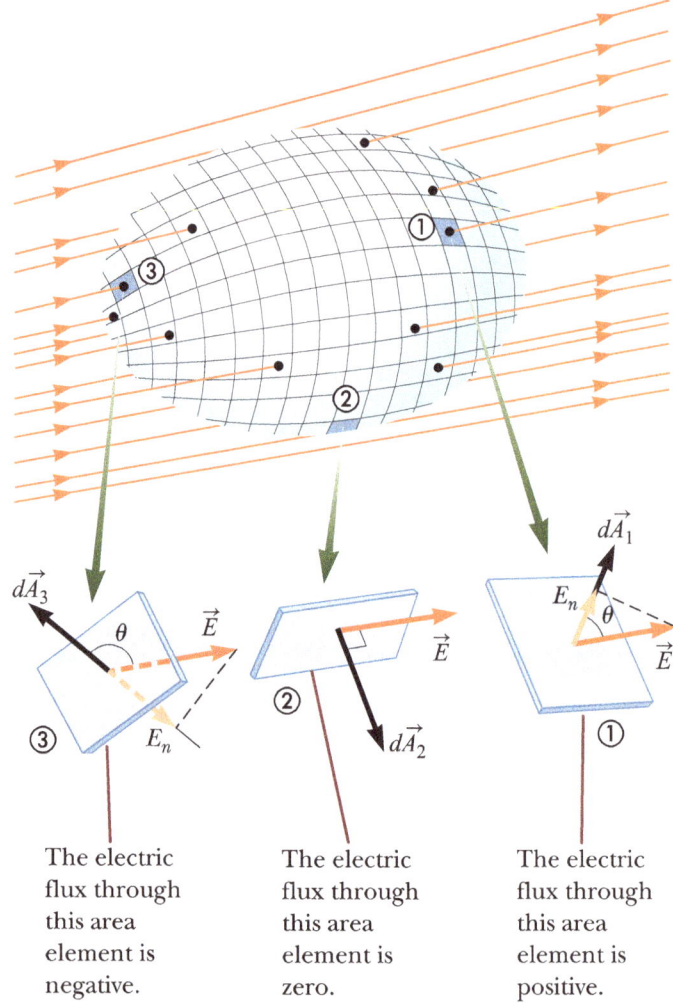

The electric flux through this area element is negative.

The electric flux through this area element is zero.

The electric flux through this area element is positive.

Figure 1.178: A closed surface in an electric field. The area vectors are, by convention, pointed normally outward to the surface.

EXAMPLE 71. Fig.1.179 shows a cube that has edge length l in a uniform electric field, $\vec{E}$, that is directed along the positive x-axis and perpendicular to the plane of one face of the cube. What is the net electric flux passing though the cube?

APPROACH To find the net flux through the cube, determine

Figure 1.179

the flux passing through each surface and add them algebraically.

SOLUTION In Fig.1.179, the electric field lines pass through two faces perpendicularly and are parallel to four other faces of the cube. So, the flux through four of the faces (③, ④, and the unnumbered faces) is zero because $\vec{E}$ is parallel to the four faces

and therefore perpendicular to $d\vec{A}$ on these faces.
The net flux through faces ① and ②:

$$\Phi_{12} = \Phi_1 + \Phi_2 = \int_{①} \vec{E} \cdot d\vec{A} + \int_{②} \vec{E} \cdot d\vec{A}$$

For face ①, $\vec{E}$ is constant and directed inward but $d\vec{A}_1$ is directed outward ($\theta = 180°$). So, the flux through surface ①,

$$\Phi_1 = \int_{①} \vec{E} \cdot d\vec{A} = \int_{①} E (\cos 180°) \, dA$$

$$= -E \int_{①} dA = -EA = -El^2 \quad (\because \cos 180° = -1)$$

For face ②, $\vec{E}$ is constant and outward and in the same direction as $d\vec{A}_2$ ($\theta = 0°$). Therefore, flux through this face:

$$\Phi_2 = \int_{②} \vec{E} \cdot d\vec{A} = \int_{②} E (\cos 0°) \, dA$$

$$= E \int_{②} dA = +EA = El^2 \quad (\because \cos 0° = 1)$$

Therefore, net flux through the cube

$$\Phi_E = \Phi_{12} + \Phi_{3456}$$
$$= -El^2 + El^2 + 0 + 0 + 0 + 0 = 0$$

EXAMPLE 72. Expression for an electric field is given by $\vec{E} = 4000x^2 \hat{i}$ V/m. Find the electric flux through the cube of side 20 cm when placed in electric field (as shown in the Fig.1.180).

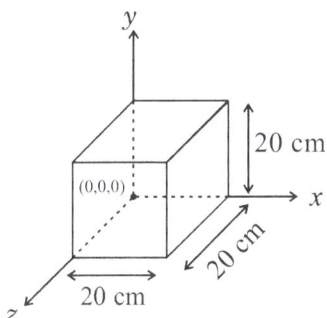

Figure 1.180

SOLUTION Electric flux (ϕ) is defined as $\phi = \vec{E} \cdot \vec{A}$. Clearly, flux will be zero if area vector is perpendicular to $\vec{E}$.
It is given that $\vec{E} = 4000x^2 \hat{i}$ V/m. It is directed along positive direction of X–axis. So, electric flux is zero corresponding to planes $y = 0$, $y = 0.2$, $z = 0$ and $z = 0.2$ m.
Now, electric flux corresponding to plane $x = 0$

$$\phi_1 = \vec{E} \cdot \vec{A} = \left[4000x^2 \hat{i} \text{ V/m} \cdot (0.20 \text{ m})^2 \left(-\hat{i}\right)\right]_{x=0} = 0$$

Electric flux corresponding to plane $x = 0.20$ m$\hat{i}$

$$\phi_1 = \vec{E} \cdot \vec{A} = \left[4000x^2 \hat{i} \text{ V/m} \cdot (0.20 \text{ m})^2 \hat{i}\right]_{x=0.20}$$

$$= 4000 \times 16 \times 10^{-4} \text{ Vm} = 640 \text{ Vcm}$$

Therefore, net electric flux linked to the given cube:

$$\phi = \phi_1 + \phi_2 + \underbrace{\phi_3}_{\text{flux passing through remaining four surfaces}}$$

$$= 0 + 640 \text{ V} \cdot \text{cm} + 0 = 640 \text{ V} \cdot \text{cm}$$

1.31 Check Point 8

1. •A person is placed in a large, hollow, metallic sphere that is insulated from ground. If a large charge is placed on the sphere, will the person be harmed upon touching the inside of the sphere?
2. •Why must hospital personnel wear special conducting shoes while working around oxygen in an operating room? What might happen if the personnel wore shoes with rubber soles?
3. ••Consider a closed triangular box resting within a horizontal electric field of magnitude $E = 7.80 \times 10^4$ N/C as shown in Fig.1.181. Calculate the electric flux through (a) the vertical rectangular surface, (b) the slanted surface, and (c) the entire surface of the box.

Figure 1.181

4. ••A uniform electric field $a\hat{i} + b\hat{j}$ intersects a surface of area A. What is the flux through this area if the surface lies (a) in the yz plane? (b) in the xz plane? (c) in the xy plane?
5. ••A point charge q is located at the centre of a uniform ring having linear charge density λ and radius a, as shown in Fig.1.182. Determine the total electric flux through a sphere centered at the point charge and having radius R where $R < a$.

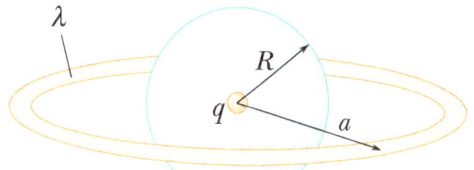

Figure 1.182

6. ••A pyramid with horizontal square base, 6.00 m on each side, and a height of 4.00 m is placed in a vertical electric field of 52.0 N/C [Fig.1.183]. Calculate the total electric flux through the pyramid's four slanted surfaces.

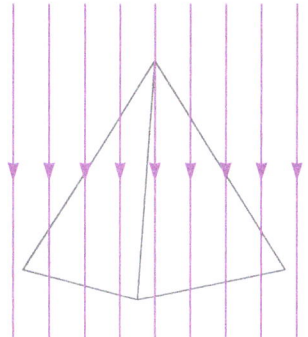

Figure 1.183

7. ••A cone with base radius R and height h is located on a horizontal table. A horizontal uniform field E penetrates the cone, as shown in Figure 1.184. Determine the electric flux that enters the left-hand side of the cone.

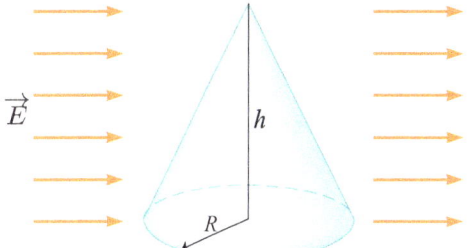

Figure 1.184

8. ●●A hemispherical surface of radius R is kept in a uniform electric field $\vec{E}$ such that $\vec{E}$ is parallel to the axis of hemi-sphere [Fig.1.185]. Find the net flux linked with hemispherical surface-

Figure 1.185

9. ●●An electric field is described by $\vec{E} = (15\hat{i} + 25\hat{j})$N/C. Find the electric flux through a surface whose area vector is $\vec{A} = (0.65\hat{i} + 0.35\hat{j})$m^2

10. ●●Calculate the electric flux through the surfaces shown in Fig.1.186

Figure 1.186

11. ●●The electric flux through the surface shown in Fig.1.187 is 25 N m^2/C. What is the electric field strength?

Figure 1.187

12. ●●A 2.0 cm × 3.0 cm rectangle lies in the xy-plane. What is the electric flux through the rectangle if
    (a) $\vec{E} = (100\,\hat{i} + 50\,\hat{k})$N/C?
    (b) $\vec{E} = (100\,\hat{i} + 50\,\hat{j})$N/C?

13. ●●A 2.0 cm × 3.0 cm rectangle lies in the xz-plane. What is the electric flux through the rectangle if
    (a) $\vec{E} = (100\,\hat{i} + 50\,\hat{k})$ N/C?
    (b) $\vec{E} = (100\,\hat{i} + 50\,\hat{j})$ N/C?

14. ●●A 3.0 cm diameter circle lies in the xz-plane in a region where the electric field is $\vec{E} = (1500\,\hat{i} + 1500\,\hat{j} - 1500\,\hat{k})$ N/C. What is the electric flux through the circle?

15. ●●A 1.0 cm × 1.0 cm × 1.0 cm box with its edges aligned with the xyz-axes, is in the electric field $\vec{E} = (350x + 150)\hat{i}$N/C, where x is in meters. What is the net electric flux through the box?

16. ●●What is the net electric flux through the two cylinders shown in Fig.1.188? Give your answer in terms of R and E.

Multiple Choice Questions

Figure 1.188

17. ●●A square surface of side L meter in the plane of the paper is placed in a uniform electric field E (volt/m) acting along the same plane at an angle θ with the horizontal side of the square as shown in Fig.1.189. The electric flux linked to the surface, in units of volt m is-

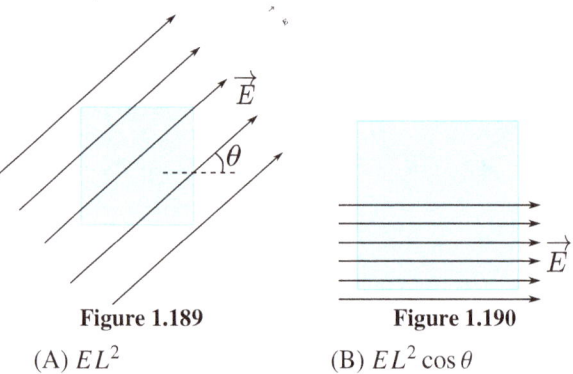

Figure 1.189   Figure 1.190

(A) EL^2 (B) $EL^2 \cos\theta$
(C) $EL^2 \sin\theta$ (D) zero

18. ●●A square surface of side L metres is in the plane of the paper. A uniform electric field $\vec{E}$(volt /m), also in the plane of the paper is limited only to the lower half of the square surface (Fig.1.190). The electric flux in SI units associated with the surface is
    (A) EL^2 (B) $EL^2/2\epsilon_0$
    (C) $EL^2/2$ (D) zero

1.32 Gauss's Law

Gauss's law is equivalent to Coulomb's law for static charges, although Gauss's law will look very different. It provides a different way to express the relationship between electric charge and electric field.

The purpose of learning Gauss's law is twofold:
- Gauss's law allows the electric fields of some continuous distributions of charge to be found much more easily than does Coulomb's law.
- Gauss's law is more general in that it also covers the case of a rapidly moving charge. For such charges the electric lines of force become compressed in a plane at right angles to the direction of motion, thus losing their spherical symmetry. Coulomb's law is not valid for moving charges.

Thus Gauss's law is ultimately a more fundamental statement about electric fields than is Coulomb's law.

Statement of Gauss's Law: Gauss's Law states that the net flux Φ_E of electric field passing through any closed surface is proportional to the net charge Q_{encl} that is enclosed by the same surface. The proportionally constant is $1/\epsilon_o$, i.e.

$$\Phi_E = \oint \vec{E}.d\vec{A} = \frac{Q_{encl}}{\epsilon_0}$$

1.32. GAUSS'S LAW

Proof: Let us consider a positive point charge q located at the centre of a sphere of radius r, as shown in Fig.1.191. The electric field created by this charge at any point the surface of the Gaussian sphere is directed outwards and be normal to the surface. It's magnitude is also constant and given by-

$$E = k\frac{q}{r^2} = \frac{1}{4\pi\epsilon_0}\frac{q}{r^2} \qquad (1.114)$$

Therefore:

$$\vec{E} \cdot d\vec{A} = E dA \cos 0° = E dA = \frac{1}{4\pi\epsilon_0}\frac{q}{r^2} dA$$

Hence, the total flux Φ_E through the entire Gaussian surface is

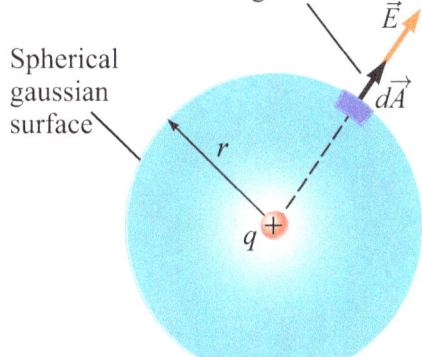

Figure 1.191

$$\Phi_E = \oint \vec{E} \cdot d\vec{A} = \frac{q}{4\pi\epsilon_\circ r^2} \oint dA$$
$$= \frac{q}{4\pi\epsilon_\circ r^2} 4\pi r^2 = \frac{q}{\epsilon_\circ}$$

or

$$\boxed{\Phi_E = \oint \vec{E} \cdot d\vec{A} = \frac{q}{\epsilon_\circ}} \qquad (1.115)$$

From Eq.(1.115), it is clear that the flux through a spherical surface of radius r is equal to the charge q inside the sphere divided by the permittivity of free space ϵ_0.

Since, entire Gaussian surface of any shape forms a solid angle of magnitude of 4π sr, therefore, this flux is linked with solid angle 4π sr. Now consider several closed Gaussian surfaces surrounding the charge as shown in Fig.1.192(a). The number of electric field lines passing through the spherical surface S_1 is the same as the number of lines passing through the non-spherical surfaces S_2 and S_3. Therefore, we conclude that the electric flux through any closed surface is independent of the shape of the surface that encloses the charge q, and it also does not depend upon the particular location of q inside the surface. The magnitude of the flux through any closed Gaussian surface surrounding the point charge q is $q/\epsilon_\circ$.

Now consider a point charge located outside a closed surface of arbitrary shape as shown in Fig.1.192(b). In such a case, any electric field line entering the surface leaves the surface at another point. The number of electric field lines entering the surface equals the number leaving the surface. Therefore, the net electric flux through a closed surface that surrounds no charge is zero.

1.32.1 General Integral Form of Gauss's Law

Now, suppose the surface encloses not just one point charge q but several charges $q_1, q_2, q_3, \cdots$. The total (resultant) electric field $\vec{E}$ at any point is the vector sum of the $\vec{E}$ fields of the individual charges. Let Q_{encl} be the total charge enclosed by the surface: $Q_{encl} = q_1 + q_2 + q_3 + \cdots$. Also let $\vec{E}$ be the total field at the position of the surface area element $d\vec{A}$, then we can write an equation like Eq.(1.115) for each charge and its corresponding field and add the results. When we do, we obtain the general statement of Gauss's law:

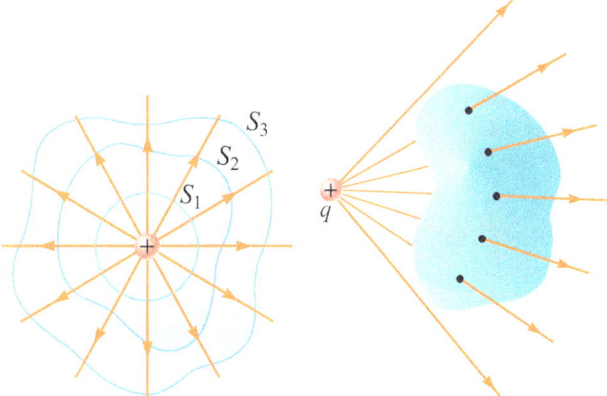

(a) The net electric flux is the same through all surfaces.

(b) The number of field lines entering the surface equals the number leaving the surface.

Figure 1.192: (a) Closed surfaces of various shapes surrounding a positive charge. (b) A point charge located outside a closed surface.

$$\boxed{\Phi_E = \oint \vec{E} \cdot d\vec{A} = \frac{Q_{encl}}{\epsilon_0}} \quad \text{(Gauss's law)} \qquad (1.116)$$

The net electric flux through any closed surface is equal to the net charge inside the surface divided by the permittivity of free space ϵ_0..

In terms of solid angle, we can say that the complete flux linked with solid angle 4π sr is Q_{encl}/ϵ_0

Note: If we have a closed surface of irregular geometry, then during the integration the value of E may be different at various locations on the surface, and the angle between $\vec{E}$ and dA may also vary as we sum the various contributions over the surface. But, interestingly, regardless of the shape of the surface the net flux through the closed surface is always Q_{encl}/ϵ_0

Key Points

1. Gauss's law is true for any closed surface, no matter what its shape or size.
2. The term Q_{encl} on the right side of Gauss's law includes the sum of all charges enclosed by the surface. The charges may be located anywhere inside the surface.
3. In the situation when the surface is so chosen that there are some charges inside and some outside the Gaussian surface, the electric field [whose flux appears on the left side of Gauss's law] is caused partly by charges inside the surface and partly by charges outside. But as Fig.1.192(b) shows, the outside charges do not contribute to the total (net) flux through the surface.
4. The closed surface that we choose for the application of Gauss's law is called the Gaussian surface. It is an imaginary surface. There need not be any material object at the position of the surface. You may choose any Gaussian surface and apply Gauss's law. However, take care not to let the Gaussian surface pass through any discrete charge. This is because electric field due to a system of discrete charges is not well defined at the location of any charge. (As you go close to the charge, the electric field increases to infinity.) However, the Gaussian surface can pass through a continuous charge distribution.

5. Gauss's law is often useful towards a much easier calculation of the electrostatic field when the system has some symmetry. This is facilitated by the choice of a suitable Gaussian surface.
6. Finally, Gauss's law is based on the inverse square dependence on distance contained in the Coulomb's law. Any violation of Gauss's law will indicate departure from the inverse square law.

EXAMPLE 73. Fig. 1.193 shows two Gaussian surfaces in uniform electric fields. In Fig.1.193(a) the Gaussian surface is a square box, and in Fig.1.193(b) the Gaussian surface is a closed cylinder. In each case, integrate $\Phi_E = \oint \vec{E} \cdot d\vec{A} = q_{encl}/\epsilon_0$ to find the amount of charge enclosed in each Gaussian surface.

SOLUTION (a) Note: We can easily determine whether the

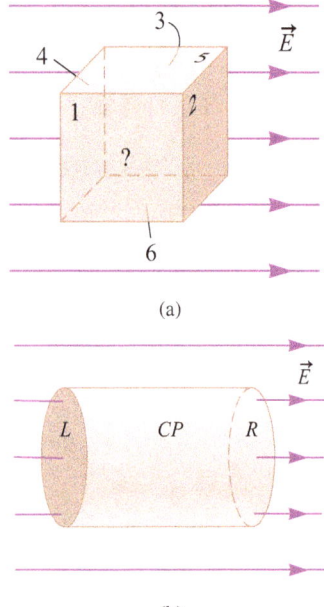

(a)

(b)

Figure 1.193

charge is positive, negative, or zero by looking at the electric field lines (Fig. 1.193). Because the electric field lines pass completely through the Gaussian surfaces, the net charge inside each surface is zero.

By Using Gauss's Law Although we already know the results, we are asked to use integration in order to practice using Gauss's law with different Gaussian surfaces.

A closed box is made up of six sides. According to Gauss's law, we must integrate over the entire closed box, so we break up the integral into six pieces, one for each side of the box. The six subscripts on the integrals correspond to the six sides of the box (Fig. 1.193(a)).

$$\Phi_E = \oint \vec{E} \cdot d\vec{A}$$
$$= \int_1 \vec{E} \cdot d\vec{A} + \int_2 \vec{E} \cdot d\vec{A} + \int_3 \vec{E} \cdot d\vec{A}$$
$$+ \int_4 \vec{E} \cdot d\vec{A} + \int_5 \vec{E} \cdot d\vec{A} + \int_6 \vec{E} \cdot d\vec{A}$$

To calculate the six dot products, we need to know the angle θ between $\vec{E}$ and $d\vec{A}$ for each side. A separate drawing of each side is helpful (Fig. 1.194). The area vector $d\vec{A}$ for each side is perpendicular to that side and points outward.

The dot product is zero whenever the angle θ between $\vec{E}$ and $d\vec{A}$ is 90°, so the dot product is zero for sides 1, 3, 5, and 6. That means these four integrals are zero, and we are left with just two

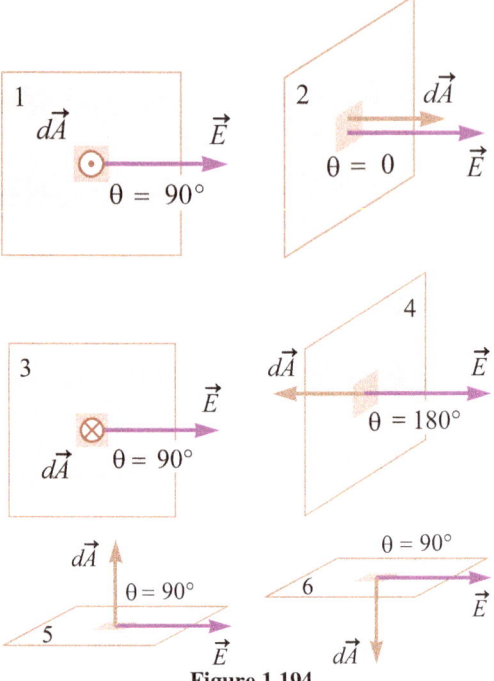

Figure 1.194

integrals.

$$\Phi_E = \oint \vec{E} \cdot d\vec{A}$$
$$= 0 + \int_2 \vec{E} \cdot d\vec{A} + 0 + \int_4 \vec{E} \cdot d\vec{A} + 0 + 0$$
$$= \int_2 \vec{E} \cdot d\vec{A} + \int_4 \vec{E} \cdot d\vec{A}$$
$$\Rightarrow \Phi_E = \int_2 \vec{E} \cdot d\vec{A} + \int_4 \vec{E} \cdot d\vec{A}$$

Write each dot product in terms of the magnitude of the vectors and the angle between them.

$$\Phi_E = \int_2 E dA \cos\theta + \int_4 E dA \cos\theta$$
$$\Rightarrow \Phi_E = \int_2 E dA \cos 0 + \int_4 E dA \cos 180°$$
$$\Rightarrow \Phi_E = \int_2 E dA - \int_4 E dA$$

The electric field is uniform, so as we integrate over dA, E is a constant that we can pull outside the integrals.

$$\Phi_E = E \int_2 dA - E \int_4 dA$$

The integrals are identical: Each equals the area A of one square side.

$$\Phi_E = EA - EA = 0$$

The net electric flux is zero, and according to Gauss's law, that means the charge enclosed by the Gaussian surface is zero.

$$\Phi_E = \frac{Q_{encl}}{\epsilon_0} = 0 \quad \Rightarrow \quad Q_{encl} = 0$$

Note: This is exactly what we predicted because the number of electric field lines entering the box is equal to the number of electric field lines leaving the box.

(b) The closed cylinder is made up of three surfaces - the left cap, the right cap, and the curved part. These are labeled L, R, and CP (Fig. 1.195). Break the integral up into three pieces, one for each surface in Fig.1.195.

$$\Phi_E = \oint \vec{E} \cdot d\vec{A}$$

1.32. GAUSS'S LAW

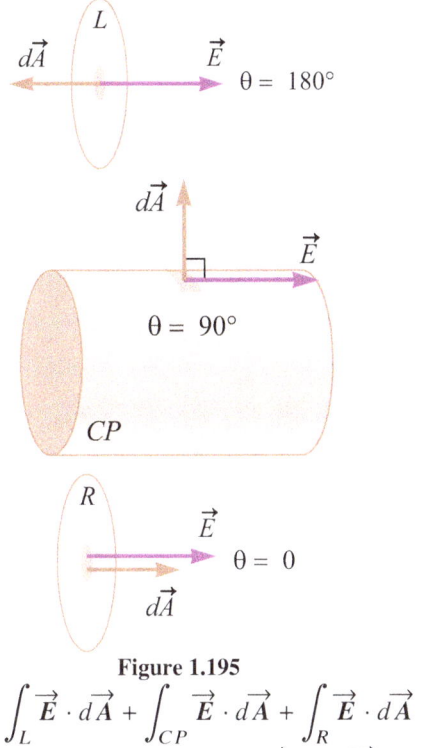

Figure 1.195

$$= \int_L \vec{E} \cdot d\vec{A} + \int_{CP} \vec{E} \cdot d\vec{A} + \int_R \vec{E} \cdot d\vec{A}$$

Figure 1.195 shows the angle θ between $\vec{E}$ and $d\vec{A}$ for each surface.

The area vector points outward for all small pieces of the curved part, while the electric field points to the right. Therefore, the angle $\theta = 90°$ for the curved part and the corresponding dot product and integral are zero, leaving two integrals.

$$\Phi_E = \int_L \vec{E} \cdot d\vec{A} + \int_R \vec{E} \cdot d\vec{A}$$

Write each dot product in terms of the magnitude of the vectors and the angle between the vectors.

$$\Phi_E = \int_L E \, dA \cos 180° + \int_R E \, dA \cos 0$$
$$= -\int_L E \, dA + \int_R E \, dA$$

As for the box, the electric field is constant and the integrals are identical. This time, each is the area of the end cap.

$$\Phi_E = -EA + EA = 0$$

The electric flux is zero, so the charge inside the Gaussian surface is zero.

$$\Phi_E = \frac{q_{encl}}{\epsilon_0} = 0 \quad \Rightarrow \quad q_{encl} = 0$$

☞ We find the same result whether we use a Gaussian cylinder or a box. The process is similar, but there is slightly less work when we use the cylinder because the Gaussian cylinder is made up of just three surfaces instead of the six surfaces of a box. Because we are free to choose a convenient Gaussian surface, it may be helpful to choose a closed cylinder instead of a box when that is possible.

EXAMPLE 74. The Gaussian surface in Figure 1.196 is a closed cylinder. An electric field points to the right throughout, made up of two uniform components with magnitude $E_L = 40.9$ N/C on the left side of the cylinder and $E_R = 81.8$ N/C on the right side. Integrate $\Phi = \oint \vec{E} \cdot d\vec{A} = q_{encl}/\epsilon_0$ to find the amount of charge enclosed in the Gaussian surface. Express your result in terms of the surface charge density (charge per unit area) σ. (Assume σ is uniform.)

SOLUTION This is similar to last problem (Fig.1.193 (b)), except the electric field is not uniform. Because more electric field

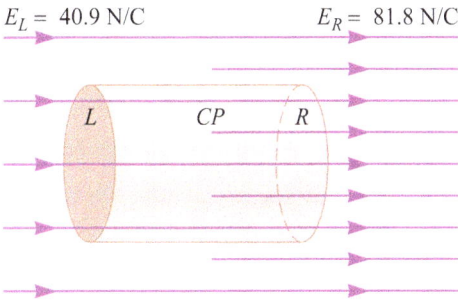

Figure 1.196

lines emerge from the Gaussian surface than enter it, we expect the net charge inside to be positive.

As like last problem, the net flux associated with the cylinder is-

$$\Phi_E = -E_L \int_L dA + E_R \int_R dA$$

The electric field is constant over each end cap, so it can be pulled outside the integrals. The magnitude of the electric field depends on the position, and the subscripts refer to the magnitudes in Fig.1.195.

$$\Phi_E = -E_L \int_L dA + E_R \int_R dA$$

The integrals are identical, equaling the area A of each end cap.

$$\Phi_E = -E_L A + E_R A = A(E_R - E_L)$$

According to Gauss's law, the electric flux is proportional to the charge inside.

$$\Phi_E = A(E_R - E_L) = \frac{Q_{encl}}{\epsilon_0}$$

The surface charge density σ is charge per unit area.

$$\sigma = \frac{Q_{encl}}{A} = \epsilon_0 (E_R - E_L)$$
$$\sigma = \left(8.85 \times 10^{-12} C^2/N \cdot m^2\right)(81.8 \text{ N/C} - 40.9 \text{ N/C})$$
$$= 3.62 \times 10^{-10} C/m^2$$

Note: As expected, the charge inside the Gaussian surface is positive.

EXAMPLE 75. A charge is placed at the centre of a cylindrical surface. Find total flux passing through lateral curved surface.

SOLUTION Total flux linked can be divided into two parts

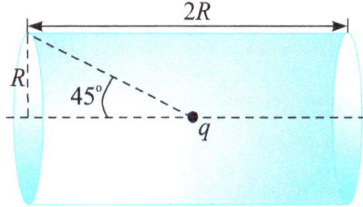

Figure 1.197

(i) Flux through longitudinal cylindrical curved surface ϕ_L
(ii) Flux through end caps ϕ_A

If end cap subtends a solid angle Ω at the centre, then

$$\Phi_L + \Phi_A = \frac{q}{\epsilon_\circ} \qquad (1.117)$$

The solid angle corresponding to semi vertical angle θ at the vertex of a cone [Fig.1.198] is given by-

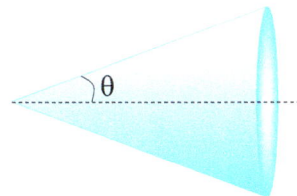

Figure 1.198

$$\Omega = 2\pi(1 - \cos\theta) \qquad (1.118)$$

Since, the electric flux corresponding to solid angle 4π sr is q/ϵ_0, therefore, the electric flux corresponding to solid angle Ω is $\dfrac{\Omega}{4\pi}\dfrac{q}{\epsilon_0}$, i.e., $\dfrac{2\pi(1-\cos\theta)}{4\pi}\dfrac{q}{\epsilon_0}$ or $\dfrac{(1-\cos\theta)}{2}\dfrac{q}{\epsilon_0}$.

The electric flux passing through left end cap of Fig.1.197,

$$\Phi_1 = \dfrac{(1-\cos 45°)}{2}\dfrac{q}{\epsilon_0} = \dfrac{1}{2}\left(1 - \dfrac{1}{\sqrt{2}}\right)\dfrac{q}{\epsilon_0}$$

Similarly, the electric flux passing through the right cap is-

$$\Phi_2 = \dfrac{1}{2}\left(1 - \dfrac{1}{\sqrt{2}}\right)\dfrac{q}{\epsilon_0}$$

Therefore, the electric flux passing through both end caps-

$$\Phi_A = \Phi_1 + \Phi_2 = \left(1 - \dfrac{1}{\sqrt{2}}\right)\dfrac{q}{\epsilon_0}$$

Substituting this value in Eq.(1.117), we get-

$$\Phi_L = \dfrac{q}{\epsilon_0} - \left(1 - \dfrac{1}{\sqrt{2}}\right)\dfrac{q}{\epsilon_0} = \dfrac{q}{\sqrt{2}\epsilon_0}$$

So, the flux passing through curved cylindrical surface is $q/\sqrt{2}\epsilon_0$.

EXAMPLE 76. A cone of base radius R and height h is located in a uniform electric field $\vec{E}$ parallel to its base [Fig.1.199]. Find the electric flux entering the cone.

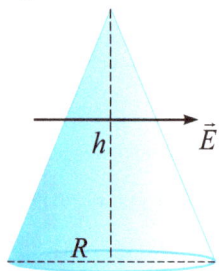

Figure 1.199

APPROACH You require to take only the projection of area of cone on a plane normal to electric field and it is a triangle of height h and base length $2R$ for the cone. To determine the required flux, just multiply this projection with electric field E.

SOLUTION Area of the triangle = $\frac{1}{2} \times 2R \times h = Rh$. Hence flux = ERh. (since, incoming flux is positive)

EXAMPLE 77. If a charge q is placed at the centre of a closed hemispherical non-conducting surface as shown in Fig.1.200, the total flux passing through the flat surface would be:

(A) 0 (B) $\dfrac{q}{2\epsilon_0}$ (C) $\dfrac{q}{4\epsilon_0}$ (D) $\dfrac{q}{2\pi\epsilon_0}$

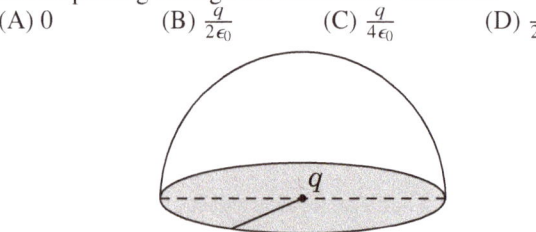

Figure 1.200

SOLUTION As the flat surface is passing through the charge, therefore field lines emitted from charge wont be able to cross the flat surface. So flux passing through the flat surface will be zero. Therefore option A is correct.

EXAMPLE 78. A charge q is surrounded by a closed surface consisting of an inverted cone of height h and base radius R, and a hemisphere of radius R as shown in the Fig.1.201. The electric flux through the conical surface is $\dfrac{nq}{6\epsilon_0}$ (in SI units). Find the value of n.

SOLUTION From Gauss law,

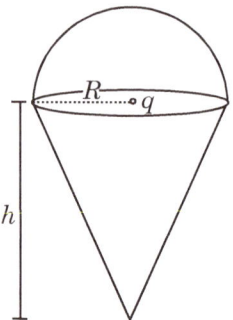

Figure 1.201

$$\phi_{\text{hemisphere}} + \phi_{\text{cone}} = \dfrac{q}{\epsilon_0} \qquad \cdots (1)$$

From Eq.1.109, the solid angle corresponding to plane angle 2θ, as shown in Fig.1.174, is

$$\Omega = 2\pi(1-\cos\theta)$$

Since, total electric flux corresponding to 4π sr solid angle is $Q_{\text{encl}}/\epsilon_0$, therefore flux corresponding to solid angle Ω is-

$$\phi = \dfrac{Q_{\text{encl}}\Omega}{4\pi\epsilon_0} = \dfrac{2\pi(1-\cos\theta)Q_{\text{encl}}}{4\pi\epsilon_0} = \dfrac{Q_{\text{encl}}}{2\epsilon_0}(1-\cos\theta)$$

For hemisphere, $2\theta = \pi$, i.e., $\theta = \pi/2$ and given that $Q_{\text{encl}} = q$, therefore

$$\phi_{\text{hemisphere}} = \dfrac{q}{2\epsilon_0}\left(1-\cos\dfrac{\pi}{2}\right) = \dfrac{q}{2\epsilon_0}$$

Substituting this value in Eq.(1), we get

$$\dfrac{q}{2\epsilon_0} + \phi_{\text{cone}} = \dfrac{q}{\epsilon_0} \implies \phi_{\text{cone}} = \dfrac{q}{2\epsilon_0}$$

It is also given that, $\phi_{\text{cone}} = \dfrac{nq}{6\epsilon_0}$, therefore

$$\dfrac{nq}{6\epsilon_0} = \dfrac{q}{2\epsilon_0} \implies n = 3$$

Alternatively, $\phi \propto$ no of electric field lines passing through the given surface. As the given point charge q, is equally distributed between the hemispherical and conical parts, therefore half of electric field lines will pass through hemisphere & other half will pass through conical surface.

1.32.2 Differential Form of Gauss's Law (Optional)

In contrast to Eq.1.116 which is called the integral form we shall seek the differential form of the Gauss theorem, which establishes the relation between the volume charge density ρ and the changes in the field intensity E in the vicinity of a given point in space.

For this purpose, we first represent the charge q in the volume V enveloped by a closed surface S in the form $Q_{\text{encl}} = \langle\rho\rangle V$, where $\langle\rho\rangle$ is the volume charge density, averaged over the volume V. Then we substitute this expression into Eq. (1.116) and divide both its sides by V, which gives

$$\dfrac{1}{V}\oint \vec{E}\cdot d\vec{A} = \langle\rho\rangle/\epsilon_0 \qquad (1.119)$$

We now make the volume V tend to zero by contracting it to the point we are interested in. In this case, $\langle\rho\rangle$ will obviously tend to the value of ρ at the given point of the field, and hence the ratio on the left-hand side of Eq. 1.119 will tend to ρ/ϵ_0.

The quantity which is the limit of the ratio of $\oint \vec{E}\cdot d\vec{A}$ to V as $V \to 0$ is called the divergence of the field $\vec{E}$ and is denoted by div $\vec{E}$. Thus, by definition,

$$\text{div}\,\vec{E} = \lim_{V\to 0}\dfrac{1}{V}\oint \vec{E}d\vec{A} \qquad (1.120)$$

The divergence of any other vector field is determined in a similar way. It follows from definition (1.120) that divergence is a scalar function of coordinates.

In order to obtain the expression for the divergence of the field

1.33. APPLICATIONS OF GAUSS'S LAW

$\vec{E}$, we must, in accordance with (1.120), take an infinitely small volume V, determine the flux of $\vec{E}$ through the closed surface enveloping this volume, and find the ratio of this flux to the volume. The expression obtained for the divergence will depend on the choice of the coordinate system (in different systems of coordinates it turns out to be different). For example, in Cartesian coordinates it is given by

$$\text{div } \vec{E} = \frac{\partial E_x}{\partial x} + \frac{\partial E_y}{\partial y} + \frac{\partial E_z}{\partial z} \quad (1.121)$$

Thus, we have found that as $V \to 0$ in Eq.1.119, its right-hand side tends to ρ/ϵ_0, while the left-hand side tends to div $\vec{E}$. Consequently, the divergence of the field $\vec{E}$ is related to the charge density at the same point through the equation

$$\boxed{\text{div } \vec{E} = \frac{\rho}{\epsilon_0}} \quad (1.122a)$$

or

$$\boxed{\frac{\partial E_x}{\partial x} + \frac{\partial E_y}{\partial y} + \frac{\partial E_z}{\partial z} = \frac{\rho}{\epsilon_\circ}} \quad (1.122b)$$

Equations (1.122a) and (1.122b) represent the Gauss theorem in the differential form. The form of many expressions and their applications can be considerably simplified if we introduce the vector differential operator $\vec{\nabla}$. In Cartesian coordinates, the operator $\vec{\nabla}$ has the form

$$\vec{\nabla} = \hat{i}\frac{\partial}{\partial x} + \hat{j}\frac{\partial}{\partial y} + \hat{k}\frac{\partial}{\partial z}$$

where $\hat{i}$, $\hat{j}$, and $\hat{k}$ are the unit vectors along the X-, Y-, and Z-axes respectively. The operator $\vec{\nabla}$ itself does not have any meaning. It becomes meaningful only in combination with a scalar or vector function by which it is symbolically multiplied. For example, if we form the scalar product of vector $\vec{\nabla}$ and vector $\vec{E} = E_x \hat{i} + E_y \hat{j} + E_z \hat{k}$, we obtain-

$$\vec{\nabla} \cdot \vec{E} = \frac{\partial}{\partial x} E_x + \frac{\partial}{\partial y} E_y + \frac{\partial}{\partial z} E_z$$

It follows from Eq.1.121 that this is just the divergence of $\vec{E}$. Thus, the divergence of the field $\vec{E}$ can be written as div $\vec{E}$ or $\vec{\nabla} \cdot \vec{E}$ (in both cases it is read as "the divergence of $\vec{E}$"). So, the Gauss theorem (Eq.1.122) in terms of $\vec{\nabla}$ operator can be written as-

$$\boxed{\vec{\nabla} \cdot \vec{E} = \frac{\rho}{\epsilon_0}} \quad (1.123)$$

The Gauss theorem in the differential form is a local theorem: the divergence of the field $\vec{E}$ at a given point depends only on the electric charge density ρ at this point. This is one of the remarkable properties of electric field. For example, the field $\vec{E}$ of a point charge is different at different points. Generally, this refers to the spatial derivatives $\partial E_x/\partial x$, $\partial E_y/\partial y$, and $\partial E_z/\partial z$ as well. However, the Gauss theorem states that the sum of these derivatives, which determines the divergence of $\vec{E}$, turns out to be equal to zero at all points of the field (outside the charge itself).

At the points of the field where the divergence of $\vec{E}$ is positive, we have the sources of the field (positive charges), while at the points where it is negative, we have sinks (negative charges). The field lines emerge from the field sources and terminate at the sinks. Eq.1.123 can also be written as-

$$\vec{\nabla} \cdot \left(\epsilon_0 \vec{E}\right) = \rho$$

or

$$\boxed{\vec{\nabla} \cdot \vec{D} = \rho} \quad (1.124)$$

In Eq.1.124, $\vec{D} = \epsilon_0 \vec{E}$ is called electric displacement vector. Detailed discussion of displacement vector is out of scope of this book.

1.33 Applications of Gauss's law

Selecting a Gaussian Surface: While applying Gauss's law we are interested in evaluating the integral

$$\Phi_E = \oint \vec{E} \cdot \vec{dA} = \oint E\, dA \cos\theta \quad (1.125)$$

Now note that, this integration may be complicated if product of E and $\cos\theta$ is not constant. Therefore we always select a Gaussian surface in such a way that product of E and $\cos\theta$ remains constant. That is we have to select some symmetrical surfaces. The types of symmetry are illustrated in Fig.1.202 and summarized in Table 1.3. If the object does not have any of these three types of symmetry, Gauss's law, even though still true, is unlikely to be helpful in calculating the electric field strength.

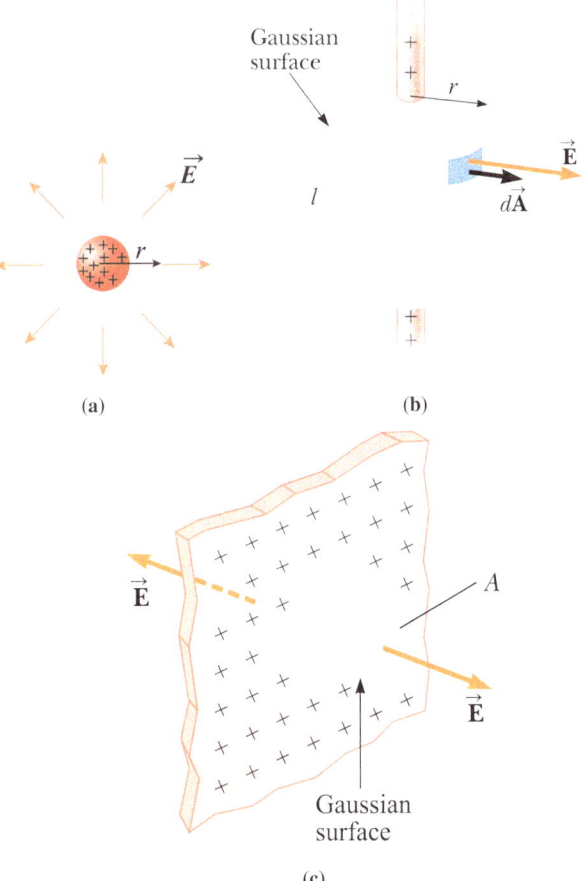

Figure 1.202: In spherical symmetry (a), the points on a sphere have constant electric field strength. For cylindrical symmetry (b), points at a common distance from a line have constant electric field strength. In planar symmetry (c), points on a plane parallel to the charged plane have constant electric field strength.

1.33.1 Gauss's Law: Problem Solving Approach

To apply Gauss's law, we simply follow the following steps-
1. Sketch the charge distribution, as well as the position(s) where the electric field is to be calculated.
2. Determine the symmetry of its electric field.

Table 1.3: Type of symmetry and the corresponding Gaussian surface that should be used

Symmetry	Type of Gaussian Surface	Objects
spherical	sphere concentric with given charged object	point charge, charged sphere, spherical shell
cylindrical	cylinder coaxial with given charged object	line of charge, charged cylinder, coaxial cylindrical shells
planar	cylinder or box perpendicular to given plane charged object	charged parallel plane(s), large flat object

3. Draw the appropriate closed Gaussian surface (see Table 1.3) based on the symmetry, choosing the size of the surface according to where the electric field is to be determined. For example, in the case of a charged sphere, a Gaussian sphere is drawn concentric with the centre of the object and is of radius equal to the radial distance where the electric field is to be computed. Sketch in the directions of $\vec{E}$ and $d\vec{A}$ on your diagram. You need not enclose all the charge within the Gaussian surface.

4. Be sure every part of the Gaussian surface satisfies one or more of the following conditions:
   (a) The value of the electric field can be argued by symmetry to be constant over the portion of the surface.
   (b) Gaussian surface is either tangent to or perpendicular to the electric field.
      i. If the electric field is everywhere tangent to a surface, the electric flux through the surface is $\Phi_E = \vec{E}.\vec{A} = 0$
      ii. If the electric field is everywhere perpendicular to a surface and has the same magnitude E at every point, the electric flux through the surface is $\Phi_E = \vec{E}.\vec{A} = EA$
   (c) The electric field is zero over the portion of the surface. Different portions of the Gaussian surface can satisfy different conditions as long as every portion satisfies at least one condition.

5. *If necessary, divide the closed surface into parts.* In some cases, the electric field is not constant over the entire surface but is known over each part. In this case, divide the Gaussian surface into different parts, ensuring that the entire closed surface is included. For example, we earlier divided the cylinder into two end caps and the curved side (see Ex.75).

6. For each part, *determine the orientation of $\vec{E}$ relative to $d\vec{A}$* and use that to find the dot product for that part.

7. Finally *integrate over the Gaussian surface to compute the electric flux*. However, if you have a surface with uniform electric field strength (and constant direction relative to the surface normal) you can take E out of the integral and simply multiply the constant E by the surface area multiplied by the $\cos\theta$ value.

8. *Calculate the net charge enclosed by the Gaussian surface.* Remember that only the charge inside the surface needs to be considered, and when there are positive and negative charges inside the surface it is the net charge that is used.

9. *Use mathematical expression of Gauss's law (with the expression for the electric flux (from step 6) and the enclosed charge (step 7)) to solve for the electric field strength-*

$$\Phi_E = \oint \vec{E} \cdot d\vec{A} = \frac{Q_{encl}}{\epsilon_0}$$

10. Finally you can check that your answer makes sense with limiting cases (i.e., does your relationship make sense just outside a surface, or at infinite distance).

EXAMPLE 79. Using Gauss's law, find the electric field at a distance r from a positive point charge q, and compare it with Coulomb's law.

APPROACH By definition, a point is a sphere of negligible radius, i.e., a sphere of zero radius, therefore we can assume a point charge as spherical and draw a spherical Gaussian surface of radius r around it (Fig.1.203). From the symmetry of the Gaussian surface, we know that at each point on it, the electric field $\vec{E}$ is perpendicular to the surface and directed outwards from the spherical center. Thus, $\vec{E}//d\vec{A}$ and $\vec{E}.d\vec{A} = EdA$, with $Q_{encl} = q$. Now, apply Gauss's law-

$$\Phi_E = \oint \vec{E} \cdot d\vec{A} = \frac{Q_{encl}}{\epsilon_0} \quad (1.126)$$

and solve for E

SOLUTION Substituting, $\vec{E} \cdot d\vec{A} = EdA$ and $Q_{encl} = q$ in Eq.

Figure 1.203

(1.126), we get

$$\oint EdA = \frac{q}{\epsilon_0}$$

$$\implies E \oint dA = \frac{q}{\epsilon_0} \implies 4\pi r^2 E = \frac{q}{\epsilon_0}$$

$$\implies \boxed{E = \frac{1}{4\pi\epsilon_0}\frac{q}{r^2} = k\frac{q}{r^2}}$$

which is simply Coulomb's law. This proves that Gauss's law and Coulomb's law are equivalent.

EXAMPLE 80. A single positive point charge, $+Q$, is placed at the exact centre of a cube of side length L (see Fig.1.204).
(a) Is it possible to calculate the electric flux through one face of the cube? If so, what is the value?
(b) Can Gauss's law be used to find the electric field on the face of the cube without considering the symmetry due to all faces? If so, what is the value?

SOLUTION (a) Yes. The enclosed charge is $+Q$, so by Gauss's law the total electric flux through all six faces of the cube is $\frac{Q}{\epsilon_0}$. By symmetry, all the faces are equal, so the electric flux through any one face is just 1/6 of this value, i.e., $Q/6\epsilon_0$.
(b) No! Since, we don't know the distribution of electric field strength across the faces, so we cannot take E out of the integral.

EXAMPLE 81. Find electric flux through square of side a, due to charge placed at distance $a/2$ from centre of a

1.33. APPLICATIONS OF GAUSS'S LAW

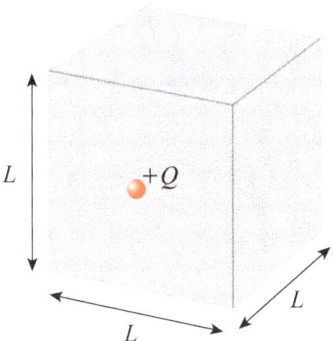

Figure 1.204: A single point charge, $+Q$, is at the exact centre of a cube. The length of each side of the cube is L.

square[Fig.1.205].
SOLUTION Let us enclose the charge q by a cubical gaussian surface with q at its centre[Fig.1.206].

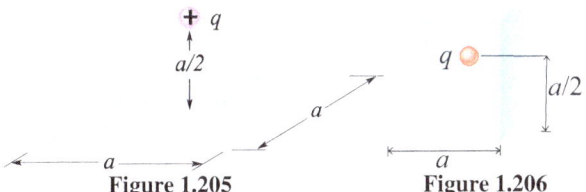

Figure 1.205   Figure 1.206

Net flux, coming out from the cubical Gaussian surface-
$$\phi = \frac{1}{\epsilon_0} Q_{encl} = \frac{1}{\epsilon_0} q$$
By symmetry all six faces have equal flux. Therefore, flux coming out from a single face, $\phi = q/6\epsilon_0$

EXAMPLE 82. By applying Gauss's law, show that the stable equilibrium of a charge under the effect of an electric field only, is impossible(Earnshaw's Theorem 1.24).
SOLUTION Suppose, in vacuum, we have a system of fixed

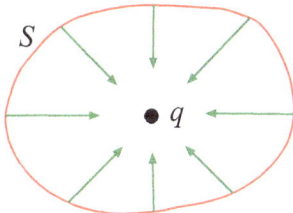

Figure 1.207

point charges in equilibrium. Now, consider, one of these charges, e.g. a charge q. Let us envelop the charge q by a small closed surface S (Fig.1.207). For the sake of definiteness, we assume that $q > 0$. For the equilibrium of this charge to be stable, it is necessary that the electric field $\vec{E}$ created by all the remaining charges of the system at all the points of the surface S be directed towards the charge q. Only in this case any small displacement of the charge q from the equilibrium position will give rise to a restoring force, and the equilibrium state will actually be stable. But such a configuration of the field $\vec{E}$ around the charge q is in contradiction to the Gauss theorem: the flux of $\vec{E}$ through the surface S is negative, while in accordance with the Gauss theorem it must be equal to zero since it is created by charges lying outside the surface S. On the other hand, the fact that E is equal to zero indicates that at some points of the surface S vector $\vec{E}$ is directed inside it and at some other points it is directed outside.
Hence it follows that in any electrostatic field a charge cannot be in stable equilibrium.

1.33.2 Cylindrical Symmetry
The following examples illustrate situations where cylindrical symmetry can be applied.
 EXAMPLE 83.   **Field of an Infinite Line Charge:** An infinitely long, straight line of uniform positive charge has a charge per unit length of λ (in units of charge per length). Find the electric field $\vec{E}$ at an arbitrary distance r from the line.
APPROACH From each incremental charge along the line, an electric field emanates equally in all directions. However, by symmetry, the superposition of the fields from all of the incremental charges results in a cancellation of fields parallel to the line of charge*. The result is a net field directed radially outward from the line. At all points at a given distance r from the line (in any direction), the field has the same magnitude.
Therefore, we match this symmetry with a Gaussian surface in the form of a cylinder of radius r and length L whose axis is the line of charge (Fig.1.208). At every point on the cylindrical curved side S_1 of the cylinder, $\vec{E}$ is parallel to the area elements dA, and it has the same magnitude everywhere. On the flat end caps S_1 and S_2 of the cylinder, $\vec{E}$ is perpendicular to $d\vec{A}$ everywhere. The net charge Q_{encl} inside the cylinder is λL. Applying Gauss's law, we obtain-
$$\Phi_E = \oint \vec{E} \cdot d\vec{A} = \frac{Q_{encl}}{\epsilon_0}$$

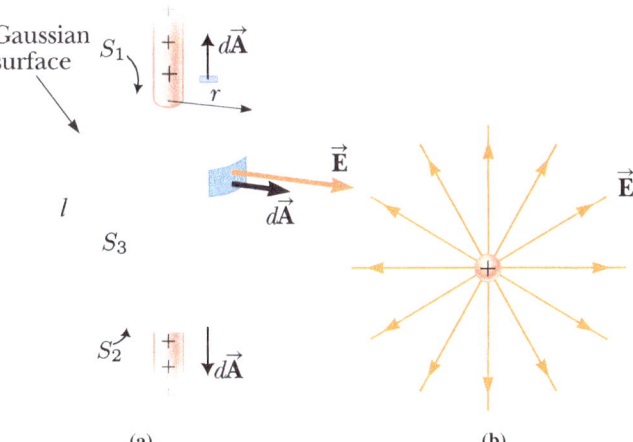

Figure 1.208: (a) An infinite line of charge surrounded by a cylindrical Gaussian surface concentric with the line. (b) An end view shows that the electric field at the cylindrical surface is constant in magnitude and perpendicular to the surface.

$$\Rightarrow \int_{S_1} \vec{E} \cdot d\vec{A} + \int_{S_2} \vec{E} \cdot d\vec{A} + \int_{S_3} \vec{E} \cdot d\vec{A} = \frac{Q_{encl}}{\epsilon_0} \quad (1.127)$$
From Fig.1.208a, we see that, $\vec{E} \perp d\vec{A}$ at end caps S_1 and S_2, therefore-
$$\int_{S_2} \vec{E} \cdot d\vec{A} = \int_{S_3} \vec{E} \cdot d\vec{A} = 0$$
and $\vec{E} \parallel \vec{A}$ at whole curved surface S_3, so -
$$\int_{S_3} \vec{E} \cdot d\vec{A} = E dA$$
Substituting these values and $Q_{encl} = \lambda L$ in Eq.(??), we get-

---
*We have already seen this in case of applications of Coulomb's law for "electric field due to an infinite line charge"

$$0 + 0 + \int_{S_3} E \, dA = \frac{\lambda L}{\epsilon_0} \implies E(2\pi rL) = \frac{\lambda L}{\epsilon_0}$$

$$\implies \boxed{E = \frac{\lambda}{2\pi\epsilon_0 r} = k\frac{2\lambda}{r}} \quad \text{(radially outward)} \quad (1.128)$$

here, $k = 1/4\pi\epsilon_0$.

Eq.1.128 gives the electric field strength, E, at a distance r from an infinite line with charge density λ. The direction of the electric field is radially away from the line of charge.

EXAMPLE 84. What if the line segment in Example 83 were not infinitely long?

SOLUTION If the line charge in Example 83 were of finite length, the electric field would not be given by Eq.1.128. A finite line charge does not possess sufficient symmetry to make use of Gauss's law because the magnitude of the electric field is no longer constant over the surface of the Gaussian cylinder: the field near the ends of the line would be different from that far from the ends. Therefore, condition (4a) of problem solving approach of Gauss's law "Subsection 1.33.1", would not be satisfied in this situation. Furthermore, $\vec{E}$ is not perpendicular to the cylindrical surface at all points: the field vectors near the ends would have a component parallel to the line. Therefore, condition (4b(ii)) of problem solving approach of Gauss's law "Subsection 1.33.1". would not be satisfied. For points close to a finite line charge and far from the ends, Equation 1.128 gives a good approximation of the value of the field.

EXAMPLE 85. The Electric Field of an Infinite Plane Sheet of Charge: Use Gauss's law to find the electric field of an infinite plane of charge with surface charge density σ (in C/m^2).

APPROACH From symmetry we can say that, as long as we are not near an edge, the electric field must extend perpendicularly away from the plane on both sides. (There is no asymmetry that would cause the field lines to bend to one side or the other as they extend away from the positive charges.) We match the symmetry of this field by considering a Gaussian surface in the form of a cylinder or pill box, of cross-sectional area A, whose axis is perpendicular to the plane and whose ends are equidistant from the plane (Fig.1.209. The net charge enclosed by the surface is $Q_{encl} = \sigma A$. By symmetry, the field emerges uniformly and perpendicularly from each end and is tangent to the curved side of the cylinder. Now, apply Gauss's law, and solve for $\vec{E}$.

SOLUTION By, Gauss's law, we have

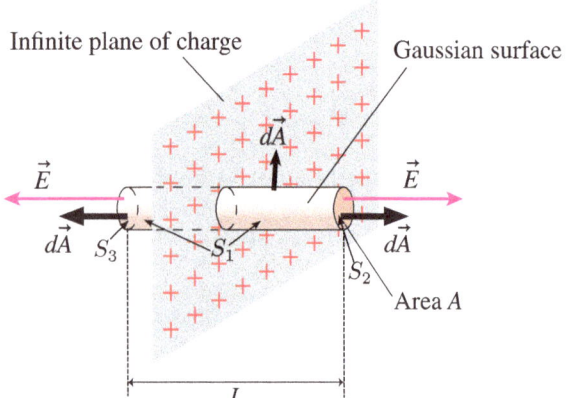

Figure 1.209: The Gaussian surface extends to both sides of a plane of charge.

$$\oint \vec{E} \cdot d\vec{A} = \frac{Q_{encl}}{\epsilon_0} = \frac{A\sigma}{\epsilon_0} \quad (1.129)$$

Now, $\oint \vec{E} \cdot d\vec{A} = \int_{S_1} \vec{E} \cdot d\vec{A} + \int_{S_2} \vec{E} \cdot d\vec{A} + \int_{S_3} \vec{E} \cdot d\vec{A}$

here, S_1 represents curved surface whereas S_2, S_3 represent two planer surfaces.

From Fig.1.209, we have -

$$\int_{S_1} \vec{E} \cdot d\vec{A} = 0 \quad \because \vec{E} \perp d\vec{A} \text{ over whole surface } S_1, \text{ and}$$

$$\int_{S_2} \vec{E} \cdot d\vec{A} + \int_{S_3} \vec{E} \cdot d\vec{A} = 2EA \quad (\because \vec{E} \parallel d\vec{A}, \vec{E} \parallel d\vec{A})$$

Note that, for S_2, both $\vec{E}$ and $d\vec{A}$ pointed towards right side, while for S_3, both $\vec{E}$ and $d\vec{A}$ pointed towards left (Fig.1.209). Substituting these values in Eq.1.129, we get -

$$\therefore \quad 2EA = \frac{A\sigma}{\epsilon_0}$$

$$\implies \boxed{E = \frac{\sigma}{2\epsilon_0}} \quad (1.130)$$

This electric field is away from the plane, in both left and right side of it.

Because the distance from the surface does not appear in the Eq.(1.130), we conclude that the field has the same constant value for all distances on either side of the plane of charge.

EXAMPLE 86. Explain why Gauss's law cannot be used to calculate the electric field near an electric dipole, a charged disc, or a triangle with a point charge at each corner.

SOLUTION The charge distributions of all these configurations do not have sufficient symmetry to make the use of Gauss's law practical. We cannot find a closed surface surrounding any of these distributions for which all portions of the surface satisfy one or more of conditions listed in "Gauss's Law: Problem solving Approach (Subsection 1.33.1)."

EXAMPLE 87. Electric Field Near a Charged Conducting Surface: Using Gauss's law, find the electric field just outside the surface of a conductor carrying a positive surface charge density σ.

APPROACH Near the conducting surface, the surface appears to be a flat (Fig.1.210, close-up), infinite plane, just as the surface of the ocean seems to be flat from our perspective standing on a beach. So, the surface charge density σ over a small part of the conductor is uniform*.

Notice that In the static case, no electric field $\vec{E}$ can exist within a conductor (because it would make the conduction charges move). So field lines extend away from the conductor, perpendicular to the surface, Fig.1.210. From symmetry considerations, we choose a cylindrical Gaussian surface as in the previous example, but in this case, we can draw two types of cylindrical Gaussian surfaces-

1. Gaussian cylinder crosses the conductor
2. One end of Gaussian cylinder is inside the conductor, whereas the other end out side the surface of the conductor.

Apply Gauss's law in both cases and solve for electric field $\vec{E}$.

SOLUTION Case 1. Let Gaussian cylinder crosses the conductor completely. 1, 2 are planer surfaces and 3 is curved surface of Gaussian cylinder

Net flux linked with the Gaussian cylinder

$$\Phi_E = \oint E \cdot dA = \frac{Q_{encl}}{\epsilon_0}$$

$$\implies \underbrace{\int_1 \vec{E} \cdot d\vec{A}}_{\vec{E} \parallel d\vec{A}} + \underbrace{\int_2 \vec{E} \cdot d\vec{A}}_{\vec{E} \parallel d\vec{A}} + \underbrace{\int_3 \vec{E} \cdot d\vec{A}}_{\vec{E} \perp d\vec{A}} = \frac{2\sigma A}{\epsilon_0}$$

---

*In general, the surface charge density σ is not constant on the surface of a conductor but depends on the shape of the conductor.

1.33. APPLICATIONS OF GAUSS'S LAW

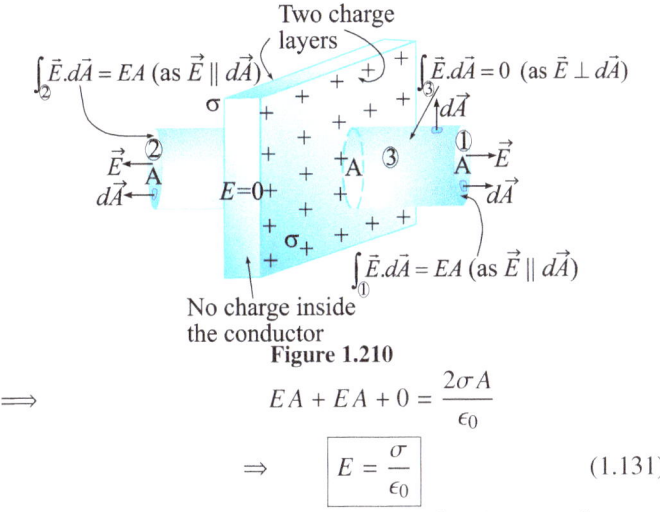

Figure 1.210

$$\Rightarrow \quad EA + EA + 0 = \frac{2\sigma A}{\epsilon_0}$$

$$\Rightarrow \quad \boxed{E = \frac{\sigma}{\epsilon_0}} \quad (1.131)$$

Thus electric field near the conducting surface is $E = \sigma/\epsilon_0$
This field is twice the value that we found for thin sheet of charge in the previous example, and it has a constant value for all distances from the infinite plane conductor.

Case 2. Let Gaussian surface does not cross the conductor i.e., one end is inside the conductor whereas other end is outside it [Fig.1.211]
Net flux linked with the Gaussian cylinder

Figure 1.211

$$\Phi_E = \oint \vec{E} \cdot d\vec{A} = \frac{Q_{encl}}{\epsilon_0}$$

$$\Rightarrow \underbrace{\int_1 \vec{E} \cdot d\vec{A}}_{\vec{E} \parallel d\vec{A}} + \underbrace{\int_2 \vec{E} \cdot d\vec{A}}_{\vec{E}=0} + \underbrace{\int_3 \vec{E} \cdot d\vec{A}}_{\vec{E} \perp d\vec{A}} = \frac{\sigma A}{\epsilon_0} \quad \Rightarrow \quad EA = \frac{\sigma A}{\epsilon_0}$$

$$\Rightarrow \quad \boxed{E = \frac{\sigma}{\epsilon_0}} \quad (1.132)$$

Thus electric field near the conducting surface is $E = \sigma/\epsilon_0$

Explanation: A solid thick conductor can be assumed as a combination of two thin plane sheets of charges having same charge densities as that of conductor.

The directions of the fields to the left, between, and right of the sheets are shown in Fig.1.212. The resultant field depends on the direction and values of E_1 and E_2 due to sheet ① and ② respectively.

Electric field inside the conductor: Inside the conductor, $\vec{E}_1$ and $\vec{E}_2$ are oppositely directed, therefore the resultant electric field-

$$E = \frac{\sigma}{2\epsilon_0} - \frac{\sigma}{2\epsilon_0} = 0$$

here, σ is the surface charge density of both sheets.
Electric field outside the conductor: Outside left or right of

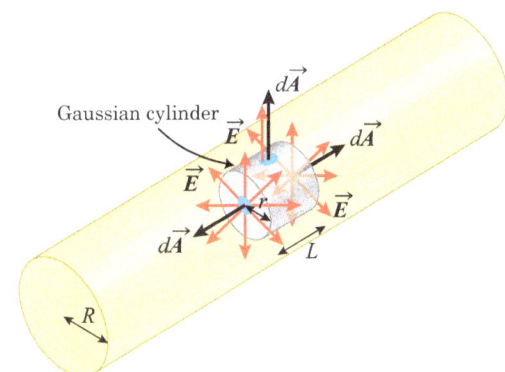

Figure 1.212

the conductor, $\vec{E}_1$ and $\vec{E}_2$ both are directed in same direction, therefore the resultant electric field-

$$E = \frac{\sigma}{2\epsilon_0} + \frac{\sigma}{2\epsilon_0} = \frac{\sigma}{\epsilon_0}$$

EXAMPLE 88. A non-conducting very long cylinder of radius R has a positive uniform volume charge density ρ throughout. Derive expressions for the electric field both (a) inside ($r < R$) and (b) outside ($r > R$) the cylinder.

SOLUTION In this case, we have cylindrical symmetry, so the appropriate closed surface is a coaxial Gaussian cylinder of radius r [Fig.1.213]. The length of the Gaussian cylinder, L, is arbitrary and will cancel out in the final answer.

Figure 1.213: The very long charged cylinder (yellow) has a uniform positive charge density. In grey is drawn the smaller Gaussian cylinder for finding the electric field inside the charged cylinder.

(a) Inside the cylinder ($r < R$) : For this part, we draw the Gaussian cylinder with radius r less than the radius, R, of the charged cylinder (i.e., the Gaussian surface is inside the physical cylinder, but coaxial with it). The situation is illustrated in Figure 1.213. From the symmetry of the situation, the positively charged cylinder results in an electric field that points radially outward and has constant electric field strength at any particular radial distance. This is a situation in which we need to divide the closed Gaussian surface into three parts, the two end caps and the curved side of the cylinder. In all cases, $d\vec{A}$ points outward from the closed Gaussian surface. This is pictured for the Gaussian surface in Figure 1.213. For the two end caps, $d\vec{A}$ is parallel to the axis of the cylinder, pointing in opposite directions at the two ends. For the curved surface of the cylinder, $d\vec{A}$ is everywhere radially outward (we only show $\vec{E}$ and $d\vec{A}$ at one point on the curved Gaussian surface, but both are similarly radially outward for all other points on the curved surface).
We find the electric flux by combining the contributions from the three parts of the surface:

$$\Phi_{E\ net} = \Phi_{E\ left\ cap} + \Phi_{E\ curved\ surface} + \Phi_{E\ right\ cap}$$

Everywhere along the curved surface, $\vec{E}$ and $d\vec{A}$ are parallel to each other, while on both end caps they are perpendicular to each other (Figure 1.213). This means that the vector dot product of $\vec{E}$ with $d\vec{A}$ yields

left end cap: $\vec{E} \cdot d\vec{A} = E\, dA\, \cos 90° = E\, dA(0) = 0$
curved wall: $\vec{E} \cdot d\vec{A} = E\, dA\, \cos 0° = E\, dA(1.00) = E\, dA$
right end cap: $\vec{E} \cdot d\vec{A} = E\, dA\, \cos 90° = E\, dA(0) = 0$

Here E represents the magnitude of the electric field, and dA is the differential area element without a vector direction. Therefore, substituting into the relationship for electric flux, we have

$$\Phi_{E\ \text{left cap}} = \oint_{\text{left cap}} \vec{E} \cdot d\vec{A} = 0,$$

$$\Phi_{E\ \text{curved wall}} = \oint_{\text{curved wall}} \vec{E} \cdot d\vec{A} = \oint_{\text{curved wall}} E\, dA$$

$$= E \oint_{\text{curved wall}} dA = E A_{\text{curved wall}},$$

$$\Phi_{E\ \text{right cap}} = \oint_{\text{right cap}} \vec{E} \cdot d\vec{A} = 0$$

The area of the curved wall can be determined if we imagine it has been unwrapped into a corresponding rectangle. One side of the rectangle is L and the other is the circumference of the surface, $2\pi r$:

$\Phi_{E\ \text{curved wall}} = E A_{\text{curved wall}} = E L (2\pi r) = 2\pi r L E$

Since the two end caps have zero electric flux, this is also the net electric flux:

$\Phi_{\text{net}} = 2\pi r L E$

Now we need to find the net enclosed charge. We multiply the volume charge density, ρ, by the volume of the Gaussian cylinder. The volume of a cylinder is the cross-sectional area times the length of the cylinder:

$Q_{\text{encl}} = \pi r^2 L \rho$

Finally, we invoke Gauss's law and solve for the electric field strength:

$$\Phi_E = \frac{Q_{\text{encl}}}{\epsilon_0} \implies 2\pi r L E = \frac{\pi r^2 L \rho}{\epsilon_0}$$

$$\implies \boxed{E = \frac{r\rho}{2\epsilon_0}} \quad (1.133)$$

The direction of the electric field points radially outward from the axis of the cylinder.
(b) Outside the charged cylinder ($r > R$) : In this case, we make a Gaussian cylinder of radius larger than that of the charged cylinder, but otherwise the solution is similar. The electric flux process is identical to that given in part (a), with the same result:

$\Phi_{\text{net}} = 2\pi r L E$

When computing the enclosed charge, we now use the volume of the charged cylinder, rather than the larger Gaussian cylinder:

$Q_{\text{encl}} = \pi R^2 L \rho$

We invoke Gauss's law and solve for the electric field strength.

$$\Phi_E = \frac{q_{\text{encl}}}{\epsilon_0} \implies 2\pi r L E = \frac{\pi R^2 L \rho}{\epsilon_0}$$

$$\implies \boxed{E = \frac{R^2 \rho}{2 r \epsilon_0}} \quad (1.134)$$

Again, the direction of the electric field is radially outward from the axis of the cylinder. So, within the cylinder, the field is directly proportional to the distance from the axis. Outside the cylinder, the field is the same as for the charged wire of infinite length. Fig.1.214 shows a plot of E vs. r.
Making sense of the result: Inside the charged cylinder, the electric field increases linearly with distance from the axis. At the exact centre of the cylinder, the electric field is zero. Outside the charged cylinder, the electric field is proportional to $1/r$, approaching zero at infinite distance. When $r = R$, we require

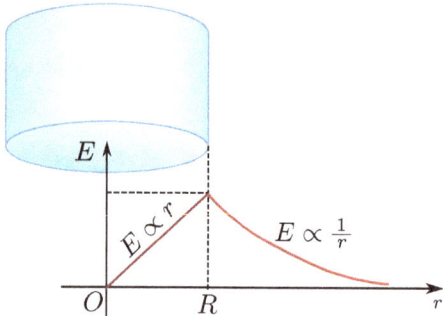

Figure 1.214: A graph of the electric field E vs. r for a uniform volume charge density throughout an infinitely long cylinder of radius R. In the graph, E is positive when $\vec{E}$ is directed outward.

that the results from the two parts be consistent, which they are. Note that this solution is for the case of a non-conducting cylinder with the charge uniformly distributed.

1.33.3 Spherical Symmetry

The following examples are based on the situations in which spherical symmetry exists.

EXAMPLE 89. Field of a Charged Conducting Sphere: We place positive charge q on a solid conducting sphere (or spherical shell) with radius R (Fig.1.215). Find $\vec{E}$ at any point inside or outside the sphere.

APPROACH For a conducting sphere, all the charge lies on its surface. So, the system has spherical symmetry. To take advantage of the symmetry, we draw a spherical Gaussian surface of radius r centred on the conductor. To calculate the field outside the conductor, we take r to be greater than the conductor's radius R; to calculate the field inside, we take r to be less than R. In either case, the point where we want to calculate $\vec{E}$ lies on the Gaussian surface. Now, we use the mathematical expression of Gauss's law (Eq.1.135) to calculate the electric field E at the required position-

$$\Phi_E = \oint \vec{E} \cdot d\vec{A} = \frac{Q_{\text{encl}}}{\epsilon_0} \quad \text{(Gauss's Law)} \quad (1.135)$$

SOLUTION (i) For a point outside the conducting sphere

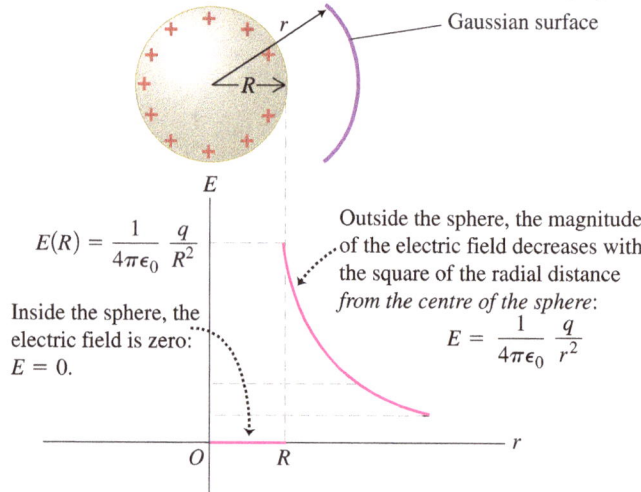

Figure 1.215: Calculating the electric field of a conducting sphere with positive charge q. Outside the sphere, the field is the same as if all of the charge were concentrated at the centre of the sphere.

($r \geq R$): In this case, the entire conductor is within the Gaussian surface, so the enclosed charge $Q_{\text{encl}} = q$

By symmetry, we conclude that the field $\vec{E}$ can only be radially outward, outside the sphere. Furthermore, for a given value of

1.33. APPLICATIONS OF GAUSS'S LAW

r, $\vec{E}$ has the same magnitude everywhere. Therefore, Eq.(1.135) can be written as

$$\oint E\, dA = \frac{q}{\epsilon_0}$$

Since, the field at each point on Gaussian surface has same magnitude, so E can be taken out of integral, so-

$$E \oint dA = \frac{q}{\epsilon_0} \implies E(4\pi r^2) = \frac{q}{\epsilon_0}$$

$$\implies \boxed{E = \frac{1}{4\pi\epsilon_0}\frac{q}{r^2} = k\frac{q}{r^2}} \quad \text{(For } r > R\text{)} \quad (1.136)$$

This is just the inverse-square-law field for a point charge q concentrated at the centre of the sphere.
Just outside the surface of the sphere, where $r = R$, Eq. (1.136) takes the form-

$$\boxed{E = \frac{1}{4\pi\epsilon_0}\frac{q}{R^2} = k\frac{q}{R^2}} \quad (1.137)$$

Note: *A point charge q can be considered to be the limiting case of a small spherical conductor whose radius tends to zero and electric field to infinity.*

Comment: Flux can be positive or negative: Note that we have chosen the charge q to be positive. If the charge is negative, the electric field is radially inward instead of radially outward, and the electric flux through the Gaussian surface is negative. The electric field magnitudes outside and at the surface of the sphere are given by the same expressions as above, except that q denotes the magnitude (absolute value) of the charge.

(ii) For a point inside the conducting sphere ($r < R$): In this case Gaussian surface does not include any charge, i.e. $Q_{encl} = 0$, therefore Eq.(1.135) gives-

$$\oint E\, dA = 0 \implies E \oint dA = 0 \implies E(4\pi r^2) = 0$$

$$\implies \boxed{E = 0} \quad (1.138)$$

So, the electric field inside the conductor is zero.

Comment: We already knew that $\vec{E} = 0$ inside the conductor, as it must be inside any solid conductor when the charges are at rest. Figure 1.215 shows E as a function of the distance r from the centre of the sphere. Note that in the limit as $R \to 0$, the sphere becomes a point charge; there is then only an 'outside', and the field is everywhere given by $E = q/4\pi\epsilon_0 r^2$. Thus we have again deduced Coulomb's law from Gauss's law.

EXAMPLE 90. A thin spherical shell of radius R has a total positive charge Q distributed uniformly over its surface. Find the electric field inside and outside the shell.

APPROACH Here, we use the method described in the last problem. By symmetry, if any field exists inside the shell, it must be radial. For any point outside or on the surface of the conducting shell, it behaves like a solid conducting sphere. So, construct spherical Gaussian surfaces for both cases and apply Gauss's law-

$$\Phi_E = \oint \vec{E} \cdot d\vec{A} = \frac{Q_{encl}}{\epsilon_0} \quad (1.139)$$

and solve above equation for E in each case.

SOLUTION (i) Inside the shell (r<R): Let us construct a spherical Gaussian surface of radius $r < R$ concentric with the shell (Fig.1.216). Since, there is no enclosed charge within the Gaussian surface, i.e., $Q_{encl} = 0$, therefore, Eq.1.139 gives-

$$\oint \vec{E} \cdot d\vec{A} = E(4\pi r^2) = 0$$

$$\implies \boxed{E = 0} \quad (1.140)$$

So, we conclude that there is no electric field inside a uniformly

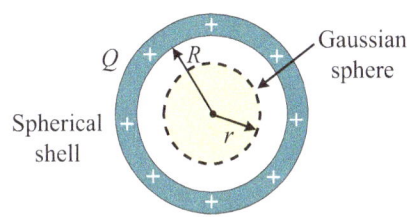

Figure 1.216: Cross sectional view of shell with Gaussian surface inside it.

charged spherical shell.

(ii) Outside the shell ($r \geq R$): Outside the shell, we construct a spherical Gaussian surface of radius $r > R$ concentric with the charged shell as shown in Fig.1.217. Symmetry suggests that E = constant on that surface and $\vec{E}$ is parallel to $d\vec{A}$, i.e. $\oint \vec{E} \cdot d\vec{A} = E(4\pi r^2)$. Since the net charge Q_{encl} inside the Gaussian surface is equal to the total charge Q on the shell, the shell is equivalent to a point charge located at the centre. Therefore, Eq.1.139 gives-

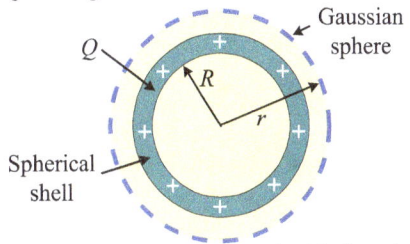

Figure 1.217: Cross sectional view of shell with Gaussian surface outside it.

$$\boxed{E = \frac{1}{4\pi\epsilon_o}\frac{Q}{r^2} = k\frac{Q}{r^2}} \quad (r > R) \quad (1.141)$$

(iii) At the surface of the shell ($r = R$): In this case, Eq.(1.141) takes the form,

$$\boxed{E = \frac{1}{4\pi\epsilon_o}\frac{Q}{r^2} = k\frac{Q}{R^2}} \quad (1.142)$$

Electric field E vs r graph will be similar to the graph obtained in the case of solid conducting sphere (Fig.1.215).

EXAMPLE 91. A solid sphere of radius R has a uniform volume charge density ρ and carries a total positive charge Q. Find and sketch the electric field at any distance r away from the sphere's center.

APPROACH The whole space can be divided into two parts:(i) $0 \leq r \leq R$ and (ii) $r \geq R$. Observe the symmetry in each case and find corresponding Q_{encl}. Apply Gauss's law and simplify for electric field E.

SOLUTION (1) For $0 \leq r \leq R$: When dealing with a spherically symmetric charge distribution, we chose a spherical Gaussian surface of radius $r < R$ concentric with the charged sphere as shown in Fig.1.218.
By symmetry, the magnitude of the electric field is constant

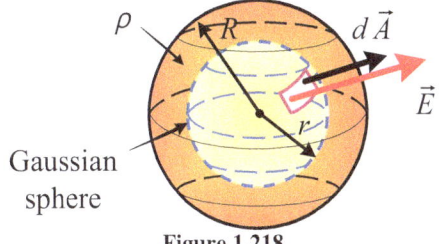

Figure 1.218

everywhere on the spherical Gaussian surface and normal to the surface at any point, i.e. $\vec{E} // d\vec{A}$. Thus:

$$\oint \vec{E} \cdot d\vec{A} = \oint E\,dA = E \oint dA = E\left(4\pi r^2\right)$$

It is important to note that the volume, say V', of the Gaussian sphere encloses a net charge $Q_{encl} = \rho V'$; that is:

$$Q_{encl} = \rho V' = \rho \left(\frac{4}{3}\pi r^3\right)$$

We can now use Gauss's law to find electric field as follows:

$$\Phi_E = \oint \vec{E} \cdot d\vec{A} = \frac{Q_{encl}}{\epsilon_\circ} \Rightarrow E\left(4\pi r^2\right) = \frac{\rho\left(\frac{4}{3}\pi r^3\right)}{\epsilon_\circ}$$

Then:
$$\boxed{E = \frac{\rho}{3\epsilon_\circ} r \quad (0 \leq r \leq R)} \quad (1.143)$$

In vector form,
$$\boxed{\vec{E} = \frac{\rho}{3\epsilon_\circ}\vec{r} \quad (0 \leq r \leq R)} \quad (1.144)$$

Using the definition $\rho = Q/\left(\frac{4}{3}\pi R^3\right)$ and $k = 1/(4\pi\epsilon_\circ)$, we get:

$$E = \frac{Q}{\frac{4}{3}\pi R^3 (3\epsilon_0)} = \frac{Q}{4\pi\epsilon_\circ R^3} r$$

$$\Rightarrow \boxed{E = k\frac{Q}{R^3} r \quad (0 \leq r \leq R)} \quad (1.145)$$

(2) For $r \geq R$ Again, because the charge distribution is spherically symmetric, we can construct a Gaussian sphere of radius $r > R$ concentric with the charged sphere, as shown in Fig.1.219.

Just as when $r < R$, $\oint \vec{E} \cdot d\vec{A} = E\left(4\pi r^2\right)$, but in this case,

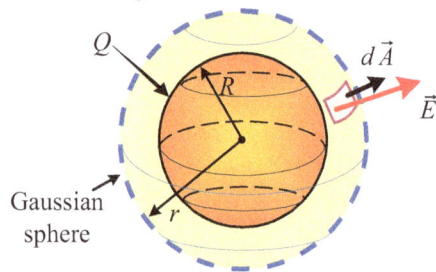

Figure 1.219

$Q_{encl} = Q$. Therefore, by Gauss's law:

$$\Phi_E = \oint \vec{E} \cdot d\vec{A} = \frac{Q_{encl}}{\epsilon_\circ} \Rightarrow E\left(4\pi r^2\right) = \frac{Q}{\epsilon_\circ}$$

i.e., $\boxed{E = \frac{1}{4\pi\epsilon_\circ}\frac{Q}{r^2} = k\frac{Q}{r^2} \quad (r \geq R)} \quad (1.146)$

Note that this is identical to the result obtained for a point charge. Therefore, we conclude that the electric field outside any uniformly charged sphere is equivalent to that of a point charge located at the centre of the sphere. At $r = R$, the two cases give identical results $E = kQ/R^2$. A plot of E versus r is shown in Fig.1.220. This figure shows the continuation of E and it is maximum at $r = R$.

EXAMPLE 92. A Uniformly charged solid non-conducting sphere of uniform volume charge density ρ and radius R has a concentric spherical cavity of radius r (Fig.1.221). Find out electric field intensity at
(i) Point A, (ii) Point B, (iii) Point C, (iv) Center of the sphere.

SOLUTION Method I:
(i) For point A: Since point A lies inside the cavity, therefore for concentric Gaussian surface through A, $Q_{encl} = 0$, so electric field intensity due to all shells will be zero.

$$\vec{E_A} = 0$$

(ii) For point B: For concentric Gaussian surface through B,

Figure 1.220

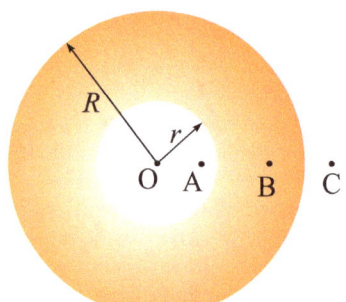

Figure 1.221

$Q_{encl} = \frac{4}{3}\pi\left((OB)^3 - r^3\right)\rho$, therefore by Gauss's theorem

$$\oint \vec{E} \cdot d\vec{A} = \frac{Q_{encl}}{\epsilon_0}$$

$$\Rightarrow E(4\pi(OB)^2) = \frac{\frac{4}{3}\pi\left((OB)^3 - r^3\right)\rho}{\epsilon_0}$$

$$\Rightarrow E = \frac{1}{4\pi\epsilon_0}\left[\frac{\frac{4}{3}\pi\left((OB)^3 - r^3\right)\rho}{(OB)^2}\right] = k\left[\frac{\frac{4}{3}\pi\left((OB)^3 - r^3\right)\rho}{(OB)^2}\right]$$

This electric field is radially away for $+ve$ charge density, therefore in vector form it can be written as

$$\vec{E} = k\left[\frac{\frac{4}{3}\pi\left((OB)^3 - r^3\right)\rho}{(OB)^2}\right]\widehat{(OB)}$$

or $\vec{E} = k\left[\frac{\frac{4}{3}\pi\left((OB)^3 - r^3\right)\rho}{(OB)^2}\right]\frac{\overrightarrow{OB}}{OB}$

So, $\vec{E}_B = \frac{k\frac{4}{3}\pi\left(OB^3 - r^3\right)\rho}{OB^3}\overrightarrow{OB} = \frac{\rho}{3\epsilon_0}\frac{[OB^3 - r^3]}{OB^3}\overrightarrow{OB}$

(iii) For point C: In this case, $Q_{encl} = \frac{4}{3}\pi\left[R^3 - r^3\right]\rho$

Now, simplifying like part (ii), we get

$$\vec{E}_C = \frac{k\frac{4}{3}\pi\left(R^3 - r^3\right)\rho}{OC^3}\overrightarrow{OC} = \frac{\rho}{3\epsilon_0}\frac{[R^3 - r^3]}{OC^3}\overrightarrow{OC}$$

Method: II We can consider that the spherical cavity is filled with charge density ρ and also $-\rho$, thereby making net charge density zero after combining. We can consider two concentric solid spheres: One of radius R and charge density ρ and other of radius r and charge density $-\rho$ [Fig.1.222].

From Eq.1.144, at A, $\vec{E}_\rho = \frac{\rho(\overrightarrow{OA})}{3\epsilon_\circ}$ and $\vec{E}_{-\rho} = \frac{-\rho(\overrightarrow{OA})}{3\epsilon_\circ}$

Applying superposition principle:

(i) $\vec{E}_A = \vec{E}_\rho + \vec{E}_{-\rho} = \frac{\rho(\overrightarrow{OA})}{3\epsilon_0} + \frac{\left[-\rho(\overrightarrow{OA})\right]}{3\epsilon_0} = 0$

(ii) Point B is inside the sphere of density ρ and outside the sphere of density $-\rho$. $\vec{E}_B = \vec{E}_\rho + \vec{E}_{-\rho}$

1.34. CHECK POINT 9

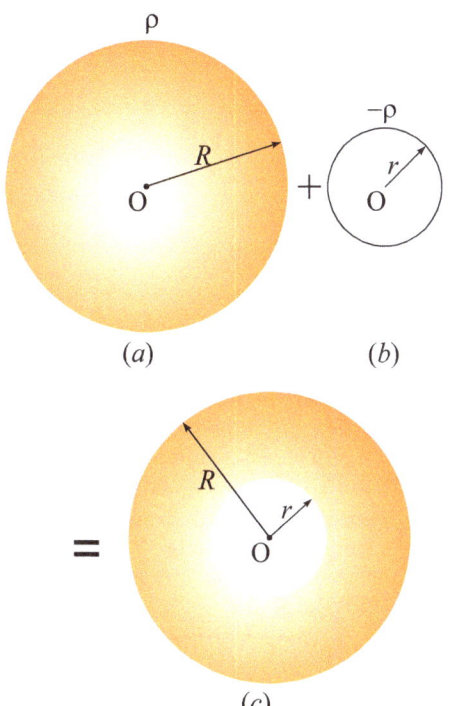

(a)　　　(b)

(c)

Figure 1.222

here, $\vec{E}_\rho = \dfrac{\rho(\overrightarrow{OB})}{3\epsilon_0}$ and $\vec{E}_{-\rho} = \dfrac{kQ_{encl}}{(OB)^2}\overrightarrow{OB}$

Therefore, $\vec{E}_B = \dfrac{\rho(\overrightarrow{OB})}{3\epsilon_0} + \dfrac{k\left[\frac{4}{3}\pi r^3(-\rho)\right]}{(OB)^3}\overrightarrow{OB}$

$= \left[\dfrac{\rho}{3\epsilon_0} - \dfrac{r^3\rho}{3\epsilon_0(OB)^3}\right]\overrightarrow{OB} = \dfrac{\rho}{3\epsilon_0}\left[1 - \dfrac{r^3}{OB^3}\right]\overrightarrow{OB}$

(iii) $\vec{E}_C = \vec{E}_\rho + \vec{E}_{-\rho} = \dfrac{K\left(\frac{4}{3}\pi R^3\rho\right)}{OC^3}\overrightarrow{OC} + \dfrac{K\left(\frac{4}{3}\pi r^3(-\rho)\right)}{OC^3}\overrightarrow{OC}$

$= \dfrac{\rho}{3\epsilon_0(OC)^3}\left[R^3 - r^3\right]\overrightarrow{OC}$

(iv) $\vec{E}_O = \vec{E}_\rho + \vec{E}_{-\rho} = 0 + 0 = 0$

EXAMPLE 93. Solve Example 92, if cavity is not concentric and centered at point P [Fig.1.223].

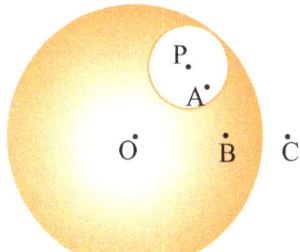

Figure 1.223

SOLUTION Again assume ρ as the charge densities of solid sphere and $(-\rho)$ as that in cavity (similar to the previous example):

(i) $\vec{E}_A = \vec{E}_\rho + \vec{E}_{-\rho}$

$\Rightarrow \vec{E}_A = \dfrac{\rho[\overrightarrow{OA}]}{3\epsilon_0} + \dfrac{-\rho[\overrightarrow{PA}]}{3\epsilon_0}$

$= \dfrac{\rho}{3\epsilon_0}\left[\overrightarrow{OA} - \overrightarrow{PA}\right] = \dfrac{\rho}{3\epsilon_0}\left[\overrightarrow{OP}\right]$

Note: From above it is clear that the electric field intensity at point A is independent of position of point A inside the cavity. Also the electric field is along the line joining the centres of the sphere and the spherical cavity.

(ii) $\vec{E}_B = \vec{E}_\rho + \vec{E}_{-\rho} = \dfrac{\rho(\overrightarrow{OB})}{3\epsilon_0} + \dfrac{k\left[\frac{4}{3}\pi r^3(-\rho)\right]}{(PB)^3}\overrightarrow{PB}$

(iii) $\vec{E}_C = \vec{E}_\rho + \vec{E}_{-\rho} = \dfrac{k\left(\frac{4}{3}\pi R^3\rho\right)}{OC^3}\overrightarrow{OC} + \dfrac{k\left(\frac{4}{3}\pi r^3(-\rho)\right)}{PC^3}\overrightarrow{PC}$

(iv) $\vec{E}_O = \vec{E}_\rho + \vec{E}_{-\rho} = 0 + \dfrac{k\left[\frac{4}{3}\pi r^3(-\rho)\right]}{[PO]^3}\overrightarrow{PO}$

EXAMPLE 94. A non-conducting solid sphere has volume charge density that varies as $\rho = \rho_0 r$, where ρ_0 is a constant and r is distance from centre. Find out electric field intensities at following positions.
(i) $r < R$　　(ii) $r \geq R$

SOLUTION Method I: (i) For $r<R$: The sphere can be considered to be made of large number of spherical shells [Fig.1.224]. Each shell has uniform charge density. So, the previous results of the spherical shell can be used. Consider a shell of radius x and thickness dx as an element. Charge on shell $dq = (4\pi x^2 dx)\rho_0 x$. Therefore, the electric field intensity at point P due to shell,

$$dE = \dfrac{k\,dq}{x^2}$$

Since all the shells will have electric field in same direction

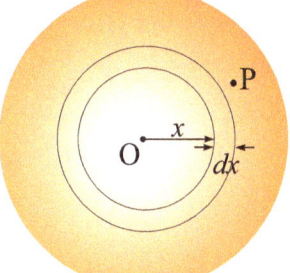

Figure 1.224

therefore, $E = \int_0^E dE = \int_0^r dE + \int_r^R dE$

Due to shells which lie between region $r < x \leq R$, electric field at point P will be zero.

$\vec{E} = \int_0^r \dfrac{k\,dq}{r^2} + 0 = \int_0^r \dfrac{k.4\pi x^2 dx \rho_0 x}{r^2} = \dfrac{4\pi k\rho_0}{r^2}\left[\dfrac{x^4}{4}\right]_0^r = \dfrac{\rho_0 r^2}{4\epsilon_0}\hat{r}$

(ii) $r \geq R$

$E = \int_0^R dE = \int_0^R \dfrac{k.4\pi x^2 dx\rho_0 x}{r^2} = \dfrac{\rho_0 R^4}{4\epsilon_0 r^2}\hat{r}$

1.34 Check Point 9

1. ●●A point charge Q is located just above the centre of the flat face of a hemisphere of radius R as shown in Figure 1.225. What is the electric flux (a) through the curved surface and (b) through the flat face?

Figure 1.225

2. ●●Fig.1.226(a) shows three charges. Draw these charges on your paper four times. Then draw two-dimensional cross sections of three-dimensional closed surfaces through which the electric flux is (a) $2q/\epsilon_0$, (b) q/ϵ_0, (c) 0, and (d) $5q/\epsilon_0$.

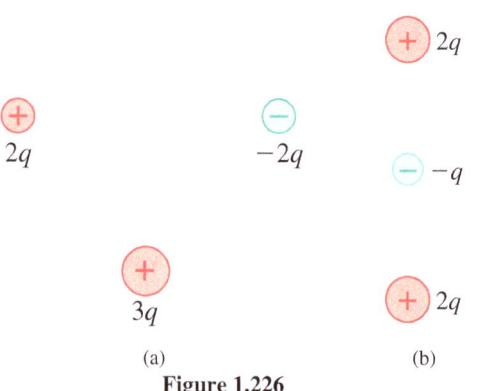

Figure 1.226

3. ••Fig.1.226(b) shows three charges. Draw these charges on your paper four times. Then draw two-dimensional cross sections of three-dimensional closed surfaces through which the electric flux is (a) $-q/\epsilon_0$, (b) q/ϵ_0, (c) $3q/\epsilon_0$, and (d) $4q/\epsilon_0$.

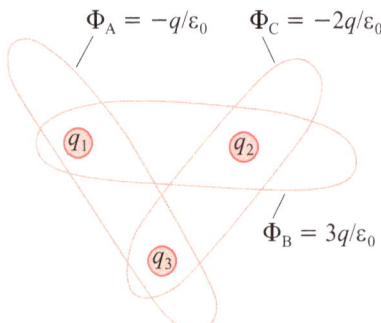

Figure 1.227

4. ••Fig.1.227 shows three Gaussian surfaces and the electric flux through each. What are the three charges q_1, q_2, and q_3?

5. ••As shown in Fig.1.228 a closed surface intersects a spherical conductor. What will be the nature of the electric flux coming out of the closed surface, if we place a negative charge at point P?

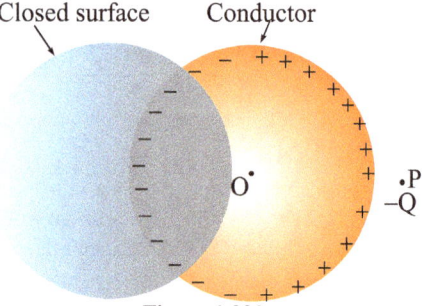

Figure 1.228

6. ••Find the electric field due to an infinitely long cylindrical charge distribution of radius R and having linear charge density λ at a distance half of the radius from its axis.

7. ••Five charges q_1, q_2, q_3, q_4, and q_5 are fixed at their positions as shown in Figure 1.229. S is Gaussian surface. The Gauss's law is given by $\int \vec{E} \cdot \vec{ds} = \frac{q}{\epsilon_0}$. Which of the following statements is correct?

(A) $\vec{E}$ on the LHS of the above equation will have a contribution from q_1, q_5, and q_3 while q on the RHS will have a contribution from q_2 and q_4 only.

(B) $\vec{E}$ on the LHS of the above equation will have a contribution from all charges while q on the RHS will have a

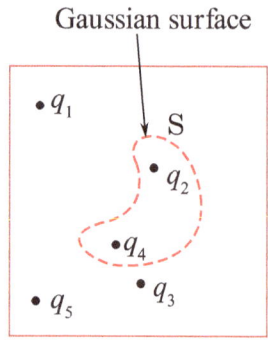

Figure 1.229

contribution from q_2 and q_4 only.

(C) $\vec{E}$ on the LHS of the above equation will have a contribution from all charges while q on the RHS will have a contribution from q_1, q_3, and q_5 only.

(D) Both $\vec{E}$ on the LHS and q on the RHS will have a contribution from q_2 and q_4 only.

8. ••What is the net electric flux through the torus (i.e., doughnut shape) of Fig.1.230

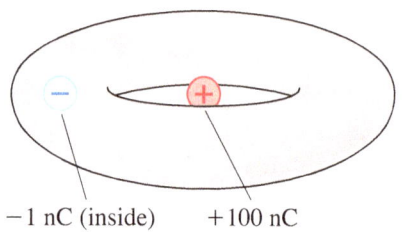

Figure 1.230

9. ••What is the net electric flux through the cylinder of Figure 1.231?

Figure 1.231

10. ••The net electric flux through an octahedron is -1000 Nm²/C. How much charge is enclosed within the octahedron?

11. ••55.3 million excess electrons are inside a closed surface. What is the net electric flux through the surface?

12. ••What is the electric flux through each face of a cube of side a (Fig.1.232) if a point charge of magnitude q is at one of its corners?

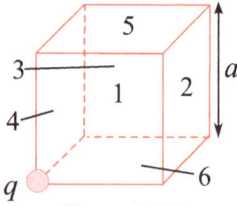

Figure 1.232

Multiple Choice Questions

13. ••What is the flux through a cube of side a if a point charge of q is at one of its corners?

(A) $2q/\epsilon_0$ (B) $q/8\epsilon_0$

(C) q/ϵ_0 (D) $q/2\epsilon_0 6a^2$

1.35. CONDUCTORS IN ELECTROSTATIC EQUILIBRIUM

14. •A charge Q is enclosed by a Gaussian spherical surface of radius R. If the radius is doubled, then the outward electric flux will -
    (A) increase four times   (B) be reduced to half
    (C) remain the same   (D) be doubled

15. ••A hollow cylinder has a charge q coulomb within it[Fig.1.233]. If ϕ is the electric flux in units of volt-meter associated with the curved surface B, the flux linked with the plane surface A in units of V.m will be-
    (A) $\frac{q}{2\epsilon_0}$ (B) $\frac{\phi}{3}$
    (C) $\frac{q}{\epsilon_0} - \phi$ (D) $\frac{1}{2}\left(\frac{q}{\epsilon_0} - \phi\right)$

Figure 1.233

16. •A charge q is located at the centre of a cube. The electric flux through any face is
    (A) $\frac{2\pi q}{6(4\pi\epsilon_0)}$ (B) $\frac{4\pi q}{6(4\pi\epsilon_0)}$
    (C) $\frac{\pi q}{6(4\pi\epsilon_0)}$ (D) $\frac{q}{6(4\pi\epsilon_0)}$

17. •A charge Q μC is placed at the centre of a cube, the flux coming out from each face will be
    (A) $\frac{Q}{6\epsilon_0} \times 10^{-6}$ (B) $\frac{Q}{6\epsilon_0} \times 10^{-3}$
    (C) $\frac{Q}{24\epsilon_0}$ (D) $\frac{Q}{8\epsilon_0}$

18. •A point charge $+q$ is placed at the centre of a cube of side l. The electric flux emerging from the cube is
    (A) $6ql^2/\epsilon_0$ (B) $q/6l^2\epsilon_0$
    (C) zero   (D) q/ϵ_0

19. ••A hollow metal sphere of radius R is uniformly charged. The electric field due to the sphere at a distance r from the centre
    (A) decreases as r increases for $r < R$ and for $r > R$
    (B) increases as r increases for $r < R$ and for $r > R$
    (C) zero as r increases for $r < R$, decreases as r increases for $r > R$
    (D) zero as r increases for $r < R$, increases as r increases for $r > R$

20. ••Two parallel infinite line charges with linear charge densities $+\lambda$C/m and $-\lambda$C/m are placed at a distance of $2R$ in free space. The electric field midway between the two line charges?
    (A) $\frac{\lambda}{2\pi\epsilon_0 R}$ N/C   (B) zero
    (C) $\frac{2\lambda}{\pi\epsilon_0 R}$ N/C   (D) $\frac{\lambda}{\pi\epsilon_0 R}$ N/C

21. ••The electric field at a distance $3R/2$ from the centre of a charged conducting spherical shell of radius R is E. The electric field at a distance $R/2$ from the centre of the sphere is
    (A) 0   (B) E (C) $E/2$ (D) $E/3$

22. ••A hollow insulated conducting sphere is given a positive charge of 10μC. What will be the electric field at the centre of the sphere if its radius is 2 meter?
    (A) 20 μCm^{-2} (B) 5 μCm^{-2}
    (C) zero   (D) 8 mCm^{-2}

23. ••Figure1.234 shows a uniformly charged hemisphere of radius R. It has volume charge density ρ. If the electric field at a point A at $2R$ distance above its centre is E then the electric field at point B which is $2R$ below its centre, is?
    (A) $\frac{\rho R}{6\epsilon_0} + E$ (B) $\frac{\rho R}{12\epsilon_0} - E$
    (C) $\frac{-\rho R}{6\epsilon_0} + E$ (D) $\frac{\rho R}{24\epsilon_0} + E$

Figure 1.234

24. ••For a spherically symmetrical charge distribution, electric field at a distance r from the centre of sphere is $\vec{E} = kr^7 \hat{r}$, where k is a constant. What will be the volume charge density at a distance r from the centre of sphere?
    (A) $\rho = 9k\epsilon_0 r^6$ (B) $\rho = 5k\epsilon_0 r^3$
    (C) $\rho = 3k\epsilon_0 r^4$ (D) $\rho = 9k\epsilon_0 r^0$

25. ••A long string with a charge of λ per unit length passes through an imaginary cube of edge a. The maximum possible flux of the electric field through the cube may be-
    (A) $\lambda a/\epsilon_o$ (B) $\sqrt{2}\lambda a/\epsilon_o$
    (C) $6\lambda a^2/\epsilon_o$ (D) $\sqrt{3}\lambda a/\epsilon_o$

26. ••The total electric flux, leaving spherical surface of radius 1 cm, and surrounding an electric dipole is-
    (A) q/ϵ_0 (B) zero
    (C) $2q/\epsilon_0$ (D) $8\pi r^2 q/\epsilon_0$

27. ••An electric dipole is placed inside a conducting shell. Mark the correct statement(s)
    (A) the flux of the electric field through the shell is zero
    (B) the electric field is zero at every point on the shell
    (C) the electric field is not zero anywhere on the shell
    (D) the electric field is zero on a circle on the shell.

1.35 Conductors in Electrostatic Equilibrium

1.35.1 Field in a Substance

1.35.1.1 Micro- and Macroscopic Fields

The real electric field in any substance (which is called the microscopic field) varies abruptly both in space and in time. It is different at different points of atoms and in the interstices. In order to find the intensity $\vec{E}$ of a real field at a certain point at a given instant, we should vectorially sum up the intensities of the fields of all individual charged particles of the substance, viz. electrons and nuclei. The solution of this problem is obviously not feasible. In any case, the result would be so complicated that it would be impossible to use it. Moreover, the knowledge of this field is not required for the solution of macroscopic problems. In many cases it is sufficient to have a simpler and rougher description which we shall be using henceforth.

Under the electric field $\vec{E}$ in a substance (which is called the macroscopic field) we shall understand the microscopic field averaged over space (in this case time averaging is worthless). This averaging is performed over what is called a *physically infinitesimal volume*, viz. the volume containing a large number of atoms and having the dimensions that are many times smaller than the distances over which the macroscopic field noticeably changes. The averaging over such volumes smoothens all irregular and rapidly varying fluctuations of the microscopic field over the distances of the order of atomic ones, but retains smooth variations of the macroscopic field over macroscopic distances. Thus, the field in the substance is

$$E = E_{\text{macro}} = \langle E_{\text{micro}} \rangle \qquad (1.147)$$

1.35.1.2 The Influence of a Substance on a Field

If any substance is introduced into an electric field, the positive and negative charges (nuclei and electrons)* are displaced, which in turn leads to a partial separation of these charges. In certain regions of the substance, uncompensated charges of different signs appear. This phenomenon is called the electrostatic induction, while the charges appearing as a result of separation are called *induced charges*.

Induced charges create an additional electric field which in combination with the initial (external) field forms the resultant field. Knowing the external field and the distribution of induced charges, we can forget about the presence of the substance itself while calculating the resultant field, since the role of the substance has already been taken into account with the help of induced charges. Thus, the resultant field in the presence of a substance is determined simply as the superposition of the external field and the field of induced charges.

1.35.2 Fields Inside and Outside a Conductor

1.35.2.1 Inside a Conductor $E = 0$

Let us place a metallic conductor into an external electrostatic field or impart a certain charge to it. In both cases, the electric field will act on all the charges of the conductor, and as a result all the negative charges (free electrons) will be displaced in the direction against the field. This displacement (current) will continue until (this practically takes a small fraction of a second) a certain charge distribution sets in, at which the electric field at all the points inside the conductor vanishes. Thus, in the static case the electric field inside a conductor is absent ($E = 0$).

Further, since $E = 0$ everywhere in the conductor, the density of excess (uncompensated) charges inside the conductor is also equal to zero at all points ($\rho = 0$). This can be easily explained with the help of the Gauss theorem. Indeed, since inside the conductor $E = 0$, the flux of E through any closed surface inside the conductor is also equal to zero. And this means that there are no excess charges inside the conductor.

Excess charges appear only on the conductor surface with a certain density σ which is generally different for different points of the surface. *It should be noted that the excess surface charge is located in a very thin surface layer (whose thickness amounts to one or two interatomic distances)*. When there is no net motion of electrons within the conductor, the conductor is said to be in electrostatic equilibrium. In electrostatic equilibrium, all conductors show following four properties:

1. The electric field is zero everywhere inside the conductor.
2. If an isolated conductor carries a charge, the charge resides on its surface.
3. The electric field just outside a charged conductor is perpendicular to the surface of the conductor and has a magnitude σ/ϵ_0, where σ is the surface charge density at that point.
4. On an irregularly shaped conductor, the surface charge density is greatest at locations where the radius of curvature of the surface is smallest.

Explanation of Above Properties on the Basis of Gauss's Law:
Here, we verify only the first three properties for the conductors. The fourth property is presented here so that we have a complete list of properties for conductors in electrostatic equilibrium, but cannot be verified until the next Chapter "Electric Potential".

---

*In conductors, only free electron can move not nuclei. So there will be a relative displacement between nuclei and electrons due to motion of free electrons only.

Property 1. We can elaborate more about the first property by considering the conducting slab in electrostatic equilibrium on the left of Fig.1.235, where the free electrons are uniformly distributed throughout the slab, i.e. $\vec{E}_{\text{int}} = 0$. When we place the slab in an external electric field $\vec{E}_{\text{ext}}$ as in the right part of Fig.1.235, the free electrons respond to the electric force $-e\vec{E}$ by moving-in the direction, opposite the field. In Fig.1.235, the direction of applied external electric field is towards right side, therefore electrons move to the left. In time, more negative and positive charges accumulate on the left and right surfaces, respectively. The resulting charge separation gives rise to an increasing electric field $\vec{E}_{\text{int}}$ within the conductor that's opposite to the applied field. As more charge moves, this internal field becomes stronger and after awhile, $\vec{E}_{\text{int}}$ will compensate $\vec{E}_{\text{ext}}$ resulting in a zero net electric field inside the conductor, i.e. $\vec{E}_{\text{net}} = \vec{E}_{\text{ext}} - \vec{E}_{\text{int}} = 0$. The time to reach this new electrostatic equilibrium is of the order 10^{-6} s. Therefore, *The electric field is zero inside a conductor in elec-*

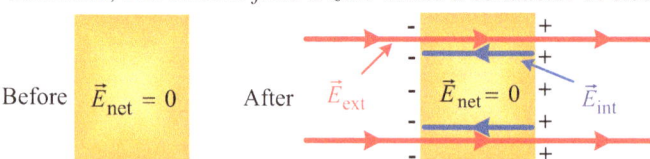

Figure 1.235: An external electric field $\vec{E}_{\text{ext}}$ creates an internal electric field $\vec{E}_{\text{int}}$ in the conductor such that the net electric field $\vec{E}_{\text{net}}$ is zero

trostatic equilibrium.

It could not be otherwise: Since a conductor contains free charges, the presence of any internal electric field would result in bulk charge motion, and we wouldn't have equilibrium. This result doesn't depend on the size or shape of the conductor, the magnitude or direction of the applied field, or even the nature of the material as long as it's a conductor. This is a macroscopic view; it considers only average fields within the material. At the atomic and molecular level, there are still strong electric fields near individual electrons and positive ions. But the average field, taken over larger distances, is zero inside a conductor in electrostatic equilibrium.

Explanation of Property 2. Figure 1.236a shows a cross-sectional view of a conductor with a Gaussian surface drawn just below the material surface. In equilibrium there's no electric field inside the conductor, and thus the field is zero everywhere on the Gaussian surface. Therefore, from Gauss's law, we have-

$$\phi_E = \oint \vec{E} \cdot d\vec{A} = \frac{Q_{\text{encl}}}{\epsilon_0}$$

$$\implies \oint 0 \, d\vec{A} = \frac{Q_{\text{encl}}}{\epsilon_0}$$

$$\implies 0 = \frac{Q_{\text{encl}}}{\epsilon_0} \implies Q_{\text{encl}} = 0$$

Thus, net electric flux ϕ, through the Gaussian surface and hence net charge enclosed within the Gaussian surface is also zero. This is true no matter where the Gaussian surface is as long as it's inside the conductor. We can move it arbitrarily close to the conductor surface and it still encloses no net charge. If there is a net charge q_C (say) on the conductor, it lies outside the Gaussian surface, and therefore we conclude: If a conductor in electrostatic equilibrium carries a net charge, that charge must reside on the conductor surface.

A Cavity Inside the Conductor: Fig.1.236(b) shows a cross-sectional view of same charged conductor with a cavity that is totally within the conductor. Can there be charge on this interior surface? To find out, we place a Gaussian surface around the cavity, infinitesimally close but entirely within the conductor. Because $\vec{E} = 0$ inside the conductor, there can be no flux through

1.35. CONDUCTORS IN ELECTROSTATIC EQUILIBRIUM

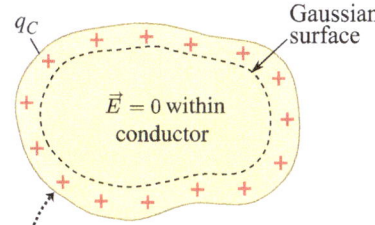

The charge q_C resides entirely on the surface of the conductor.

(a)

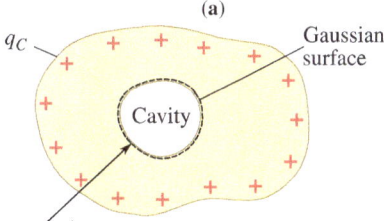

Because $\vec{E} = 0$ at all points within the conductor, the electric field at all points on the Gaussian surface must be zero.

(b)

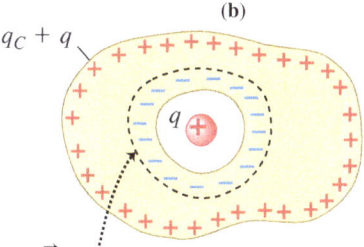

For $\vec{E}$ to be zero at all points on the Gaussian surface, the surface of the cavity must have a total charge $-q$.

(c)

Figure 1.236: (a) A cross sectional view of a solid conductor of arbitrary shape. The broken line represents a gaussian surface that can be as close to the surface of the conductor as we wish (b) The same conductor with an internal cavity (c) An isolated charge q placed in the cavity

this new Gaussian surface. Therefore, from Gauss' law, we have-

$$\phi_E = \oint \vec{E} \cdot d\vec{A} = \frac{Q_{encl}}{\epsilon_0}$$

$$\implies \oint 0\, d\vec{A} = \frac{Q_{encl}}{\epsilon_0} \implies 0 = \frac{Q_{encl}}{\epsilon_0}$$

$$\implies Q_{encl} = 0$$

So, we conclude that there is no net charge on the cavity walls; all the excess charge remains on the outer surface of the conductor, as in Fig.1.236b.

Suppose that, by some process, the excess charges could be "frozen" into position on the conductor surface of Fig.1.236b, perhaps by embedding them in a thin plastic coating, and suppose that the conductor could then be removed completely. This is equivalent to enlarging the cavity of Fig.1.236b until it consumes the entire conductor, leaving only the charges. The electric field pattern would not change at all; it would remain zero inside the thin shell of charge and would remain unchanged for all external points. The electric field is set up by the charges and not by the conductor. The conductor simply provides a pathway so that the charges can change their positions.

Now, Suppose we place a small body with a charge q inside a cavity within a conductor (Fig.1.236(c)). The conductor is uncharged and is insulated from the charge q. Again $\vec{E} = 0$ everywhere on the Gaussian surface. If we assume that the charge on the cavity wall is x, then by Gausses law. we have-

$$\phi_E = \oint \vec{E} \cdot d\vec{A} = \frac{Q_{encl}}{\epsilon_0} \implies \oint 0\, d\vec{A} = \frac{Q_{encl}}{\epsilon_0}$$

$$\implies 0 = \frac{Q_{encl}}{\epsilon_0} \implies Q_{encl} = 0 \implies q + x = 0 \implies x = -q$$

So according to Gauss's law there must be a charge $-q$ distributed on the surface of the cavity, drawn there by the charge q inside the cavity. Let charge y appears on the outer surface of conductor, then by charge conservation principle, we have

$$x + y = 0 \implies y = -x \implies y = -(-q) = q$$

Therefore, we conclude that the charge $+q$ must appear on the outer surface of the conductor.

If the conductor originally had a charge q_C, then charge conservation principle gives-

$$x + y = q_C \implies y = q_C - x = q_C - (-q) = q_C + q$$

i.e., the total charge on the outer surface must be $q_C + q$ after the charge q is inserted into the cavity.

Explanation of Property 3. There can't be an electric field within a conductor in electrostatic equilibrium, but there may be a field right at the conductor surface (Fig.1.237). Such a field must be perpendicular to the surface; otherwise, charge would move along the surface in response to the field's parallel component, and we wouldn't have equilibrium.

Figure 1.237: The electric field at the surface of a charged conductor is perpendicular to the conductor surface.

1.35.2.2 The Field Near a Conductor Surface

Here we will show that the electric field intensity in the immediate vicinity of the surface of a conductor is connected with the local charge density at the conductor surface through a simple relation. This relation can be established with the help of the Gauss theorem.

Suppose that the region of the conductor surface we are in-

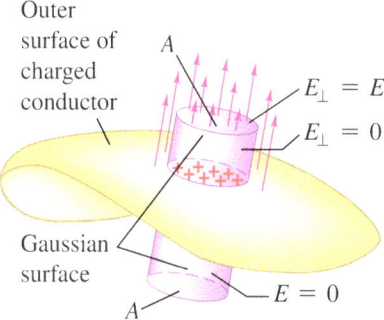

Figure 1.238: The field just outside a charged conductor is perpendicular to the surface, and its perpendicular component $E_\perp$ is equal to σ/ϵ_0.

terested in borders on a vacuum. The field lines are normal to the conductor surface. Hence for a closed surface we shall take a small cylinder and arrange it as is shown in Fig.1.238. Then the flux of $\vec{E}$ through this surface will be equal only to the flux through the "outer" endface of the cylinder. (the fluxes through the lateral surface and the inner endface are equal to zero). Thus we obtain $\vec{E} \cdot \vec{A} = E_\perp A = Q_{encl}/\epsilon_0 = \sigma \vec{A}/\epsilon_0$, where $E_\perp$ is the projection of vector $\vec{E}$ onto the outward normal (with respect to the conductor), A is the cross-sectional area of the cylinder and σ is the local surface charge density of the conductor. Cancelling both sides of this expression by A, we get

$$E_\perp = \frac{\sigma}{\epsilon_0} \quad (1.148)$$

If $\sigma > 0$, then $E_\perp > 0$, i.e., vector $\vec{E}$ is directed from the conductor surface (coincides in direction with the outward normal). If $\sigma < 0$, then $E_\perp < 0$, and vector $\vec{E}$ is directed towards the conductor surface.

Relation (1.148) may lead to the erroneous conclusion that the field $\vec{E}$ in the vicinity of a conductor depends only on the local charge density σ. This is not so. The intensity $\vec{E}$ is determined by all the charges of the system under consideration as well as the value of σ itself.

1.35.3 Mechanical Pressure (or Surface Density of Force) on the Surface of a Charged Conductor

Let us consider the case when a charged region of the surface of a conductor borders, is in a vacuum. It's surface charge density is σ. A small patch of charge on a conducting surface, is shown in Figure 1.239.

Total electric field anywhere outside the surface is -

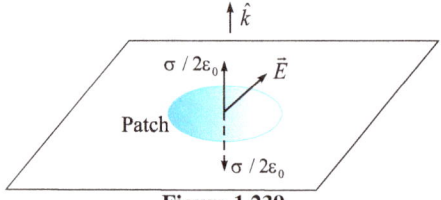

Figure 1.239

$$\vec{E} = \vec{E}_{patch} + \vec{E}' \quad (1.149)$$

where $\vec{E}_{patch}$ is the electric field due to charge on the patch, and $\vec{E}'$ is the electric field due all other charges. Since by Newton's third law, the patch cannot exert a force on itself, the force on the patch must come solely from $\vec{E}'$. Assuming the patch to be a flat surface, from Gauss's law, the electric field due to the patch is

$$\vec{E} = \begin{cases} \frac{+\sigma}{2\epsilon_0}\hat{k} & z > 0 \\ -\frac{\sigma}{2\epsilon_0}\hat{k} & z < 0 \end{cases} \quad (1.150)$$

By superposition principle, the electric field above the conducting surface is

$$\vec{E}_{above} = \left(\frac{\sigma}{2\epsilon_0}\right)\hat{k} + \vec{E}' \quad (1.151)$$

Similarly, below the conducting surface, the electric field is

$$\vec{E}_{below} = -\left(\frac{\sigma}{2\epsilon_0}\right)\hat{k} + \vec{E}' \quad (1.152)$$

Notice that $\vec{E}'$ is continuous across the boundary. This is due to the fact that if the patch were removed, the field in the remaining "hole" exhibits no discontinuity. Using the two equations above, we find

$$\vec{E}' = \frac{1}{2}\left(\vec{E}_{above} + \vec{E}_{below}\right) = \vec{E}_{avg} \quad (1.153)$$

But, for a thick conductor which is parallel to xy-plane,

$$\vec{E}_{above} = \left(\frac{\sigma}{\epsilon_0}\right)\hat{k} \text{ and } \vec{E}_{below} = 0,$$

therefore, above equation gives-

$$\vec{E}_{avg} = \frac{1}{2}\left(\frac{\sigma}{\epsilon_0}\hat{k} + 0\right) = \frac{\sigma}{2\epsilon_0}\hat{k} \quad (1.154)$$

Thus, the force acting on the patch is

$$\vec{F} = q\vec{E}_{avg} = (\sigma A)\frac{\sigma}{2\epsilon_0}\hat{k} = \frac{\sigma^2 A}{2\epsilon_0}\hat{k} \quad (1.155)$$

where A is the area of the patch. This is precisely the force needed to drive the charges on the surface of a conductor to an equilibrium state where the electric field just outside the conductor takes on the value σ/ϵ_0 and vanishes inside. *Note that irrespective of the sign of σ, the force tends to pull the patch into the field.*

Using the above result, we may define electrostatic pressure on the patch as

$$P = \frac{F}{A} = \frac{\sigma^2}{2\epsilon_0} = \frac{1}{2}\epsilon_0\left(\frac{\sigma}{\epsilon_0}\right)^2 = \frac{1}{2}\epsilon_0 E^2 \quad (1.156)$$

where E is the magnitude of the field just above the patch. The pressure is being transmitted via the electric field.

This electrostatic pressure on the patch is also called the surface density of force on the surface of charged conductor.

EXAMPLE 95. Figure 1.240 shows a cross-sectional view of a thick spherical conductor. The conductor is neutral, and a small charged sphere ($q = +29.5$ μC) hangs from an insulating thread. The sphere is not in the centre of the conductor; instead, it is closer to the left side as shown in Fig. 1.240.

(a) Find the charge q_{wall} on the wall of the cavity and the charge

Figure 1.240: Cross-sectional view

q_{out} on the outer surface of the conductor. Start by sketching the electric field for all regions-inside the cavity, inside the body of the conductor, and outside the conductor.

(b) If the radius of the conductor is $R = 14.4$ cm, what is the magnitude of the electric field just outside the conductor?

(a) APPROACH The positive charge on the small sphere attracts electrons in the conductor. These electrons move close to the walls of the cavity. If the positively charged sphere were in the centre of the cavity, the electrons would be uniformly distributed on the cavity wall. However, the electrons are more concentrated on the left side of the cavity because the positive sphere is closer to the left side (Fig.1.241). The electric field in the body of any conductor in electrostatic equilibrium is zero. By choosing a Gaussian sphere that is concentric with the conductor and embedded in it, we can determine the amount of charge on the walls of the cavity.

SOLUTION The electric flux through the Gaussian surface (Fig. 1.241) is zero because the electric field in the body of the conductor is zero. Therefore, the net charge inside the Gaussian sphere is zero, so the total charge on the inside wall of the cavity is negative, equal in magnitude to the charge on the small sphere inside.

$$\Phi_E = \oint \vec{E} \cdot d\vec{A} = 0 = \frac{Q_{encl}}{\epsilon_0}$$

$$\Rightarrow \quad Q_{encl} = 0 = q + q_{wall}$$
$$\Rightarrow \quad q_{wall} = -q = -29.5 \, \mu C$$

1.35. CONDUCTORS IN ELECTROSTATIC EQUILIBRIUM

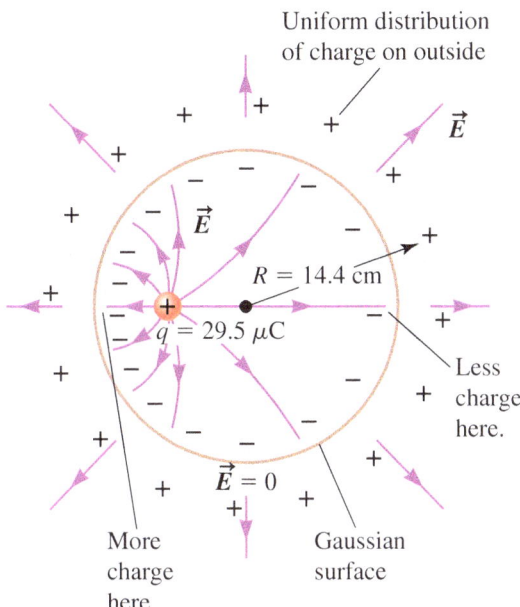

Figure 1.241

Because the conductor is neutral, therefore by charge conservation, the charge on its surface must be positive and equal to the charge on the cavity wall.

$$q_{\text{out}} = -q_{\text{wall}} = -(-29.5\mu C) = +29.5\mu C$$

The positive charge q_{out} on the outer surface of the conductor is uniformly distributed, as excess charge always is on the surface of a spherical conductor. To see why this is so, imagine that free electrons move toward the cavity wall, leaving positively charged ions in place.

(b) APPROACH The electric field just outside any conductor depends only on the surface charge density, so we need to find the surface charge density so that we can find the electric field.

SOLUTION The charge is uniformly distributed on a sphere of radius R. To find surface charge density, we divide the charge q_{out} by the surface area of the sphere.

$$\sigma = \frac{q_{\text{out}}}{A} = \frac{q_{\text{out}}}{4\pi R^2} = \frac{29.5 \times 10^{-6} C}{4\pi (14.4 \times 10^{-2}\, m)^2}$$

$$\Rightarrow \quad \sigma = 1.13 \times 10^{-4} C/m^2$$

The magnitude of the electric field just outside a conductor is given by-

$$E = \frac{\sigma}{\epsilon_0} = \frac{1.13 \times 10^{-4} C/m^2}{8.85 \times 10^{-12} C^2/N \cdot m^2} = 1.28 \times 10^7\, N/C$$

EXAMPLE 96. The Fig.1.242 shows a cross section of a spherical metal shell of inner radius R. A point charge of $-5.0\,\mu C$ is located at a distance $R/2$ from the centre of the shell. If the shell is electrically neutral, what are the (induced) charges on its inner and outer surfaces?

ANSWER Total induced charge on inner surface = $5.0\,\mu C$. It's distribution is non-uniform. The total induce charge on outer surface = $-5.0\,\mu C$. This charge will be distributed uniformly.

EXAMPLE 97. An uncharged spherical conductor centered at the origin has a cavity of some weird shape carved out of it (Fig.1.243). Somewhere within the cavity is a charge q. What is the field outside the sphere?

SOLUTION The answer is -

$$\vec{E} = \frac{1}{4\pi\epsilon_0} \frac{q}{r^2} \hat{r}$$

The conductor conceals from us all information concerning the nature of the cavity, revealing only the total charge it contains. How can this be? The charge $+q$ induces an opposite charge $-q$ on the wall of the cavity, which distributes itself in such a

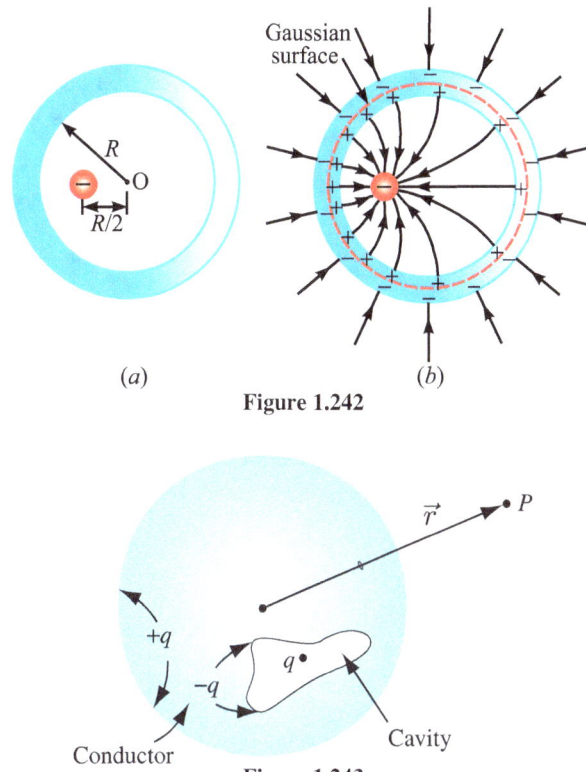

(a)   (b)
Figure 1.242

Figure 1.243

way that its field cancels that of q, for all points exterior to the cavity. Since the conductor carries no net charge, this leaves $+q$ to distribute itself uniformly over the surface of the sphere. (It is *uniform* because the asymmetrical influence of the point charge $+q$ is negated by that of the induced charge $-q$ on the inner surface.) For points outside the sphere, then, the only thing that survives is the field of the leftover $+q$ uniformly distributed over the outer surface.

1.35.4 Faraday Cage Protection

We have already seen that the electric field is zero inside any cavity within a conductor unless there is a charge in the cavity. This conclusion has an important practical application. For example, suppose we need to exclude the electric field from the region in Fig.1.244(a) enclosed within dashed lines. We can do so by surrounding this region with the neutral conducting box of Fig.1.244(b).

This region of space is now a cavity inside a conductor, thus

Figure 1.244: The electric field can be excluded from a region of space by surrounding it with a conducting box.

the interior electric field is zero. The use of a conducting box to exclude electric fields from a region of space is called *screening*. Solid metal walls are ideal, but in practice wire screen or wire mesh provides sufficient screening for all the sensitive equipment. In such case, the exterior field is now very complicated. Such shielding enclosures are called Faraday cages, after Michael Faraday. In 1816, Faraday used them to demonstrate that no charge was present inside a conducting shell.

Applications: *We can use these ideas to explain how antistatic bags work and why your cell phone doesn't work in a closed elevator. The antistatic bag is coated with a conductor, and sensitive equipment is sealed inside. If the bag comes in contact with a charged object, the outside of the bag becomes charged but the inside remains uncharged, and the electric field inside the bag is zero. The antistatic bag is a Faraday cage. In order for your cell phone to work, it must be able to receive signals that consist of changing electric and magnetic fields. If your cell phone is inside a closed elevator, the elevator acts as a Faraday cage. The electric field inside the elevator is zero, so your cell phone cannot receive signals. The same physics tells you that one of the safest places to be in a lightning storm is inside a car; if the car is struck by lightning, the charge tends to remain on the metal skin of the vehicle, and little or no electric field is produced inside the passenger compartment.*

EXAMPLE 98. A conducting sphere of radius R carries a net positive charge $2Q$. A conducting spherical shell of inner radius R_1 ($R_1 > R$) and outer radius R_2 carries a net negative charge $-Q$. This shell is concentric with the conducting sphere. Find the magnitude of the electric field at a distance r away from the common centre when: (a) $r < R$, (b) $R < r < R_1$, (c) $R_1 < r < R_2$, and (d) $r > R_2$.

SOLUTION The charge distributions under consideration are characterized by being spherically symmetrical around the common centre c. This suggests that a spherical Gaussian surface of radius r is to be constructed in each case such as S_1, S_2, S_3, and S_4 that are displayed in Fig.1.245. In addition, we use the fact that the electric field inside a conductor is zero and all the excess charge will lie entirely on the outer surface of the isolated conductor.

(a) In this region the Gaussian sphere S_1 of Fig. 1.245 satisfies

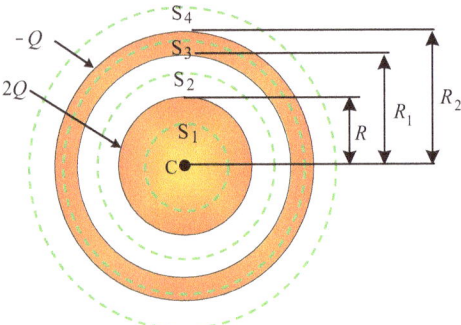

Figure 1.245

the condition $r < R$. Because there is no charge inside the conductor in this region, i.e. $Q_{encl} = 0$; then, $E_1 = 0$.

(b) In this region the Gaussian sphere S_2 of Fig. 1.245 satisfies the condition $R < r < R_1$. Because $Q_{encl} = 2Q$ inside this surface and because $\oint \vec{E}_2 \cdot d\vec{A} = E_2 (4\pi r^2)$, we can use Gauss's law to find:

$$E_2 = \frac{1}{4\pi\epsilon_o}\frac{2Q}{r^2} = k\frac{2Q}{r^2} \quad (R < r < R_1)$$

(c) In this region, the Gaussian sphere S_3 of Fig. 1.245 satisfies the condition $R_1 < r < R_2$. Because the electric field inside an equilibrium conductor is zero, i.e. $E_3 = 0$; then, based on Gauss's law, the net charge Q_{encl} must be zero. From this argument, we find that an induced charge $-2Q$ must be established on the inner surface of the shell to cancel the charge $+2Q$ on the solid sphere.

In addition, because the net charge on the whole shell is $-Q$, we conclude that its outer surface must carry an induced charge $+Q$.

(d) In this region, the Gaussian sphere S_4 of Fig. 1.245 satisfies the condition $r > R_2$. Because $Q_{encl} = 2Q - Q = Q$ inside this surface and because $\oint \vec{E}_4 \cdot d\vec{A} = E_4 (4\pi r^2)$, we can use Gauss's law to find:

$$E_4 = \frac{1}{4\pi\epsilon_o}\frac{Q}{r^2} = k\frac{Q}{r^2} \quad (r > R_2)$$

Figure 1.246 shows a graphical representation of the variation of

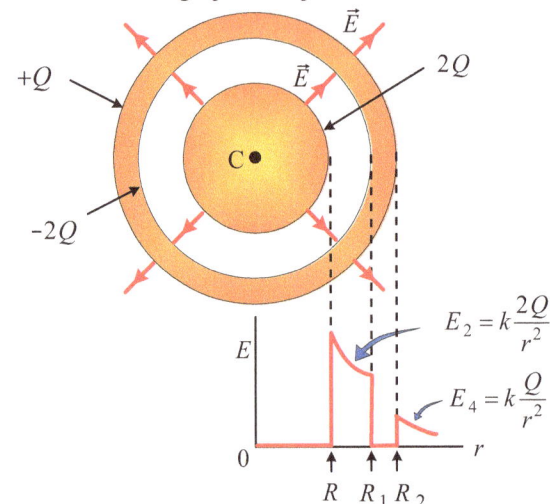

Figure 1.246

the electric field E with r. In addition, the Fig.1.246 shows the final distribution of the charge on the two conductors.

EXAMPLE 99. **Electric Field Due to Plane Plus Point Charge:** A very large (and thin) non-conducting plane slab carries a uniform surface charge density σ (assume positively charged). A single positive point charge, $+Q$, is located a distance d above the plane. What is the electric field at a point between the point charge and the plane and at a distance b from the plane?

APPROACH The situation is drawn in Fig.1.247. We will use the superposition principle to solve this problem.

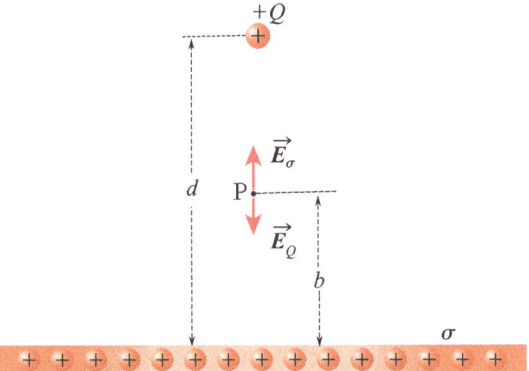

Figure 1.247: Situation of EXAMPLE 99 with the electric field at point P shown separately due to the non-conducting plane of charge $\vec{E}_\sigma$ and the point charge $\vec{E}_Q$

SOLUTION The electric field at point P due to the charged plane is-

$$E_\sigma = \frac{\sigma}{2\epsilon_0}$$

The direction of this electric field is perpendicular to the plane and away from it (due to the positive charge), as shown in Figure 1.247

The electric field at point P due to the point charge Q is-

1.35. CONDUCTORS IN ELECTROSTATIC EQUILIBRIUM

$$E_Q = k\frac{Q}{(d-b)^2}$$

The direction of this electric field is away from positive point charge Q and directed perpendicularly towards the charged plane. The net electrical field at P is given by-

$$\vec{E}_{net} = \vec{E}_\sigma + \vec{E}_Q = \left(\frac{\sigma}{2\epsilon_0} - k\frac{Q}{(d-b)^2}\right)\hat{k}$$

The net electric field could be either upward or downward, depending on the relative magnitudes of the point charge and the surface charge density.

Note: While the superposition technique can be powerful, you have to be careful not to use it when it is not valid. If you have a conductor, the presence of another charged object will change the distribution of charges on the conductor, so we could not assume that the plane is still uniformly charged.

EXAMPLE 100. Why is it normally not possible to use the technique of superposition along with Gauss's law when one of the objects is a conductor?
(A) Conductors have zero interior electric fields.
(B) Charges on conductors are all on the outer surface.
(C) The electric fields on conductors (just outside) are perpendicular to the surface.
(D) Charges in conductors are free to move, which will often destroy the symmetry needed for Gauss's law.
ANSWER: (D) While the first three statements are true, that is not why we can't use the Gauss's law plus superposition technique. When we have two charged objects, and one of them is a conductor, the presence of the other object will distort the symmetry of the charge distribution and we can no longer make assumptions such as a constant-magnitude of electric field on a surface.

EXAMPLE 101. Two conducting plates A and B are placed parallel to each other. A is given a charge Q_1 and B a charge Q_2. Find the distribution of charges on the four surfaces.
SOLUTION Consider a Gaussian surface as shown in Fig.1.248. Two faces of this closed surface lie completely inside the conductor where the electric field is zero. The flux through these faces is, therefore, zero. The other parts of the closed surface which are outside the conductor are parallel to the electric field and hence the flux on these parts is also zero. The total flux of the electric field through the closed surface is, therefore, zero. So, from Gauss's law, the total charge inside this closed surface should be zero. The charge on the inner surface of A should be equal and opposite to the inner surface of B.

The distribution should be like the one shown in figure 1.249. To

Figure 1.248

find the value of q, consider the field at a point P inside the plate A. Suppose, the surface area of each plate is A. Let σ_1, σ_2, are respective surface charge densities on the outer, inner surfaces of the plate A and σ_3, σ_4 are respectively the surface charge densities on inner and outer surface of the plate B.
Now, the electric field at point P inside the plate A-
(i) due to the charge $Q_1 - q$ on the outer surface of plate A, is given by-

$$E_1 = \frac{\sigma_1}{2\epsilon_0}$$

Since, $\sigma_1 = (Q_1 - q)/A$, therefore,

Figure 1.249

$$E_1 = \frac{Q_1 - q}{2A\epsilon_0} \quad \text{(downwards)},$$

(ii) due to the charge $+q$ on the inner surface of plate A,

$$E_2 = \frac{\sigma_2}{2\epsilon_0}$$

Since, $\sigma_2 = q/A$, therefore,

$$E_2 = \frac{q}{2A\epsilon_0} \quad \text{(upwards)},$$

(iii) Due to the charge $-q$ on the inner surface of plate B,

$$E_3 = \frac{\sigma_3}{2\epsilon_0}$$

Since, $\sigma_3 = q/A$, therefore,

$$E_3 = \frac{q}{2A\epsilon_0} \quad \text{(downwards)},$$

(iv) due to the charge $Q_2 + q$ on the outer surface of the plate B,

$$E_4 = \frac{\sigma_4}{2\epsilon_0}$$

Since, $\sigma_3 = (Q_2 + q)/A$, therefore,

$$E_4 = \frac{Q_2 + q}{2A\epsilon_0} \quad \text{(upward)},$$

If we consider, upward direction as positive direction of electric field, then, net electric field at point P due to all the four charged surfaces is (in the downward direction)

$$E = -E_1 + E_2 - E_3 + E_4$$
$$= -\frac{Q_1 - q}{2A\epsilon_0} + \frac{q}{2A\epsilon_0} - \frac{q}{2A\epsilon_0} + \frac{Q_2 + q}{2A\epsilon_0}$$
$$= -\frac{Q_1}{2A\epsilon_0} + \frac{2q}{2A\epsilon_0} + \frac{Q_2}{2A\epsilon_0}$$

or $\quad E = \frac{Q_2 - Q_1}{2A\epsilon_0} + \frac{2q}{2A\epsilon_0}$

As the point P is inside the conductor, this field should be zero. Hence,

$$E = \frac{Q_2 - Q_1}{2A\epsilon_0} + \frac{2q}{2A\epsilon_0} = 0$$

$$\implies \boxed{q = \frac{Q_1 - Q_2}{2}} \quad (1.157)$$

It is the electric charge on the inner surface of plate A (which has the net electric charge Q_1). The electric charge on the inner surface of the plate B, is $-q$, i.e.,

$$\boxed{-q = -\left(\frac{Q_1 - Q_2}{2}\right)} \quad (1.158)$$

Thus, the charge on outer surface of the plate A-

$$\boxed{Q_1 - q = \frac{Q_1 + Q_2}{2}} \quad (1.159)$$

and the charge on the outer surface of the plate B-

$$\boxed{Q_2 + q = \frac{Q_1 + Q_2}{2}} \quad (1.160)$$

From Eq.1.159 and 1.160, we see that the charge distribution on the outer surfaces of plates A and B remains identical. Final charge distribution is shown in Fig.1.250

EXAMPLE 102. Figure 1.251 shows three large metallic plates with charges $-Q$, $3Q$ and Q respectively. Determine the final charges on all the surfaces.

Figure 1.250

Figure 1.251

APPROACH In Fig.1.252 we have numbered each surface of the given plates. Now, assume that the charge on surface 2 is x. Following conservation of charge, we see that surfaces 1 has charge $(-Q - x)$ i.e., $-(Q + x)$. Similarly, we can also find the charge distribution on other plates. Since, in electrostatic equilibrium, the net electric field inside a metal plate is always zero, therefore, the net electric field at P should be zero, i.e., $E_P = 0$.

SOLUTION Resultant electric field at point P inside the left

Figure 1.252

most plate -

$$\frac{Q+x}{2A\epsilon_0} + \frac{x}{2A\epsilon_0} + \frac{3Q}{2A\epsilon_0} + \frac{Q}{2A\epsilon_0} = 0$$

(Since electric fields due to all surface charges are directed in same direction (left))

$\Rightarrow \quad 2x + 5Q = 0 \quad \Rightarrow x = -5Q/2$

Therefore, charge on surface ①

Figure 1.253

$$= -(Q + x) = -\left(Q - \frac{5Q}{2}\right) = +\frac{3Q}{2}$$

Since, the front surfaces of two parallel metal plates always acquire equal and opposite electric charges, therefore, electric charge on surface ③ will be $-x$, i.e., $+5Q/2$.

By charge conservation, the electric charge on surface ④

$$= 3Q - \frac{5Q}{2} = \frac{Q}{2}$$

Therefore, the electric charge on the opposite front surface ⑤

will be $-Q/2$.

Again by charge conservation, the electric charge on the surface ⑥-

$$= Q - \left(-\frac{Q}{2}\right) = \frac{3Q}{2}$$

Final charge distribution is shown in Fig.1.253.

EXAMPLE 103. An isolated conducting sheet of area A and carrying a charge Q is placed in a uniform electric field $\vec{E}$, such that electric field is perpendicular to sheet and covers all the sheet[Fig.1.254]. Find out charges appearing on its two surfaces.

SOLUTION Let the electric charge on the left surface of the

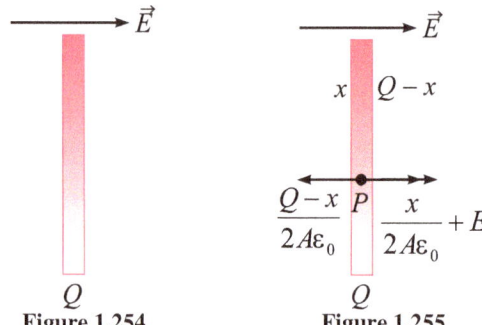

Figure 1.254   **Figure 1.255**

plate is x, then by charge conservation, the electric charge on the right surface of the plate will be $Q - x$ [see Fig.1.255].

Now, consider a point P inside the metal plate. Find the net electric field at P and equate it with zero because in electrostatic static equilibrium the net electric field inside the plate should always be zero, i.e., $E_P = 0$

From Fig.1.255, the net electric field at P–

$$E_P = \frac{x}{2A\epsilon_0} - \frac{Q-x}{2A\epsilon_0} + E = 0$$

$$\Rightarrow \quad 2x + 2A\epsilon_0 E - Q = 0 \quad \Rightarrow \quad x = \frac{Q}{2} - A\epsilon_0 E$$

So charge on one side is $\frac{Q}{2} - A\epsilon_0 E$ and other side

$$= Q - x = Q - \left(\frac{Q}{2} - A\epsilon_0 E\right) = \frac{Q}{2} + A\epsilon_0 E$$

Remark Solve this question for $Q = 0$ without using the above answers and match that answers with the answers that you will get by putting $Q = 0$ in the above answers.

1.36 Check Point 10

1. ●●The electric field strength just above one face of a copper penny is 2000 N/C. What is the surface charge density on this face of the penny?
2. ●●A spark occurs at the tip of a metal needle if the electric field strength exceeds 3.0×10^6 N/C, the field strength at which air breaks down. What is the minimum surface charge density for producing a - spark?
3. ●●The conducting box in Fig.1.256 has been given an excess negative charge. The surface density of excess electrons at the centre of the top surface is 5.0×10^{10} electrons/m². What are the electric field strengths E_1 to E_3 at points 1 to 3?

Figure 1.256

4. ●●A thin, horizontal, 10 cm diameter copper plate is charged

to 3.5 nC. If the electrons are uniformly distributed on the surface, what are the strength and direction of the electric field
   (a) 0.1 mm above the centre of the top surface of the plate?
   (b) at the plate's centre of mass?
   (c) 0.1 mm below the centre of the bottom surface of the plate?
5. ●●Figure 1.257 shows a hollow cavity within a neutral conductor. A point charge Q is inside the cavity. What is the net electric flux through the closed surface that surrounds the conductor?

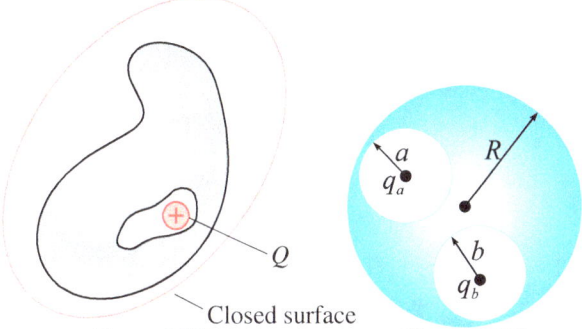

Figure 1.257   Figure 1.258

6. ●●Two spherical cavities, of radii a and b, are hollowed out from the interior of a (neutral) conducting sphere of radius R (Fig.1.258). At the centre of each cavity a point charge is placed-call these charges q_a and q_b.
   (a) Find the surface charge densities σ_a, σ_b, and σ_R.
   (b) What is the field outside the conductor?
   (c) What is the field within each cavity?
   (d) What is the force on q_a and q_b?
   (e) Which of these answers would change if a third charge, q_c, were brought near the conductor?
7. ●●Electronic charge e may be determined by Millikan's oil drop method. Oil drops of radius r acquire a terminal speed v_1 with downward electric field E and a speed v_2 with the upward electric field. Find the value of e in terms of E, v_1, v_2, r and η, the viscosity of oil in air.

Multiple Choice Questions

8. ●●A soap bubble has radius R, charge Q, surface tension T. Find the excess pressure in it.
   (A) $\dfrac{32\pi^2 R^2 \epsilon_0 T - q^2}{32\pi^2 R^4 \epsilon_0}$ (B) $\dfrac{64\pi^2 R^3 \epsilon_0 T - q^2}{32\pi^2 R^4 \epsilon_0}$
   (C) $\dfrac{128\pi^2 R^3 \epsilon_0 T - q^2}{32\pi^2 R^4 \epsilon_0}$ (D) none of these

1.37 Questions and Exercises
1.37.1 Conceptual Questions
1. Why does a plastic ruler that has been rubbed with a cloth have the ability to pick up small pieces of paper? Why is this difficult to do on a humid day?
2. A popular classroom demonstration consists of rubbing a plastic rod with fur to give the rod charge, and then placing the rod near an empty soda can that is on its side (Figure 1.259). Explain why the can will roll toward the rod.

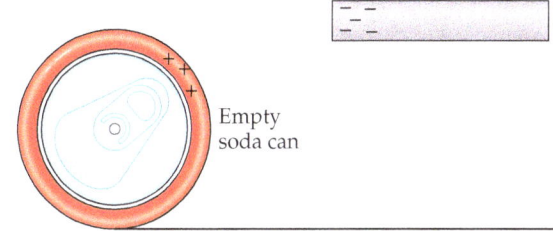

Figure 1.259

3. A metal sphere is positively charged. Is it possible for the sphere to electrically attract another positively charged ball? Explain your answer
4. We are not normally aware of the gravitational or electric force between two ordinary objects. What is the reason in each case? Give an example where we are aware of each one and why.
5. What are the two major reasons for introducing the field concept?
6. Explain why the test charges we use when measuring electric fields must be small.
7. When determining an electric field, must we use a *positive* test charge, or would a negative one do as well? Explain.
8. Why do electric field lines point away from positive charges and toward negative charges?
9. An electron initially moving horizontally near Earth's surface enters a uniform electric field and is deflected upward. What can you say about the direction of the electric field (assuming no other interaction)? What can you say about the direction of the electric field if the electron is deflected downward?
10. Consider the electric field at the three points indicated by the letters A, B, and C in Fig. 1.260. First draw an arrow at each point indicating the direction of the net force that a positive test charge would experience if placed at that point, then list the letters in order of *decreasing* field strength (strongest first). Explain.

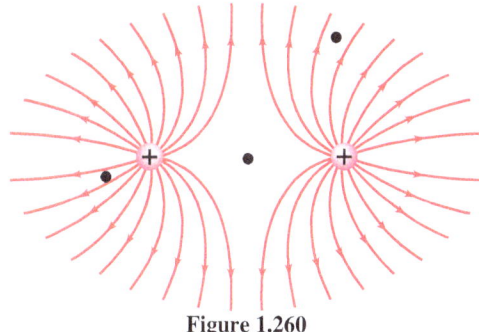

Figure 1.260

11. Given two point charges, Q and $2Q$, a distance l apart, is there a point along the straight line that passes through them where $E = 0$ when their signs are (a) opposite, (b) the same? If yes, state roughly where this point will be.
12. Consider a small positive test charge located on an electric field line at some point, such as point P in Fig.1.261. Is the direction of the velocity and/or acceleration of the test charge along this line? Discuss.

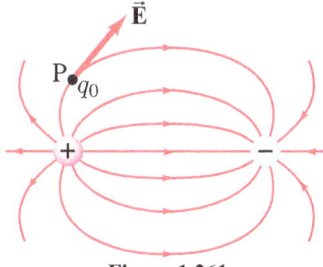

Figure 1.261

13. Is it possible for a dipole that is initially moving in a straight line to be deflected by a uniform electric field?
14. An electron is travelling in circular motion at constant speed

due to the effect of an electric field. Is it possible for the electric field to be uniform?

15. A point charge is surrounded by a spherical Gaussian surface of radius r. If the sphere is replaced by a cube of side r, will Φ_E be larger, smaller, or the same? Explain.

16. Eight electrons are the only charged particles inside an isolated balloon, producing a field line flux through its surface. (a) If eight additional electrons are set outside the balloon to form the eight corners of a cube, how does the field line flux through the balloon change? (b) How does the flux change if the eight additional electrons are all placed at one location outside the balloon instead of being distributed at the corners of a cube?

17. Under what conditions is the electric field magnitude nonzero inside the cavity enclosed by a hollow conducting sphere?

18. What is the distinction between field line flux and electric flux? Why is electric flux the preferred variable for analyzing electric fields?

19. Objects are composed of atoms which are composed of charged particles (protons and electrons); however, we rarely observe the effects of the electrostatic force. Explain why we do not observe these effects.

20. Occasionally, people who gain static charge by shuffling their feet on the carpet will have their hair stand on end. Why does this happen?

1.37.2 Problems

Discrete Charge Distributions

1. •How many electrons are required to yield a total charge of 1.00 C?
2. •A plastic rod is rubbed against a wool shirt, thereby acquiring a charge of $-0.80\ \mu C$. How many electrons are transferred from the wool shirt to the plastic rod?
3. •What is the total charge of all of the protons in 1.00 kg of carbon (C)? (Given: atomic number of C is 6 and it's molar mass is 12 gram)
4. ••The faraday is a unit of charge frequently encountered in electrochemical applications and named for the British physicist and chemist Michael Faraday. It consists of exactly 1 mole of elementary charges (i.e., 1 mol of electrons). Calculate the number of coulombs in 1 faraday (1F).
5. ••Suppose a cube of aluminum which is 1.00 cm on a side accumulates a net charge of +2.50 pC. (a) What percentage of the electrons originally in the cube was removed? (b) By what percentage has the mass of the cube decreased because of this removal?

Electric Force

6. ••What is the magnitude of the electric force between two protons separated by 2.00×10^{-15} m?
7. •A point charge $q_1 = 4.0\ \mu C$ is at the origin and a point charge $q_2 = 6.0\ \mu C$ is on the x axis at $x = 3.0$ m. (a) Find the electric force on charge q_2. (b) Find the electric force on q_1. (c) How would your answers for Parts (a) and (b) differ if q_2 were $-6.0\ \mu C$?
8. •Three point charges are on the x-axis: $q_1 = -6.0\mu C$ is at $x = -3.0$ m, $q_2 = 4.0\ \mu C$ is at the origin, and $q_3 = -6.0\ \mu C$ is at $x = 3.0$ m. Find the electric force on q_1.
9. ••A 2.0 μC point charge and a 4.0 μC point charge are a distance L apart. Where should a third point charge be placed so that the electric force on that third charge is zero?
10. ••A $-2.0\ \mu C$ point charge and a 4.0 μC point charge are a distance L apart. Where should a third point charge be placed so that the electric force on that third charge is zero?
11. ••A point particle that has a charge of $-2.5\ \mu C$ is located at the origin. A second point particle that has a charge of 6.0 μC is at $x = 1.0$ m, $y = 0.50$ m. A third point particle, an electron, is at a point with coordinates (x, y). Find the values of x and y such that the electron is in equilibrium.
12. •••Five identical point charges, each having charge Q, are equally spaced on a semicircle of radius R as shown in Figure 1.262. Find the force (in terms of k, Q, and R) on a charge q located equidistant from the five other charges.
13. •••The structure of the NH_3 molecule is approximately that of an equilateral tetrahedron, with three H^+ ions forming the base and an N^{3-} ion at the apex of the tetrahedron. The length of each side is 1.64×10^{-10} m. Calculate the electric force that acts on each ion.

The Electric Field

14. •A point charge of 4.0 μC is at the origin. What are the magnitude and direction of the electric field on the x axis at (a) $x = 6.0$ m and (b) $x = -10$ m ? (c) Sketch the function E_x versus x for both positive and negative values of x. (Remember that E_x is negative when $\vec{E}$ points in the $-x$ direction.)
15. ••Two point charges, each $+4.0\mu C$, are on the x axis; one point charge is at the origin and the other is at $x = 8.0$ m. Find the electric field on the x axis at (a) $x = -2.0$ m, (b) $x = 2.0$ m, (c) $x = 6.0$ m, and (d) $x = 10$m. (e) At what point on the x axis is the electric field zero? (f) Sketch E_x versus x for -3.0 m $< x < 11$ m.
16. ••A point charge of $+5.0\ \mu C$ is located on the x axis at $x = -3.0$ cm, and a second point charge of $-8.0\ \mu C$ is located on the x axis at $x = +4.0$ cm. Where should a third charge of $+6.0\ \mu C$ be placed so that the electric field at the origin is zero?
17. ••A $-5.0\ \mu C$ point charge is located at $x = 4.0$ m, $y = -2.0$ m and a 12 μC point charge is located at $x = 1.0$ m, $y = 2.0$ m. (a) Find the magnitude and direction of the electric field at $x = -1.0$ m, $y = 0$. (b) Calculate the magnitude and direction of the electric force on an electron that is placed at $x = -1.0$ m, $y = 0$.
18. ••Two point particles, each having a charge q, sit on the base of an equilateral triangle that has sides of length L as shown in Figure 1.263. A third point particle that has a charge equal to $2q$ sits at the apex of the triangle. Where must a fourth point particle that has a charge equal to q be placed in order that the electric field at the centre of the triangle be zero? (The centre is in the plane of the triangle and equidistant from the three vertices.)

Figure 1.262   Figure 1.263

19. •••Two equal positive point charges $+q$ are on the y axis; one is at $y = +a$ and the other is at $y = -a$. The electric field at the origin is zero. A test charge q_0 placed at the origin will therefore be in equilibrium. (a) Discuss the stability of the equilibrium for a positive test charge by considering small displacements from equilibrium along the x axis and small displacements along the y axis. (b) Repeat Part (a) for a negative test charge. (c) Find the magnitude and sign of a

1.37. QUESTIONS AND EXERCISES

charge q_0 that when placed at the origin results in a net force of zero on each of the three charges.

20. ●● Two equal positive charges q are on the y axis; one point charge is at $y = +a$ and the other is at $y = -a$. (a) Show that on the x axis the x component of the electric field is given by $E_x = 2kqx/(x^2 + a^2)^{3/2}$. (b) Show that near the origin, where x is much smaller than a, $E_x \approx 2kqx/a^3$. (c) Show that for values of x much larger than a, $E_x \approx 2kq/x^2$. Explain why a person might expect this result even without deriving it by taking the appropriate limit.

21. ●● Two positive point charges $+q$ are on the y axis at $y = +a$ and $y = -a$. A bead of mass m and charge $-q$ slides without friction along a taut thread that runs along the x axis. Let x be the position of the bead. (a) Show that for $x \ll a$, the bead experiences a linear restoring force (a force that is proportional to x and directed toward the equilibrium position at $x = 0$) and therefore undergoes simple harmonic motion. (b) Find the period of the motion.

22. ● The acceleration of a particle in an electric field depends on the charge-to-mass ratio of the particle. (a) Compute q/m for a proton, and find its acceleration in a uniform electric field that has a magnitude of 100 N/C. (b) Find the time it takes for a proton initially at rest in such a field to reach a speed of $0.01c^*$ (where c is the speed of light). Given: mass of proton = 1.673×10^{-27} kg and charge of a proton = 1.602×10^{-19} C.

23. ●● An electron is released from rest in a weak electric field $\vec{E} = (-1.50 \times 10^{-10} \text{ N/C})\,\hat{j}$. After the electron has travelled a vertical distance of $1.0\,\mu\text{m}$, what is its speed? (Do not neglect the gravitational force on the electron.)

24. ●● A 2.00 g charged particle is released from rest in a region that has a uniform electric field $\vec{E} = (300\,\text{N/C})\hat{i}$. After travelling a distance of 0.500 m in this region, the particle has a kinetic energy of 0.120 J. Determine the charge of the particle.

25. ●● A charged particle leaves the origin with a speed of 3.00×10^6 m/s at an angle of 35° above the x axis. A uniform electric field given by $\vec{E} = -E_0\hat{j}$ exists throughout the region. Find E_0 such that the particle will cross the x axis at $x = 1.50$ cm if the particle is (a) an electron, and (b) a proton.

Dipoles

26. ●● Two point charges, $q_1 = 2.0$ pC and $q_2 = -2.0$ pC, are separated by 4.0 μm. (a) What is the magnitude of the dipole moment of this pair of charges? (b) Sketch the pair and show the direction of the dipole moment.

27. ●● An electron (charge $-e$, mass m) and a positron (charge $+e$, mass m) revolve around their common centre of mass under the influence of their attractive coulomb force. Find the speed v of each particle in terms of e, m, k, and their separation distance L.

28. ●● A point particle of mass m and charge q is constrained to move vertically inside a narrow, friction less cylinder (Figure 1.264). At the bottom of the cylinder is a point charge Q having the same sign as q. (a) Show that the particle whose mass is m will be in equilibrium at a height $y_0 = (kqQ/mg)^{1/2}$ (b) Show that if the particle is displaced from its equilibrium position by a small amount and released, it will exhibit simple harmonic motion with angular frequency $\omega = (2g/y_0)^{1/2}$

29. ●● During the Millikan experiment used to determine the charge on the electron, a charged polystyrene microsphere is released in still air in a known vertical electric field. The charged microsphere will accelerate in the direction of the net force until it reaches terminal speed. The charge on the microsphere is determined by measuring the terminal speed. During one such experiment, the microsphere has radius of $r = 5.50 \times 10^7$ m, and the field has a magnitude $E = 6.00 \times 10^4$ N/C. The magnitude of the drag force on the sphere is given by $F_D = 6\pi\eta rv$, where v is the speed of the sphere and η is the viscosity of air ($\eta = 1.8 \times 10^{-5}$ N·s/m²). Polystyrene has density 1.05×10^3 kg/m³. (a) If the electric field is pointing down and the polystyrene microsphere is rising with a terminal speed of 1.16×10^{-4} m/s, what is the charge on the sphere? (b) How many excess electrons are on the sphere? (c) If the direction of the electric field is reversed but its magnitude remains the same, what is the new terminal speed?

30. ●●● In problem 29, there is a description of the Millikan experiment used to determine the charge on the electron. During the experiment, a switch is used to reverse the direction of the electric field without changing its magnitude, so that one can measure the terminal speed of the microsphere both as it is moving upward and as it is moving downward. Let v_u represent the terminal speed when the particle is moving up, and v_d the terminal speed when moving down. (a) If we let $u = v_u + v_d$, show that $q = 3\pi\eta ru/E$, where q is the microsphere's net charge. For the purpose of determining q, what advantage does measuring both v_u and v_d have over measuring only one terminal speed? (b) Because charge is quantized, u can only change by steps of magnitude N, where N is an integer. Using the data from Problem 29, calculate Δu.

Continuous Charge Distribution

31. ●● Two infinite nonconducting sheets of charge are parallel to each other, with sheet A in the $x = -2.0$ m plane and sheet B in the $x = +2.0$ m plane. Find the electric field in the region $x < -2.0$ m, in the region $x > +2.0$ m, and between the sheets for the following situations. (a) When each sheet has a uniform surface charge density equal to $+3.0\,\mu\text{C/m}^2$ and (b) when sheet A has a uniform surface charge density equal to $+3.0\,\mu\text{C/m}^2$ and sheet B has a uniform surface charge density equal to $-3.0\,\mu\text{C/m}^2$. (c) Sketch the electric field line pattern for each case.

32. ● A non-conducting disc of radius R lies in the $z = 0$ plane with its centre at the origin. The disc has a uniform surface charge density σ. Find the value of z for which $E_z = \sigma/(4\epsilon_0)$. Note that at this distance, the magnitude of the electric field strength is half the electric-field strength at points on the x axis that are very close to the disc.

33. ● A ring that has radius a lies in the $z = 0$ plane with its centre at the origin. The ring is uniformly charged and has a total charge Q. Find E_z on the z axis at (a) $z = 0.2a$, (b) $z = 0.5a$, (c) $z = 0.7a$, (d) $z = a$, and (e) $z = 2a$. (f) Use your results to plot E_z versus z for both positive and negative values of z. (Assume that these distances are exact.)

34. ● A line charge that has a uniform linear charge density λ lies along the x axis from $x = x_1$ to $x = x_2$ where $x_1 < x_2$. Show that the x component of the electric field at a point on the y-axis is given by $E_x = \dfrac{k\lambda}{y}(\cos\theta_2 - \cos\theta_1)$, where $\theta_1 = \tan^{-1}(x_1/y)$, $\theta_2 = \tan^{-1}(x_2/y)$ and $y \neq 0$.

35. ●● A thin hemispherical shell of radius R has a uniform surface charge σ. Find the electric field at the centre of the base

---

*When the speed of an electron approaches the speed of light c, relativistic kinematics must be used to calculate its motion, but at speeds of $0.01c$ or less, non-relativistic kinematics is sufficiently accurate for most purposes.

of the hemispherical shell.
Gauss's Law

36. ••An imaginary right circular cone (Figure 1.265) that has a base angle θ and base radius R is in a charge free region that has a uniform electric field $\vec{E}$ (field lines are vertical and parallel to the cone's axis). What is the ratio of the number of field lines per unit area penetrating the base to the number of field lines per unit area penetrating the conical surface of the cone? Use Gauss's law in your answer. (The field lines in the figure are only a representative sample.)

Figure 1.264

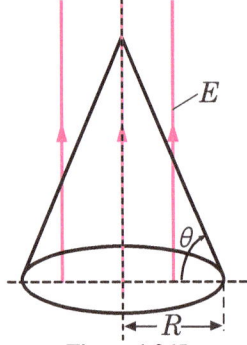
Figure 1.265

37. ••A non-conducting solid sphere of radius R has a volume charge density that is proportional to the distance from the centre. That is, $\rho = Ar$ for $r \leq R$, where A is a constant. (a) Find the total charge on the sphere. (b) Find the expressions for the electric field inside the sphere ($r < R$) and outside the sphere ($r > R$) (c) Sketch the magnitude of the electric field as a function of the distance r from the sphere's centre.

38. •••A non-conducting spherical shell of inner radius R_1 and outer radius R_2 has a uniform volume charge density ρ. (a) Find the total charge on the shell. (b) Find expressions for the electric field everywhere.

Gauss's Law Applications in Cylindrical Symmetry Situations

39. ••An infinitely long non-conducting solid cylinder of radius a has a uniform volume charge density of ρ_0. Show that the electric field is given by the following expressions: $E_a = \rho_0 a/(2\epsilon_0)$ for $0 \leq r < a$ and $E_a = \rho_0 a^2/(2\epsilon_0 r)$ for $r > a$, where r is the distance from the long axis of the cylinder.

40. ••Consider two infinitely long, coaxial thin cylindrical shells. The inner shell has a radius a_1 and has a uniform surface charge density of σ_1, and the outer shell has a radius a_2 and has a uniform surface charge density of σ_2. (a) Use Gauss's law to find expressions for the electric field in the three regions: $0 \leq r < a_1, a_1 < r < a_2$, and $r > a_2$, where r is the distance from the axis.
(b) What is the ratio of the surface charge densities σ_2/σ_1 and their relative signs if the electric field is to be zero everywhere outside the largest cylinder? (c) For the case in Part (b), what would be the electric field between the shells?

Electric Charge and Field at Conductor Surfaces

41. •An uncharged penny is in a region that has a uniform electric field of magnitude 1.60 kN/C directed perpendicular to its faces. (a) Find the charge density on each face of the penny, assuming the faces are planes. (b) If the radius of the penny is 1.00 cm, find the total charge on one face.

42. •A thin metal slab has a net charge of zero and has square faces that have 12 cm long sides. It is in a region that has a uniform electric field that is perpendicular to its faces. The total charge induced on one of the faces is 1.2 nC. What is the magnitude of the electric field?

43. •••If the magnitude of an electric field in air is as great as 3.0×10^6 N/C, the air becomes ionized and begins to conduct electricity. This phenomenon is called **dielectric breakdown**. A charge of 18 μC is to be placed on a conducting sphere. What is the minimum radius of a sphere that can hold this charge without breakdown?

44. •••A quantum-mechanical treatment of the hydrogen atom shows that the electron in the atom can be treated as a smeared-out distribution of negative charge of the form $\rho(r) = -\rho_0 e^{-2r/a}$. Here r represents the distance from the centre of the nucleus and a represents the first *Bohr radius* which has a numerical value of 0.0529 nm. Recall that the nucleus of a hydrogen atom consists of just one proton and treat this proton as a positive point charge. (a) Calculate ρ_0, using the fact that the atom is neutral. (b) Calculate the electric field at any distance r from the nucleus.

45. •••A thin, non-conducting, uniformly charged spherical shell of radius R (Figure 1.266a) has a total positive charge of Q. A small circular plug is removed from the surface. (a) What are the magnitude and direction of the electric field at the centre of the hole? (b) The plug is now put back in the hole (Figure 1.266b). Using the result of Part (a), find the electric force acting on the plug. (c) Using the magnitude of the force, calculate the "electrostatic pressure" (force/unit area) that tends to expand the sphere.

Figure 1.266

46. ••A uniformly charged, infinitely long line of negative charge has a linear charge density of $-\lambda$ and is located on the z axis. A small positively charged particle that has a mass m and a charge q is in a circular orbit of radius R in the xy plane centred on the line of charge. (a) Derive an expression for the speed of the particle. (b) Obtain an expression for the period of the particle's orbit.

47. ••A uniformly charged non-conducting solid sphere of radius R has its centre at the origin and has a volume charge density of ρ. (a) Show that at a point within the sphere a distance r from the centre $\vec{E} = \dfrac{\rho}{3\epsilon_0}r\hat{r}$. (b) Material is removed from the sphere leaving a spherical cavity that has a radius $b = R/2$ and its centre at $x = b$ on the x axis (Figure 1.267). Calculate the electric field at points 1 and 2 shown in Figure 1.267.
Hint: Model the sphere-with-cavity as two uniform spheres of equal positive and negative charge densities.

48. ••Show that the electric field throughout the cavity of Problem 47(b) is uniform and is given by $\vec{E} = \dfrac{\rho}{3\epsilon_0}b\hat{i}$

49. ••The cavity in Problem 47(b) is now filled with a uniformly charged non-conducting material with a total charge of Q. Calculate the new values of the electric field at points 1 and 2 shown in Figure 1.267

50. ••An electric dipole that has a dipole moment of $\vec{p}$ is located at a perpendicular distance R from an infinitely long line charge that has a uniform linear charge density λ. Assume

1.37. QUESTIONS AND EXERCISES

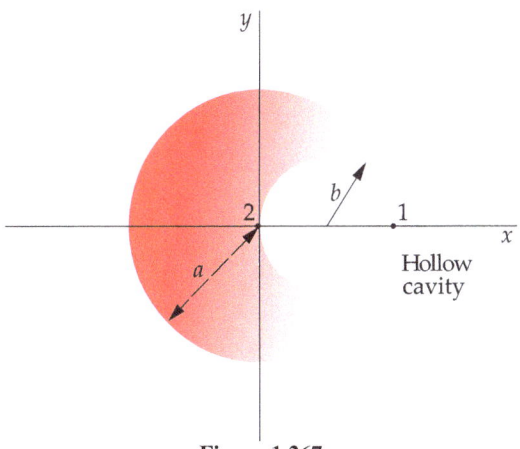

Figure 1.267

that the dipole moment is in the same direction as the field of the line of charge. Determine an expression for the electric force on the dipole.

1.37.3 Multiple Choice Questions
1.37.3.1 Level 1
Charge & its Properties

1. The ratio of electric force to gravitational force acting between two electrons will be
   (A) 1×10^{36} (B) 2×10^{39}
   (C) 6×10^{45} (D) 4×10^{42}

2. F_g and F_e represent the gravitational and electrostatic force respectively between two electrons situated at some distance. The ratio of F_g to F_e is of the order of -
   (A) 10^{36} (B)10^1 (C) 10^0 (D) 10^{-43}

3. One quantum of charge should be at least be equal to the charge in coulomb:
   (A) 1.6×10^{-17}C. (B) 1.6×10^{-19}C.
   (C) 1.6×10^{-10}C (D). 4.8×10^{-10}C.

4. The unit of charge is coulomb in SI system and esu of charge (or stat coulomb) in C.G.S. system 1 coulomb equals
   (A) 3×10^9 esu (B) $\left(1/3 \times 10^9\right)$ esu
   (C) $\left(1/3 \times 10^8\right)$ esu (D) $\left(9 \times 10^9\right)$ esu

5. The relative strengths of gravitational, electromagnetic and strong nuclear forces are-
   (A) $1 : 10^{39} : 10^{36}$ (B) $1 : 10^{36} : 10^{39}$
   (C) $1 : 10^{-26} : 10^{-39}$ (D) $1 : 10^{-39} : 10^{-36}$

6. An electron at rest has a charge of 1.6×10^{-19}C. It starts moving with a velocity $v = c/2$, where c is the speed of light, then the new charge on it is -
   (A) 1.6×10^{-19} coulomb
   (B) $1.6 \times 10^{-19} \sqrt{1-\left(\frac{1}{2}\right)^2}$ coulomb
   (C) $1.6 \times 10^{-19} \sqrt{\left(\frac{2}{1}\right)^2 - 1}$ coulomb
   (D) $\frac{1.6 \times 10^{-19}}{\sqrt{1-\left(\frac{1}{2}\right)^2}}$ coulomb

7. If a glass rod is rubbed with silk, it acquires a positive charge because-
   (A) Protons are added to it.
   (B) Protons are removed from it.
   (C) Electrons are added to it.
   (D) Electrons are removed from it.

8. Which one of the following is the unit of electric charge?
   (A) Coulomb (B) Newton
   (C) Volt (D) Coulomb/Volt

9. An accelerated or decelerated charge produces-
   (A) Electric field only
   (B) Magnetic field only
   (C) Localized electric and magnetic fields
   (D) Electric and magnetic fields that are radiated

10. Which one of the following statement regarding electrostatics is wrong ?
    (A) Charge is quantized
    (B) Charge is conserved
    (C) There is an electric field near an isolated charge at rest
    (D) A stationary charge produces both electric and magnetic fields

11. The dielectric constant for water is-
    (A) 1 (B) 40 (C) 81 (D) 0.3

12. In M.K.S. System, $1/4\pi\epsilon_0$ equals-
    (A) 9×10^9 N.m^2/C^2
    (B) 1 N.m^2/C^2
    (C) 1 dyne .cm^2/ stat C^2
    (D) 9×10^9 dyne ×cm^2/ stat C^2

13. A stationary electric charge produces-
    (A) Only electric fields
    (B) Only magnetic field
    (C) Both electric and magnetic field
    (D) Neither electric nor magnetic field

14. Charges reside on the-
    (A) Outer surface of the charged conductor
    (B) Inner surface of the charged conductor
    (C) Inner as well as outer surface of the charged conductor
    (D) None of the above

15. An isolated solid metallic sphere is charged with $+Q$ charge .The distribution of their $+Q$ charge on the sphere will be-
    (A) uniform but on the surface alone
    (B) non uniform but on the surface alone
    (C) uniform inside the volume
    (D) non uniform inside the volume

Coulomb's Law

16. Two similar charge of $+Q$, as shown in Figure1.268 are placed at A and B. $-q$ charge is placed at point C midway between A and B. $-q$ charge will oscillate if
    (A) It is moved towards A.
    (B) It is moved towards B.
    (C) It is moved upwards AB.
    (D) Distance between A and B is reduced.

Figure 1.268

17. When the distance between two charged particle is halved, the force between them becomes-
    (A) One fourth (B) One half
    (C) Double (D) Four times

18. The force between two point charges in vacuum is 15N, if a brass plate is introduced between the two charges, then force between them will-
    (A) Become zero (B) Remain the same
    (C) Become 30 N (D) Become 60 N

19. The force between an α particle and an electron separated by a distance of 1 Å is-

(A) 2.3×10^{-8} N attractive
(B) 2.3×10^{-8} N Repulsive
(C) 4.6×10^{-8} N attractive
(D) 4.6×10^{-8} repulsive

20. Two charges are at distance (d) apart in air. Coulomb force between them is F. If a dielectric material of dielectric constant (κ) is placed between them, the coulomb force now becomes.
    (A) F/κ (B) $F\kappa$ (C) F/κ^2 (D) $\kappa^2 F$

21. Two point charges in air at a distance of 20 cm. from each other interact with a certain force. At what distance from each other should these charges be placed in oil of relative permittivity 5 to obtain the same force of interaction –
    (A) 8.94×10^{-2} m (B) 0.894×10^{-2} m
    (C) 89.4×10^{-2} m (D) 8.94×10^{2} m

22. A certain charge Q is divided at first into two parts, (q) and $(Q - q)$. Later on the charges are placed at a certain distance. If the force of interaction between the two charges is maximum then-
    (A) $Q/q = 4/1$ (B) $Q/q = 2/1$
    (C) $Q/q = 3/1$ (D) $Q/q = 5/1$

23. A unit charge is one which when placed in vacuum one cm from an equal charge of the same kind will repel it with a force of-
    (A) 1 N (B) 1 dyne (C) 2 dyne (D) 4 dyne

24. The permittivity ϵ_0 of vacuum is 8.86×10^{-12} C²/N – m² and the dielectric constant of water is 81. The permittivity of water in C²/Nm² is-
    (A) $81 \times 8.86 \times 10^{-12}$ (B) 8.86×10^{-12}
    (C) $(8.86 \times 10^{-12})/81$ (D) $81/(8.86 \times 10^{-12})$

25. The force between two point charges placed in vacuum at distance 1 mm is 18 N. If a glass plate of thickness 1 mm and dielectric constant 6, be kept between the charges then new force between them would be-
    (A) 18 N (B) 10^8 N
    (C) 3 N (D) 3×10^{-6} N

26. Two similar and equal charges repel each other with force of $1.6\,N$, when placed 3 m apart. Strength of each charge is -
    (A) 40 μC (B) 20 μC
    (C) 4 μC (D) 2 μC

27. There are two charges +1 micro-coulomb and +5 micro-coulomb, the ratio of force on them will be -
    (A) 10^{43} (B) 1 : 1 (C) 10° (D) 10^{-43}

Superposition Principle

28. A charge Q is divided in two parts Q_1 and Q_2 and these charges are placed at distance R. there will be maximum repulsion between them, when-
    (A) $Q_2 = (Q/R), Q_1 = Q - (Q/R)$
    (B) $Q_2 = (Q/3), Q_1 = (2Q/3)$
    (C) $Q_2 = (Q/4), Q_1 = (3Q/4)$
    (D) $Q_1 = Q_2 = Q/2$

29. The three charges each of 5×10^{-6} coloumb are placed at vertex of an equilateral triangle of side 10 cm. The force exerted on the charge of $1\mu C$ placed at centre of triangle in newton will be-
    (A) 13.5 (B) zero (C) 4.5 (D) 6.75

30. A point charge q_1 exerts a force F upon another charge q_2. If one other charge q_3 be placed quite near to charge q_2, then the froce that charge q_1 exerts on the charge q_2 will be
    (A) F (B) > F (C) (C) < F (D) zero

31. A mass particle (mass = m and charge = q) is placed bewteen two point charges of charge q separtion between these two charge is 2 L. The frequency of oscillation of mass particle, if it is displaced for a small distance along the line joining the charges-
    (A) $\dfrac{q}{2\pi}\sqrt{\dfrac{1}{m\pi\epsilon_0 L^3}}$ (B) $\dfrac{q}{2\pi}\sqrt{\dfrac{4}{m\pi\epsilon_0 L^3}}$
    (C) $\dfrac{q}{2\pi}\sqrt{\dfrac{1}{4m\pi\epsilon_0 L^3}}$ (D) $\dfrac{q}{2\pi}\sqrt{\dfrac{1}{16\pi\epsilon_0 m L^3}}$

32. Two small balls having equal positive charge Q (Coulomb) on each are suspended by two insulating strings of equal length 'L' metre, from a hook fixed to a stand. The whole set up is taken in a satellite in to space where there is no gravity (state of weightlessness) Then the angle (θ) between the two strings is-
    (A) $0°$ (B) $90°$
    (C) $180°$ (D) $0° < q < 180°$

33. ABC is a right angle triangle AB = 3 cm, BC = 4 cm charges +15, +12, −12 esu are placed at A, B and C respectively. The magnitude of the force experienced by the charge at B in dyne is-
    (A) 125 (B) 35 (C) 22 (D) 0

34. Equal charges of each $2\mu C$ are placed at a point $x = 0, 2, 4,$ and 8 cm on the x-axis. The force experienced by the charge at $x = 2$ cm is equal to -
    (A) 5 Newton (B) 10 Newton
    (C) 0 Newton (D) 15 Newton

35. Three equal charges (q) are placed at corners of a equilateral triangle. The force on any charge is-
    (A) Zero (B) $\sqrt{3}\dfrac{Kq^2}{a^2}$
    (C) $\dfrac{Kq^2}{\sqrt{3}a^2}$ (D) $3\sqrt{3}\dfrac{Kq^2}{a^2}$

36. Two identical charges of magnitude 'q' are placed at $(-a, 0)$ and $(a, 0)$. Same nature charged particle is placed at origin. It executes S.H.M. If it is displaced-
    (A) in x-direction
    (B) in y-direction
    (C) at an angle of $45°$ from the x-axis
    (D) along perpendicular to the plane.

37. Two equal negative charge $(-q)$ are fixed at the points $(0, a)$ and $(0, -a)$ on the y-axis. A positive charge (Q) is released from rest at the point $(2a, 0)$ on the x-axis. The charge Q will -
    (A) execute simple harmonic motion about the origin.
    (B) move to the origin and remains at rest
    (C) move to infinity
    (D) execute oscillatory but not simple harmonic motion

38. Five point charges, each of value +q coulomb, are placed on five vertices of a regular hexagon of side L meter. The magnitude of the force on a point charge of value $-q$ coulomb placed at the centre of the hexagon is-
    (A) $\dfrac{kq^2}{L^2}$ (B) $\sqrt{5}\dfrac{kq^2}{L^2}$
    (C) $\sqrt{3}\dfrac{kq^2}{L^2}$ (D) Zero

Electric Field

39. A pendulum bob of mass 80 mg and carrying a charge of 2×10^{-8} C. is at rest in a horizontal uniform electric field of $20,000$ V m^{-1}. Find the tension in the thread of pendulum-
    (A) 8.8×10^{-2} N (B) 8.8×10^{-3} N
    (C) 8.8×10^{-4} N (D) 8.8×10^{-5} N

40. Two charges 4q and q are placed 30 cm. apart. At what point the value of electric field will be zero

(A) 10 cm away from q and between the charge
(B) 20 cm away from q and between the charge
(C) 10 cm away from q and out side the line joining the charge.
(D) 10 cm away from $4q$ and out side the line joining them.

41. One unit of electric field intensity is newtons/coulomb. The other unit of this can be
    (A) Vm (B) Vm^2
    (C) V/m (D) V/m^2

42. If $Q = 2$ C coulomb and force on it is $F = 100$ N. Then the value of field intensity will be-
    (A) 100 N/C (B) 50 N/C
    (C) 200 N/C (D) 10 N/C

43. Four equal but like charge are placed at four corners of a square. The electric field intensity at the centre of the square due to any one charge is E then the resultant electric field intensity at centre of square will be
    (A) Zero (B) 4E (C) E (D) 1/2 E

44. Two charges $9e$ and $3e$ are placed at a distance r. The distance of the point where the electric field intensity will be zero is
    (A) $\left(\dfrac{r}{1+\sqrt{3}}\right)$ from $9e$ charge
    (B) $\left(\dfrac{r}{1+\sqrt{1/3}}\right)$ from $9e$ charge
    (C) $\left(\dfrac{r}{1-\sqrt{3}}\right)$ from $3e$ charge
    (D) $\left(\dfrac{r}{1+\sqrt{1/3}}\right)$ from $3e$ charge.

45. An electric field can deflect-
    (A) X-rays (B) Neutrons
    (C) α-particles (D) γ - rays

46. Which one of the following relations is correct-
    (A) 1 N/C = 10^8 Volt /m
    (B) 1 N/C = 10^{-6} V/m
    (C) 1 N/C = 1 V/m
    (D) 1 N/C = 10^{-8} V/m

47. If mass of the electron = 9.1×10^{-31} kg. Charge on the electron = 1.6×10^{-19} coulomb and g = 9.8 m/s^2. Then the intensity of the electric field required to balance the weight of an electron is-
    (A) 5.6×10^{-9} N/C (B) 5.6×10^{-11} N/C
    (C) 5.6×10^{-8} N/C (D) 5.6×10^{-7} N/C

48. Six charges $+Q$ each are placed at the corners of a regular hexagon of side (a), the electric field at the centre of hexagon is-
    (A) Zero (B) $\dfrac{1}{4\pi\epsilon_0}\cdot\dfrac{6Q^2}{a^2}$
    (C) $\dfrac{1}{4\pi\epsilon_0}\cdot\dfrac{Q^2}{a^2}$ (D) $\dfrac{1}{4\pi\epsilon_0}\cdot\dfrac{6Q^2}{a\sqrt{2}}$

49. Two charged spheres A and B are charged with the charges of +10 and +20 coulomb respectively and separated by a distance of 80 cm. The electric field at a point on the line joining the centers of the two spheres will be zero at a distance ... from sphere A.
    (A) 20 cm (B) 33 cm (C) 55 cm (D) 60 cm

50. Four charges $+q, +q, -q$ and $-q$ are placed respectively at the corners A, B, C and D of a square of side (a), arranged in the given order. Calculate the intensity at (O) the centre of the square
    (A) $\dfrac{4\pi\epsilon_0 \cdot a^2}{4\sqrt{2}q}$ (B) $\dfrac{4\sqrt{2}q}{4\pi\epsilon_0 \cdot a^2}$
    (C) $\dfrac{\pi\epsilon_0 \cdot a^2}{4\sqrt{2}q}$ (D) $\dfrac{4\sqrt{2}q}{\pi\epsilon_0 \cdot a^2}$

51. A ring of radius (R) carries a uniformly distributed charge $+Q$. A point charge $-q$ is placed on the axis of the ring at a distance 2R from the centre of the ring and released from rest. The particle-
    (A) Becomes in rest condition immediately.
    (B) Executes simple harmonic motion
    (C) Motion is not SHM
    (D) Come at the centre of ring immediately.

52. A small circular ring has a uniform charge distribution. On a far-off axial point distance x from the centre of the ring, the electric field is proportional to-
    (A) x^{-1} (B) $x^{-3/2}$ (C) x^{-2} (D) $x^{5/4}$

Electric flux and Gauss Laws

53. The tangent drawn at a point on a line of electric force shows the-
    (A) intensity of gravity field
    (B) intensity of magnetic field
    (C) intensity of electric field
    (D) direction of electric field

54. Which of the following statements concerning the electrostatics is correct-
    (A) electric line of force never intersect each other
    (B) electric lines of force start from positive charge and end at the negative charge
    (C) electric lines of force start or ends perpendicular to the surface of a charged metal.
    (D) all of the above

55. When no charge is confined with in the Gauss's surface, it implies that-
    (A) $E = 0$
    (B) $\vec{E}$ and $d\vec{A}$ are parallel
    (C) $\vec{E}$ and $d\vec{A}$ are mutually perpendicular
    (D) $\vec{E}$ and $d\vec{A}$ are inclined at some angle

56. If electric field flux coming out of a closed surface is zero, the electric field at the surface will be-
    (A) zero
    (B) same at all places
    (C) dependent upon the location of points
    (D) infinite

57. If three electric di-poles are placed in some closed surface, then the electric flux emitting from the surface will be-
    (A) zero (B) positive
    (C) negative (D) None

58. For which of the following fields, Gauss's law is valid-
    (A) fields following inverse square law
    (B) uniform field
    (C) all types of field
    (D) this law has no concern with the field

59. A charge of Q coloumb is located at the centre of a cube. If the corner of the cube is taken as the origin, then the flux coming out from the faces of the cube in the direction of X - axis will be-
    (A) $4\pi Q$ (B) $Q/6\epsilon_0$
    (C) $Q/3\epsilon_0$ (D) $Q/4\epsilon_0$

60. A rectangular surface of 2 metre width and 4 metre length, is placed in an electric field of intensity 20 newton/C, there is an angle of 60° between the perpendicular to surface and electrical field intensity. Then total flux emitted from the surface will be- (In volt-metre)
    (A) 80 (B) 40 (C) 20 (D) 160

61. A charge q is inside a closed surface and charge $-q$ is outside. The out going electric flux is-
(A) $-q/\epsilon_0$ (B) zero (C) q/ϵ_0 (D) $2q/\epsilon_0$

62. If the electric field is uniform, then the electric lines of forces are-
(A) Divergent (B) Convergent
(C) Circular (D) Parallel

63. Electric lines of forces-
(A) Exist everywhere
(B) Are imaginary
(C) Exist only in the immediate vicinity of electric charges
(D) None of the above

64. Which one of the following diagrams Fig.1.269 shows the correct lines of force ?

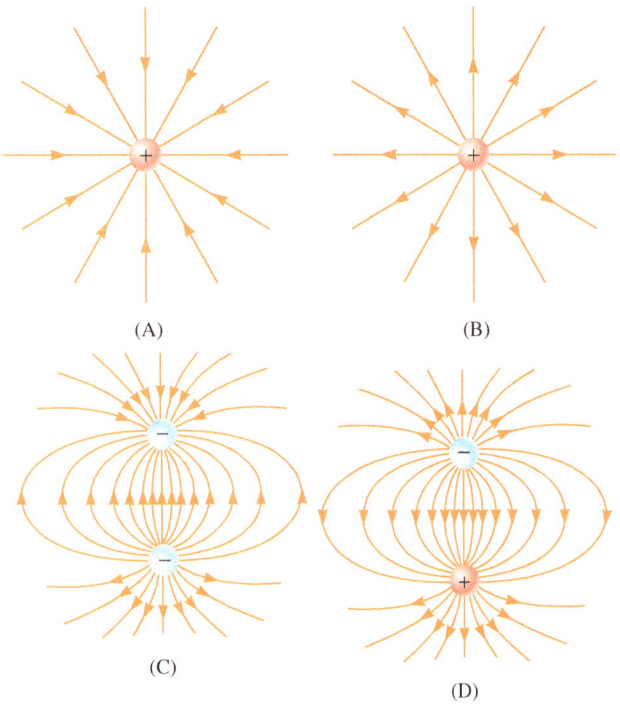

Figure 1.269

65. In Fig.1.270 shown the electric lines of force emerging from a charged body. If the electric fields at A and B are E_A and E_B are respectively, If the distance between A and B is r then
(A) $E_A > E_B$ (B) $E_A < E_B$
(C) $E_A = E_B$ (D) $E_A = (E_B)/r^2$

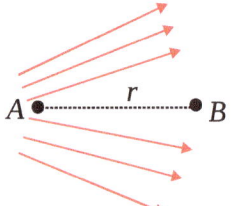

Figure 1.270

Application of Gauss law

66. Three charges $q_1 = 1\ \mu C, q_2 = 2\ \mu C$ and $q_3 = -3\ \mu C$ and four surfaces S_1, S_2, S_3 and S_4 are shown in Fig.1.271. The flux emerging through surface S_2 in N.m^2/C is-
(A) $36\pi \times 10^3$ (B) $-36\pi \times 10^3$
(C) $36\pi \times 10^9$ (D) $-36\pi \times 10^9$

67. A surface enclosed an electric dipole, the flux through the surface is -

Figure 1.271

(A) Infinite (B) Positive
(C) Negative (D) Zero

68. Total flux coming out of some closed surface is-
(A) q/ϵ_0 (B) ϵ_0/q (C) $q\epsilon_0$ (D) $\sqrt{\frac{q}{\epsilon_0}}$

69. A square of side 20 cm. is enclosed by a concentric spherical surface of radius 80 cm. Four charges $+2 \times 10^{-6}$ C, -5×10^{-6} C, -3×10^{-6} C, $+6 \times 10^{-6}$ C, are located at the four corners of a square, then out going total flux from spherical surface in N m^2/C, will be-
(A) zero (B) $(16\pi) \times 10^{-6}$
(C) $(8\pi) \times 10^{-6}$ (D) $(36\pi) \times 10^{-6}$

70. The flux emerging out from any one face of the cube will be -
(A) $\frac{q}{6\epsilon_0}$ (B) $\frac{q}{3\epsilon_0}$ (C) $\frac{q}{\epsilon_0}$ (D) $\frac{q}{4\epsilon_0}$

71. The electric field inside a spherical shell of uniform surface charge density is -
(A) Zero
(B) Constant, different from zero
(C) Proportional to the distance from the centre
(D) None of the above

72. A cubical box of side 1 m is immersed a uniform electric field of strength 10^4 N/C. The flux through the cube is-
(A) 10^4 (B) 6×10^4 (C) 2×10^4 (D) Zero

73. A charge (q) is located at the centre of a cube. The electric flux through any face of the cube is-
(A) q/ϵ_0 (B) $q/2\epsilon_0$ (C) $q/4\epsilon_0$ (D) $q/6\epsilon_0$

74. A large isolated metal sphere of radius (R) carries a fixed charge. A small charge is placed at a distance (r) from its surface experiences a force which is -
(A) Proportional to R
(B) Independent of R and
(C) Inversely proportional to $(R + r)^2$
(D) inversely proportional to r^2

75. A hollow sphere of charge does not produce an electric field at any-
(A) Interior point (B) Outer point
(C) Surface point (D) None of the above

76. A spherical conductor of radius 50 cm has a surface charge density of 8.85×10^{-6} C/m^2. The electric field near the surface in N/C is-
(A) 8.85×10^{-6} (B) 8.85×10^6
(C) 1×10^6 (D) Zero

77. The electric field intensity at a point located at distance $r(r < R)$ from the centre of a spherical conductor (radius R) of charge Q will be-
(A) kQR/r^3 (B) kQr/R^3
(C) kQ/r^2 (D) zero.

Electric Dipole

78. If an electric dipole is kept in a uniform electric field, then it will experience
(A) a force
(B) a couple and moves
(C) a couple and rotates
(D) a force and moves.

79. An electric dipole consists of two opposite charges each of

magnitude 1×10^{-6}C separated by a distance 2 cm. The dipole is placed in an external field of 10×10^5 N/C. The maximum torque on the dipole is -
(A) 0.2×10^{-3} N.m (B) 1.0×10^{-3} N.m
(C) 2×10^{-3} N.m (D) 4×10^{-3} N.m

80. The ratio of the electric field due to an electric dipole on its axis and on the perpendicular bisector of the dipole is -
(A) 1 : 2 (B) 2 : 1 (C) 1 : 4 (D) 4 : 1

81. The region surrounding a stationary electric dipole has-
(A) electric field only
(B) magnetic field only
(C) both electric and magnetic fields
(D) neither electric non magnetic field

1.37.3.2 Level 2

1. 5×10^5 lines of electric flux are entering in a closed surface and 4×10^5 lines come out of the surface the charge enclosed by the surface is -
(A) 0.885×10^{-6}C (B) 8.85×10^{-6}C
(C) -8.85×10^{-7}C (D) 8.85×10^{-8}C

2. A cylinder of radius (R) and length (L) is placed in a uniform electrical field (E) parallel to the axis of the cylinder the total flux for the surface of the cylinder is given by-
(A) $2\pi R^2 E$ (B) $\pi R^2 E$
(C) $\dfrac{\pi R^2 + \pi R^2}{E}$ (D) zero

3. A hemisphere (radius R) is placed in electric field as shown in Fig.1.272 Total outgoing flux is -
(A) $\pi R^2 E$ (B) $2\pi R^2 E$
(C) $4\pi R^2 E$ (D) $(\pi R^2 E)/2$

Figure 1.272

4. Three identical charges each of 1 μC are kept on the circumference of a circle of radius 1 metre forming equilateral triangle. The electric intensity at the centre of the circle in N/C is-
(A) 9×10^3 (B) 13.5×10^3 (C) 27×10^3 (D) Zero

5. The number of electrons, falling on spherical conductor (radius = 0.1 m) to produce 0.036 N/C electric field at the surface of conductor, is-
(A) 2.7×10^5 (B) 2.5×10^2
(C) 2.6×10^5 (D) 2.4×10^5

6. A particle of mass 6 μg carrying a charge of 10^{-9}C, is placed in the electric field of strength E = 6×10^5 V/m, the acceleration acquired by the particle is-
(A) 10^2 m/s^2 (B) 10^3 m/s^2
(C) 10^5 m/s^2 (D) 10^{20} m/s^2

7. Two small balls having equal positive charge Q on each are suspended by two insulating strings at equal length L meter, from a hook fixed to a stand. The whole set up is taken in a satellite into space where there is no gravity. Then the angle θ between two strings and tension in each string is-
(A) $0, \dfrac{kq^2}{L^2}$ (B) $\pi, \dfrac{kq^2}{2L^2}$
(C) $\pi, \dfrac{kq^2}{4L^2}$ (D) $\dfrac{\pi}{2}, \dfrac{kq^2}{2L^2}$

8. The magnitude of the electric field strength (E) such that an electron placed in the field would experience an electrical force equal to its weight is [assume g = 10 m/see^2]
(A) 5.68×10^{-11} N/C vertically up.
(B) 5.68×10^{-11} N/C vertically down.
(C) 5.68×10^{-10} N/C vertically up.
(D) 5.68×10^{-10} N/C vertically down.

9. The electric field at the surface of a charged spherical conductor is 10 kV/m. The electric field at a distance equal to the diameter from its centre will be -
(A) 2.5 V/m (B) 2.5 kV/m
(C) 5.0 kV/m (D) 5.0 V/m

10. An Electron is situated 3×10^{-9} m from one α - particle and 4×10^{-9} m from another α - particle. The magnitude of force on the electron, when two α - particles are 5×10^{-9} m apart is -
(A) 5.64×10^{-11} N (B) 56.4×10^{-11} N
(C) 0.564×10^{-11} N (D) 564×10^{-11} N.

11. Two large metal plates, each of area A, carry charges $+q$ and $-q$ and face each other. the plates are separated by a small distance d the electric field between the plates would be-
(A) $\dfrac{2q}{\epsilon_0 A}$ (B) $\dfrac{qA}{\epsilon_0 A}$ (C) $\dfrac{q}{\epsilon_0 A}$ (D) $\dfrac{A}{q\epsilon_0}$

12. Two parallel plates of infinite dimensions are uniformly charged [Fig.1.273]. The surface charge density on one plate is σ_A and on the other is σ_B, field intensity at point C will be-
(A) $\propto (\sigma_A - \sigma_B)$ (B) $\propto (\sigma_A + \sigma_B)$
(C) zero (D) $2\sigma_A$

Figure 1.273

13. A charged particle of mass m is kept in equilibrium in the electric field between the plates of Millikan oil drop experiment. If the direction of the electric field between the plate is reversed. then acceleration of the charged particle will be-
(A) Zero (B) $g/2$ (C) g (D) $2g$

14. Two identical small balls, each of mass, are suspended by two light inelastic conducting threads each of length l from the same fixed point support. If the distance d between two balls is very small as compared to l, then d is equal to-
(A) $\left(\dfrac{2kq^2}{mg}\right)^{1/3}$ (B) $\left(\dfrac{2klq^2}{mg}\right)^{2/3}$
(C) $\left(\dfrac{kq^2}{2mg}\right)^{2/3}$ (D) none of these

15. A metal sphere A of radius R has a charge of Q on it The field at a point B outside the sphere is E. Now another sphere of radius R having a charge $-3Q$ is placed at point B. The total field at a point midway between A and B due to both sphere is-
(A) $4E$ (B) $8E$ (C) $12E$ (D) $16E$

16. A uniformly charged rod with charge per unit length λ is bent in to the shape of a semicircle of radius R. The electric field at the centre is -
(A) $2k\lambda/R$ (B) $k\lambda/2R$ (C) Zero (D) None

17. A thin stationary ring of radius 1 m has a positive charge 10 μC uniformly distributed over it. A particle of mass 0.9 gram and having a negative charge of 1μC is placed on the axis at a distance of 1 cm from the centre of the ring and

released then time period of oscillation of particle will be-
(A) 0.6 s   (B) 0.2 s   (C) 0.3 s   (D) 0.4 s

18. Three charges $+3q$, $+q$ and Q are placed on a straight line with equal separation. In order to make the net force on q to be zero, the value of Q will be-
(A) $+3q$ (B) $+2q$ (C) $-3q$ (D) $-4q$

19. As per diagram 1.274 point charge $+q$ is placed at the origin O. Work done in taking another point charge $-Q$ from the point A [coordinates $(0, a)$] to another point B coordinates $(a, 0)$] along the straight path AB is-

(A) Zero   (B) $\left(\dfrac{-qQ}{4\pi\epsilon_0}\dfrac{1}{a^2}\right)\sqrt{2}a$

(C)   (D) $\left(\dfrac{qQ}{4\pi\epsilon_0}\dfrac{1}{a^2}\right)\sqrt{2}a$

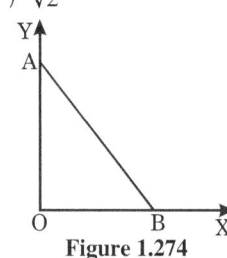
Figure 1.274

20. The electric field due to an electric dipole at a distance r from its centre in axial position is E. if the dipole is rotated through an angle of 90° about its perpendicular axis, the electric field at the same point will be-
(A) E (B) $E/4$ (C) $E/2$ (D) $2E$

21. Two electric dipoles of moment p and $64\ p$ are placed in opposite direction on a line at a distance of 25 cm. The electric field will be zero at point between the dipoles whose distance from the dipole of moment p is-
(A) 5 cm   (B) $\frac{25}{9}$ cm   (C) 10 cm   (D) $\frac{4}{13}$ cm

22. When an electric dipole of dipole moment $\vec{p}$, is placed in a uniform electric field $\vec{E}$ then at what angle between $\vec{p}$ and $\vec{E}$ the value of torque will be maximum -
(A) 90°   (B) 0°   (C) 180°   (D) 45°

23. An electric dipole has the magnitude of its charge as q and its dipole moment is p. It is placed in a uniform electric field E. If its dipole moment is along the direction of the field, the force on it is-
(A) $2qE$ (B) qE (C) 0   (D) $-qE$

24. Two opposite and equal charges 4×10^{-8} coulomb when placed 2×10^{-2} m away, form a dipole. If this dipole is placed in an external electric field 4×10^3 newton/coulomb, the value of maximum torque and the work done in rotating it through 180° will be-
(A) 64×10^{-4} Nm and 64×10^{-4} J
(B) 32×10^{-4} Nm and 32×10^{-4} J
(C) 64×10^{-4} Nm and 32×10^{-4} J
(D) 32×10^{-4} Nm and 64×10^{-4} J

1.37.3.3 Level 3

1. If an electron enters into a space between the plates of a parallel plate capacitor at an angle α with the plates and leaves at an angle β with the plates. The ratio of its kinetic energy while entering the capacitor to that while leaving will be -
(A) $\left(\dfrac{\sin\beta}{\sin\alpha}\right)^2$ (B) $\left(\dfrac{\cos\beta}{\cos\alpha}\right)^2$
(C) $\left(\dfrac{\cos\alpha}{\cos\beta}\right)^2$ (D) $\left(\dfrac{\sin\alpha}{\sin\beta}\right)^2$

2. Force between two identical charges placed at a distance of r in vacuum is F. Now a liquid of dielectric constant $\kappa = 4$ is inserted between these two charges. The thickness of the liquid is $r/2$. The force between the charges will now become -
(A) $F/4$ (B) $F/2$ (C) $3F/5$ (D) $4F/9$

3. In a certain region of surface there exists a uniform electric field of $2 \times 10^3 \hat{k}$ V/m. A rectangular coil of dimensions 10 cm × 20 cm is placed in xy plane. The electric flux through the coil is-
(A) Zero   (B) 30 V.m   (C) 40 V.m   (D) 50 V.m

4. The electric flux from a cube of edge l is ϕ. What will be its value if edge of cube is made $2\ l$ and charge enclosed is halved -
(A) $\phi/2$ (B) 2ϕ (C) 4ϕ (D) ϕ

5. Each of the two point charges are doubled and their distance is halved. Force of interaction becomes n times, where n is -
(A) 4   (B) 1   (C) 1/16   (D) 16

6. Two point charges repel each other with a force of 100 N. One of the charges is increased by 10% and other is reduced by 10%. The new force of repulsion at the same distance would be -
(A) 100 N   (B) 121 N
(C) 99 N   (D) None of these

7. A spherical charged conductor has σ as the surface density of charge. The electric field on its surface is E. If the radius of the sphere is doubled keeping the surface density of charge unchanged, what will be the electric field on the surface of the new sphere -
(A) $E/4$ (B) $E/2$ (C) E (D) $2E$

8. Three equal and similar charges are placed at $(-a, 0, 0)$, $(0, 0, 0)$ and $(+a, 0, 0)$. What is the nature of equilibrium of the charge at the origin-
(A) Stable when moved along the Y-axis
(B) Stable when moved along Z-axis
(C) Stable when moved along X-axis
(D) Unstable in all of the above cases

9. The electric field strength due to a ring of radius R at a distance x from its centre on the axis of ring carrying charge Q is given by
$$E = \frac{1}{4\pi\epsilon_0}\frac{Qx}{(R^2+x^2)^{3/2}}$$
The value of x corresponding to maximum electric field, is-
(A) R (B) $R/2$ (C) $R/\sqrt{2}$ (D) $\sqrt{2}R$

10. Two point charges Q and $-3Q$ are placed certain distance apart. If the electric field at the location of Q be $\vec{E}$, then at the location of $-3Q$ will be-
(A) $3\vec{E}$ (B) $-3\vec{E}$ (C) $\vec{E}/3$ (D) $-\vec{E}/3$

11. A and B are two points on the axis and the perpendicular bisector respectively of an electric dipole. A and B are far away from the dipole and at equal distances from it. The fields at A and B are $\vec{E}_A$ and $\vec{E}_B$ respectively such that-
(A) $\vec{E}_A = \vec{E}_B$ (B) $\vec{E}_A = 2\vec{E}_B$
(C) $\vec{E}_A = -2\vec{E}_B$ (D) $\vec{E}_A = \vec{E}_B/2$

12. A long string with a charge of λ per unit length passes through an imaginary cube of edge l. The maximum possible flux of the electric field through the cube will be
(A) $\lambda l/\epsilon_0$ (B) $\sqrt{2}\lambda l/\epsilon_0$ (C) $6\lambda l^2/\epsilon_0$ (D) $\sqrt{3}\lambda l/\epsilon_0$

13. A charge Q is placed at each of two opposite corners of a square and a charge q is placed at other two opposite corners of the square. If the resultant electric field at the position of Q is zero, then -

(A) $Q = -\frac{q}{2\sqrt{2}}$ (B) $Q = -2\sqrt{2}q$
(C) $Q = -2q$ (D) $Q = 2\sqrt{2}q$

14. The electric field outside a charged long straight wire is given by $E = -\frac{5000}{r}\text{Vm}^{-1}$. It is radially inward. The value of $V_B - V_A$ is -
[Given $r_B = 60$ cm and $r_A = 30$ cm]
(A) $5000\log_e 2$ V   (B) 0 V   (C) 2 V   (D) 2500 V

15. An electron moves with velocity $\vec{v}$ in x - direction. An electric field acts on it in y - direction. The force on the electron acts in -
(A) +ve direction of Y - axis
(B) −ve direction of Y - axis
(C) +ve direction of Z - axis
(D) −ve direction of Z - axis

16. Two identical simple pendulums A and B, are suspended from the same point. The bobs are given positive charges, with A having more charge than B. They diverge and reach equilibrium, with A and B making angles θ_1 and θ_2 with the vertical respectively. T_A and T_B are tensions in A and B respectively. Which of the following is correct-
(A) $\theta_1 > \theta_2$ (B) $\theta_1 < \theta_2$
(C) $\theta_1 = \theta_2$ (D) $T_A > T_B$

Statements Type Question:
Each of the questions given below consist of Statement - I and Statement- II. Use the following Key to choose the appropriate answer.
(A) If both Statement-I and Statement-II are true, and Statement-II is the correct explanation of Statement-I.
(B) If both Statement -I and Statement -II are true but Statement- II is not the correct explanation of Statement-I.
(C) If Statement -I is true but Statement -II is false.
(D) If Statement -I is false but Statement -II is true.

17. Statement I: Force between two charges decreases when air separating the charges is replaced by water.
Statement II : Medium intervening the charges has no effect on force.

18. Statement I: The number of lines of force emanating from 1μC charge in vacuum is 1.13×10^5*
Statement II: This follow from Gauss's theorem in electrostatics.

1.37.3.4 Level 4 (Previous Years JEE Main & Advanced Questions)
Section A: JEE Main

1. Two points are separated by a certain distance. A charge Q is placed at each of these two points. Find the third charge which is placed at the mid point of the line joining the charges so that the system is in equilibrium - [2002]
(A) $-Q/4$ (B) $-Q/2$ (C) $-Q/3$ (D) $-Q$

2. A charged particle q is placed at the centre O of cube of length L (ABCDEFGH). Another same charge q is placed at a distance L from O. Then the electric flux through BCFG is[Fig.1.275]- [2002]
(A) L (B) zero   (C) $3L$ (D) $\frac{q}{3\pi\epsilon_0 L}$

3. Three charges $-q_1$, $+q_2$ and $-q_3$ are placed as shown in Fig.1.276. The x - component of the force on $-q_1$ is proportional to – [2003]
(A) $\frac{q_2}{b^2} + \frac{q_3}{a^2}\sin\theta$ (B) $\frac{q_2}{b^2} + \frac{q_3}{a^2}\cos\theta$
(C) $\frac{q_2}{b^2} + \frac{q_3}{a^2}\cos\theta$ (D) $\frac{q_2}{b^2} - \frac{q_3}{a^2}\cos\theta$

4. If the electric flux entering and leaving an enclosed surface respectively is ϕ_1 and ϕ_2, the electric charge inside the surface will be – [2003]

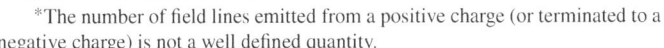
*The number of field lines emitted from a positive charge (or terminated to a negative charge) is not a well defined quantity.

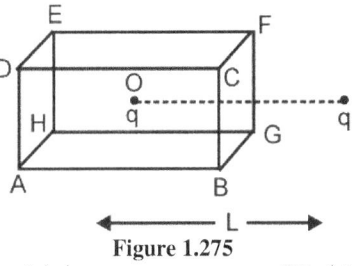
Figure 1.275

(A) $(\phi_1 + \phi_2)/\epsilon_0$ (B) $(\phi_2 - \phi_1)/\epsilon_0$
(C) $(\phi_1 + \phi_2)\epsilon_0$ (D) $(\phi_2 - \phi_1)\epsilon_0$

5. Two spherical conductors B and C having equal radii and carrying equal charges on them repel each other with a force F when kept apart at some distance. A third spherical conductor having same radius as that of B but uncharged is brought in contact with B, then brought in contact with C and finally removed away from both. The new force of repulsion between B and C is -[2004]
(A) $F/4$ (B) $3F/4$ (C) $F/8$ (D) $3F/8$

6. Four charges equal to $-Q$ are placed at the four corners of a square and a charge q is at its centre. If the system is in equilibrium the value of q is - [2004]
(A) $-\frac{Q}{4}(1+2\sqrt{2})$ (B) $\frac{Q}{4}(1+2\sqrt{2})$
(C) $-\frac{Q}{2}(1+2\sqrt{2})$ (D) $\frac{Q}{2}(1+2\sqrt{2})$

7. A charged oil drop is suspended in a uniform field of 3×10^4 V/m so that it neither falls nor rises. The charge on the drop will be (Take the mass of the charge $= 9.9 \times 10^{-15}$ kg and $g = 10$ m/s^2) - [2004]
(A) 3.3×10^{-18}C   (B) 3.2×10^{-18}C
(C) 1.6×10^{-18}C   (D) 4.8×10^{-18}C

8. A charged ball B hangs from a silk thread S which makes an angle θ with a large charged conducting sheet P, as shown in the Figure 1.277. The surface charge density σ of the sheet is proportional to- [2005]
(A) $\cos\theta$ (B) $\cot\theta$ (C) $\sin\theta$ (D) $\tan\theta$

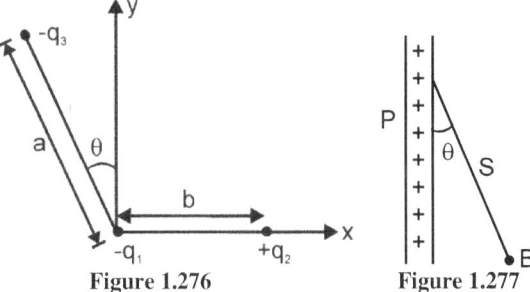
Figure 1.276   **Figure 1.277**

9. Two point charges $+8q$ and $-2q$ are located at $x = 0$ and $x = L$ respectively. The location of a point on the x axis at which the net electric field due to these two point charges is zero is - [2005]
(A) $2L$ (B) $L/4$ (C) $8L$ (D) $4L$

10. An electric dipole is placed at an angle of 30° to a non-uniform electric field. the dipole will experience - [2006]
(A) a torque as well as a translational force
(B) a torque only
(C) a translational force only in the direction of the field
(D) a translational force only in a directin normal to the direction of the field

11. If g_E and g_M are the accelerations due to gravity on the surfaces of the earth and the moon respectively and if Millikan's oil drop experiment could be performed on the two surfaces, one will find the ratio (electronic charge on

the moon/ electronic charge on the earth) to be- [2007]
(A) 1   (B) 0   (C) g_E/g_M (D) g_M/g_E

12. A thin spherical shell of radius R has charge Q spread uniformly over its surface. Which of the following graphs [Fig.1.278] most closely represents the electric field $E(r)$ produced by the shell in the range $0 \leq r < \infty$, where r is the distance from the centre of the shell? [2008]

Questions 13 and 14 consist of Statement-I and Statement-

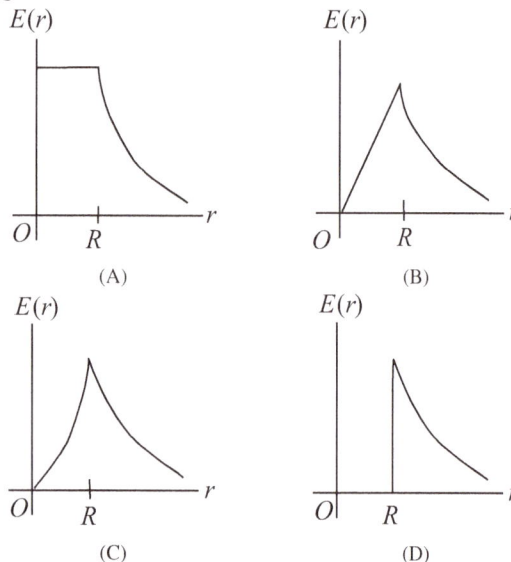

Figure 1.278

II. Use the following Key to choose the appropriate answer.
(A) If both Statement-I and Statement-II are true, and Statement - II is the correct explanation of Statement- I.
(B) If both Statement -I and Statement -II are true but Statement - II is not the correct explanation of Statement -I.
(C) If Statement -I is true but Statement -II is false.
(D) If Statement -I is false but Statement -II is true.

13. **Statement-1:** For a mass M kept at the centre of a cube of side 'a', the flux of gravitational field passing through its sides is $4\pi GM$.
Statement-2: If the direction of a field due to a point source is radial and its dependence on the distance 'r' from the source is given as $\frac{1}{r^2}$, its flux through a closed surface depends only on the strength of the source enclosed by the surface and not on the size or shape of the surface. [2008]

14. **Statement-1:** For a charged particle moving from point P to point Q, the net work done by an electrostatic field on the particle is independent of the path connecting point P to point Q.
Statement-2: The net work done by a conservative force on an object moving along a closed loop is zero. [2009]

15. Let $P(r) = \frac{Q}{\pi R^4}r$ be the charge density distribution for a solid sphere of radius R and total charge Q. For point 'P' inside the sphere at distance r_1 from the centre of the sphere, the magnitude of electric field is - [2009]
(A) 0   (B) $\frac{Q}{4\pi\epsilon_0 r_1^2}$ (C) $\frac{Qr_1^2}{4\pi\epsilon_0 R^4}$ (D) $\frac{Qr_1^2}{3\pi\epsilon_0 R^4}$

16. A charge Q is placed at each of the opposite corners of a square. A charge q is placed at each of the other two corners. If the net electrical force on Q is zero, then Q/q equals - [2009]
(A) $-2\sqrt{2}$ (B) -1 (C) 1 (D) $-\frac{1}{\sqrt{2}}$

17. A thin semi-circular ring of radius r has a positive charge q distributed uniformly over it[Fig.1.279]. The net field $\vec{E}$ at the centre O is [2010]
(A) $\frac{q}{4\pi^2\epsilon_0 r^2}j$ (B) $-\frac{q}{4\pi^2\epsilon_0 r^2}j$
(C) $-\frac{q}{2\pi^2\epsilon_0 r^2}j$ (D) $\frac{q}{2\pi^2\epsilon_0 r^2}j$

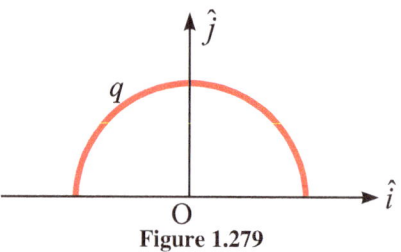

Figure 1.279

18. Let there be a spherically symmetric charge distribution with charge density varying as $\rho(r) = \rho_0\left(\frac{5}{4} - \frac{r}{R}\right)$ upto $r = R$, and $r(r) = 0$ for $r > R$, where r is the distance from the origin. The electric field at a distance $r(r < R)$ from the origin is given by [2010]
(A) $\frac{4\pi\rho_0 r}{3\epsilon_0}\left(\frac{5}{3} - \frac{r}{R}\right)$ (B) $\frac{\rho_0 r}{4\epsilon_0}\left(\frac{5}{3} - \frac{r}{R}\right)$
(C) $\frac{4\rho_0 r}{3\epsilon_0}\left(\frac{5}{4} - \frac{r}{R}\right)$ (D) $\frac{\rho_0 r}{3\epsilon_0}\left(\frac{5}{4} - \frac{r}{R}\right)$

19. Two identical charged spheres suspended from a common point by two massless strings of length l are initially a distance $d(d << 1)$ apart because of their mutual repulsion. The charge begins to leak from both the spheres at a constant rate. As a result the charges approach each other with a velocity v. Then v as a function of distance x between them - [2011]
(A) $v \propto x^{-1}$ (B) $v \propto x^{1/2}$ (C) $v \propto x$ (D) $v \propto x^{-1/2}$

20. In a uniformly charged sphere of total charge Q and radius R, the electric field E is plotted as a function of distance from the centre. The graph which would correspond to the above will be[Fig.1.280]- [2012]

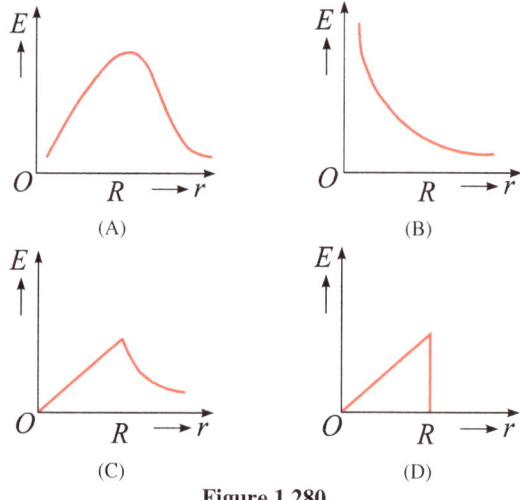

Figure 1.280

21. Let $[\epsilon_0]$ denote the dimensional formula of the permittivity of vacuum. If M = mass, L = length, T = time and A = electric current, then $[\epsilon_0]$ is: [2013]
(A) $[M^{-1}L^{-3}T^2A]$ (B) $[M^{-1}L^{-3}T^4A^2]$
(C) $[M^{-1}L^2T^{-1}A^{-2}]$ (D) $[M^{-1}L^2T^{-1}A]$

22. Two charges, each equal to q, are kept at $x = -a$ and $x = a$ on the x - axis. A particle of mass m and charge $q_0 = q/2$ is placed at the origin. If charge q_0 is given a small displacement $(y << a)$ along the y - axis, the net force acting on the particle is proportional to: [2013]

(A) y (B) −y (C) 1/y (D) −1/y

23. The region between two concentric spheres of radii 'a' and 'b', respectively (Figure 1.281), has volume charge density $\rho = A/r'$, where A is a constant and r is the distance from the centre. At the centre of the spheres is a point charge Q. The value of A such that the electric field in the region between the spheres will be constant, is: [2016]

(A) $\frac{2Q}{\pi a^2}$ (B) $\frac{e}{2\pi a^2}$
(C) $\frac{Q}{2\pi(b^2-a^2)}$ (D) $\frac{2Q}{\pi(a^2-b^2)}$

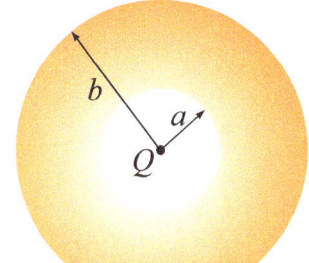

Figure 1.281

24. An electric dipole has a fixed dipole moment $\vec{p}$, which makes angle θ with respect to x-axis. When subjected to an electric field $\vec{E}_1 = E\hat{i}$, it experiences a torque $\vec{\tau}_1 = \tau\hat{k}$. When subjected to another electric field $\vec{E}_2 = E_1\hat{i}$ it experiences a torque $\vec{\tau}_2 = -\vec{\tau}_1$. The angle θ is: [2017]
(A) 90° (B) 30° (C) 45° (D) 60°

25. Two identical conducting spheres A and B, carry equal charge. They are separated by a distance much larger than their diameter, and the force between them is F. A third identical conducting sphere, C, is uncharged. Sphere C is first touched to A, then to B, and then removed. As a result, the force between A and B would be equal to [9 Jan 2018]
(A) 3F/4 (B) F/2 (C) F (D) 3F/8

26. A solid ball of radius R has a charge density ρ given by $\rho = \rho_0\left(1-\frac{r}{R}\right)$ for $0 \le r \le R$. The electric field outside the ball is: [15 April 2018]

(A) $\frac{\rho_0 R^3}{\epsilon_0 r^2}$ (B) $\frac{4\rho_0 R^3}{3\epsilon_0 r^2}$
(C) $\frac{3\rho_0 R^3}{4\epsilon_0 r^2}$ (D) $\frac{\rho_0 R^3}{12\epsilon_0 r^2}$

27. A charge Q is placed at a distance $a/2$ above the centre of the square surface of edge a as shown in the figure 1.282. The electric flux through the square surface is: [15 April 2018]

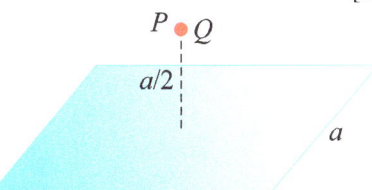

Figure 1.282

(A) $Q/3\epsilon_0$ (B) $Q/6\epsilon_0$ (C) $Q/2\epsilon_0$ (D) Q/ϵ_0

28. Three charges $+Q, q, +Q$ are placed respectively, at distance, $d/2$ and d from the origin, on the x-axis. If the net force experienced by $+Q$, placed at $x = 0$, is zero, then value of q is: [2019]
(A) $-Q/4$ (B) $+Q/2$ (C) $+Q/4$ (D) $-Q/2$

29. Charge is distributed within a sphere of radius R with a volume charge density $p(r) = \frac{A}{r^2}e^{-2r/a}$ where A and a are constants. If Q is the total charge of this charge distribution, the radius R is: [2019]

(A) $a\log\left(1-\frac{Q}{2\pi aA}\right)$ (B) $\frac{a}{2}\log\left(\frac{1}{1-\frac{Q}{2\pi aA}}\right)$
(C) $a\log\left(\frac{1}{1-\frac{Q}{2\pi aA}}\right)$ (D) $\frac{a}{2}\log\left(1-\frac{Q}{2\pi aA}\right)$

30. Four point charges $-q, +q, +q$ and $-q$ are placed on y-axis at $y = -2d, y = -d, y = +d$ and $y = +2d$, respectively. The magnitude of the electric field E at a point on the x-axis at $x = D$, with $D \gg d$, will behave as: [9 April 2019, (II)]
(A) $E \propto 1/D^3$ (B) $E \propto 1/D$
(C) $E \propto 1/D^4$ (D) $E \propto 1/D^2$

31. The bob of a simple pendulum has mass 2g and a charge of $5.0^{1/4}$C. It is at rest in a uniform horizontal electric field of intensity 2000 V/m. At equilibrium, the angle that the pendulum makes with the vertical is: (take $g = 10$ m/s^2) [8 April 2019(I)]
(A) $\tan^{-1}(2.0)$ (B) $\tan^{-1}(0.2)$
(C) $\tan^{-1}(5.0)$ (D) $\tan^{-1}(0.5)$

32. For a uniformly charged ring of radius R, the electric field on its axis has the largest magnitude at a distance h from its centre. Then value of h is: [9 Jan. 2019 (I)]
(A) $R/\sqrt{5}$ (B) $R/\sqrt{2}$ (C) R (D) $R\sqrt{2}$

33. Shown in the Fig.1.283, is a shell made of a conductor. It has inner radius a and outer radius b, and carries charge Q. At its centre is a dipole having dipole moment $\vec{p}$ as shown. In this case: [12 April 2019(I)]
(A) surface change density on the inner surface is uniform and equal to $\frac{Q/2}{4\pi a^2}$
(B) electric field outside the shell is the same as that of a point charge at the centre of the shell.
(C) surface charge density on the outer surface depends on $|\vec{p}|$
(D) surface charge density on the inner surface of the shell is zero everywhere.

34. An electric dipole is formed by two equal and opposite charges q with separation d. The charges have same mass m. It is kept in a uniform electric field E. If it is slightly rotated from its equilibrium orientation, then its angular frequency ω is: [8 April 2019, (II)]
(A) $\sqrt{\frac{qE}{md}}$ (B) $\sqrt{\frac{2qE}{md}}$ (C) $2\sqrt{\frac{qE}{md}}$ (D) $\sqrt{\frac{qE}{2md}}$

35. Charges Q_1 and Q_2 are at points A and B of a right angle triangle OAB (Fig.1.284). The resultant electric field at point O is perpendicular to the hypotenuse, then Q_1/Q_2 is proportional to: [06 Sep 2020 (I)]
(A) x_1^3/x_2^3 (B) x_2/x_1 (C) x_1/x_2 (D) x_2^2/x_1^2

  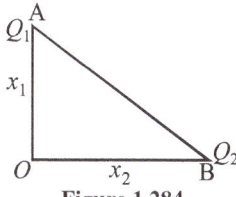

Figure 1.283   Figure 1.284

36. Consider the force F on a charge 'q' due to a uniformly charged spherical shell of radius R carrying charge Q distributed uniformly over it. Which one of the following statements is true for F, if 'q' is placed at distance r from the centre of the shell? [06 Sep. 2020 (II)]

(A) $F = \frac{1}{4\pi\epsilon_0} \frac{Qq}{R^2}$ for $r < R$
(B) $\frac{1}{4\pi\epsilon_0} \frac{Qq}{R^2} > F > 0$ for $r < R$
(C) $F = \frac{1}{4\pi\epsilon_0} \frac{Qq}{R^2}$ for $r > R$
(D) $F = \frac{1}{4\pi\epsilon_0} \frac{Qq}{R^2}$ for all r

37. Two charged thin infinite plane sheets of uniform surface charge density σ_+ and σ_-, where $|\sigma_+| > |\sigma_-|$, intersect at right angle. Which of the diagram of Fig.1.285 best represents the electric field lines for this system?
[04 Sep. 2020 (I)]

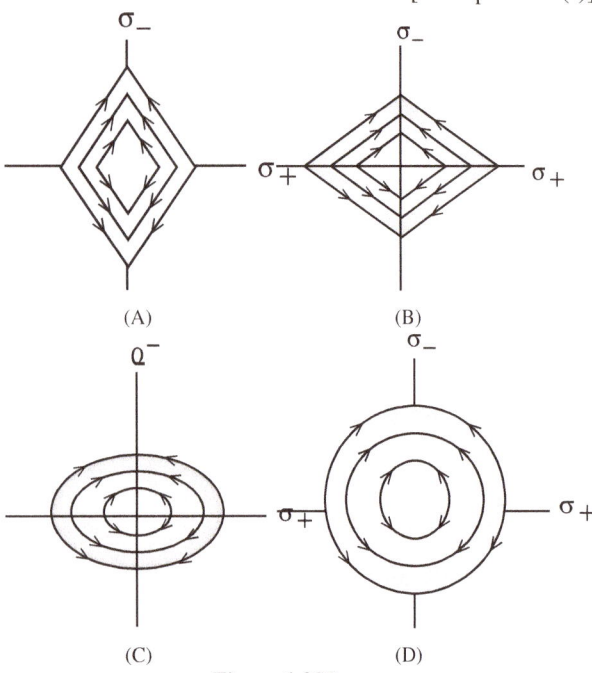

Figure 1.285

38. A particle of charge q and mass m is subjected to an electric field $E = E_0(1 - ax^2)$ in the x-direction, where a and E_0 are constants. Initially the particle was at rest at $x = 0$. Other than the initial position the kinetic energy of the particle becomes zero when the distance of the particle from the origin is: [04 Sep 2020 (II)]
(A) a (B) $\sqrt{2/a}$ (C) $\sqrt{3/a}$ (D) $\sqrt{1/a}$

39. A small point mass carrying some positive charge on it, is released from the edge of a table [Fig.1.286]. There is a uniform electric field in this region in the horizontal direction. Which of the given options then correctly describe the trajectory of the mass in Fig.1.287? (Curves are drawn schematically and are not to scale). [02 Sep 2020 (II)]

Figure 1.286

40. Consider a sphere of radius R which carries a uniform charge density ρ. If a sphere of radius $R/2$ is carved out of it as shown in Fig.1.288, then the ratio $|\vec{E}_A|/|\vec{E}_B|$ of magnitude of electric field $\vec{E}_A$ and $\vec{E}_B$, respectively, at points A and B due to the remaining portion is:
[9 Jan. 2020 (I)]
(A) $\frac{21}{34}$ (B) $\frac{18}{34}$ (C) $\frac{17}{54}$ (D) $\frac{18}{54}$

Figure 1.287

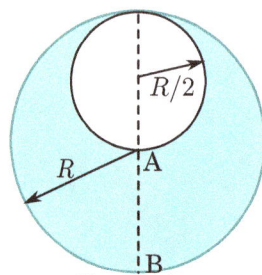

Figure 1.288

41. An electric dipole of moment $\vec{p} = (\hat{i} - 3\hat{j} + 2\hat{k}) \times 10^{-29}$ C.m is at the origin $(0,0,0)$. The electric field due to this dipole at $\vec{r} = +\hat{i} + 3\hat{j} + 5\hat{k}$ (note that $\vec{r} \cdot \vec{p} = 0$) is parallel to: [9 Jan. 2020 (I)]
(A) $(+\hat{i} - 3\hat{j} - 2\hat{k})$ (B) $(-\hat{i} + 3\hat{j} - 2\hat{k})$
(C) $(+\hat{i} + 3\hat{j} - 2\hat{k})$ (D) $(-\hat{i} - 3\hat{j} + 2\hat{k})$

42. A particle of mass m and charge q is released from rest in a uniform electric field. If there is no other force on the particle, the dependence of its speed v on the distance x travelled by it is correctly given by[Fig.1.289] (graphs are schematic and not drawn to scale) [8 Jan. 2020 (II)]

Figure 1.289

43. Two infinite planes each with uniform surface charge density $+\sigma$ are kept in such a way that the angle between them is $30°$. The electric field in the region shown between them is given by: [7 Jan. 2020, (I)]
(A) $\frac{\sigma}{2\epsilon_0}\left[(1+\sqrt{3})\hat{y} - \frac{\hat{x}}{2}\right]$ (B) $\frac{\sigma}{\epsilon_0}\left[\left(1+\frac{\sqrt{3}}{2}\right)\hat{y} + \frac{\hat{x}}{2}\right]$
(C) $\frac{\sigma}{2\epsilon_0}\left[(1+\sqrt{3})\hat{y} + \frac{\hat{x}}{2}\right]$ (D) $\frac{\sigma}{2\epsilon_0}\left[\left(1-\frac{\sqrt{3}}{2}\right)\hat{y} - \frac{\hat{x}}{2}\right]$

44. A particle of mass m and charge q has an initial velocity $\vec{v} = v_0\hat{j}$. If an electric field $\vec{E} = E_0\hat{i}$ and magnetic field $\vec{B} = B_0\hat{i}$ act on the particle, its speed will double after a time: [7 Jan 2020, (II)]
(A) $\frac{2mv_0}{qE_0}$ (B) $\frac{3mv_0}{qE_0}$ (C) $\frac{\sqrt{3}mv_0}{qE_0}$ (D) $\frac{\sqrt{2}mv_0}{qE_0}$

45. Two identical electric point dipoles have dipole moments $\vec{p}_1 = p\hat{i}$ and $\vec{p}_2 = -p\hat{i}$ and are held on the x axis at distance 'a' from each other. When released, they move along x axis with the direction of their dipole moments remaining unchanged. If the mass of each dipole is 'm', their speed when they are infinitely far apart is: [06 Sep. 2020 (II)]

(A) $\frac{p}{a}\sqrt{\frac{1}{\pi\epsilon_0 ma}}$
(B) $\frac{p}{a}\sqrt{\frac{1}{2\pi\epsilon_0 ma}}$
(C) $\frac{p}{a}\sqrt{\frac{2}{\pi\epsilon_0 ma}}$
(D) $\frac{p}{a}\sqrt{\frac{2}{2\pi\epsilon_0 ma}}$

46. In finding the electric field using Gauss law the formula $|\vec{E}| = \frac{q_{encl}}{\epsilon_0|A|}$ is applicable. In the formula ϵ_0 is permittivity of free space, A is the area of Gaussian surface and q_{enc} is charge enclosed by the Gaussian surface. This equation can be used in which of the following situation? [8 Jan 2020, (I)]
    (A) Only when the Gaussian surface is an equipotential surface.
    (B) Only when the Gaussian surface is an equipotential surface and $|\vec{E}|$ is constant on the surface.
    (C) Only when $|\vec{E}|$ = constant on the surface.
    (D) For any choice of Gaussian surface.

47. An electric field $\vec{E} = 4x\hat{i} - (y^2 + 1)\hat{j}$ N/C passes through the box shown in Fig.1.290. The flux of the electric field through surfaces $ABCD$ and $BCGF$ are marked as ϕ_1 and ϕ_{11} respectively. The difference $(\phi_1 - \phi_{11})$ is _____ Nm^2/C. [9 Jan 2020 (II)]

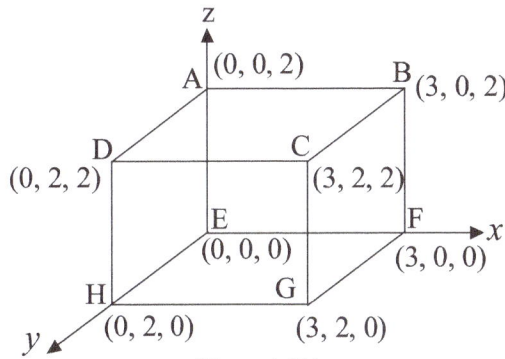

Figure 1.290

48. A uniformly charged disc of radius R having surface charge density σ is placed in the xy plane with its centre at the origin. Find the electric field intensity along the z-axis at a distance Z from origin: [27 Aug.2021 (I)]
    (A) $E = \frac{\sigma}{2\epsilon_0}\left(\frac{1}{Z^2+R^2}\right) + \frac{1}{Z^2}$
    (B) $E = \frac{2\epsilon_0}{2\sigma}\left(\frac{1}{(Z^2+R^2)^{1/2}} + Z\right)$
    (C) $E = \frac{\sigma}{2\epsilon_0}\left(1 + \frac{Z}{(Z^2+R^2)^{1/2}}\right)$
    (D) $E = \frac{\sigma}{2\epsilon_0}\left(1 - \frac{Z}{(Z^2+R^2)^{1/2}}\right)$

49. Figure 1.291 shows a rod AB, which is bent in a 120° circular arc of radius R. A charge $(-Q)$ is uniformly distributed over rod AB. What is the electric field $\vec{E}$ at the centre of curvature O? [27-Aug-2021 (I)]
    (A) $\frac{3\sqrt{3}Q}{8\pi^2\epsilon_0 R^2}(\hat{i})$
    (B) $\frac{3\sqrt{3}Q}{8\pi\epsilon_0 R^2}(\hat{i})$
    (C) $\frac{3\sqrt{3}Q}{8\pi^2\epsilon_0 R^2}(-\hat{i})$
    (D) $\frac{3\sqrt{3}Q}{16\pi^2\epsilon_0 R^2}(\hat{i})$

50. A vertical electric field of magnitude 4.9×10^5 N/C just prevents a water droplet of a mass 0.1 g from falling. The value of charge on the droplet will be: (Given $g = 9.8$ m/s^2) [24 June 22 Shift I]
    (A) 1.6×10^{-9} C
    (B) 2.0×10^{-9} C
    (C) 3.2×10^{-9} C
    (D) 0.5×10^{-9} C

51. In the Fig.1.292, a very large plane sheet of positive charge is shown. P_1 and P_2 are two points at distance l and $2l$ from

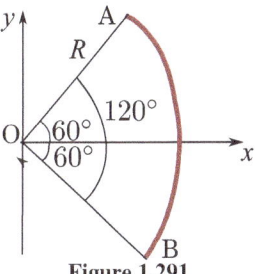

Figure 1.291

the charge distribution. If σ is the surface charge density, then the magnitude of electric fields E_1 and E_2 at P_1 and P_2 respectively are: [25 June 22 Shift I]

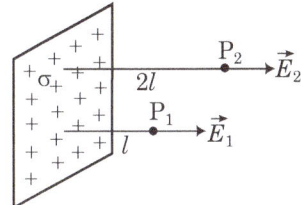

Figure 1.292

(A) $E_1 = \sigma/\epsilon_0$, $E_2 = \sigma/2\epsilon_0$
(B) $E_1 = 2\sigma/\epsilon_0$, $E_2 = \sigma/\epsilon_0$
(C) $E_1 = E_2 = \sigma/2\epsilon_0$
(D) $E_1 = E_2 = \sigma/\epsilon_0$

52. **Assertion (A)**: Non-polar materials do not have any permanent dipole moment.
    Reason (R): When an non-polar material is placed in an electric field. the centre of the positive charge distribution of it's individual atom or molecule coincides with the centre of the negative charge distribution.
    In the light of above statements, choose the most appropriate answer from the options given below. [26 June 22 Shift I]
    (A) Both (A) and (R) are correct and (R) is the correct explanation of (A).
    (B) Both (A) and (R) are correct and (R) is not the correct explanation of (A).
    (C) (A) is correct but (R) is not correct.
    (D) (A) is not correct but (R) is correct.

53. Sixty four conducting drops each of radius 0.02 m and each carrying a charge of 5 μC are combined to form a bigger drop at constant temperature. The ratio of surface density of bigger drop to the smaller drop will be: [26 June 22 Shift II]
    (A) 1 : 4 (B) 4 : 1 (C) 1 : 8 (D) 8 : 1

54. A force of 10 N acts on a charged particle placed between two plates of a charged capacitor. If one plate of capacitor is removed, then the force acting on that particle will be: [27 June 22 Shift I]
    (A) 5 N (B) 10 N (C) 20 N (D) Zero

55. If a charge q is placed at the centre of a closed hemispherical non-conducting surface as shown in Fig.1.293, the total flux passing through the flat surface would be: [27 June 22 Shift II]

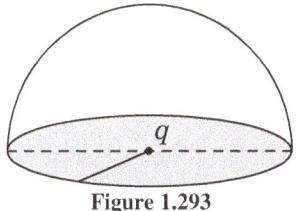

Figure 1.293

(A) 0 (B) $\frac{q}{2\epsilon_0}$ (C) $\frac{q}{4\epsilon_0}$ (D) $\frac{q}{2\pi\epsilon_0}$

56. Three identical charged balls each of charge 2 C are suspended from a common point P by silk threads of 2 m each (as shown in Fig.1.294). They form an equilateral triangle of side 1 m. The ratio of net electric force on a charged ball to the force between any two charged balls will be: [27 June 22 Shift II]
    (A) 1 : 1 (B) 1 : 4 (C) $\sqrt{3}$: 2 (D) $\sqrt{3}$: 1

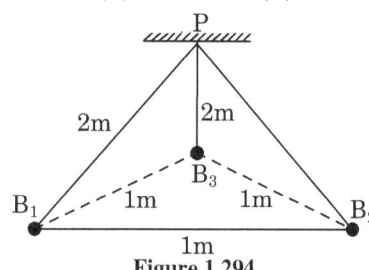

Figure 1.294

57. Given below are two statements: [28 June 22 Shift I]
    Statement-I: A point charge is brought in an electric field. The value of electric field at a point near to the charge may increase if the charge is positive.
    Statement-II: An electric dipole is placed in a non-uniform electric field. The net electric force on the dipole will not be zero.
    Choose the correct answer from the options given below:
    (A) Both statement-I and statement-II are true.
    (B) Both statement-I and statement-I are false.
    (C) Statement-I is true but statement-II is false.
    (D) Statement-I is false but statement-II is true.

58. The three charges $q/2$, q and $q/2$ are placed at the corners A, B and C of a square of side 'a' as shown in Fig.1.295. The magnitude of electric field ($\vec{E}$) at the corner D of the square, is: [28 June 22 Shift I]

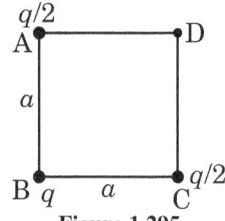

Figure 1.295

(A) $\frac{q}{4\pi\epsilon_0 a^2}\left(\frac{1}{\sqrt{2}} + \frac{1}{2}\right)$ (B) $\frac{q}{4\pi\epsilon_0 a^2}\left(1 + \frac{1}{\sqrt{2}}\right)$
(C) $\frac{q}{4\pi\epsilon_0 a^2}\left(1 - \frac{1}{\sqrt{2}}\right)$ (D) $\frac{q}{4\pi\epsilon_0 a^2}\left(\frac{1}{\sqrt{2}} - \frac{1}{2}\right)$

59. Two point charges A and B of magnitude $+8 \times 10^{-6}$ C and -8×10^{-6} C respectively are placed at a distance d apart. The electric field at the middle point O between the charges is 6.4×10^4 NC^{-1}. The distance 'd' between the point charges A and B is: [28 June 22 Shift II]
    (A) 2.0 m (B) 3.0 m (C) 1.0 m (D) 4.0 m

60. Two point charges Q each are placed at a distance d apart. A third point charge q is placed at a distance x from mid-point on the perpendicular bisector. The value of x at which charge q will experience the maximum Coulomb's force is: [29 June 23 Shift II]
    (A) d (B) $d/2$ (C) $d/\sqrt{2}$ (D) $\frac{d}{2\sqrt{2}}$

61. If two charges q_1 and q_2 are separated with distance 'd' and placed in a medium of dielectric constant κ. What will be the equivalent distance between charges in air for the same electrostatic force? [24 Jan 2023 Shift I]
    (A) $d\sqrt{k}$ (B) $k\sqrt{d}$ (C) $1.5d\sqrt{k}$ (D) $2d\sqrt{k}$

62. A point charge of 10 C is placed at the origin. At what location on the X–axis should a point charge of 40 μC be placed so that the net electric field is zero at $x = 2$ cm on the X–axis? [25 Jan 2023 Shift II]
    (A) $x = 6$ cm (B) $x = 4$ cm
    (C) $x = 8$ cm (D) $x = -4$ cm

63. A point charge $q_1 = 4q_0$ is placed at origin. Another point charge $q_2 = -q_0$ is placed at $x = 12$ cm. Charge of proton is q_0. The proton is placed on x-axis so that the electrostatic force on the proton in zero. In this situation, the position of the proton from the origin is ___ cm. [29 Jan 2023 Shift I]

64. A point charge 2×10^{-2} C is moved from P to S in a uniform electric field of 30 NC^{-1} directed along positive x-axis as shown in Fig.1.296. If coordinates of P and S are (1 m, 2 m, 0 m) and (0 m, 0 m, 0 m) respectively, the work done by electric field will be [29 Jan 2023 Shift II]
    (A) 1200 mJ (B) 600 mJ
    (C) –600 mJ (D) –1200 mJ

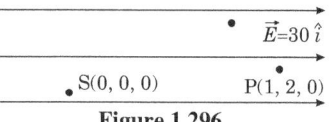

Figure 1.296

65. Electric field in a certain region is given by $\vec{E} = \left(\frac{A}{x^2}\hat{i} + \frac{B}{y^3}\hat{j}\right)$. The SI unit of A and B are: [30 Jan 2023 Shift I]
    (A) Nm^{3}C^{-1}; Nm^{2}C^{-1} (B) Nm^{2}C^{-1}; Nm^{3}C^{-1}
    (C) Nm^{3}C; Nm^{2}C (D) Nm^{2}C; Nm^{3}C

66. As shown in the Fig.1.297, a point charge Q is placed at the centre of conducting spherical shell of inner radius a and outer radius b. The electric field due to charge Q in three different regions I, II and III is given by : ($I: r < a$, $II: a < r < b, III: r > b$) [30 Jan 2023 Shift II]
    (A) $E_I = 0, E_{II} = 0, E_{III} \neq 0$
    (B) $E_I \neq 0, E_{II} = 0, E_{II} \neq 0$
    (C) $E_I \neq 0, E_{II} = 0, E_{II} = 0$
    (D) $E_I = 0, E_{II} = 0, E_{II} = 0$

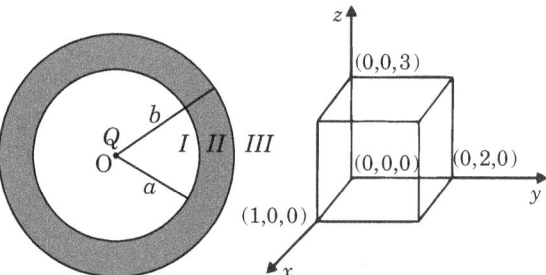

Figure 1.297   **Figure 1.298**

67. As shown in Fig.1.298, a cuboid lies in a region with electric field $\vec{E} = 2x^2\hat{i} - 4y\hat{j} + 6\hat{k}$ N/C. The magnitude of charge within the cuboid is $n\epsilon_0$ C. The value of n is ___ (if dimension of cuboid is $1 \times 2 \times 3$ m^3) [30 Jan 2023 Shift II]

68. Expression for an electric field is given by $\vec{E} = 4000x^2\hat{i}$ V/m. The electric flux through the cube of side 20 cm when placed in electric field (as shown in the Fig.1.299) is ___ V·cm. [31 Jan 2023 Shift I]

69. Let σ be the uniform surface charge density of two infinite thin plane sheets shown in Fig.1.300. Then the electric fields in three different region E_I, E_{II} and E_{III} are:

1.37. QUESTIONS AND EXERCISES   105

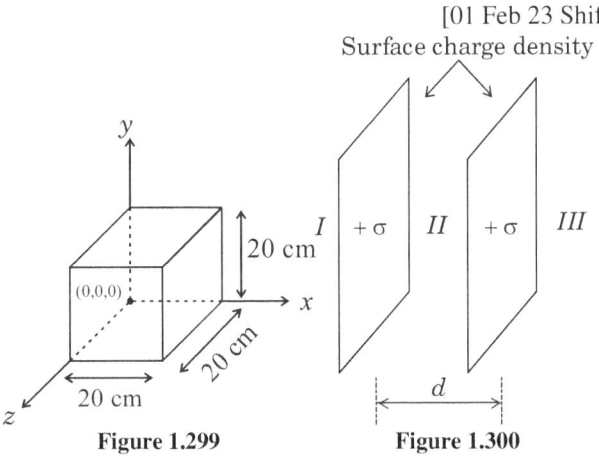

Figure 1.299   Figure 1.300

(A) $\vec{E}_I = \frac{2\sigma}{\epsilon_0}\hat{n}, \vec{E}_{II} = 0, \vec{E}_{III} = \frac{2\sigma}{\epsilon_0}\hat{n}$
(B) $\vec{E}_I = 0, \vec{E}_{II} = \frac{\sigma}{\epsilon_0}\hat{n}, \vec{E}_{III} = 0$
(C) $\vec{E}_I = \frac{\sigma}{2\epsilon_0}\hat{n}, \vec{E}_{II} = 0, \vec{E}_{III} = \frac{\sigma}{2\epsilon_0}\hat{n}$
(D) $\vec{E}_I = -\frac{\sigma}{\epsilon_0}\hat{n}, \vec{E}_{II} = 0, \vec{E}_{III} = \frac{\sigma}{\epsilon_0}\hat{n}$

70. In Fig.1.301, two equal positive point charges are separated by a distance $2a$. The distance of a point from the centre of the line joining two charges on the equatorial line (perpendicular bisector) at which force experienced by a test charge q_0 becomes maximum is $a/\sqrt{x}$. The value of x is ____.
[01 Feb 23 Shift I]

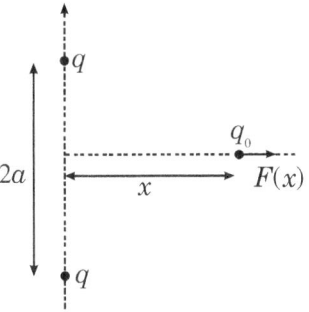

Figure 1.301

71. A cubical volume is bounded by the surfaces $x = 0$, $x = a$, $y = 0$, $y = a$, $z = 0$, $z = a$. The electric field in the region is given by $\vec{E} = E_0 x \hat{i}$. Where $E_0 = 4 \times 10^4$ NC^{-1}m^{-1}. If $a = 2$ cm, the charge contained in the cubical volume is $Q \times 10^{-14}$ C. The value of Q is ____. (Take $\epsilon_0 = 9 \times 10^{-12}$ C^2/Nm^2) [01 Feb 23 Shift II]

72. A dipole comprises of two charged particles of identical magnitude q and opposite in nature. The mass 'm' of the positive charged particle is half of the mass of the negative charged particle. The two charges are separated by a distance 'l'. If the dipole is placed in a uniform electric field '$\vec{E}$'; such a way that dipole axis makes a very small angle with the electric field, '$\vec{E}$'. The angular frequency of the oscillations of the dipole when released is given by:
[06 April 23 Shift II]

(A) $\sqrt{\frac{3qE}{2ml}}$ (B) $\sqrt{\frac{4qE}{ml}}$ (C) $\sqrt{\frac{4qE}{3ml}}$ (D) $\sqrt{\frac{8qE}{ml}}$

73. Which of the following graphs, shown in Fig.1.303, represents the graphical variation of the electric field due to a uniformly charged insulating solid sphere of radius R at a distance r from the centre O, as depicted in Fig.1.302?
[08 April 23 Shift I]

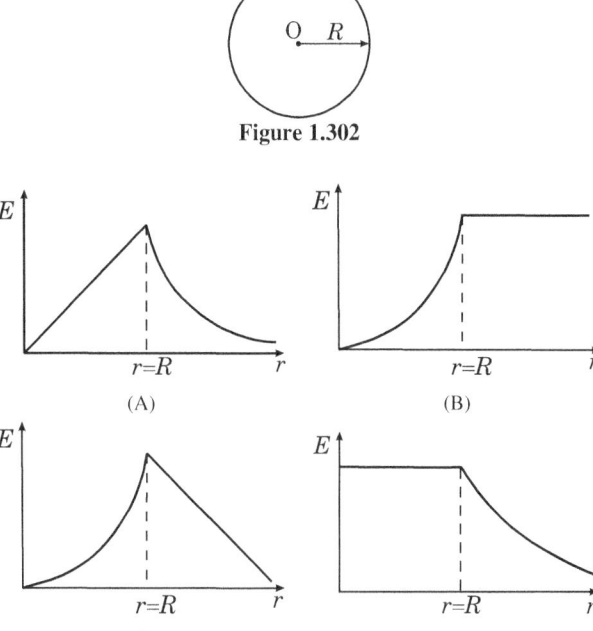

Figure 1.302

Figure 1.303

74. Given below are two statement: one is labelled as Assertion A and the other is labelled as Reason R.
[12 April 23 Shift I]

Assertion A: If an electric dipole of dipole moment 30×10^{-5} cm is enclosed by a closed surface, the net flux coming out of the surface will be zero.

Reason R: Electric dipole consists of two equal and opposite charges.

In the light of above, statements, choose the correct answer from the options given below:
(A) Both A and R are true and R is the correct explanation of A.
(B) A is true but R is false.
(C) Both A and R true but R is not the correct explanation of A.
(D) A is false but R is true.

75. A thin infinite sheet charge and an infinite line charge of respective charge densities $+\sigma$ and $+\lambda$ are placed parallel at 5 m distance from each other. Points 'A' and 'B' are at $(3/\pi)$ m and $(4/\pi)$ m perpendicular distance from line charge towards sheet charge, respectively. 'E_A' and 'E_B' are the magnitudes of resultant electric field intensities at point 'A' and 'B', respectively. If $E_A/E_B = 4/a$ for $2|\sigma| = |\lambda|$. Then the value of a is [13 April 23 Shift I]

76. Two charges each of magnitude 0.01 C and separated by a distance of 0.4 mm constitute an electric dipole. If the dipole is placed in a uniform electric field $|\vec{E}|$ of 10 dyne/C making 30° angle with $\vec{E}$, the magnitude of torque acting on dipole is- [13 April 23 Shift I]
(A) 4.0×10^{-10} Nm   (B) 2.0×10^{-10} Nm
(C) 1.0×10^{-8} Nm   (D) 15×10^{-9} Nm

77. A 10 μC charge is divided into two parts and placed at 1 cm distance so that the repulsive force between them is maximum. The charges of the two parts are:
[13 April 23 Shift II]
(A) 9 μC, 1 μC   (B) 5 μC, 5 μC
(C) 7 μC, 3 μC   (D) 8 μC, 2 μC

78. The electric field due to a short electric dipole at a large distance (r) from centre of dipole on the equatorial plane

varies with distance as: [15 April 23 Shift I]
(A) r (B) $1/r$ (C) $1/r^3$ (D) $1/r^2$

Section B: JEE Advanced

1. Five point charges, each of value $+q$, are placed on five vertices of a regular hexagon of side L. The magnitude of the force on a point charge of value $-q$ placed at the centre of the hexagon is [1992]
   (A) $\kappa q^2/L^2$ (B) $\kappa q^2/4L^2$ (C) $\kappa q^2/2L^2$ (D) $\kappa q^2/8L^2$

2. Two point charges $+q$ and $-q$ are held fixed at $(-d, 0)$ and $(d, 0)$ respectively of $a(x, y)$ coordinate system, then [1995]
   (A) The electric field $\vec{E}$ at all points on the x-axis has the same direction
   (B) $\vec{E}$ at all points on the Y-axis is along $\hat{i}$
   (C) Work has to be done in bringing a test charge from infinity to the origin
   (D) The dipole moment is $2qd$ directed along $\hat{i}$

3. A metallic solid sphere is placed in a uniform electric field. The lines of force follow the path(s) shown in Fig.1.304 as [1996]
   (A) 1 (B) 2 (C) 3 (D) 4

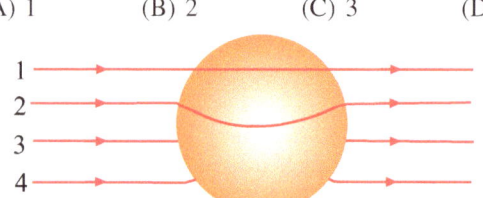

Figure 1.304

4. An electron of mass m_e, initially at rest, moves through a certain distance in a uniform electric field in time t_1. A proton of mass m_p, also, initially at rest, takes time t_2 to move through an equal distance in this uniform electric field. Neglecting the effect of gravity, the ratio t_2/t_1 is nearly equal to [1997]
   (A) 1 (B) $(m_p/m_e)^{1/2}$ (C) $(m_e/m_p)^{1/2}$ (D) 1836

5. A non-conducting solid sphere of radius R is uniformly charged. The magnitude of the electric field due to the sphere at a distance r from its centre [1998]
   (A) increases as r increases, for $r < R$
   (B) decreases as r increases, for $0 < r < \infty$
   (C) increases as r increases, for $R < r < \infty$
   (D) is discontinuous at $r = R$

6. Three positive charges of equal value q are placed at the vertices of an equilateral triangle. The resulting lines of force should be sketched as in Fig.1.305 [2001]

7. A point charge 'q' is placed at a point inside a hollow conducting sphere. Which of the following electric lines of force pattern is correct[Fig.1.306]? [2003]

8. In Fig.1.307, charges q_1 and $-q_1$ are inside a Gaussian surface. Where as charge q_2 is outside the surface. Electric field on the Gaussian surface will be [2004]
   (A) only due to q_2
   (B) zero on the Gaussian surface
   (C) uniform on the Gaussian surface
   (D) due to all

9. Six charges of equal magnitude are placed at six corners of a regular hexagon [Fig.1.308]. Find arrangement the charges in order PQRSTU which produce double electric field at centre as compared to electric field produce by single charges $+q$ at R- [2004]
   (A) $+ + + - - -$ (B) $+ - + + -$
   (C) $- + + + -$ (D) $- + + - -$

10. Three large charged sheets having surface charge density as

Figure 1.305

Figure 1.306

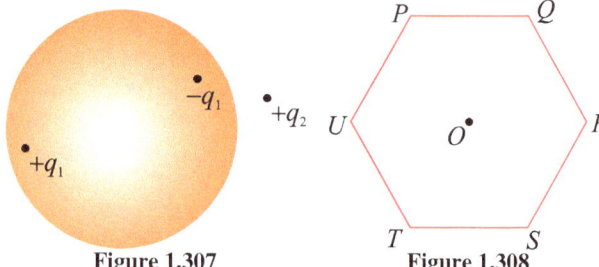

Figure 1.307   **Figure 1.308**

shown in the Figure1.309. The sheets are placed parallel to XY plane. Then electric field at point P- [2005]
(A) $\frac{-4\sigma}{\epsilon_0} k$ (B) $\frac{4\sigma}{\epsilon_0} k$ (C) $\frac{2\sigma}{\epsilon_0} k$ (D) $-\frac{2\sigma}{\epsilon_0} k$

Figure 1.309

11. Consider a neutral conducting sphere. A positive point charge is placed outside the sphere. The net charge on the sphere is then, [2007]
(A) negative and distributed uniformly over the surface of the sphere

(B) negative and appears only at the point on the sphere closest to the point charge
(C) negative and distributed non-uniformly over the entire surface of the sphere
(D) zero

12. A disc of radius $a/4$ having a uniformly distributed charge 6C is placed in the x-y plane with its centre at $(-a/2, 0, 0)$. A rod of length 'a' carrying a uniformly distributed charge 8C is placed on the x - axis from $x = a/4$ to $x = 5a/4$ [Fig.1.310]. Two point charges $-7C$ and $3C$ are placed at $(a/4, -a/4, 0)$ and $(-3a/4, 3a/4, 0)$ respectively. Consider a cubical surface formed by six surfaces $x = \pm a/2$, $y = \pm a/2$, $z = \pm a/2$. The electric flux this cubical surface is- [2009]
(A) $-2C/\epsilon_0$ (B) $2C/\epsilon_0$ (C) $10C/\epsilon_0$ (D) $12C/\epsilon_0$

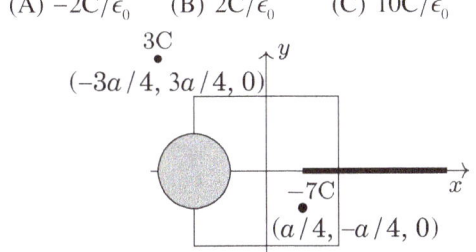

Figure 1.310

13. Under the influence of the Coulomb field of charge $+Q$, a charge $-q$ is moving around it in an elliptical orbital. Find out the correct statement(s) - [2009]
(A) The angular momentum of the charge $-q$ is constant
(B) The linear momentum of the charge $-q$ is constant
(C) The angular velocity of the charge $-q$ is constant
(D) The linear speed of the charge $-q$ is constant

14. A few electric field lines for a system of two charges Q_1 and Q_2 fixed at two different points on the x-axis are shown in the Fig.1.311. These lines suggest that [2010]
(A) $|Q_1| > |Q_2|$
(B) $|Q_1| < |Q_2|$
(C) at a finite distance to the left of Q_1 the electric field is zero
(D) at a finite distance to the right of Q_2 the electric field is zero

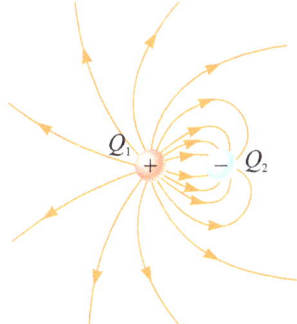

Figure 1.311

15. A uniformly charged thin spherical shell of radius R carries uniform surface charge density of σ per unit area. It is made of two hemispherical shells, held together by pressing them with force F (Figure 1.312). F is proportional to [2010]
(A) $\frac{1}{\epsilon_0}\sigma^2 R^2$ (B) $\frac{1}{\epsilon_0}\sigma^2 R$ (C) $\frac{1}{\epsilon_0}\frac{\sigma^2}{R}$ (D) $\frac{1}{\epsilon_0}\frac{\sigma^2}{R^2}$

16. A tiny spherical oil drop carrying a net charge q is balanced in still air with a vertical uniform electric field of strength $\frac{81\pi}{7} \times 10^5 \text{Vm}^{-1}$. When the field is switched off, the drop is

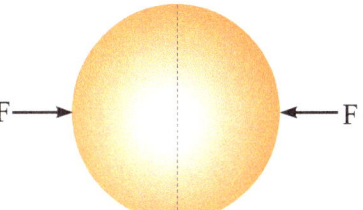

Figure 1.312

observed to fall with terminal velocity 2×10^{-3} ms^{-1} Given g = 9.8 ms^{-2}, viscosity of the air = 1.8×10^{-5}Nsm^{-2} and the density of oil = 900 kg m^{-3}, the magnitude of q is [2010]
(A) 1.6×10^{-19}C   (B) 3.2×10^{-19}C
(C) 4.8×10^{-19}C   (D) 8.0×10^{-19}C

17. Consider an electric field $E = E_0 x$, where E_0 is a constant. The flux through the shaded area (as shown in the Fig.1.313) due to this field is [2011]
(A) $2E_0 a^2$ (B) $\sqrt{2}E_0 a^2$ (C) $E_0 a^2$ (D) $\frac{E_0 a^2}{\sqrt{2}}$

Figure 1.313

18. A cubical region of side a has its centre at the origin. It encloses three fixed point charges [Fig. 1.314], $-q$ at $(0, -a/4, 0)$, $+3q$ at $(0,0,0)$ and $-q$ at $(0, +a/4, 0)$. Choose the correct options(s) [2012]
(A) The net electric flux crossing the plane $x = +a/2$ is equal to the net electric flux crossing the plane $x = -a/2$
(B) The net electric flux crossing the plane $y = +a/2$ is more than the net electric flux crossing the plane $y = -a/2$.
(C) The net electric flux crossing the entire region is q/ϵ_0.
(D) The net electric flux crossing the plane $z = +a/2$ is equal to the net electric flux crossing the plane $x = +a/2$.

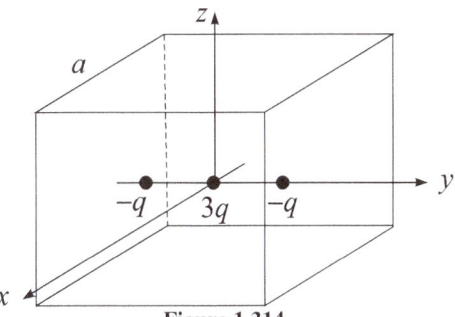

Figure 1.314

19. Two non-conducting solid spheres of radii R and $2R$, having uniform volume charge densities ρ_1 and ρ_2 respectively, touch each other . The net electric field at a distance $2R$ from the centre of the smaller sphere, along the line joining the centres of the spheres, is zero. The ratio (ρ_1/ρ_2) can be [2013]
(A) -4 (B) $-(32/25)$ (C) $(32/25)$ (D) 4

20. Let $E_1(r)$, $E_2(r)$ and $E_3(r)$ be the respective electric fields at a distance r from a point charge Q, an infinitely long

wire with constant linear charge density λ, and an infinite plane with uniform surface charge density σ. If $E_1(r_0) = E_2(r_0) = E_3(r_0)$ at a given distance r_0, then [2014]
(A) $Q = 4\sigma\pi r_0^2$
(B) $r_0 = \frac{\lambda}{2\pi\sigma}$
(C) $E_1\left(\frac{r_0}{2}\right) = 2E_2\left(\frac{r_0}{2}\right)$
(D) $E_2\left(\frac{r_0}{2}\right) = 4E_3\left(\frac{r_0}{2}\right)$

21. Charges Q, $2Q$ and $4Q$ are uniformly distributed in three dielectric solid spheres 1, 2 and 3 of radii $R/2$, R and $2R$ respectively, as shown in Fig.1.315. If magnitudes of the electric fields at point P at a distance R from the centre of spheres 1, 2 and 3 are E_1, E_2 and E_3 respectively, then [2014]
(A) $E_1 > E_2 > E_3$
(B) $E_3 > E_1 > E_2$
(C) $E_2 > E_1 > E_3$
(D) $E_3 > E_2 > E_1$

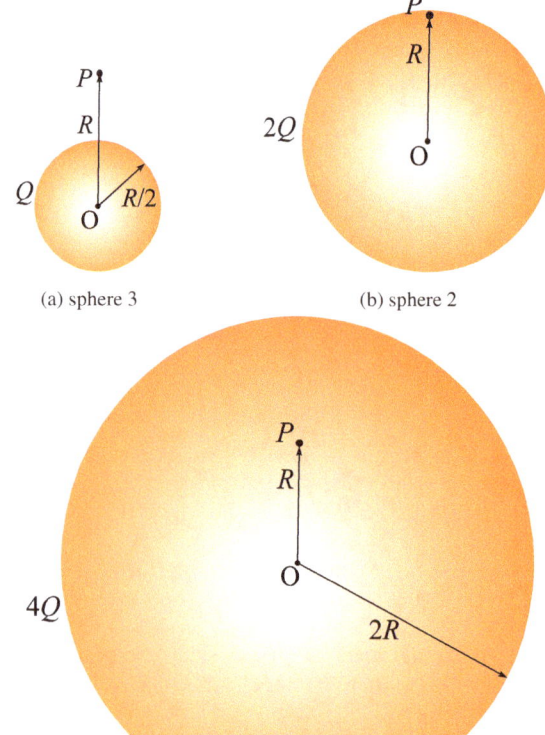

(a) sphere 3   (b) sphere 2

(c) sphere 3

Figure 1.315

22. A long cylindrical shell carries positive surface charge σ in the upper half and negative surface charge $-\sigma$ in the lower half. The electric field lines around the cylinder will look like Fig.1.316 given in : (figures are schematic and not drawn to scale) [2015]

23. A solid sphere of radius R has a charge Q distributed in its volume with a charge density $\rho = \kappa r^a$, where κ and a are constants and r is the distance from its centre. If the electric field at $r = \frac{R}{2}$ is $1/8$ times that at $r = R$, find the value of a. [2009]

24. Four point charges, each of $+q$, are rigidly fixed at the four corners of a square planar soap film of side 'a'. The surface tension of the soap film is y. The system of charges and planar film are in equilibrium, and $a = k\left[\frac{q^2}{\gamma}\right]^{1/N}$, where '$k$' is a constant. Then N is [2011]

25. An infinitely long solid cylinder of radius R has a uniform volume charge density ρ. It has a spherical cavity of radius $R/2$ with its centre on the axis of the cylinder, as shown in the Figure 1.317. The magnitude of the electric field at the point P, which is at a distance $2R$ from the axis of the

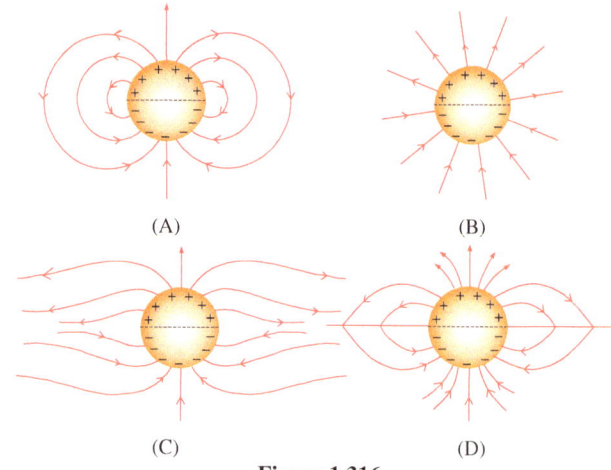

Figure 1.316

cylinder, is given by the expression $\frac{23\rho R}{16k\epsilon_0}$. The value of k is [2013]

26. An infinitely long thin non-conducting wire is parallel to the z-axis and carries a uniform line charge density λ. It pierces a thin non-conducting spherical shell of radius R in such a way that the arc PQ subtends an angle 120° at the centre O of the spherical shell, as shown in the Figure 1.318. The permittivity of free space is ϵ_0. Which of the following statements is (are) true? [2018]
(A) The electric flux through the shell is $\sqrt{3}R/\lambda\epsilon_0$
(B) The z-component of the electric field is zero at all the points on the surface of the shell
(C) The electric flux through the shell is $\sqrt{2}R/\lambda\epsilon_0$
(D) The electric field is normal to the surface of the shell at all points

Figure 1.317   **Figure 1.318**

27. The electric field E is measured at a point P $(0, 0, d)$ generated due to various charge distributions and the dependence of E on d is found to be different for different charge distributions. List-I contains different relations between E and d. List-II describes different electric charge distributions, along with their locations. Match the functions in List-I with the related charge distributions in List-II. [2018]

List I	List II
P. E is independent of d	1. A point charge Q at the origin
Q. $E \propto 1/d$	2. A small dipole with point charges Q at $(0,0,1)$ and $-Q$ at $(0,0,-1)$. Take $2 < d$
R. $E \propto 1/d^2$	3. An infinite line charge coincident with the x-axis, with uniform linear charge density λ.
S. $E \propto 1/d^x$	4. Two infinite wires carrying uniform linear Charge density parallel to the x-axis. The one along $(y = 0, z = 1)$ has a charge density $+\lambda$ and the one along $(y = 0, z = -1)$ has a charge density $-\lambda$. Take $2\lambda << d$
	5. Infinite plane charge coincident with the xy-plane with uniform surface charge density

(A) P → 5; Q → 3,4; R → 1; S → 2
(B) P → 5; Q → 3; R → 1,4; S → 2
(C) P → 5; Q → 3; R → 1,2; S → 4
(D) P → 4; Q → 2,3; R → 1; S → 5

28. An electric dipole with moment $\frac{p_0}{\sqrt{2}}(\hat{i} + \hat{j})$ is held fixed at the origin O in the presence of a uniform electric field of magnitude E_0. If the potential is constant on a circle of radius R centered at the origin as shown in Fig.1.319, then the correct statement(s) is/are: (ϵ_0 is permittivity of free space. $R \gg$ dipole size) [2019]

Figure 1.319

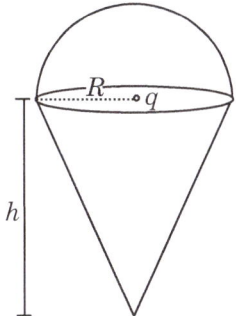

Figure 1.320

(A) The magnitude of total electric field on any two points of the circle will be same
(B) Total electric field at point B is $\vec{E}_B = 0$
(C) Total electric field at point A is $\vec{E}_A = \sqrt{2}E_0(\hat{i} + \hat{j})$
(D) $R = \left(\frac{p_0}{4\pi\epsilon_0 E_0}\right)^{1/3}$

29. A charged shell of radius R carries a total charge Q. Given ϕ as the flux of electric field through a closed cylindrical surface of height h, radius r and with its centre same as that of the shell. Here, centre of the cylinder is a point on the axis of the cylinder which is equidistant from its top and bottom surfaces. Which of the following option(s) is/are correct? [ϵ_0 is the permittivity of free space] [2019]
    (A) If $h > 2R$ and $r = 3R/5$ then $\phi = Q/5\epsilon_0$
    (B) If $h > 2R$ and $r > R$ then $\phi = Q/\epsilon_0$
    (C) If $h < 8R/5$ and $r = 3R/5$ then $\phi = 0$
    (D) If $h > 2R$ and $r > 4R/5$ then $\phi = Q/5\epsilon_0$

30. Two identical non-conducting solid spheres of same mass and charge are suspended in air from a common point by two non-conducting, massless strings of same length. At equilibrium, the angle between the strings is α. The spheres are now immersed in a dielectric liquid of density 800 kg.m^{-3} and dielectric constant 21. If the angle between the strings remains the same after the immersion, then [2020]
    (A) electric force between the spheres remains unchanged
    (B) electric force between the spheres reduces
    (C) mass density of the spheres is 840 kgm^{-3}
    (D) the tension in the strings holding the spheres remains unchanged

31. A circular disc of radius R carries surface charge density $\sigma(r) = \sigma_0\left(1 - \frac{r}{R}\right)$, where σ_0 is a constant and r is the distance from the centre of the disc. Electric flux through a large spherical surface that encloses the charged disc completely is ϕ_0. Electric flux through another spherical surface of radius $R/4$ and concentric with the disc is ϕ. Then the ratio ϕ_0/ϕ is ____. [2020]

32. A charge q is surrounded by a closed surface consisting of an inverted cone of height h and base radius R, and a hemisphere of radius R as shown in the Fig.1.320. The electric flux through the conical surface is $\frac{nq}{6\epsilon_0}$ (in SI units). The value of n is ____. [2022]

1.38 Answer Keys and Solutions
1.38.1 Check Point 1
1. Rub a glass rod with silk and use it to charge an electroscope. The electroscope will end up with a net positive charge. Bring the pocket comb close to the electroscope. If the electroscope leaves move farther apart, then the charge on the comb is positive, the same as the charge on the electroscope. If the leaves move close to each other, then the charge on the comb is negative, opposite the charge on the electroscope.

2. The shirt or blouse becomes charged as a result of rubbing against the dryer sides and other clothes. When you put on the charged object (shirt), it causes charge separation within the molecules of your skin, which results in attraction between the shirt and your skin.

3. Fog or rain droplets tend to form around ions because water is a polar molecule, with a positive region and a negative region. The charge centers on the water molecule will be attracted to the ions (positive to negative).

4. The negatively charged electrons in the paper are attracted to the positively charged rod and move towards it within their molecules. The attraction occurs because the negative charges in the paper are closer to the positive rod than are the positive charges in the paper, and therefore the attraction between the unlike charges is greater than the repulsion between the like charges.

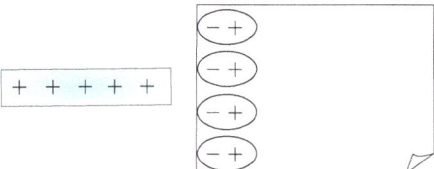

Figure 1.321: Polarization in the paper atoms.

5. A plastic ruler that has been rubbed with a cloth is charged. When brought near small pieces of paper, it will cause separation of charge in the bits of paper, which will cause the paper to be attracted to the ruler. On a humid day, polar water molecules will be attracted to the ruler and to the sep-

arated charge on the bits of paper, neutralizing the charges and thus eliminating the attraction.

6. The net charge on a conductor is the difference between the total positive charge and the total negative charge in the conductor. The "free charges" in a conductor are the electrons that can move about freely within the material because they are only loosely bound to their atoms. The "free electrons" are also referred to as "conduction electrons." A conductor may have a zero net charge but still have substantial free charges.

7. Most of the electrons are strongly bound to nuclei in the metal ions. Only a few electrons per atom (usually one or two) are free to move about throughout the metal. These are called the "conduction electrons". The rest are bound more tightly to the nucleus and are not free to move. Furthermore, in the cases shown in Figures 1.4a and 1.4b, not all of the conduction electrons will move. In Figure 1.4a, electrons will move until the attractive force on the remaining conduction electrons due to the incoming charged rod is balanced by the repulsive force from electrons that have already gathered at the left end of the neutral rod. In Figure 1.4a, conduction electrons will be repelled by the incoming rod and will leave the stationary rod through the ground connection until the repulsive force on the remaining conduction electrons due to the incoming charged rod is balanced by the attractive force from the net positive charge on the stationary rod.

8. No, the basic physics of electric charges would not have been affected at all by an opposite assignment of positive and negative labels. The use of + and − signs, as opposed to labels such as A and B, has the distinct advantage that it gives zero net charge to an object that contains equal amounts of positive and negative charge.

9. Initially, the bits of paper are uncharged and are attracted to the comb by polarization effects. When one of the bits of paper comes into contact with the comb, it acquires charge from the comb. Now the piece of paper and the comb have charge of the same sign, and hence there is a repulsive force between them.

10. (a) -4.4×10^{11} C (b) 2.4×10^8 C

11. 0.275 C

12. **APPROACH** Since, total charge on a body having N electrons is given by, $Q = Ne$. So, first of all, find the total number of electrons (N) in 3.10 g of copper and then use above relation.

    SOLUTION As, the number of electrons in a copper (Cu) atom is 29 (the atomic number of copper). Therefore, total number of electrons in 3.10 gram of copper is given by-
    N = No. of Cu atoms in 3.10 gram of Cu × 29
    i.e., $N = \dfrac{\text{Given mass of copper}}{\text{molar mass of copper}} \times N_A \times 29$

    Here, N_A is avogadro's number and it's value is 6.02×10^{23}, molar mass of the copper is 63.5 g, therefore,
    $N = \dfrac{3.10 \text{ g}}{63.5 \text{ g}} \times 6.02 \times 10^{23} \times 29 = 8.53 \times 10^{23}$ electrons

    So, the total charge on the copper penny,
    $Q = Ne = 8.53 \times 10^{23} \times 1.6 \times 10^{-19}$ C $= 1.37 \times 10^5$ C

    Since, electrons are negatively charged particles, so the nature of this charge will be negative, i.e., $Q = -1.37 \times 10^5$ C.

1.38.2 Check Point 2

1. For point charges, the two charges must have opposite signs.
2. Yes, when the charge on one body (q_1) is much greater than that on the other (q_2) and they are close enough to each other so that force of attraction between q_1 and induced charge on the other exceeds the force of repulsion between q_1 and q_2 plus similar induced charge. However, two similar point charges can never attract each other because no induction will take place here.

3. Yes, as charging a body means addition or removal of electrons and electron has a mass.

4. All electric appliances may end with some charge due to faulty connections. In such a situation charge will be accumulated on the appliance. When the user touches the appliance, he may get a shock. By providing the third hole for grounding all accumulated charge is discharged to the ground and the appliance is safe.

5. Consider any two equal charges q, with an electric force $F = kq^2/r^2$. If we transfer a charge δ from one to the other, the charges become $(q+\delta)$ and $(q-\delta)$, so the force becomes $F' = k(q+\delta)(q-\delta)/r^2$. But $(q+\delta)(q-\delta) = q^2 - \delta^2$, which is less than q^2, so the force decreases.

6. Total number of atoms in the sphere of mass 1 gram
$$N = \dfrac{\text{mass of copper sphere}}{\text{molar mass of copper}} \times \text{Avogadro number}$$
$$= \dfrac{1}{63.54} \times 6.023 \times 10^{23}$$
So the total nuclear charge
$$Q_1 = \text{number of atoms} \times \text{atomic number}$$
i.e., $Q_1 = N \times 29$
$$= \dfrac{6.023 \times 10^{23}}{63.54} \times 1.6 \times 10^{-19} \times 29$$

Now the charge on the sphere = total nuclear charge − total electronic charge
$$= \dfrac{6.023 \times 10^{23}}{63.54} \times 1.6 \times 10^{-19} \times \dfrac{29 \times 1}{100}$$
$$= 4.298 \times 10^2 \text{ C}$$

Hence, force of interaction between these two spheres,
$$F = k \cdot \dfrac{[4.398 \times 10^2]^2}{1^2} \text{ N}$$
$$= 9 \times 10^9 \times 10^4 \times 19.348 \text{ N} = 1.74 \times 10^{15} \text{ N}$$

7. Given situation is shown in Fig.1.322.
Since it is given that the magnitude of force between $+q$ and $+Q$ is F, therefore, magnitude of force between $2q$ and Q is $2F$; between $-3q$ and Q is $3F$ and between $-4q$ and Q is $4F$.

Charge $+q$ at A applies repulsion F on charge $+Q$ at O. It

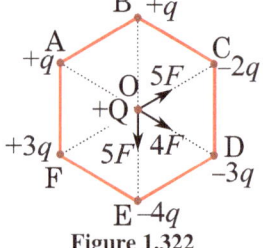

Figure 1.322

is directed from O to D.
Charge $-q$ at D applies attraction $3F$ on charge $+Q$ at O. It is also directed from O to D.
So, resultant of these two forces is $4F$ and it is along OD.
Similarly, Charge $+q$ at B applies repulsion of F on charge $+Q$ at O. It is directed from O to E.
Charge $-4q$ at E applies attraction of $4F$ on charge $+Q$ at O. It is also directed from O to E.

So, resultant of these two forces is $5F$ and it is along OE.
Charge $+3q$ at F applies repulsion of $3F$ on charge $+Q$ at O. It is directed from O to C.
Charge $-2q$ at C applies attraction of $2F$ on charge $+Q$ at O. It is also directed from O to C.
So, resultant of these two forces is $5F$ and it is along OC.
These opposite pair resultants are shown in Fig.1.322.
Now, if F_1 is the resultant of $5F$ acting along OC and $5F$ along OE, then
$$F_1 = \sqrt{(5F)^2 + (5F)^2 + 2(5F)(5F)\cos 120°} = 5F$$
$$[\because \angle COE = 120°]$$
This resultant of two equal force bisects the angle F_1 bisects the angle $\angle COE$ and hence it will be along OD, i.e., along third force of magnitude $4F$. The resultant of these two forces acting along the same direction is
$$F_{\text{net}} = F_1 + 4F = 5F + 4F = 9F$$
Therefore, $n = 9$

8. Consider the equilibrium of charge at A,
   F = force due to other charge at distance $L = \dfrac{kq^2}{L^2}$

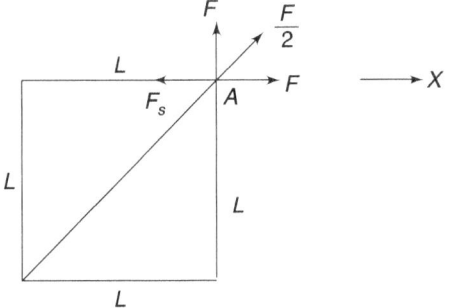

Figure 1.323

Force due to diagonally opposite charge,
$$\dfrac{kq^2}{(\sqrt{2}L)^2} = \dfrac{F}{2}$$
The resultant force in X direction,
$$F_x = F + \dfrac{F}{2}\cos 45° = F\left(1 + \dfrac{1}{2\sqrt{2}}\right)$$
This gets balanced by the spring force.
$$k \cdot \dfrac{L}{2} = F\left(1 + \dfrac{1}{2\sqrt{2}}\right)$$
$$kL = \dfrac{q^2}{4\pi \epsilon_0 L^2}\left(\dfrac{2\sqrt{2}+1}{\sqrt{2}}\right)$$
$$\therefore \quad k = \dfrac{q^2}{4\pi \epsilon_0 L^3}\left(\dfrac{2\sqrt{2}+1}{\sqrt{2}}\right)$$

9. **APPROACH 1.** Suppose the charge on one object is q and on the other is $Q - q$. Since, the electric force between these charges depends on the magnitude of q, therefore, for maximum force between the charges, we have-
$$\dfrac{dF}{dq} = 0$$
So, first find the expression for electric force F and then use above expression and solve for q.
SOLUTION The force between the objects is
$$F = k\dfrac{q(Q-q)}{r^2} = k\dfrac{(qQ - q^2)}{r^2}$$
where r is the separation between them.
Now, $\dfrac{dF}{dq} = \dfrac{d}{dq}\left(k\dfrac{(qQ-q^2)}{r^2}\right) = \left(k\dfrac{(Q-2q)}{r^2}\right)$

For maximum force, $\dfrac{dF}{dq} = 0$, i.e.,
$$\left(k\dfrac{(Q-2q)}{r^2}\right) = 0 \implies q = Q/2$$
Thus, the charge should be divided equally on the two objects.

10. The equilibrium will be unstable.
    Explanation We take the case when the central charge is of opposite sign to the two other charges. If this charge is moved slightly along the line joining the three charges, the attraction due to the nearer charge increases, whilst that due to the more distant charge diminishes, with the result that the charge moves still further from the equilibrium position. Its equilibrium is therefore unstable.
    If the central charge is of the same sign as the other two, and it moves slightly along the line joining the charges, forces will arise that tend to return it to its equilibrium position. However, if it moves at right angles to the line joining the charges, the resultant of the repulsions will no longer be zero and will act in the direction in which it has moved. As a result the charge will tend to move further from its equilibrium position. The equilibrium is thus unstable.
    This result, which we have obtained for an elementary case, is always valid. If only Coulomb forces of interaction are present in a system of free electric charges, the equilibrium is always unstable.

11. $Q = q/\sqrt{3}$ (the equilibrium is unstable).

12. (C) : In case I:
$$F = -\dfrac{1}{4\pi\varepsilon_0}\dfrac{Q^2}{r^2} \quad (1)$$
In Case II : $Q_A = Q - \dfrac{Q}{4}, Q_B = -Q + \dfrac{Q}{4}$
$$\therefore \quad F' = \dfrac{1}{4\pi\varepsilon_0}\dfrac{\left(Q - \frac{Q}{4}\right)\left(-Q + \frac{Q}{4}\right)}{r^2}$$
$$\Rightarrow F' = \dfrac{1}{4\pi\varepsilon_0}\dfrac{\left(\frac{3}{4}Q\right)\left(\frac{-3}{4}Q\right)}{r^2} = -\dfrac{1}{4\pi\varepsilon_0}\dfrac{9}{16}\dfrac{Q^2}{r^2} \quad (2)$$
From equations (1) and (2), $F' = \dfrac{9}{16}F$

13. (B): A hydrogen atom consists of an electron and a proton.
    $\therefore$ Charge on one hydrogen atom
    $= q_e + q_p = -e + (e + \Delta e) = \Delta e$
    Since a hydrogen atom carries a net charge Δe

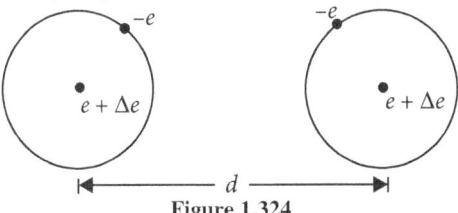

Figure 1.324

$\therefore$ Electrostatic force,
$$F_e = \dfrac{1}{4\pi\varepsilon_o}\dfrac{(\Delta e)^2}{d^2} \quad (1)$$
will act between two hydrogen atoms. The gravitational force between two hydrogen atoms is given as
$$F_g = \dfrac{Gm_h m_h}{d^2} \quad (2)$$
Since, the net force on the system is zero, therefore
$$F_e = F_g$$
Using Eq. 1 and 2, we get

$$\frac{(\Delta e)^2}{4\pi\varepsilon_o d^2} = \frac{Gm_h^2}{d^2}$$

or $\quad (\Delta e)^2 = 4\pi\varepsilon_0 G m_h^2$

$\qquad = 6.67 \times 10^{-11} \times (1.67 \times 10^{-27})^2 \left[1/(9 \times 10^9)\right]$

or $\quad \Delta e \approx 10^{-37}\,\text{C}$

14. (B)
$$T\cos\theta = mg \qquad (1)$$
$$T\sin\theta = \frac{kq^2}{x^2} \qquad (2)$$

From Equations (1) and (2), we have

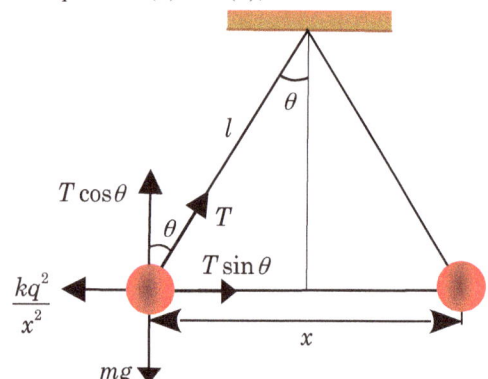

Figure 1.325

$$\tan\theta = \frac{kq^2}{x^2 mg}$$

Since θ is small, therefore, $\tan\theta = \sin\theta = \frac{x}{2l}$

so, $\quad \frac{x}{2l} = \frac{kq^2}{x^2 mg} \Rightarrow q^2 = x^3 \frac{mg}{2lk}$

or $\quad q \propto x^{3/2}$

$\Rightarrow \quad \frac{dq}{dt} \propto \frac{3}{2}\sqrt{x}\frac{dx}{dt} = \frac{3}{2}(\sqrt{x})v$

Since, $\frac{dq}{dt}$ = constant, therefore, $v \propto \frac{1}{\sqrt{x}}$

15. (A) The given situation is drawn in Fig.1.326. A, B and C are three small charged spheres each has charge q and mass m. They are suspended from three strings AS, BS and CS respectively to the common point of suspension S. Length of each string is l, i.e., $AS = BS = CS = l$ and the angle of these strings from vertical downward direction is θ. The separation between charges is same and at any instant it is x. These charges are in a same horizontal plane and the triangle ABC is an equilateral triangle of side x. $\angle CAB = \angle ABC = \angle BCA = 60°$. O is the centroid of the triangle, therefore, $OA = OB = OC$. SO is perpendicular to the plane of triangle ABC. If we extrapolate median BO behind O, it perpendicularly bisects side CA at point D. So, $DA = x/2$.

Since, lines joining the vertices of an equilateral triangle to it's centroid bisects the angle, therefore $\angle DOA = 60°/2 = 30°$. Also in right angle $\triangle ADO$, we have
$$\cos 30° = \frac{DA}{OA}$$
$\Rightarrow \quad OA = \frac{DA}{\cos 30°} = \frac{x/2}{\sqrt{3}/2} = \frac{x}{\sqrt{3}} \qquad (1)$

In Fig.1.326, the positive directions of X, Y and Z axes are also shown in separate diagram.

Now, consider the sphere A. The forces acting on it are:

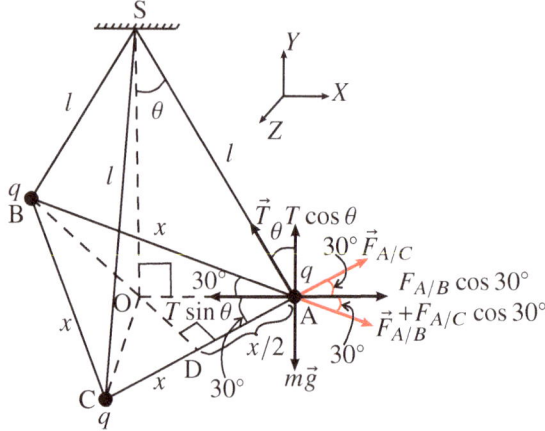

Figure 1.326

(a) **Gravitational force** $m\vec{g}$: It is acting vertically downwards along negative direction of Y axis:
$$m\vec{g} = mg\hat{j}$$

(b) **Tension** $\vec{T}$: It is acting along $\overrightarrow{AS}$ in XOY plane
$$\vec{T} = -T\sin\theta\hat{i} + T\sin\theta\hat{j}$$

(c) **Electric force** $\vec{F}_{A/B}$: It is applied by sphere B along $\overrightarrow{BA}$ in XOZ plane:
$$\vec{F}_{A/B} = F_{A/B}\cos 30°\hat{i} + F_{A/B}\sin 30°\hat{k}$$

(d) **Electric force** $\vec{F}_{A/C}$: It is applied by sphere B along $\overrightarrow{CA}$ in XOZ plane:
$$\vec{F}_{A/C} = F_{A/C}\cos 30°\hat{i} - F_{A/C}\sin 30°\hat{k}$$

All forces and their components are shown in Fig.1.326 except Z components of $\vec{F}_{A/B}$ and $\vec{F}_{A/C}$ which will be outward and inward respectively to the plane pf paper.

Since, charges and distances of B and C from A are identical, therefore they apply same electric force on A. It is given as:
$$F_{A/B} = F_{A/C} = k\frac{q^2}{x^2} \qquad (2)$$

Net force on the charged sphere A,
$$\vec{F} = m\vec{g} + \vec{T} + \vec{F}_{A/B} + \vec{F}_{A/C}$$
$$= -mg\hat{j} + (-T\sin\theta\hat{i} + T\sin\theta\hat{j})$$
$$\quad + \left(F_{A/B}\cos 30°\hat{i} + F_{A/B}\sin 30°\hat{k}\right)$$
$$\quad + \left(F_{A/C}\cos 30°\hat{i} - F_{A/C}\sin 30°\hat{k}\right)$$
$$\Rightarrow \vec{F} = \left(2F_{A/B}\cos 30° - T\sin\theta\right)\hat{i} + (T\cos\theta - mg)\hat{j} \quad (3)$$
$$[\because \ F_{A/B} = F_{AC}]$$

Now, according to problem, the length of side of triangle $\triangle ABC$ is slowly increasing, i.e., the acceleration of each sphere is zero. So, the sphere A is in dynamic equilibrium and hence net force on it must be zero. Therefore from Eq.(3), we get-
$$\vec{F} = \left(2F_{A/B}\cos 30° - T\sin\theta\right)\hat{i} + (T\cos\theta - mg)\hat{j} = 0$$

On comparing both sides, we get

$\quad 2F_{A/B}\cos 30° - T\sin\theta = 0 \quad$ and $\quad T\cos\theta - mg = 0$

or $\quad T\sin\theta = 2F_{A/B}\cos 30° = F_{A/B}\sqrt{3} \qquad (4)$

and $\qquad T\cos\theta = mg \qquad (5)$

Dividing Eq.(4) by Eq.(5), we get
$$\tan\theta = \frac{F_{A/B}\sqrt{3}}{mg} \qquad (6)$$

Also in right angle $\triangle SOA$, we have
$$\tan\theta = \frac{OA}{OS} = \frac{OA}{\sqrt{(AS)^2 - (OA)^2}} = \frac{x/\sqrt{3}}{\sqrt{l^2 - (x/2)^2}} \qquad (7)$$

Since, $x << l$, therefore $l^2 - (x/2)^2 \approx l^2$.
Substituting it in Eq.(7), we get-
$$\tan\theta = \frac{x}{l\sqrt{3}} \qquad (8)$$
Comparing the values of $\tan\theta$ obtained from Eq.(6) and (8), we get
$$\frac{F_{A/B}\sqrt{3}}{mg} = \frac{x}{l\sqrt{3}}$$
$$F_{A/B} = \frac{xmg}{3l} \qquad \ldots (9)$$
Putting the value of $F_{A/B}$ from Eq.(2) in (9), we get
$$k\frac{q^2}{x^2} = \frac{xmg}{3l} \implies q = \sqrt{\left[\frac{mg}{3lk}\right]}x^{3/2}$$
Differentiating both sides with respect to time 't', we get
$$\frac{dq}{dt} = \sqrt{\left[\frac{mg}{3lk}\right]}\frac{3}{2}x^{1/2}\frac{dx}{dt}$$
It is also given that, $\frac{dx}{dt} = \frac{a}{\sqrt{x}}$, therefore, above equation gives
$$\frac{dq}{dt} = \sqrt{\left[\frac{mg}{3lk}\right]}\frac{3}{2}x^{1/2}\frac{a}{\sqrt{x}} = \sqrt{\frac{3mga^2}{4lk}} \qquad (10)$$
Since, $k = 1/4\pi\epsilon_0$, therefore above equation can also be written as
$$\frac{dq}{dt} = \sqrt{\frac{3\pi\epsilon_0 mga^2}{l}}$$

16. **(D)** : Let m be mass of each ball and q be charge on each ball. Force of repulsion,
$$F = \frac{1}{4\pi\epsilon_0}\frac{q^2}{r^2}$$

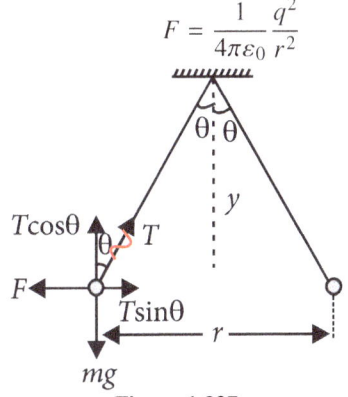

Figure 1.327

In equilibrium, we have
$$T\cos\theta = mg \qquad (1)$$
$$T\sin\theta = F \qquad (2)$$
Dividing Eq.(2) by (1), we get,
$$\tan\theta = \frac{F}{mg} = \frac{\frac{1}{4\pi\epsilon_0}\frac{q^2}{r^2}}{mg}$$
From Fig.1.327,
$$\frac{r/2}{y} = \frac{\frac{1}{4\pi\epsilon_0}\frac{q^2}{r^2}}{mg} \qquad (3)$$
From Fig.1.328, we have
$$\tan\theta' = \frac{\frac{1}{4\pi\epsilon_0}\frac{q^2}{r'^2}}{mg} \implies \frac{r'/2}{y/2} = \frac{\frac{1}{4\pi\epsilon_0}\frac{q^2}{r'^2}}{mg} \qquad (4)$$
Divide Eq.(4) by (3), we get
$$\frac{2r'}{r} = \frac{r^2}{r'^2} \implies r'^3 = \frac{r^3}{2} \implies r' = \frac{r}{\sqrt[3]{2}}$$

17. **(C)** According to Coulomb's law, the force of repulsion be-

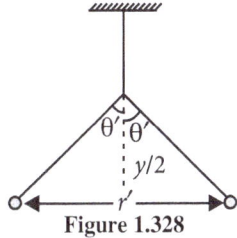

Figure 1.328

tween the two positive ions each of charge q, separated by a distance d is given by -
$$F = \frac{1}{4\pi\epsilon_0}\frac{(q)(q)}{d^2} = \frac{q^2}{4\pi\epsilon_0 d^2}$$
$$\implies q^2 = 4\pi\epsilon_0 F d^2$$
$$\text{or} \quad q = \sqrt{4\pi\epsilon_0 F d^2} \qquad (1)$$
Since, $q = ne$ where, n = number of electrons missing from each ion, e = magnitude of charge on electron
$$\therefore \quad n = \frac{q}{e} = \sqrt{\frac{4\pi\epsilon_0 F d^2}{e^2}}$$

18. **(D)**: $F_m = F_0/\kappa$, i.e., decreases to κ^{-1} times.
Note that it is the net force on either charged particles due to polarisation effect of medium. The individual force acting between the charges remains unchanged.

19. **(A)** The situation is as shown in the Fig.1.329.

Figure 1.329

Let two equal charges Q each placed at points A and B at a distance r apart. C is the centre of AB where charge q is placed.
For equilibrium, net force on charge $Q = 0$
$$\therefore \quad \frac{1}{4\pi\epsilon_0}\frac{QQ}{r^2} + \frac{1}{4\pi\epsilon_0}\frac{Qq}{(r/2)^2} = 0$$
$$\text{or} \quad \frac{1}{4\pi\epsilon_0}\frac{Q^2}{r^2} = -\frac{1}{4\pi\epsilon_0}\frac{4Qq}{r^2}$$
$$\text{or} \quad Q = -4q \quad \text{or} \quad q = -Q/4$$

20. **(C)** Net force on each of the charge due to the other charges is zero. However, disturbance in any direction other than along the line on which the charges lie, will not make the charges return.

21. A, B, C

22. **(C)** Is pushed outside the triangle. Like charges repel. The forces from the two charges at vertices adjacent to the fourth charge are equal but opposite and cancel. The net force is thus the same as the contribution from the charge at the opposite vertex; that repulsion will push the fourth charge away, out of the triangle.

1.38.3 Check Point 3

1. **Approach** We can use symmetry to significantly reduce the number of actual calculations. If we place a small test charge at point P, the forces from the top left and the bottom right will be of equal magnitude but opposite direction and will cancel. The same is true for the charges at the top right and the bottom left. Therefore, we only need to consider the two central charges. Since the top one is positive and the bottom one is negative, a positive test charge placed at P would experience downward forces from both charges. We simply calculate the equal electric fields using the point charge relationship and then use the superposition principle

to combine.

SOLUTION $E = k\dfrac{Q}{a^2} + k\dfrac{Q}{a^2} = 2k\dfrac{Q}{a^2}$ (downwards)

2. If $\vec{r}_0$ is the position vector of position O (1 m, 2 m, 4 m) of charge and $\vec{r}_P$ is the position vector of position P (0, −4 m, 3 m) with respect to origin (0, 0, 0), then
$$\vec{r}_0 = (\hat{i} + 2\hat{j} + 4\hat{k}) \text{ m and } \vec{r}_P = (-4\hat{j} + 3\hat{k}) \text{ m}$$
Therefore, $\vec{r}_{OP} = \vec{r}_P - \vec{r}_0 = (-\hat{i} - 6\hat{j} - \hat{k})$ m
and $\vec{r}_{OP} = \sqrt{(-1)^2 + (-6)^2 + (-1)^2} = \sqrt{38}$ m
So, $E = k\dfrac{Q}{|\vec{r}_{OP}|^2} = \dfrac{9.0 \times 10^9 \times 10^{-6}}{38}$
$\approx 0.24 \times 10^3 \text{N/C} = 240 \text{ N/C}$

3. Here, $r - r_0 = (8\hat{i} - 5\hat{j}) - (2\hat{i} + 3\hat{j}) = 6\hat{i} - 8\hat{j}$
Therefore, $|r - r_0| = \sqrt{6\hat{i} - 8\hat{j}} = 10 \, m$
So, Required field strength,
$$|\vec{E}| = \dfrac{1}{4\pi\varepsilon_0}\dfrac{q}{|\vec{r} - \vec{r}_0|^2} = 9.0 \times 10^9 \dfrac{50 \times 10^{-6}}{10^2}$$
$= 4.5 \times 10^3$ V/m = 4.5 kV/m

4. **APPROACH** follow the approach of Example 43. If $\vec{E}$ is the electric field at point P, then, the electric force on test charge placed at point P, is given by $\vec{F} = q_0 \vec{E}$.
Answer $\vec{E} = \dfrac{qz.2\sqrt{2}}{ql^3}\hat{k}, \vec{F} = q_0\vec{E} = \dfrac{qq_0z.2\sqrt{2}}{ql^3}\hat{k}$

5. Given: Charges Q_1 and Q_2 are located on vertices of a right angled triangle OAB,
The resultant electric field at O due to Q_1 and Q_2 is perpen-

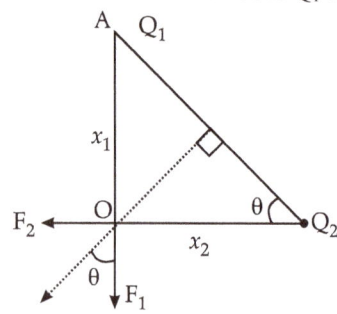

Figure 1.330

dicular to side AB.
To find $\dfrac{Q_1}{Q_2}$: From the Fig.1.330,
$$\tan\theta = \dfrac{x_1}{x_2} = \dfrac{F_2}{F_1}$$
$\Rightarrow \dfrac{x_1}{x_2} = \dfrac{Q_2/x_2^2}{Q_1/x_1^2} = \dfrac{Q_2}{Q_1}\cdot\dfrac{x_1^2}{x_2^2} \Rightarrow \dfrac{Q_1}{Q_2} = \dfrac{x_1}{x_2}$

6. Due to repulsion of test charge q_0 (assumed not too small), placed at point P, the suspended charged ball repel away from q_0, therefore the separation between the two charges increases. As a result of it, the measured value of electric field $E_{measured} = \dfrac{F}{q_0}$ will be less than the actual value E_{act}.

7. $k\dfrac{4Q}{x^2} = k\dfrac{9Q}{(d-x)^2}$
$\Rightarrow \dfrac{4}{x^2} = \dfrac{9}{(d-x)^2}$
$\Rightarrow \dfrac{2}{x} = \pm\dfrac{3}{(d-x)}$
$\Rightarrow 2d - 2x = \pm 3x$
On simplifying above expression, we get
$$x = \dfrac{2d}{5} \text{ and } x = -2d$$
The only acceptable answer is, $x = 2d/5$

8. Net intensity of electric field at origin is given by-
$$E_O = \dfrac{kq}{r_1^2} + \dfrac{kq}{r_2^2} + \dfrac{kq}{r_3^2} + \ldots \infty$$
$= kq \times 10^4 \left[\dfrac{1}{1} + \dfrac{1}{4} + \dfrac{1}{16} + \dfrac{1}{64} + \ldots \infty\right]$
$= 9 \times 10^9 \times 20 \times 10^{-6} \times 10^4 \times \left(\dfrac{1}{1 - 1/4}\right)$
$= 24 \times 10^8$ N/C

9. **APPROACH** Apply the principle of superposition of electric fields and the expansion-
$$\ln(1+x) = x - \dfrac{x^2}{2} + \dfrac{x^3}{3} - \dfrac{x^4}{4} + \ldots$$
For, $x = 1$, above expression gives-
$$\ln 2 = 1 - \dfrac{1}{2} + \dfrac{1}{3} - \dfrac{1}{4} + \ldots$$

SOLUTION Net intensity of electric field at origin is given by-
$$E_O = \dfrac{kq}{a^2} - \dfrac{kq}{2a^2} + \dfrac{kq}{3a^2} - \dfrac{kq}{3a^2} + \dfrac{kq}{4a^2} - \dfrac{kq}{5a^2} - \ldots$$
$= \dfrac{kq}{a^2}\left[1 - \dfrac{1}{2} + \dfrac{1}{3} - \dfrac{1}{4} + \ldots\right]$
$= \dfrac{kq}{a^2}[\ln 2]$

10. (B) From Fig.1.331, in right angle $\triangle OLP$ we have-
$$\tan\theta = \dfrac{LP}{OL} = \dfrac{2\sqrt{2}}{2} = \sqrt{2} \qquad \cdots (1)$$
Now, from Eq.1.53, we have

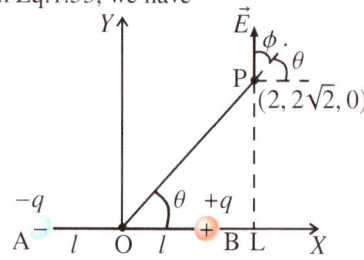

Figure 1.331

$$\tan\phi = \dfrac{\tan\theta}{2} = \dfrac{\sqrt{2}}{2} = \dfrac{1}{\sqrt{2}} = \dfrac{1}{\tan\theta} = \cot\theta$$
[since from Eq.(1), $\tan\theta = \sqrt{2}$]
$\Rightarrow \tan\phi = \tan(90° - \theta)$
$\Rightarrow \theta + \phi = 90°$
i.e., $\vec{E}$ is along positive y-axis.

11. (C) Direction of dipole moment will be opposite to the electric field

1.38.4 Check Point 4

1. **APPROACH** From Eq.1.69, the total electric field at the centre of the uniformly charged circular arc of radius R, subtending an angle α at the centre, is given by-
$$E = \dfrac{kQ}{R^2}\dfrac{\sin\alpha/2}{\alpha/2} \qquad (1)$$
Here, Q is the charge on circular arc. For positive charge, the direction of field is away from the circular arc along the symmetry axis i.e., net electric field vector passing through the mid point of the arc at at angle $\alpha/2$ at the center. Similarly, for negative charge, the direction of $\vec{E}$ is towards the circular arc at angle $\alpha/2$ from the line dividing the circular arc.
In this problem, each charged quarter-circle produces a field of magnitude

$$E = \frac{kQ}{R^2} \frac{\sin \pi/4}{\pi/4} = \frac{2\sqrt{2}kQ}{\pi R^2}$$

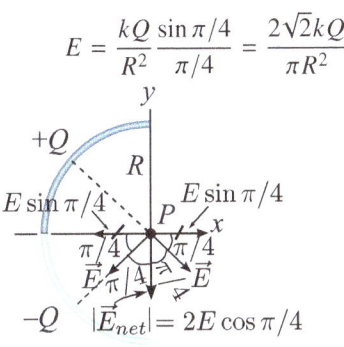

Figure 1.332

From Fig.1.332, it is clear that the x components of these electric fields are opposite to each other and hence get cancelled with each other. The components along negative direction of y-axis add up to each other and give the net electric field $\vec{E}_{net}$.

SOLUTION Net electric field at point P,

$$E_{net} = 2E\cos\pi/4 = 2\left(\frac{2\sqrt{2}kQ}{\pi R^2}\right)\cos\pi/4 = \frac{4kQ}{\pi R^2}$$

On substituting the given values in above expression, we get-

$$E = \frac{(9.0 \times 10^9 \text{ N}\cdot\text{m}^2/\text{C}^2)\, 4\,(4.50 \times 10^{-12}\text{C})}{\pi\,(5.00 \times 10^{-2} \text{ m})^2}$$

$$= 20.6 \text{ N/C}$$

(b) By symmetry, the net field points vertically downward in the $-\hat{j}$ direction, or $-90°$ counter-clockwise from the $+x$ axis.

2. **APPROACH** From symmetry, we see that the net field at P is twice the field caused by the upper semicircular charge $+q$ (and that it points downward).

The electric field at centre P, due to each halves, can be obtained by using expression-

$$\vec{E} = -\frac{kq}{R^2}\frac{\sin\alpha/2}{\alpha/2}\hat{j} \quad \text{(with } \alpha = \pi\text{)}$$

Therefore, the net electric field at point P due to both halves can be given as-

$$\vec{E}_{net} = 2\vec{E} = -\frac{kq}{R^2}\frac{\sin\alpha/2}{\alpha/2}\hat{j}$$

Now, substitute the values in above expression and solve for $\vec{E}_{net}$.

SOLUTION (a) With $R = 8.50 \times 10^{-2}$m, $\alpha = \pi$ and $q = 1.50 \times 10^{-8}$C, $|\vec{E}_{net}| = 23.8$N/C.

(b) The net electric field $\vec{E}_{net}$ points in the $-\hat{j}$ direction, or $-90°$ clockwise from the $+x$ axis.

3. $4\lambda a^2$
   Hint: Follow the approach used in solved Ex. 55.

4. $2\sqrt{2}\lambda a^2$
   Hint: Follow the approach used in solved Ex.55

5. From Eq.1.78, we have

$$E_{(a)} = \frac{Q}{2\pi\epsilon_0 R^2}\left(1 - \frac{x}{(x^2+R^2)^{1/2}}\right),$$

$$E_{(b)} = \frac{Q}{2\pi\epsilon_0 (2R)^2}\left(1 - \frac{x}{(x^2+(2R)^2)^{1/2}}\right),$$

and

$$E_{(c)} = \frac{Q}{2\pi\epsilon_0 (2R)^2}\left(1 - \frac{x}{(x^2+(2R)^2)^{1/2}}\right)$$

$$-\frac{Q}{2\pi\epsilon_0 R^2}\left(1 - \frac{x}{(x^2+R^2)^{1/2}}\right)$$

Clearly, $E_{(a)} > E_{(b)} > E_{(c)}$. So, the correct order is a > b > c.

6. From Eq.1.78,

$$E_x = \frac{Q}{2\pi\epsilon_0 R^2}\left(1 - \frac{x}{(x^2+R^2)^{1/2}}\right)$$

Since, charge Q is placed at distance h from the centre of charged disc, therefore putting $x = h$ in above expression, we get

$$E_h = \frac{Q}{2\pi\epsilon_0 R^2}\left(1 - \frac{h}{(h^2+R^2)^{1/2}}\right)$$

Therefore, $F = QE_h = \frac{Q^2}{2\pi\epsilon_0 R^2}\left(1 - \frac{h}{(h^2+R^2)^{1/2}}\right)$

The direction of this force is directly away from the charged plate.

7. The angle swept by radius vector is called azimuthal angle. Since, λ is cosine function of ϕ, so, ring is not uniformly charged. A part of ring is positively charged and remaining part of ring is negatively charged. The distribution of charge on ring is shown in Fig. 1.333.

Since, $\lambda = \lambda_0 \cos\phi$, therefore, first and fourth quadrants are

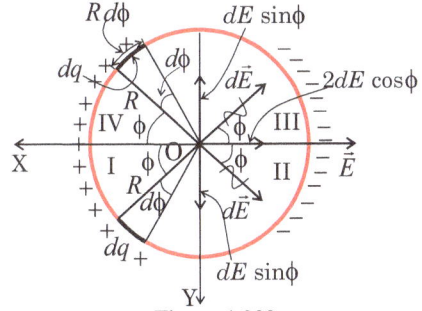

Figure 1.333

positively charged (because from fourth to first $-\pi/2 < \phi < +\pi/2$) but second and third quadrants are negatively charged (because from second to third $\pi/2 < \phi < +3\pi/2$).

In Fig.1.333, we have considered two symmetrical elementary charges dq in fourth and first quadrants. Each of these charges produces an electric field $d\vec{E}$ at centre of the ring. Vertical components of each of these fields are oppositely directed and each has equal magnitude of $dE\sin\phi$. So, they get cancelled. The horizontal components are directed in same direction and each has magnitude $dE\cos\phi$. So, they get added and their resultant is $2dE\cos\phi$. It is along the negative direction of X–axis.

Note that $dq = \lambda R d\phi = \lambda_0 \cos\phi R d\phi$.

Now, if we consider similar elementary $-ve$ charges in second and third quadrants, they also produce the field along $-ve$ X–axis. Their resultant will also be $2dE\cos\phi$. So, resultant of above field is $4dE\cos\phi$. It is directed along negative direction of X–axis.

Therefore, net field at the centre of ring can be obtained by integrating above resultant from limits $\phi = 0$ to $\phi = \pi/2$.

$$E_{net} = \int_{\phi=0}^{\pi/2} 4\,dE\cos\phi$$

Here, $dE = k\frac{dq}{R^2}$ and $dq = \lambda_0 \cos\phi R d\phi$, therefore,

$$dE = k\frac{\lambda_0 \cos\phi\, R d\phi}{R^2}$$

So, $E_{net} = 4\int_{\phi=0}^{\pi/2} k\dfrac{\lambda_0 \cos^2\phi \, Rd\phi}{R^2}$

$= \dfrac{4k\lambda_0}{R}\int_{\phi=0}^{\pi/2} \cos^2\phi \, Rd\phi$

Since, $\cos 2\phi = 2\cos^2\phi - 1 \implies \cos^2\phi = \tfrac{1}{2}(\cos 2\phi + 1)$

Therefore, $\int_{\phi=0}^{\pi/2} \cos^2\phi \, Rd\phi = \int_0^{\pi/2} \tfrac{1}{2}(\cos 2\phi + 1)$

$= \dfrac{1}{2}\int_0^{\pi/2}(\cos 2\phi + 1)d\phi$

$= \dfrac{1}{2}\left[\dfrac{\sin 2\phi}{2}\right]_0^{\pi/2} + \dfrac{1}{2}[\phi]_0^{\pi/2} = \dfrac{\pi}{4}$

Therefore, $E_{net} = \dfrac{4k\lambda_0}{R}\dfrac{\pi}{4}$

Since, $k = 1/4\pi\varepsilon_0$, therefore, $E_{net} = \dfrac{\lambda_0}{4\epsilon_0 R}$

(b) Let us take differential length element of the ring at an azimuthal angle ϕ from the x-axis, the element subtends an angle $d\phi$ at the centre, and carries charge $dq = \lambda R d\phi = (\lambda_0 \cos\phi)R\, d\phi$.

Taking the plane of ring as $x-y$ plane and centre of the ring as origin O, locations of field point $\vec{r}$, of charge element $\vec{R}$ and of field point relative to charge element $\vec{r}$ are shown in the Fig.1.334.

Electric field strength at the field point due to considered

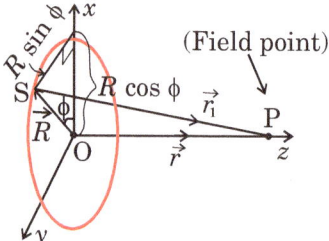

Figure 1.334

charge element

$$d\vec{E} = \dfrac{1}{4\pi\varepsilon_0}\dfrac{dq}{r_1^3}\vec{r}_1$$

Using $\vec{r}_1 = \vec{r} - \vec{R} = z\hat{k} - (R\cos\phi\,\hat{i} + R\sin\phi\,\hat{j})$
and $\vec{r}_1^3 = (R^2 + z^2)^{3/2}$

$d\vec{E} = \dfrac{1}{4\pi\varepsilon_0}\dfrac{(\lambda_0\cos\phi)R\,d\phi}{(R^2+z^2)^{3/2}}\{z\hat{k} - (R\cos\phi\,\hat{i} + R\sin\phi\,\hat{j})\}$

So, the net electric field strength

$\vec{E} = \int d\vec{E} = \dfrac{\lambda_0 R}{4\pi\varepsilon_0(R^2+z^2)^{3/2}}\left[z\hat{k}\int_0^{2\pi}\cos\phi\,d\phi\right.$

$\left. -R\hat{i}\int_0^{2\pi}\cos^2\phi\,d\phi - R\hat{j}\int_0^{2\pi}\sin 2\phi\,d\phi\right]$

Taking into account

$\int_0^{2\pi}\cos\phi\,d\phi = 0, \quad \int_0^{2\pi}\sin 2\phi\,d\phi = 0$

and $\int_0^{2\pi}\cos^2\phi\,d\phi = \pi$

We get $\vec{E} = \dfrac{1}{4\pi\varepsilon_0}\dfrac{\pi\lambda_0 R^2(-\hat{i})}{(R^2+z^2)^{3/2}}$

Hence, $E = \dfrac{\lambda_0 R^2}{4\varepsilon_0(R^2+z^2)^{3/2}}$

For $z \gg R$, $E_x = \dfrac{p}{4\pi\varepsilon_0 z^3}$, where $p = \lambda_0 \pi R^2$

8. (a) $E = \dfrac{\lambda\sqrt{2}}{4\pi\varepsilon_0 R}$; (b) $E = 0$

9. (A) The electric field E just outside the earth's surface is same as if the entire charge q were concentrated at it's centre. Thus,

$E = \dfrac{1}{4\pi\varepsilon_0}\dfrac{q}{R^2} = \dfrac{1}{4\pi\varepsilon_0}\dfrac{4\pi R^2 \sigma}{R^2} = \dfrac{\sigma}{\varepsilon_0}$

Substituting the given value:

$E = \dfrac{(-1.6\times 10^{-19})\,C/m^2}{8.9\times 10^{-12}\,C^2/N.m^2} = -1.8108\,N/C$

The minus sign indicates that E is radially inward.

10. Electric field due to an arc at its centre is $(k\lambda/R)2\sin\theta/2$, where $k = 1/4\pi\varepsilon_0$
    θ = angle subtended by the wire at the centre,
    λ = Linear density of charge.

Let E be the electric field due to remaining portion. Since intensity at the centre due to the circular wire is zero. Applying principle of superposition.

$\dfrac{k\lambda}{R}2\sin\dfrac{\theta}{2} - E = 0$

$\implies E = \dfrac{1}{4\pi\varepsilon_0 R}\dfrac{Q}{2\pi R}2\sin\dfrac{\theta}{2} = \dfrac{Q}{4\pi^2\varepsilon_0 R^2}\sin\dfrac{\theta}{2}$

1.38.5 Check Point 5

1. (a) Charge ① is negative. (b) Charge ② is positive. (c) Since, the number of field lines terminating on charge ① is greater than that of emitting from ②, therefore the magnitude of charge ① is greater than the magnitude of charge 2.

2. $A > B > C$.

3. (a) 2 : 1 (b) 1 : 3

4. (a) charges A and C are positive and charge B is negative
   (b) $|q_A| : |q_B| : |q_C| = 5 : 10 : 15 = 1 : 2 : 3$
   i.e., $|q_A| < |q_B| < |q_C|$

5. See Fig.1.335

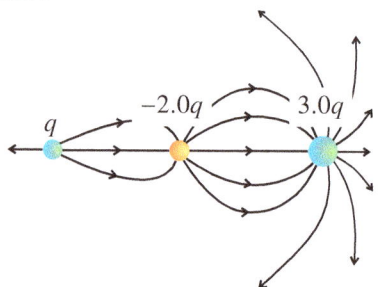

Figure 1.335

6. See Fig.1.336
7. (B), (D)
8. (B) The electric field depends on the density of electric field lines. The field will be greater where the density of field lines is greater.

1.38.6 Check Point 6

1. (A) FBD of the charged particle is shown in Fig.1.337. As the charged particle is floating in air (neglecting the buoyant force due to air) we obtain-

$mg = qE \implies q = \dfrac{mg}{E}$

2. (B) Since, the electric field is a conservative field, therefore, the work done by electric field is independent of the path followed and is equal to-

1.38. *ANSWER KEYS AND SOLUTIONS* 117

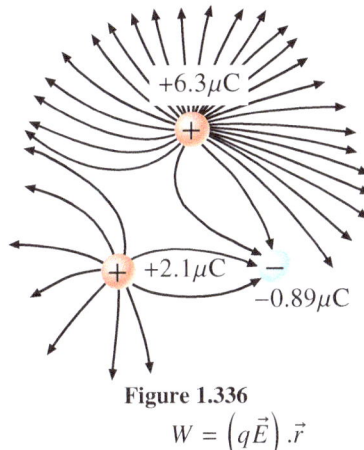

Figure 1.336

$$W = \left(q\vec{E}\right).\vec{r}$$

where $\vec{r}$ = displacement from P to S.
Here, $\vec{r} = -a\hat{i} - b\hat{j}$, while $\vec{E} = E\hat{i}$
Therefore, work $= -(qE\hat{i}).(a\hat{i} + b\hat{j}) = -qaE$

3. **(C)** Let x be the small displacement given to the pendulum such that the angle θ is small. The forces acting at A are [Fig.1.338]-
   (i) tension $\vec{T_0}$ along the thread

Figure 1.337   Figure 1.338

(ii) weight $m\vec{g}$ acting vertically downwards.
(iii) electrical force qE vertically upwards.
The resultant force, vertically downwards, is $(mg - qE)$.
If g_{eff} is the effective acceleration of bob, then

$$mg_{\text{eff}} = mg - qE \implies g_{\text{eff}} = g - \frac{qE}{m}$$

So, time period, $T = 2\pi\sqrt{\frac{l}{g_{\text{eff}}}} = 2\pi\left[\frac{l}{g - \frac{qE}{m}}\right]^{1/2}$

4. **(B)** FBD of given situation is shown in Fig.1.339
   From Fig.1.339, the resultant of vertical downward force

Figure 1.339

$m\vec{g}$ and horizontal force $q\vec{E}$ is $\sqrt{(mg)^2 + (qE)^2}$. If g_{eff} is the effective acceleration, then

$$mg_{\text{eff}} = \sqrt{(mg)^2 + (qE)^2} \implies g_{\text{eff}} = \sqrt{g^2 + \left(\frac{qE}{m}\right)^2}$$

Therefore, $T = 2\pi\sqrt{\frac{l}{g_{\text{eff}}}} = 2\pi\sqrt{\frac{l}{\sqrt{g^2 + \left(\frac{qE}{m}\right)^2}}}$.

5. **(D)** The torque exerted by an electric field E on a dipole of moment p is given by

$$\tau = pE\sin\theta$$

where θ is the angle which the dipole moment is making with the electric-field.
Corresponding to maximum torque, $\theta = 90°$, therefore-
$$\tau_{\max} = pE$$
Here, $p = q(2l) = 1 \times 10^{-6} \times 0.02$ C/m and $E = 10^5$ N/C
Therefore, $\tau_{\max} = 1 \times 10^{-6} \times 0.02 \times 10^5 = 2 \times 10^{-3}$ N.m

6. **(B)** When the dipole is given a small angular displacement θ from its mean position, the magnitude of restoring torque on it, is

$$\tau = -pE\sin\theta$$

For small angular displacement $\sin\theta \approx \theta$, therefore-
$$\tau = -pE\theta$$
If α is the angular acceleration of dipole, then $\tau = I\alpha$

Therefore, $I\alpha = -pE\theta \implies \alpha = -\frac{pE\theta}{I} = -\omega^2\theta$

where $\omega^2 = \frac{pE}{I}$

Therefore, $T = 2\pi\sqrt{\frac{I}{pE}}$

1.38.7 Check Point 7

1. **APPROACH** Time period of a simple pendulum is given by-

$$T = 2\pi\sqrt{\frac{L}{|g_{\text{eff}}|}} \qquad (1)$$

here, g_{eff} is the effective gravitational acceleration. Now, find the value of g_{eff} in given situations and by using Eq.1, calculate the value of time period T.

(a) In this case, from FBD [Fig.1.340a], the effective value of g can be given by

$$mg_{\text{eff}} = mg + qE = mg + mg \quad \left(\because E = \frac{mg}{q}\right)$$

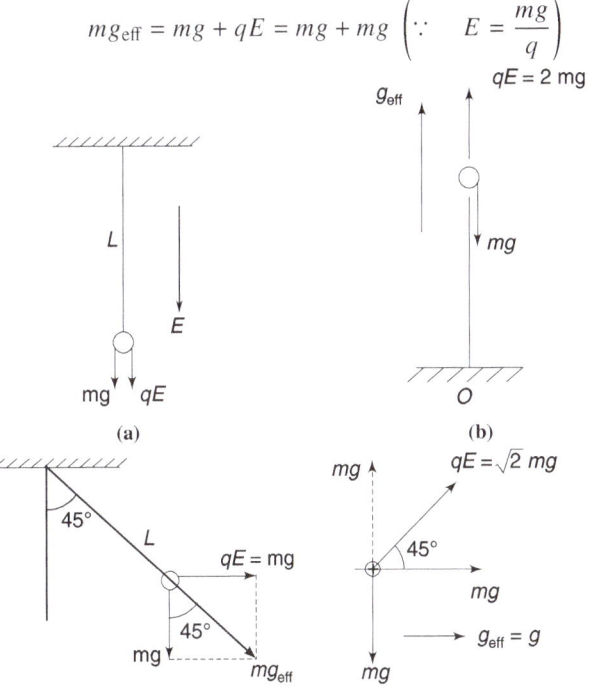

Figure 1.340: Different situations of pendulum

$$\implies g_{\text{eff}} = 2g$$

Substituting this value of g_{eff} in Eq.1, we get

$$T = 2\pi\sqrt{\frac{L}{2g}}$$

(b) In this case, from FBD (Fig.1.340b), the effective

acceleration is given by-
$$mg_{eff} = mg - qE = mg - 2mg \quad \left(\because E = \tfrac{2mg}{q}\right)$$
$$\Rightarrow \quad g_{eff} = -g$$
Here, $-ve$ sign just says that the effective gravitational acceleration is directed in upward direction. So, the pendulum get inverted and oscillates about the fixed point as shown in Fig.1.340b. On substituting this value of g_{eff} in Eq.1, we get
$$T = 2\pi\sqrt{\dfrac{L}{g}}$$

(c) Equilibrium position is shown in Fig.1.340c
In this case, the effective gravitational acceleration is given by-
$$mg_{eff} = \sqrt{(mg)^2 + qE^2} = mg\sqrt{2} \quad \left(\because E = \tfrac{mg}{q}\right)$$
$$\Rightarrow \quad g_{eff} = g\sqrt{2}$$
On substituting this value of g_{eff} in Eq.1, we get
$$T = 2\pi\sqrt{\dfrac{L}{g\sqrt{2}}}$$

(d) In this case, the equilibrium is shown in Fig.1.340d (thread is horizontal). From, this figure, we have-
$$mg_{eff} = \sqrt{(mg - qE\sin 45°)^2 + (qE\cos 45°)^2}$$
$$= \sqrt{\left(mg - \dfrac{qE}{\sqrt{2}}\right)^2 + \left(\dfrac{qE}{\sqrt{2}}\right)^2}$$
$$= \sqrt{\left(mg - \dfrac{qE}{\sqrt{2}}\right)^2 + \left(\dfrac{qE}{\sqrt{2}}\right)^2} = mg$$
$$\left[\because E = \dfrac{\sqrt{2}mg}{q}\right]$$
$$\Rightarrow \quad g_{eff} = g$$
Substituting this value of g_{eff} in Eq.1, we get
$$T = 2\pi\sqrt{\dfrac{L}{g}}$$

2. (a) At every point on the path, the electric force on the particle and it's displacement always remains tangential to the path, i.e., both remains in the same direction. So, work done by the electric force on the particle-
$$W = \int_A^B q\vec{E}\cdot d\vec{l} = qE\cdot \pi R$$
Now, by Work-Energy theorem, we have-
$$W = K_B - K_A$$
Therefore, $K_B - K_A = qE\pi R$
$$\Rightarrow \quad K_B = \pi qER \qquad (\because K_A = 0)$$
(b) Suppose, at mid point the speed of the particle is v [Fig.1.341], then by work-energy theorem, we have
$$\tfrac{1}{2}mv^2 - 0 = qE\dfrac{\pi R}{2} \Rightarrow \dfrac{v^2}{R} = \dfrac{\pi qE}{m}$$

Figure 1.341

Therefore, radial acceleration of the particle,
$$a_r = \dfrac{v^2}{R} = \dfrac{\pi qE}{m}$$

This acceleration arises due to constraining forces.
Tangential acceleration, $a_t = \dfrac{F_t}{m} = \dfrac{qE}{m}$
∴ Resultant acceleration,
$$a = \sqrt{a_r^2 + a_t^2} = \dfrac{qE}{m}\sqrt{1+\pi^2}$$

3. Electrostatic force,
$$F = qE \Rightarrow ma_0 = q(-bx+a) \quad [a_0 = \text{accleration}]$$
$$\Rightarrow \quad a_0 = v\dfrac{dv}{dx} = -\dfrac{bq}{m}x + \dfrac{aq}{m} \qquad \cdots (1)$$
$$\int_0^v v\,dv = -\dfrac{bq}{m}\int_0^x x\,dx + \dfrac{aq}{m}\int_0^x dx$$
Now, $\dfrac{v^2}{2} = -\dfrac{bqx^2}{2m} + \dfrac{aqx}{m}$
$v = 0$ when $\dfrac{bqx^2}{2m} = \dfrac{aqx}{m} \Rightarrow x = 2a/b$
From Eq.(1), we have: $a_0 = \dfrac{q}{m}(-bx+a)$
at $x = 2a/b$, acceleration: $a_0 = \dfrac{q}{m}(-2a+a) = -\dfrac{qa}{m}$

4. Charge Q resides on each of the blocks, which repel as point charges. In equilibrium, we have-
$$F = \dfrac{1}{4\pi\epsilon_0}\dfrac{Q^2}{L^2} = k(L - L_i)$$
here, $k \to$ spring constant, $\dfrac{1}{4\pi\varepsilon_0} = 9.0\times 10^9$ N.m²/C².
Solving for Q, we get
$$Q = L\sqrt{\dfrac{k(L-L_i)}{1/4\pi\epsilon_0}}$$
Substituting the given values in above expression, we get
$$Q = (0.500\text{ m})\sqrt{\dfrac{(100\text{ N/m})(0.500\text{ m} - 0.400\text{ m})}{9.0\times 10^9\text{ N}\cdot\text{m}^2/\text{C}^2}}$$
$$= 1.67\times 10^{-5}\text{C}$$

5. (A): Force experienced by a charged particle in an electric field, $F = qE$
As $F = ma$, therefore, $ma = qE \Rightarrow a = \dfrac{qE}{m} \qquad \cdots (1)$
As electron and proton both fall from same height at rest. Therefore, for both of them, initial velocity = 0.
Now, from Newton's second equation of motion $y = v_0 t + \tfrac{1}{2}at^2$, with initial velocity $v_0 = 0$, we get
$$h = \dfrac{1}{2}at^2$$
Substituting the value of a from Eq.(1) in above expression, we get
$$h = \dfrac{1}{2}\dfrac{qE}{m}t^2 \Rightarrow t = \sqrt{\dfrac{2hm}{qE}} \Rightarrow t \propto \sqrt{m}$$
Since, 'q' is same for electron and proton and electron has smaller mass than proton, so electron will take smaller time.

6. (B) Given situation is shown in Fig.1.342.
Acceleration, $a = \dfrac{6-0}{1} = 6$ ms⁻²

Figure 1.342

For $t = 0$ to $t = 1$ s, $x_1 = \tfrac{1}{2}\times 6(1)^2 = 3$ m

1.38. ANSWER KEYS AND SOLUTIONS

For $t = 1$ s to $t = 2$ s, $x_2 = 6 \times 1 - \frac{1}{2} \times 6(1)^2 = 3$ m
For $t = 2$ s to $t = 3$ s, $x_3 = 0 - \frac{1}{2} \times 6(1)^2 = -3$ m
Total displacement $x = x_1 + x_2 + x_3 = 3$ m
Average velocity = $3/3 = 1$ m s^{-1}
Total distance travelled = 9 m
Average speed = $9/3 = 3$ ms^{-1}

7. (A) Since, $v^2 = 0^2 + 2ay = 2(F/m)y = 2\left(\frac{qE}{m}\right)y$,
Therefore-
$$KE = \frac{1}{2}mv^2 = \frac{1}{2}m\left[2\frac{(qE)}{m}y\right] \implies KE = qEy$$

8. (C) **APPROACH** Equation of path always represents a relation between space coordinates, i.e., between x and y coordinates in two dimensions.
Given situation can be divided in two parts as shown in Fig.1.343:
1. Region *I* for which $0 \leq x \leq d$ and
2. the region *II* for $x > d$.
Charged particle enters in region first at O with velocity v_0 directed along positive X–direction. In first region ($0 \leq x \leq d$), electric field is directed along negative Y–direction, so in this region, the path of charged particle is parabolic. When it emerges out from this region at P(x_0, y_0), it has X and Y components of velocities v_x and v_y respectively. Now, with these velocities it enters in second region ($x > d$), where electric field is absent. So, in second region it will move in straight line path with constant velocities v_x and v_y obtained at boundary point P of region first.
To find equation of path of charged particle in region *II*, find the position and velocity components of particle at P. Now, consider a position Q(x, y) of the particle at any instant t after emerging from P and apply relevant equation of motion to obtain a relation between x and y.
SOLUTION Given that the mass of the charged particle is m, charge on the particle is q, initial velocity of particle before entering the region of electric field is v_0 and it is directed along positive direction of X–axis, electric field E in the region $x = 0$ to $x = d$ is directed along negative Y– axis.
To find : The equation of the particle for the region $x > d$.
Let after travelling through the electric field in parabolic

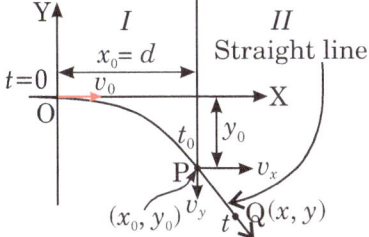

Figure 1.343

path, the charged particle emerges out the field at point P (x_0, y_0). The components of velocity of the particle at point P will change as shown in Fig.1.343.
Motion of charged particle from O to P
Along X axis Given: $x_0 = d$, $v_{0x} = v_0$, $a_x = 0$, time = t_0.
If at point P, X–component of velocity is v_x, then
$$v_x = v_0 + a_x t = v_0 + 0(t) = v_0$$
and $\quad x_0 = v_0 t_0 \implies t_0 = x_0/v_0 = d/v_0 \quad \cdots (1)$
Along Y axis Given: $v_{0y} = 0$, $a_y = -\frac{qE}{m} \quad \cdots (2)$
If at point P, Y–component of velocity is v_y, then
$$v_y = 0 + a_y t_0 = -\frac{qEd}{mv_0} \quad \cdots (3)$$
and displacement along Y–axis,

$$y_0 = v_{0y}t_0 + \frac{1}{2}a_y t_0^2 = -\frac{1}{2}\frac{qEd^2}{mv_0^2} \quad \cdots (4)$$

[since from Eq.(1), $t_0 = d/v_0$ and from (2), $a_y = -qE/m$]
In the region $x > d$, $a_x = a_y = 0$. Therefore, the equation of motion of the charged particle will be a straight line. Let after emerging through electric field at point P, the charged particle reaches to point Q (x, y) in time t.
Motion of charged particle from P to Q
Along X axis: $(x - d) = v_0 t \implies t = (x - d)/v_0 \quad \cdots (5)$
Along Y axis: $y - y_0 = v_y t \implies y = y_0 + v_y t \quad \cdots (6)$
Substituting the values of y_0, v_y and t in Eq.(6), we get

$$y = -\frac{1}{2}\frac{qEd^2}{mv_0^2} - \frac{qEd}{mv_0}\frac{(x-d)}{v_0} = -\frac{qEd}{mv_0^2}\left(\frac{d}{2} + x - d\right)$$
$$\implies y = \frac{qEd}{mv_0^2}\left(\frac{d}{2} - x\right)$$

9. (C) Given : Charge on a point mass released from the edge of a table is $+q$, initial velocity of the point mass is $v_0 = 0$. To find: The trajectory of the point mass in presence of horizontal electrical and vertical gravitational field [Fig.1.344].
Components of acceleration of the point mass, m :

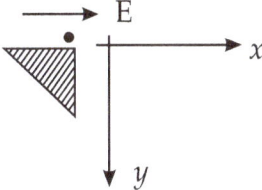

Figure 1.344

$$a_x = \frac{qE}{m}, a_y = g$$

Resultant acceleration of the point mass:
$$a = \sqrt{\left(\frac{qE}{m}\right)^2 + g^2} = \text{constant}$$

As initial velocity of the point mass was $v_0 = 0$ and the acceleration is constant its trajectory will be straight line.

10. (D) Consider an elementary segment AB, subtending a very small angle 2θ at the centre of the given ring of radius R [Fig.1.345].
From Fig.1.345, we have $AB = R(2\theta)$

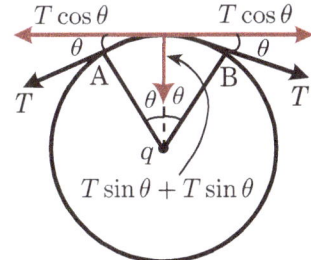

Figure 1.345

Charge on AB is, $dQ = \frac{Q}{2\pi R}(2R\theta) = \frac{Q\theta}{\pi}$

$$2T\sin\theta = \frac{dQ \cdot q}{4\pi\epsilon_0 R^2} = \frac{Qq\theta}{4\pi^2\epsilon_0 R^2}$$

Since, for a very small value of θ, $\sin\theta \approx \theta$, therefore
$$2T\theta = \frac{Qq\theta}{4\pi^2\epsilon_0 R^2} \quad \text{or} \quad T = \frac{Qq}{8\pi^2\epsilon_0 R^2}$$

11. (C) Because electric field inside the conductor is zero and electric field lines are perpendicular to Gaussian surface.

1.38.8 Check Point 8

1. The answer depends on whether the person is initially (a) uncharged or (b) charged.
   (a) No. If the person is uncharged, the electric field inside the sphere is zero. The interior wall of the shell carries no charge. The person is not harmed by touching this wall.
   (b) If the person carries a (small) charge q, the electric field inside the sphere is no longer zero. Charge $-q$ is induced on the inner wall of the sphere. The person will get a (small) shock when touching the sphere, as all the charge on his body jumps to the metal.

2. Conducting shoes are worn to avoid the build up of a static charge on them as the wearer walks. Rubber-soled shoes acquire a charge by friction with the floor and could discharge with a spark, possibly causing an explosive burning situation, where the burning is enhanced by the oxygen.

3. (a) From Fig.1.346, we have
$$A' = (10.0 \text{ cm})(30.0 \text{ cm}) = 300 \text{ cm}^2 = 0.0300 \text{ m}^2$$
$$\Phi_{E,\,A'} = EA'\cos\theta = (7.80 \times 10^4)(0.0300)\cos 180°$$
$$= -2.34 \text{ kN}\cdot\text{m}^2/\text{C}$$

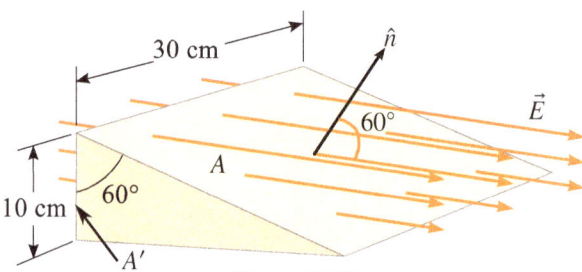

Figure 1.346

(b) The electric flux through the inclined plane is given by-
$$\Phi_{E,\,A} = EA\cos\theta = (7.80 \times 10^4)(A)\cos 60.0°$$
$$A = (30.0 \text{ cm})(w) = (30.0 \text{ cm})\left(\frac{10.0 \text{ cm}}{\cos 60.0°}\right)$$
$$= 600 \text{ cm}^2 = 0.0600 \text{ m}^2$$
$$\Rightarrow \Phi_{E,A} = (7.80 \times 10^4)(0.0600)\cos 60.0°$$
$$= +2.34 \text{ kNm}^2/\text{C}$$

(c) The bottom and the two triangular sides all lie parallel to $\vec{E}$, so $\Phi_E = 0$ for each of these. Thus,
$$\Phi_{E,\,\text{total}} = -2.34 \text{ kN}\cdot\text{m}^2/\text{C} + 2.34 \text{ kN}\cdot\text{m}^2/\text{C} + 0 + 0 + 0$$
$$= 0$$

4. (a) $\Phi_E = \vec{E}\cdot\vec{A} = (a\hat{i} + b\hat{j})\cdot A\hat{i} = aA$
   (b) $\Phi_E = (a\hat{i} + b\hat{j})\cdot A\hat{j} = bA$
   (c) $\Phi_E = (a\hat{i} + b\hat{j})\cdot A\hat{k} = 0$

5. Only the charge inside radius R contributes to the total flux.
$$\Phi_E = q/\epsilon_0$$

6. Electric flux passing through the base is given by [Fig1.347]-
$$\Phi_E = EA\cos\theta$$
$$= (52.0)(36.0)\cos 180° = -1.87 \text{ kNm}^2/\text{C}$$
Note the same number of electric field lines go through the base as go through the pyramid's surface (not counting the base). For the slanting surfaces,
$$\Phi_E = +1.87 \text{ kNm}^2/\text{C}$$

7. The flux entering the closed surface equals the flux exiting the surface. The flux entering the left side of the cone is $\Phi_E = \int \vec{E}\cdot d\vec{A} = ERh$. This is the same as the flux that exits the right side of the cone. Note that for a uniform field only the cross sectional area matters, not shape.

8. The hemisphere seems like a much more complicated calculation; you may be tempted to break up the surface into many

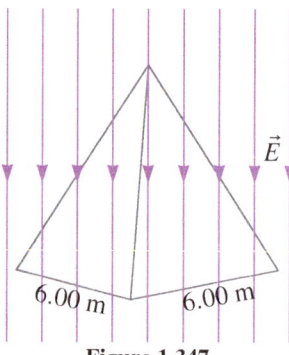

Figure 1.347

small pieces and integrate using Equation $\Phi_E = \oint \vec{E}\cdot d\vec{A}$. This would work, but there is a much simpler way. Imagine shining a light parallel to the electric field so that the shadow cast by the hemisphere is a disk of radius R. now by applying equation, $\Phi_E = \vec{E}\cdot\vec{A}$, we can find the electric flux from the area of the shadow, A.

SOLUTION Net electric flux through the hemispherical surface-
$$\Phi_E = \oint \vec{E}\cdot d\vec{A}$$
$$= E(\text{Area of surface perpendicular to}\vec{E})$$
$$= E\cdot\pi R^2$$

9. Electric flux, $\Phi_E = \vec{E}\cdot\vec{A} = (15\hat{i} + 25\hat{j})\cdot(0.65\hat{i} + 0.35\hat{j})$
$$= (9.75 + 8.75) \text{ m}^2/\text{C} = 18.5 \text{ m}^2/\text{C}$$

10. In Fig.1.186(a), the electric field is uniform over the entire surface. The electric field vectors make an angle of 30° with the planar surface. Because the normal $\hat{n}$ to the planar surface is at an angle of 90° with the surface, the angle between $\hat{n}$ and $\vec{E}$ is $\theta = 60°$.
The electric flux is
$$\Phi_E = \vec{E}\cdot\vec{A} = EA\cos\theta$$
$$= (200 \text{ N/C})(1.0 \times 10^{-2} \text{ m}^2)\cos 60°$$
$$= 1.0 \text{ N m}^2/\text{C}$$
In Fig.1.186(b) the electric field vectors make an angle of 30° below the surface. Because the normal $\hat{n}$ to the planar surface is at an angle of 90° relative to the surface, the angle between $\hat{n}$ and $\vec{E}$ is $\theta = 120°$.
The electric flux is
$$\Phi_E = \vec{E}\cdot\vec{A} = EA\cos\theta$$
$$= (180 \text{ N/C})(15 \times 10^{-2} \text{ m})^2\cos 120°$$
$$= -2.3 \text{ N m}^2/\text{C}$$

11. **APPROACH** The electric field is uniform over the entire surface. Please refer to Fig.1.348. The electric field vectors make an angle of 60° above the surface. Because the normal $\hat{n}$ to the planar surface is at an angle of 90° relative to the surface, the angle between $\hat{n}$ and $\vec{E}$ is $\theta = 30°$.
SOLUTION The electric flux is
$$\Phi_E = \vec{E}\cdot\vec{A} = EA\cos\theta$$
$$E = \frac{\Phi_E}{A\cos\theta} = \frac{25 \text{ N m}^2/\text{C}}{(10\times 10^{-2} \text{ m})(20\times 10^{-2} \text{ m})\cos 30°}$$
$$= 1.4\times 10^3 \text{ N/C}$$

12. The electric field is uniform over the rectangle in the xy plane. (a) The area vector is perpendicular to the xy plane and points in the $\hat{k}$ direction. Thus
$$\vec{A} = (2.0 \text{ cm}\times 3.0 \text{ cm})\hat{k} = (6.0\times 10^{-4} \text{ m}^2)\hat{k}$$
The electric flux through the rectangle is

$$\Phi_E = \vec{E} \cdot \vec{A} = (100\hat{i} + 50\hat{k}) \cdot \left(6.0 \times 10^{-4}\hat{k}\right) \text{Nm}^2/\text{C}$$
$$= 3.0 \times 10^{-2} \text{ N m}^2/\text{C}$$

(b) The electric flux is
$$\Phi_E = \vec{E} \cdot \vec{A} = (100\hat{i} + 50\hat{j}) \cdot \left(6.0 \times 10^{-4}\hat{k}\right) \text{Nm}^2/\text{C}$$
$$= 0.0 \text{ N m}^2/\text{C}$$

In (b), $\vec{E}$ is in the plane of the rectangle, which is why the flux is zero.

13. The electric field over the rectangle in the xz plane is uniform.

    (a) The area vector is perpendicular to the xz plane and points in the $\hat{j}$ direction. Thus
    $$\vec{A} = (2.0 \text{ cm} \times 3.0 \text{ cm})\hat{j} = \left(6.0 \times 10^{-4} \text{ m}^2\right)\hat{j}$$
    The electric flux through the rectangle is
    $$\Phi_E = \vec{E} \cdot \vec{A} = (100\hat{i} + 50\hat{k}) \text{N/C} \cdot \left(6.0 \times 10^{-4} \text{ m}^2\right)\hat{j}$$
    $$= \left(600 \times 10^{-4} \text{ N m}^2/\text{C}\right)(\hat{i} \cdot \hat{j})$$
    $$+ \left(300 \times 10^{-4} \text{ N m}^2/\text{C}\right)(\hat{k} \cdot \hat{j}) = 0.0 \text{ N m}^2/\text{C}$$

    (b) The flux is
    $$\Phi_E = \vec{E} \cdot \vec{A} = (100\hat{i} + 50\hat{j}) \text{ N/C} \cdot \left(6.0 \times 10^{-4} \text{ m}^2\right)\hat{j}$$
    $$= \left(600 \times 10^{-4} \text{ N m}^2/\text{C}\right)(\hat{i} \cdot \hat{j})$$
    $$+ \left(300 \times 10^{-4} \text{ N m}^2/\text{C}\right)\hat{j} \cdot \hat{j}$$
    $$= 0 \text{ N m}^2/\text{C} + \left(3.0 \times 10^{-2} \text{ N m}^2/\text{C}\right)$$
    $$= 3.0 \times 10^{-2} \text{ N m}^2/\text{C}$$

14. The area vector of the circle is
    $$\vec{A} = \pi r^2 \hat{k} = \pi(0.015 \text{ m})^2 \hat{j} = \left(7.07 \times 10^{-4} \text{ m}^2\right)\hat{j}$$
    Thus, the flux through the area of the circle is
    $$\Phi_E = \vec{E} \cdot \vec{A} = (1500\hat{i} + 1500\hat{j} - 1500\hat{k}) \text{ N/C}$$
    $$\cdot \left(7.07 \times 10^{-4} \text{ m}^2\right)\hat{j}$$
    Using $\hat{i} \cdot \hat{j} = \hat{j} \cdot \hat{k} = 0$ and $\hat{j} \cdot \hat{j} = 1$, we find
    $\Phi_E = (1500 \text{ N/C})\left(7.07 \times 10^{-4} \text{ m}^2\right) = 1.1 \text{ Nm}^2/\text{C}$

15. The electric field is uniform, and we take the area vectors to point outward from the box. In the Figure 1.348, the box is positioned with its edges aligned with the xyz axes, and the electric field is evaluated at the input face and the exit face. The area vectors of the six box faces are-

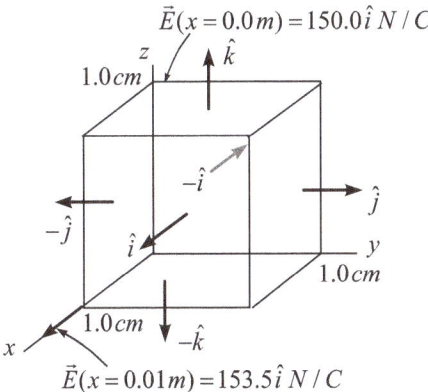

Figure 1.348

$\vec{A}_1 = \left(1.0 \times 10^{-2} \text{ m}\right)^2 \hat{i} = \left(1.0 \times 10^{-4} \text{ m}^2\right) \hat{i}$,
$\vec{A}_2 = -\left(1.0 \times 10^{-4} \text{ m}^2\right) \hat{i}$, $\vec{A}_3 = \left(1.0 \times 10^{-4} \text{ m}^2\right) \hat{j}$,
$\vec{A}_4 = -\left(1.0 \times 10^{-4} \text{ m}^2\right) \hat{j}$, $\vec{A}_5 = \left(1.0 \times 10^{-4} \text{ m}^2\right) \hat{k}$,
and $\vec{A}_6 = -\left(1.0 \times 10^{-4} \text{ m}^2\right) \hat{k}$.

$$\Phi_E = \sum_{i=1}^{6} \vec{E} \cdot \vec{A}_i$$
$$= \vec{E}_{(x=0.01 \text{ m})} \cdot \vec{A}_1 + \vec{E}_{(x=0.0 \text{ m})} \cdot \vec{A}_2$$
$$= (350(0.01) + 150) \hat{i} \text{N/C} \cdot \left(1.0 \times 10^{-4}\right) \hat{i} \text{m}^2$$
$$+ (350(0.0) + 150) \hat{i} \text{N/C} \cdot \left(1.0 \times 10^{-4}\right)(-\hat{i}) \text{ m}^2$$
$$= (153.5 \text{ N/C})\left(1.0 \times 10^{-4} \text{ m}^2\right)$$
$$-(150 \text{ N/C})\left(1.0 \times 10^{-4} \text{ m}^2\right)$$
$$= 3.5 \times 10^{-4} \text{ N m}^2/\text{C}$$

16. The electric field through the two cylinders is uniform. Let $A = \pi R^2$ be the area of the end of the cylinder and let the area vector point outward from the cylinder ends. E is the electric field strength.

    (a) There's no flux through the side walls of the cylinder because $\vec{E}$ is parallel to the wall. On the right end, where $\vec{E}$ points outward, $\Phi_{\text{right}} = EA\cos(0°) = \pi R^2 E$. The field points inward on the left, so $\Phi_{\text{left}} = EA\cos(180°) = -\pi R^2 E$. Altogether, the net flux is $\Phi_E = 0 \text{ N m}^2/\text{C}^2$

    (b) The only difference from part (a) is that $\vec{E}$ points outward on the left end, making $\Phi_{\text{left}} = EA\cos(0°) = \pi R^2 E$. Thus the net flux through the cylinder is $\Phi_e = 2\pi R^2 E \text{ N m}^2/\text{C}^2$.

17. (D): Electric flux,
    $$\Phi_E = \int \vec{E} \cdot d\vec{A} = \int E dA \cos\theta = \int E dA \cos 90° = 0$$
    (because field lines are parallel to the surface.)

18. (D)

1.38.9 Check Point 9

1. (a) With δ very small, all points on the hemisphere are nearly at a distance R from the charge [Fig.1.349], so the field everywhere on the curved surface is kQ/R^2 radially outward (normal to the surface). Therefore, the flux is this field strength times the area of half a sphere:
   $$\Phi_{\text{curved}} = \int \vec{E} \cdot d\vec{A} = E_{\text{local}} A_{\text{hemisphere}}$$

Figure 1.349

$$= \left(k\frac{Q}{R^2}\right)\left(\frac{1}{2}4\pi R^2\right) = \frac{1}{4\pi\epsilon_0}Q(2\pi) = \frac{+Q}{2\epsilon_0}$$

(b) The closed surface encloses zero charge so Gauss's law gives-
$$\Phi_{\text{curved}} + \Phi_{\text{flat}} = 0 \quad \text{or} \quad \Phi_{\text{flat}} = -\Phi_{\text{curved}} = \frac{-Q}{2\epsilon_0}$$

2. Required Gaussian surfaces are shown in Fig.1.350 For any closed surface that encloses a total charge Q_{encl}, the net electric flux through the closed surface is $\Phi_e = Q_{\text{encl}}/\epsilon_0$.

3. The required Gaussian surfaces are shown in Fig.1.351 For any closed surface that encloses a total charge Q_{encl}, the net electric flux through the closed surface is $\Phi_E = Q_{\text{encl}}/\epsilon_0$.

4. Please refer to Figure 1.227. For any closed surface that encloses a total charge Q_{encl}, the net electric flux through the surface is $\Phi_E = Q_{\text{encl}}/\varepsilon_0$. We can write three equations from the three closed surfaces in the Fig.1.227:

122   Electric Charge and Field

Figure 1.350

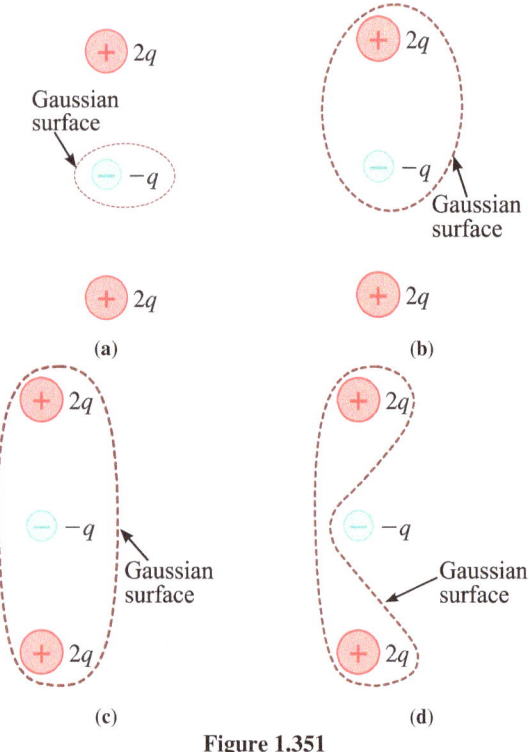

Figure 1.351

$$\Phi_A = -\frac{q}{\epsilon_0} = \frac{q_1 + q_3}{\epsilon_0} \implies q_1 + q_3 = -q$$

$$\Phi_B = \frac{3q}{\epsilon_0} = \frac{q_1 + q_2}{\epsilon_0} \implies q_1 + q_2 = 3q$$

$$\Phi_C = \frac{-2q}{\epsilon_0} = \frac{q_2 + q_3}{\epsilon_0} \implies q_2 + q_3 = -2q$$

Subtracting third equation from the first gives
$$q_1 - q_2 = +q$$
Adding second equation to this equation,
$$2q_1 = +4q \quad q_1 = 2q$$
That is, $q_1 = +2q, q_2 = +q$, and $q_3 = -3q$.

5. Point charge Q induces charge on conductor as shown in Figure 1.228.
Net charge enclosed by closed surface is negative so flux is negative.

6. From Eq.1.133, electric field inside a cylindrical geometry is given by-
$$E = \frac{r\rho}{2\varepsilon_0}$$

Since, $\rho = \frac{\lambda}{\pi R^2}$, therefore $E = \frac{\lambda}{2\pi\epsilon_0}\frac{r}{R^2}$
For $r = R/2$, we have-
$$E = \frac{\lambda}{2\pi\epsilon_0}\frac{R/2}{R^2} = \frac{\lambda}{4\pi\,\epsilon_0\,R}$$

7. (B) In Gauss's law, $\oint \vec{E}\cdot d\vec{A} = \frac{Q_{encl}}{\varepsilon_0}$,
$\vec{E}$ on the LHS, the electric field vector, $\vec{E}$, will have a contribution from all charges while Q_{encl} on the RHS will have a contribution from enclosed charges q_2 and q_4 only.

8. Please refer to Figure 1.230. For any closed surface that encloses a total charge Q_{encl}, the net electric flux through the closed surface is $\Phi_E = Q_{encl}/\epsilon_0$. For the closed surface of the torus, Q_{encl} includes only the $-1nC$ charge. Thus, the net flux through the torus is due only to this charge:
$$\Phi_E = \frac{-1\times 10^{-9}\text{C}}{8.85\times 10^{-12}\text{C}^2/\text{Nm}^2} = -0.11\text{ kNm}^2/\text{C}$$
Here, negative sign indicates that the flux is inward.

9. Please refer to Figure 1.231. For any closed surface that encloses a total charge Q_{encl}, the net electric flux through the closed surface is $\Phi_E = Q_{encl}/\epsilon_0$. The cylinder encloses the $+1$ nC charge only as both the $+100$ nC and the -100 nC charges are outside the cylinder. Thus,
$$\Phi_E = \frac{1\times 10^{-9}\text{C}}{8.85\times 10^{-12}\text{ C}^2/\text{Nm}^2} = 0.11\text{ kNm}^2/\text{C}$$
This is outward flux.

10. For any closed surface enclosing a total charge Q_{encl}, the net electric flux through the surface is
$$\Phi_E = \frac{Q_{encl}}{\epsilon_0}$$
$$\implies Q_{encl} = \epsilon_0\Phi_E$$
$$= (8.85\times 10^{-12}\text{C}^2/\text{Nm}^2)(-1000\text{Nm}^2/\text{C})$$
$$= -8.85\text{ nC}$$

11. For any closed surface that encloses a total charge Q_{encl}, the net electric flux through the surface is
$$\Phi_E = \frac{Q_{encl}}{\epsilon_0} = \frac{(55.3\times 10^6)(-1.60\times 10^{-19}\text{C})}{8.85\times 10^{-12}\text{C}^2/\text{Nm}^2}$$
$$= -1.00\text{ N m}^2/\text{C}$$

12. **APPROACH** our given information is that the cube side length is a and the cube has a point charge q at one corner (Fig.1.232). Our task here is to use Gauss's law, $\oint \vec{E}.d\vec{A} = \frac{Q_{encl}}{\epsilon_0}$, to solve the problem of how much electric flux passes through each face of the cube. We know how to calculate the electric field surrounding the point charge, and we know the locations and orientations of the faces. We also know that a cube possesses symmetries that we can exploit to simplify the problem.

SOLUTION Because in Fig.1.352a, the point charge q

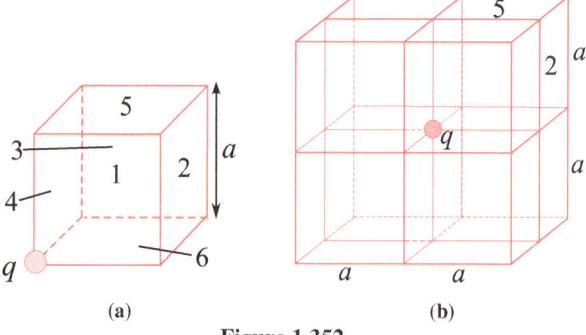

Figure 1.352

is at the corner of the cube, the cube does not enclose

1.38. ANSWER KEYS AND SOLUTIONS

it completely and so cannot be our Gaussian surface. However, we can enclose the point charge q by attaching seven additional cubes, each of side length a, to our original cube, creating a cube of side length $2a$ and having q at its geometric centre as shown in Fig.1.352b. This large cube is our Gaussian surface. We can apply Gauss's law to get the electric flux through any face of this surface and then use symmetry to get the flux through any face of our original cube of side length a.

From Gauss's law, we know that the flux through the large cube enclosing a point charge q is q/ϵ_0 Because the *point charge q* is at the geometric center, symmetry tells us that the electric flux is the same through all six faces of the large cube. As Figure 1.352b shows, each face of the large cube is composed of four squares, each of side length a: One face of the large cube contains four faces of our original cube. Thus, if we apply Gauss's law and symmetry, the flux through each of these smaller faces must be-
$$\Phi_E = \frac{1}{(6)(4)}\frac{q}{\epsilon_0} = \frac{q}{24\epsilon_0}.$$
This result gives us the flux passing through each of the faces 2, 3 and 5 of our original cube shown in Fig.1.352a. Since, the electric field lines cannot cross through the faces 1, 4, and 6 of the original cube because they pass through the source charge q. Therefore the electric flux passing through each of the faces 1, 4, and 6 is still zero.

13. (B) As explained in previous answer total eight identical cubes are required so that the given charge q appears at the centre of the bigger cube. Thus, the electric flux passing through the given cube is-
$$\Phi_E = \frac{1}{8}\left(\frac{q}{\epsilon_0}\right) = \frac{q}{8\epsilon_0}$$

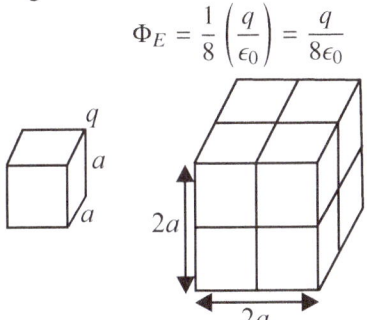

Figure 1.353

14. (C) : According to Gauss's law, $\phi_E = \frac{Q_{\text{encl}}}{\varepsilon_0}$

From the Gauss's law given above, it is evident that the electric flux depends solely on the charge enclosed by the Gaussian surface and not on the radius of the Gaussian surface. Therefore, when the radius of the Gaussian surface is doubled, the electric flux will remain the same.

15. (D): Let Φ_A, Φ_B and ϕ_C are the electric flux linked with surfaces A, B and C respectively. According to Gauss theorem,
$$\Phi_A + \Phi_B + \Phi_C = \frac{q}{\varepsilon_0}$$
Since, $\Phi_A = \Phi_C$, therefore $2\Phi_A + \Phi_B = \frac{q}{\varepsilon_0}$
or $2\Phi_A = \frac{q}{\varepsilon_0} - \Phi_B$ or, $2\Phi_A = \frac{q}{\varepsilon_0} - \Phi$ (Given: $\Phi_B = \Phi$)
Therefore, $\Phi_A = \frac{1}{2}\left(\frac{q}{\varepsilon_0} - \Phi\right)$

16. (B): The total flux through the cube $\Phi_{\text{total}} = \frac{q}{\varepsilon_0}$, therefore, the electric flux through any face
$$\Phi_{\text{face}} = \frac{q}{6\varepsilon_0} = \frac{4\pi q}{6(4\pi\varepsilon_0)}$$

17. (A): For complete cube, $\Phi_E = \frac{Q_{\text{encl}}}{\epsilon_0} = \frac{Q}{\epsilon_0} \times 10^{-6}$.

Therefore, corresponding to each face, $\Phi_E = \frac{1}{6}\frac{Q}{\epsilon_0} \times 10^{-6}$

18. (D): Electric flux emerging from the cube does not depend on size of cube. Total flux = q/ϵ_0

19. (C): In a uniformly charged hollow conducting sphere,
(i) For $r < R, \vec{E} = 0$
(ii) For $r > R, \vec{E} = \frac{1}{4\pi\varepsilon_0}\frac{Q}{|r^2|}\hat{r}; \vec{E}$ decreases

20. (D): Electric field due to an infinite line charge, [Fig.1.354]

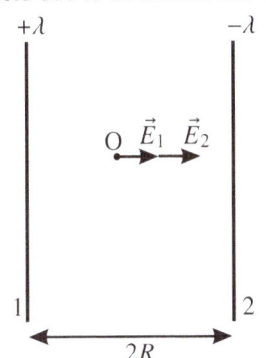

Figure 1.354

$$E = \frac{\lambda}{2\pi\varepsilon_0 r}$$
Net electric field at mid-point O,
$$\vec{E}_0 = \vec{E}_1 + \vec{E}_2$$
As, $E_1 = E_2 = \frac{\lambda}{2\pi\varepsilon_0 R}$
Therefore, $E_0 = 2E_1 = \frac{\lambda}{\pi\varepsilon_0 R}$ NC^{-1}

21. (A): Electric field inside the charged spherical shell is zero as there is no charge inside it.

22. (C): Field inside a conducting sphere = 0.

23. APPROACH Apply the principle of superposition of electric fields.

(B) SOLUTION Electric field due to a uniformly charged

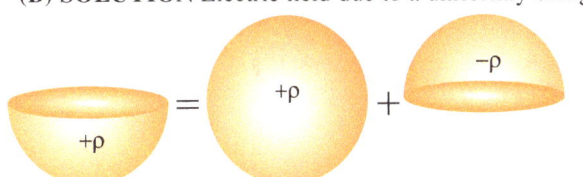

Figure 1.355

sphere, at any external point at distance r from it's center is given by-
$$E_1 = \frac{1}{4\pi\epsilon_0}\frac{Q}{r^2} = \frac{1}{4\pi\varepsilon_0}\frac{\rho(4\pi R^3)/3}{r^2}$$
Since, $r = 2R$, therefore, $E_1 = \frac{\rho R}{12\epsilon_0}$

The given electric field at a distance $2R$, above the hemisphere is-
$$E_2 = E$$
i.e., a hemispherical part creates an electric field E at distance $2R$ above it's centre.

Now, for a point at distance $r = 2R$, below the hemispherical section, we model the hemisphere as a super position of a sphere of charge density $+\sigma$ and an upper hemisphere of charge density $-\sigma$.

So, the net electric field at lower point at distance $2R$,

from the centre of the hemisphere will be equal to the vector sum of field created by complete sphere and a hemisphere of negative charge density as shown in Fig.1.355, i.e.,

$$\vec{E}_{net} = \vec{E}_1 + \vec{E}_2 \implies E_{net} = E_1 - E_2 = \frac{\rho R}{12\varepsilon_0} - E$$

24. (A) By using Gauss law [Fig.1.356] -

$$\oint \vec{E} \cdot d\vec{A} = \frac{Q_{encl}}{\epsilon_0} \implies (E)\left(4\pi r^2\right) = \frac{\int \rho \left(4\pi r^2 dr\right)}{\epsilon_0}$$

$$\left(kr^7\right)\left(4\pi r^2\right) = \frac{\int_0^r \rho \left(4\pi r^2 dr\right)}{\epsilon_0} \implies k\varepsilon_0 r^9 = \int_0^r \rho r^2 dr$$

$$\implies k\epsilon_0 r^9 = \rho \frac{r^3}{3} \implies \rho \propto r^6$$

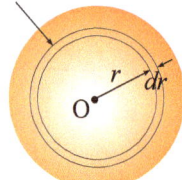

Figure 1.356

Remark : Students can also check dimensionally that $\rho \propto r^6$.

25. (D) The maximum length of the string which can fit into the cube is $a\sqrt{3}$, equal to its body diagonal. The total charge inside the cube is $\lambda a\sqrt{3}$, and hence the total flux through the cube is-

$$\Phi = \frac{Q_{encl}}{\varepsilon_0} = \frac{\lambda a\sqrt{3}}{\varepsilon_o}.$$

26. (C) Outgoing electric flux due to $+q$ charge of dipole $= +q/\epsilon_0$
    Incoming electric flux due to $-q$ charge of dipole $= -q/\epsilon_0$
    Therefore, net electric flux leaving the spherical surface due to dipole:

$$\Phi_{net} = \frac{q}{\epsilon_0} - \frac{q}{\epsilon_0} = 0$$

Alternatively, we can directly find the answer by applying Gauss's theorem:

$$\Phi_E = \frac{Q_{encl}}{\epsilon_0} = \frac{0}{\epsilon_0} = 0$$

(because for a dipole inside a sphere, $Q_{encl} = +q - q = 0$)

27. (A) Inside the surface the total charge is zero. So flux must be zero.

1.38.10 Check Point 10

1. **APPROACH** A copper penny is a conductor. Assume the penny to be a flat disc of radius much, much greater than the distance from the surface at which we are measuring the electric field.
   SOLUTION The excess charge on a conductor resides on the surface. The electric field just outside the surface of a charged conductor is

$$\vec{E} = \left(\frac{\sigma}{\varepsilon_0}, \text{ perpendicular to surface}\right)$$

$$\implies \sigma = \varepsilon_0 E = \left(8.85 \times 10^{-12} C^2/Nm^2\right)(2000 \text{ N/C})$$
$$= 17.7 \times 10^{-9} C/m^2$$

☞ Because the actual surface of a penny is not flat, the surface charge density will not be uniform. The result above gives the average surface charge density, far from the edges of the coin.

2. In electrostatic equilibrium of a conductor, the excess charge always resides on the outer surface. The electric field just above the surface of a charged conductor is

$$\vec{E} = \left(\frac{\sigma}{\epsilon_0}, \text{ perpendicular to surface}\right)$$

Therefore,

$$\sigma = \epsilon_0 E = \left(8.85 \times 10^{-12} C^2/Nm^2\right)\left(3.0 \times 10^6 \text{ N/C}\right)$$
$$= 2.7 \times 10^{-5} C/m^2$$

☞ It is the air molecules just above the surface that "break down" when the E-field becomes strong enough to accelerate stray charges to approximately 15 eV between collisions, thus causing collisional ionization. It does not make any difference whether E points toward or away from the surface.

3. The excess charge on a conductor resides on the outer surface. In Fig.1.256, point 1 is just above the surface of a charged conductor, hence

$$\vec{E} = \left(\frac{\sigma}{\epsilon_0}, \text{ perpendicular to surface}\right)$$

$$\implies E = \frac{(5.0 \times 10^{10})(1.60 \times 10^{-19} C/m^2)}{8.85 \times 10^{-12} C^2/Nm^2} = 0.90 \text{ kN/C}$$

At point 2 the electric field strength is zero because this point lies inside the conductor. The electric field strength at point 3 is zero because there is no excess charge on the interior surface of the box. This can be quickly seen by considering a Gaussian surface just inside the interior surface of the box as shown in Figure1.256.

4. The copper plate is a conductor. The excess charge resides on the surface of the plate. Ignore the charge that resides on the edge of the plate because the plate's thickness is much, much less than the radius.
   (a) One-half of the charge is located on the top surface and one-half on the bottom surface of the copper plate, so the surface charge density is

$$\sigma = \frac{q}{A} = \frac{(3.5/2 nC)}{\pi(0.10/2 \text{ m})^2} = 2.23 \times 10^{-7} C/m^2$$

Thus, the electric field just above the surface of the plate is

$$E = \frac{\sigma}{\epsilon_0} = \frac{2.23 \times 10^{-7} C/m^2}{8.85 \times 10^{-12} C^2/Nm^2}$$
$$= 25.2 \times 10^3 \text{ N/C} \approx 25 \text{ kN}$$

Because the charge on the plate is positive, the direction of the electric field is away from the plate. Thus the electric field is $\vec{E} = 25$ kN/C upward from the plate.
(b) The centre of mass of the plate is in the interior of the plate, so $E = 0.0$ N/C because the electric field within a conductor is zero.
(c) The electric field $E = 25.2 \times 10^3$ N/C, away from the plate, which is downward. Thus $\vec{E} = 25.2$ kN/C downward from the plate.

5. For any closed surface that encloses a total charge Q_{encl}, the net electric flux through the closed surface is $\Phi_E = Q_{encl}/\epsilon_0$. In the present case (Figure 1.257), the conductor is neutral and there is a point charge Q inside the cavity. Thus $Q_{encl} = Q$ and the flux is, $\Phi_E = Q/\epsilon_0$

6. (a) $\sigma_a = -\frac{q_a}{4\pi a^2}$; $\sigma_b = -\frac{q_b}{4\pi b^2}$; $\sigma_R = \frac{q_a + q_b}{4\pi R^2}$
   (b) $E_{out} = \frac{1}{4\pi\epsilon_0}\frac{(q_a+q_b)}{r^2}\hat{r}$, where $\hat{r}$ = vector from centre of large sphere.
   (c) $E_a = \frac{1}{4\pi\epsilon_0}\frac{q_a}{r_a^2}\hat{r}_a$, $E_b = \frac{1}{4\pi\epsilon_0}\frac{q_b}{r_b^2}\hat{r}_b$,
   where $\hat{r}_a$ and $\hat{r}_b$ are outward unit vectors from centres of cavities of radii a and b respectively.
   (d) Zero
   (e) σ_R changes (but not σ_a or σ_b); $E_{outside}$ changes (but not

E_a or E_b); force on q_a and q_b still zero.

7. Equations of motion are-
$$-\underbrace{eE}_{\substack{\text{upward}\\\text{electric force}}} + \underbrace{mg}_{\substack{\text{downward}\\\text{weight}}} - \underbrace{6\pi\eta r v_1}_{\substack{\text{upward}\\\text{viscous force}}} = 0 \quad \text{(downward field)}$$

$$\underbrace{eE}_{\substack{\text{downward}\\\text{electric force}}} + \underbrace{mg}_{\substack{\text{downward}\\\text{weight}}} - \underbrace{6\pi\eta r v_2}_{\substack{\text{upward}\\\text{viscous force}}} = 0 \quad \text{(upward field)}$$

On subtracting first equation from second, we get
$$2eE - 6\pi\eta r (v_2 - v_1)$$
On simplifying above equation for e, we get
$$e = \frac{3\pi\eta r}{E}(v_2 - v_1)$$

8. (C) If T is the surface tension of the liquid of soap bubble and σ is the surface charge density, then, in electrostatic equilibrium of soab bubble, we have
$$P_{\text{gas}} + P_{\text{el}} = P_0 + \frac{4T}{R} \implies P_{\text{gas}} - P_0 = \frac{4T}{R} - P_{\text{el}}$$
Here, P_{gas} is the outward gas pressure, P_0 is the inward atmospheric pressure and P_{el} is the outward electric pressure on the surface of soap bubble.

Since, due to surface charge density σ on the soap bubble, the outward electric pressure $P_{\text{el}} = \sigma^2/2\epsilon_0$, therefore the excess pressure on the soap bubble is given by,
$$P_{\text{excess}} = P_{\text{gas}} - P_0 = \frac{4T}{R} - \frac{\sigma^2}{2\epsilon_0}$$
Since, $\sigma = \dfrac{Q}{4\pi R^2}$ therefore-
$$P_{\text{excess}} = \frac{4T}{R} - \frac{Q^2}{32\pi^2 R^4 \epsilon_0}$$

1.38.11 Conceptual Questions

1. A plastic ruler that has been rubbed with a cloth is charged. When brought near small pieces of paper, it will cause separation of charge (polarization) in the bits of paper, which will cause the paper to be attracted to the ruler. A small amount of charge is able to create enough electric force to be stronger than gravity. Thus the paper can be lifted. On a humid day this is more difficult because the water molecules in the air are polar. Those polar water molecules will be attracted to the ruler and to the separated charge on the bits of paper, neutralizing the charges and thus reducing the attraction.

2. Because the can is grounded, the presence of the negatively charged plastic rod induces a positive charge on it. The positive charges induced on the can are attracted, via the Coulomb interaction, to the negative charges on the plastic rod. Unlike charges attract, so the can will roll toward the rod.

3. It is possible only if one of them has very large positive charge as compared to other. Because a metal sphere is a conductor, the proximity of a positively charged ball (not necessarily a conductor), will induce a redistribution of charges on the metal sphere with the surface nearer the positively charged ball becoming negatively charged. Because the negative charges on the metal sphere are closer to the positively charged ball than are the positive charges on the metal sphere, the net force will be attractive.

4. For the gravitational force, we don't notice it because the force is very weak, due to the very small value of G, the gravitational constant, and the relatively small value of ordinary masses. For the electric force, we don't notice it because ordinary objects are electrically neutral to a very high degree. We notice our weight (the force of gravity) due to the huge mass of the Earth, making a significant gravity force. We notice the electric force when objects have a net static charge (like static cling from the clothes dryer), creating a detectable electric force.

5. It is impossible to deal with the interactions of moving charged particles without the field concept, and it is often easier to deal with fields than with distributions of charge.

6. The test charge creates its own electric field, The measured electric field is the sum of the original electric field plus the field of the test charge. If the test charge is small, then the field that it causes is small. Therefore, the actual measured electric field is not much different than the original field.

7. A negative test charge could be used. For the purposes of defining directions, the electric field might then be defined as the opposite of the force on the test charge, divided by the test charge. Equation $E = F/q_0$ might be changed to $\vec{E} = -\vec{F}/q_0, q_0 < 0$

8. The direction of the field is defined to be the direction of the force on a positively charged test particle. Positive charges always move away from other +ve charges and towards −ve charges.

9. A negatively charged particle placed in a uniform electric field is accelerated in the direction opposite the direction of $\vec{E}$, and so the upward acceleration of the electron tells you that $\vec{E}$ must have a component that is directed vertically downward. (There might also be a component parallel to the initial motion of the electron. This would change the electron's speed but not its direction.) If the electron is accelerated downward, $\vec{E}$ must have a component that is directed vertically upward.

10. See Fig.1.357. At point A, the direction of the net force on a positive test charge would be down and to the left, parallel to the nearby electric field lines. At point B, the direction of the net force on a positive test charge would be up and to the right, parallel to the nearby electric field lines. At point C, the net force on a positive test charge would be 0. In order of decreasing field strength, the points would be ordered A, B, C.

11. The two charges are located as shown in Fig.1.358.
    (a) If the signs of the charges are opposite, then the point on the line where $E = 0$ will lie to the left of Q. In that region the electric fields from the two charges will point in opposite directions, and the point will be closer to the smaller charge.
    (b) If the two charges have the same sign, then the point on

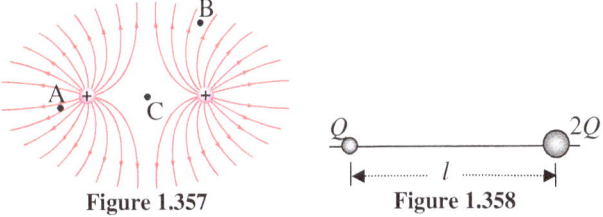

Figure 1.357   **Figure 1.358**

the line where $E = 0$ will lie between the two charges, closer to the smaller charge. In this region, the electric fields from the two charges will point in opposite directions.

12. We assume that there are no other forces (like gravity) acting on the test charge. The direction of the electric field line shows the direction of the force on the test charge. The acceleration is always parallel to the force by Newton's second law, so the acceleration lies along the field line. If the particle is at rest initially and then is released, the initial velocity will also point along the field line, and the particle will start to move along the field line. However, once the particle has a velocity, it will not follow the field line unless

the line is straight. The field line gives the direction of the acceleration, or the direction of the change in velocity.

13. No, because the forces exerted on the two charged regions of the dipole are equal in magnitude but opposite in direction, making the vector sum of forces zero.

14. No. The acceleration is not constant (it is centripetal and always changing direction). Therefore the electric force exerted on the electron must be nonconstant, implying the electric field is not uniform.

15. The electric flux depends only on the charge enclosed by the Gaussian surface $\left(\Phi = \frac{Q_{encl}}{\varepsilon_0}\right)$, not on the shape of the surface. Φ_E will be the same for the cube as for the sphere.

16. (a) Because the amount of charge enclosed by the balloon does not change, the field line flux through its surface does not change. (b) Again the field line flux does not change because the enclosed charge does not change. (However, in either case the field line pattern changes when the electrons are placed outside the balloon.)

17. If there are no charged particles inside the cavity, then the electric field is zero inside the cavity, no matter what the charge distribution is outside the sphere. The electric field inside the cavity can be non-zero when there are charged particles inside the cavity. In either case, the electric field is zero within the conducting material of the sphere itself.

18. The two are proportional to each other. The field line flux has an arbitrary value, depending on how many lines are drawn to represent a certain amount of charge. The electric flux is uniquely defined by $\Phi = \oint \vec{E}.d\vec{A} = Q_{encl}/\varepsilon_0$

19. The net charge on large objects is always very close to zero. Hence the most obvious force is the gravitational force.

20. The accumulation of static charge gives the individual hairs a charge. Since like charges repel and because the electrostatic force is inversely proportional to the charges separation distance squared, the hairs arrange themselves in a manner in which they are as far away from each other as possible. In this case that configuration is when the hairs are standing on end.

1.38.12 Problems

1. **APPROACH** Since, total charge is given therefore, to determine the number of electrons, apply the principle of quantization of charge: $q = ne$
   SOLUTION Here, charge of each electron is $e = 1.602 \times 10^{-19}$ C, $q = 1$ C, therefore-
   $$n = \frac{q}{e} = \frac{1\,\text{C}}{1.602 \times 10^{-19}\text{C}} = 6.18 \times 10^{18}\,\text{electrons}$$

2. The charge acquired by the plastic rod is an integral number of electronic charges, that is, $q = n(-e)$. Relate the charge acquired by the plastic rod to the number of electrons transferred from the wool shirt:
   $$q = n(-e) \implies n = \frac{q}{-e}$$
   Substitute numerical values and evaluate n:
   $$n = \frac{-0.80\mu C}{-1.602 \times 10^{-19}C} = 5.0 \times 10^{12}\,\text{electrons}$$

3. Number of moles in 1 kg., i.e., 10^3 gram of carbon:
   $$n = \frac{w}{M} = \frac{10^3}{12}$$
   Therefore, number of C atoms,
   $$N = nN_A = \frac{10^3}{12} \times 6.022 \times 10^{23}$$
   Since, each carbon atom has 6 protons in it's nucleus, therefore, number of protons in 10^3 gram carbon
   $$N_p = 6N = 3.011 \times 10^{26}$$

And the total charge of protons
$$Q_p = N_p e = 6Ne = 3.011 \times 10^{26} \times 1.6 \times 10^{-19}$$
$$= 4.82 \times 10^7 C$$

4. **APPROACH** 1 mol of electrons = 6.02×10^{23} electrons, i.e., N_A (Avogadro's number) electrons.
   Now, apply the principle of quantization of charge $q = ne$, with $n = N_A = 6.02 \times 10^{23}$ electrons.
   SOLUTION 1 faraday = $N_A e$
   $$= \left(6.02 \times 10^{23}\right) \times (1.602 \times 10^{-19}C) = 96470\,C$$

5. **APPROACH** (a) The percentage of the electrons originally in the cube that was removed can be found from the ratio of the number of electrons removed to the number of electrons originally in the cube. (b) The percentage decrease in the mass of the cube can be found from the ratio of the mass of the electrons removed to the mass of the cube.
   SOLUTION (a) Express the ratio of the electrons removed to the number of electrons originally in the cube:
   $$\frac{N_{rem}}{N_{ini}} = \frac{\frac{Q_{accumulated}}{e}}{N_{\text{electrons per atom}} N_{atoms}} \quad (1.161)$$
   The number of atoms in the cube is the ratio of the mass of the cube to the mass of an aluminum atom:
   $$N_{atoms} = \frac{m_{cube}}{m_{Al\,atom}} = \frac{\rho_{Al} V_{cube}}{m_{Al\,atom}}$$
   The mass of an aluminum atom is its molar mass divided by Avogadro's number:
   $$m_{Al\,atom} = \frac{M_{Al}}{N_A}$$
   Substituting and simplifying yields:
   $$N_{atoms} = \frac{\rho_{Al} V_{cube}}{\frac{M_{Al}}{N_A}} = \frac{\rho_{Al} V_{cube} N_A}{M_{Al}}$$
   Substitute for N_{atoms} in equation (1.161) and simplify to obtain:
   $$\frac{N_{rem}}{N_{ini}} = \frac{\frac{Q_{accumulated}}{e}}{N_{\text{electrons per atom}} \left(\frac{\rho_{Al} V_{cube} N_A}{M_{Al}}\right)}$$
   $$= \frac{Q_{accumulated} M_{Al}}{N_{\text{electrons per atom}} \rho_{Al} V_{cube} e N_{per\,atom}}$$
   Substitute numerical values and evaluate $\frac{N_{rem}}{N_{ini}}$, you get-
   $$\frac{N_{rem}}{N_{ini}} \approx 1.99 \times 10^{-15}\%$$
   (b) Express the ratio of the mass of the electrons removed to the mass of the cube:
   $$\frac{m_{rem}}{m_{cube}} = \frac{N_{rem} m_{electron}}{\rho_{Al} V_{cube}}$$
   From (a), the number of electrons removed is given by:
   $$N_{rem} = \frac{Q_{accumulated}}{e}$$
   Substituting and simplifying yields:
   $$\frac{m_{rem}}{m_{cube}} = \frac{\frac{Q_{accumulated}}{e} m_{electron}}{\rho_{Al} V_{cube}}$$
   $$= \frac{Q_{accumulated} m_{electron}}{e \rho_{Al} V_{cube}}$$
   Substituting numerical values and evaluating for $\frac{m_{rem}}{m_{cube}} \times 100$, yields
   $$\frac{m_{rem}}{m_{cube}} \times 100 = \frac{(2.50\,\text{pC})(9.109 \times 10^{-31}\text{kg})}{(1.602 \times 10^{-19}\text{C})\left(2.70\frac{g}{cm^3}\right)(1.00\,\text{cm}^3)}$$
   $$\approx 5.26 \times 10^{-19}\%$$

6. **APPROACH** The force on one proton is $F = \dfrac{kq_1q_2}{r^2}$ away from the other proton.
   SOLUTION The magnitude of above force,
   $$\left(8.99 \times 10^9 \text{ N} \cdot \text{m/C}^2\right) \left(\dfrac{1.60 \times 10^{-19}\text{C}}{2 \times 10^{-15} \text{ m}}\right)^2 = 57.5 \text{ N}$$

7. **APPROACH** We can find the electric forces the two charges exert on each by applying Coulomb's law and Newton's 3rd law. Note that $\hat{r}_{1,2} = \hat{i}$ because the vector pointing from q_1 to q_2 is in the positive x direction. Figure 1.359 shows the situation for Parts (a) and (b).
   SOLUTION (a) Use Coulomb's law to express the force

   Figure 1.359

   that q_1 exerts on q_2:
   $$\vec{F}_{1,2} = \dfrac{kq_1q_2}{r_{1,2}^2}\hat{r}_{1,2}$$
   Substitute numerical values and evaluate $\vec{F}_{2,1}$:
   $$\vec{F}_{1,2} = \dfrac{(8.988 \times 10^9 N \cdot m^2/C^2)\,(4.0\mu C)(6.0\mu C)}{(3.0 m)^2}\hat{i}$$
   $$= (24 mN)\hat{i}$$
   (b) Because these are action and reaction forces, we can apply Newton's 3rd law to obtain:
   $$\vec{F}_{2,1} = -\vec{F}_{1,2} = -(24 mN)\hat{i}$$
   (c) If q_2 is $-6.0\mu C$, the force between q_1 and q_2 is attractive and both force vectors are reversed:
   $$\vec{F}_{1,2} = \dfrac{(8.988 \times 10^9 N \cdot m^2/C^2)\,(4.0\ \mu C)(-6.0\mu C)}{(3.0 \text{ m})^2}\hat{i}$$
   $$= -(24 \text{ mN})\hat{i}$$
   and $\vec{F}_{2,1} = -\vec{F}_{1,2} = (24 \text{ mN})\hat{i}$

8. **APPROACH** q_2 exerts an attractive electric force $\vec{F}_{2,1}$ on point charge q_1 and q_3 exerts a repulsive electric force $\vec{F}_{3,1}$ on point charge q_1 [Fig.1.360]. We can find the net electric force on q_1 by adding these forces (that is, by using the superposition principle).
   SOLUTION Express the net force acting on q_1:

   Figure 1.360

   $$\vec{F}_1 = \vec{F}_{2,1} + \vec{F}_{3,1} \qquad \cdots (1)$$
   Express the force that q_2 exerts on q_1
   $$\vec{F}_{2,1} = \dfrac{k|q_1||q_2|}{r_{2,1}^2}\hat{i}$$
   Express the force that q_3 exerts on q_1:
   $$\vec{F}_{3,1} = \dfrac{k|q_1||q_3|}{r_{3,1}^2}(-\hat{i})$$
   Substituting these values of $\vec{F}_{2,1}$ and $\vec{F}_{3,1}$ in Eq.(1), we get:
   $$\vec{F}_1 = \dfrac{k|q_1||q_2|}{r_{2,1}^2}\hat{i} - \dfrac{k|q_1||q_3|}{r_{3,1}^2}\hat{i}$$
   $$= k|q_1|\left(\dfrac{|q_2|}{r_{2,1}^2} - \dfrac{|q_3|}{r_{3,1}^2}\right)\hat{i}$$
   Now, substituting numerical values in above equation and solving for $\vec{F}_1$, we get:
   $$\vec{F}_1 = \left(8.988 \times 10^9 \text{ N} \cdot \text{m}^2/\text{C}^2\right)(6.0\mu C)$$
   $$\times \left(\dfrac{4.0\mu C}{(3.0 \text{ m})^2} - \dfrac{6.0\mu C}{(6.0 \text{ m})^2}\right)\hat{i}$$
   $$= (1.5 \times 10^{-2} \text{ N})\hat{i}$$

9. **APPROACH** The third point charge should be placed at the location at which the forces on the third point charge due to each of the other two point charges cancel. There can be no such place except on the line between the two point charges. Denote the 2.0 μC and 4.0 μC point charges by the numerals 2 and 4, respectively, and the third point charge by the numeral 3. Assume that the 2.0μC point charge is to the left of the 4.0 μC point charge, let the $+x$ direction be to the right. Then the 4.0 μC point charge is located at $x = L$.
   SOLUTION Apply the condition for translational equilibrium to the third point charge:
   $$\vec{F}_{4,3} + \vec{F}_{2,3} = 0 \implies F_{4,3} = F_{2,3} \qquad \cdots (1)$$
   Letting the distance from the third point charge to the 4.0μC point charge be x, express the force that the 4.0μC point charge exerts on the third point charge:
   $$F_{4,3} = \dfrac{kq_3q_4}{(L-x)^2}$$
   The force that the 2.0μC point charge exerts on the third charge is given by:
   $$F_{2,3} = \dfrac{kq_3q_2}{x^2}$$
   Substitute in equation (1) to obtain:
   $$\dfrac{kq_3q_4}{(L-x)^2} = \dfrac{kq_3q_2}{x^2} \implies \dfrac{q_4}{(L-x)^2} = \dfrac{q_2}{x^2}$$
   Substituting $q_4 = 2q_2$ in above equation, we get
   $$\dfrac{2}{(L-x)^2} = \dfrac{1}{x^2} \implies \left(\dfrac{x}{L-x}\right)^2 = \dfrac{1}{2} \implies \dfrac{x}{L-x} = \pm\dfrac{1}{\sqrt{2}}$$
   Taking +ve sign, we get
   $$\dfrac{x}{L-x} = \dfrac{1}{\sqrt{2}} \implies \sqrt{2}x = L - x$$
   $$\implies x = \dfrac{L}{\sqrt{2}+1} = \left(\sqrt{2}-1\right)L = 0.41L$$
   Similarly, taking $-$ve sign, we get
   $$\dfrac{x}{L-x} = -\dfrac{1}{\sqrt{2}} \implies \sqrt{2}x = -(L-x)$$
   $$\implies x = -\dfrac{L}{\sqrt{2}-1} = -\left(\sqrt{2}+1\right)L = 2.41L$$
   The root corresponding to the negative sign is extraneous because it corresponds to a position to the left of the 2.0 μC point charge and is, therefore, not a physically meaningful root. Hence the third point charge should be placed between the point charges and a distance equal to 0.41 L away from the 2.0 μC charge.

10. **APPROACH** The third point charge should be placed at the location at which the forces on the third point charge due to each of the other two point charges cancel. There can be no such place between the two point charges. Beyond the 4.0 μC point charge, and on the line containing the two point charges, the force due to the 4.0 μC point charge overwhelms the force due to the -2.0 μC point charge. Beyond the -2.0 μC point charge, and on the line containing the two point charges, however, we can find a place where these forces cancel because they are equal in magnitude and oppositely directed. Denote the -2.0 μC and 4.0 μC point charges by the numerals 2 and 4, respectively, and the third point charge by the numeral 3. Let the $+x$ direction be to the right with the origin at the position of the -2.0 μC point charge and the 4.0 μC point charge be located at $x = L$.
    SOLUTION Apply the condition for translational equilib-

rium to the third point charge:
$$\vec{F}_{4,3} + \vec{F}_{2,3} = 0 \implies F_{4,3} = F_{2,3} \quad (1)$$
Letting the distance from the third point charge to the 2.0 μC point charge be x, express the force that the 4.0 μC point charge exerts on the third point charge:
$$F_{4,3} = \frac{kq_3q_4}{(L+x)^2}$$
The force that the -2.0μC point charge exerts on the third point charge is given by:
$$F_{2,3} = \frac{kq_3q_2}{x^2}$$
Substitute for $F_{4,3}$ and $F_{2,3}$ in Eq.(1) to obtain:
$$\frac{kq_3q_4}{(L+x)^2} = \frac{kq_3q_2}{x^2} \implies \frac{q_4}{(L+x)^2} = \frac{q_2}{x^2}$$
Substituting $q_4 = 2q_2$ and rewriting this equation explicitly as a quadratic equations yields:
$$x^2 - 2Lx - L^2 = 0$$
On simplifying above equation, we get-
$$x = \frac{2L \pm \sqrt{4L^2 + 4L^2}}{2} = L \pm \sqrt{2}L$$
The root corresponding to the positive sign between the terms is extraneous because it corresponds to a position to the right of the 2.0 μC point charge and is, therefore, not a physically meaningful root. Hence the third point charge should be placed a distance equal to $0.41L$ from the -2.0 μC charge on the side away from the 4.0 μC charge.

11. **APPROACH** The positions of the point particles are shown in Fig.1.361. It is apparent that the electron must be located along the line joining two point particles. Moreover, because it is negatively charged, it must be closer to the particle with a charge which is smaller in magnitude i.e., -2.5 μC than to the particle with a charge of 6.0 μC, as is indicated in the Fig.1.361. We can find the x and y coordinates of the electron's position by equating the two electrostatic forces acting on it and solving for its distance from the origin. We can use similar triangles to express this radial distance in terms of the x and y coordinates of the electron.
SOLUTION Express the condition that must be satisfied

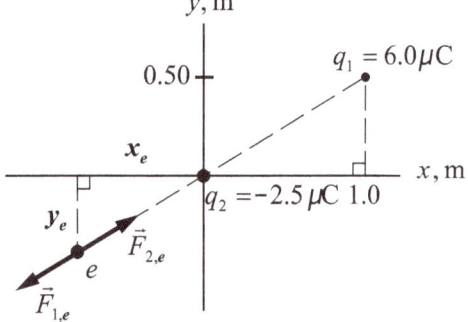

Figure 1.361

if the electron is to be in equilibrium:
$$F_{1,e} = F_{2,e}$$
here, $F_{1,e}$ is the force applied by q_1 on charge e whereas $F_{2,e}$ is the force applied by q_2 on e.
Let r represents the distance from the origin to the electron, express the magnitude of the force that the particle whose charge is q_1 exerts on the electron:
$$F_{1,e} = \frac{kq_1e}{(r+\sqrt{1.25})^2}$$

Express the magnitude of the force that the particle whose charge is q_2 exerts on the electron:
$$F_{2,e} = \frac{k|q_2|e}{r^2}$$
Substitute and simplify to obtain:
$$\frac{q_1}{(r+\sqrt{1.25})^2} = \frac{|q_2|}{r^2}$$
Substitute for q_1 and q_2 and simplifying for r, we get:
$$\frac{6}{(r+\sqrt{1.25})^2} = \frac{2.5}{r^2}$$
Solving for r yields:
$$r = 2.036 \text{ m and } r = -0.4386 \text{ m}.$$
Because $r < 0$ is unphysical, we'll consider only the positive root.
Use the similar triangles in the diagram to establish the proportion involving the y coordinate of the electron:
$$\frac{y_e}{0.50} = \frac{2.036}{1.12} \implies y_e = 0.909 \text{ m}$$
Use the similar triangles in the diagram to establish the proportion involving the x coordinate of the electron:
$$\frac{x_e}{1.0} = \frac{2.036}{1.12} \implies x_e = 1.82 \text{m}$$
Therefore, the coordinates of the electron's position are:
$$(x_e, y_e) = (-1.8 \text{ m}, -0.91 \text{ m})$$
Substituting for q_1 and q_2 and simplifying for r we get-
$$\frac{6}{(r+\sqrt{1.25})^2} = \frac{2.5}{r^2}$$
Solving for r yields: $r = 2.036$ m and $r = -0.4386$ m
Because $r < 0$ is unphysical, we'll consider only the positive root. Use the similar triangles in the diagram to establish the proportion involving the y coordinate of the electron:
$$\frac{y_e}{0.50} = \frac{2.036}{1.12} \implies y_e = 0.909 \text{ m}$$
Use the similar triangles in the diagram to establish the proportion involving the x coordinate of the electron:
$$\frac{x_e}{1.0} = \frac{2.036}{1.12} \implies x_e = 1.82 \text{ m}$$
The coordinates of the electron's position are :
$$(x_e, y_e) = (-1.8 \text{ m}, -0.91 \text{ m})$$

12. **APPROACH** Apply Coulomb's law and the principle of superposition of forces to find the net force acting on the central charge q.
SOLUTION The given situation is shown in Fig.1.362. A, B, C, D and E are five equally spaced points on the semicircle of radius R. At each of these points a charge Q is placed. O os the centre of this semicircle where another point charge q is placed. Since each charge is at same distance from central charge at O, therefore magnitude of electric force applied by each point charge on the charge at O will be same and if it is F, then
$$F = k\frac{Qq}{R^2} \quad \cdots (1)$$

Figure 1.362

If $\vec{F}_{A,O}$, $\vec{F}_{B,O}$, $\vec{F}_{C,O}$, $\vec{F}_{D,O}$, and $\vec{F}_{E,O}$ are electric forces

applied by charge Q at A, B, C, D and E respectively on the charge q at O, then
$\vec{F}_{AO} = -F\hat{j}$, $\vec{F}_{BO} = F\cos 45°\hat{i} - F\sin 45°\hat{j}$, $\vec{F}_{CO} = F\hat{i}$, $\vec{F}_{DO} = F\cos 45°\hat{i} + F\sin 45°\hat{j}$, $\vec{F}_{EO} = F\hat{j}$.
Net force on the charge q at O:
$$\vec{F}_{net} = \vec{F}_{AO} + \vec{F}_{BO} + \vec{F}_{CO} + \vec{F}_{DO} + \vec{F}_{EO}$$
Substituting the values of all forces, we get
$$\vec{F}_{net} = -F\hat{j} + (F\cos 45°\hat{i} - F\sin 45°\hat{j}) + F\hat{i}$$
$$+ (F\cos 45°\hat{i} + F\sin 45°\hat{j}) + F\hat{j}$$
$$\vec{F}_{net} = F(2\cos 45° + 1)\hat{i} = F(\sqrt{2} + 1)\hat{i} = \quad \cdots (2)$$
Substituting the value of F from Eq.(1) in Eq.(2), we get
$$\vec{F}_{net} = k\frac{Qq}{R^2}(\sqrt{2} + 1)\hat{i}$$

13. **APPROACH** Let the H$^+$ ions be in the x-y plane with one H$^+$ at $(0,0,0)$, one H$^+$ at $(a,0,0)$, and one H$^+$ at $\left(\frac{a}{2}, \frac{a\sqrt{3}}{2}, 0\right)$. These charges are denoted by q_1, q_2, and q_3 respectively in Fig.1.363. The N^{-3} ion, with charge q_4 in our notation, is then at $\left(\frac{a}{2}, \frac{a}{2\sqrt{3}}, a\sqrt{\frac{2}{3}}\right)$ where $a = 1.64 \times 10^{-10}$ m. To simplify our calculations we'll set $ke^2/a^2 = C = 8.56 \times 10^{-9}$ N. We can apply Coulomb's law and the principle of superposition of forces to find the net force acting on each ion.

SOLUTION Express the net force acting on point charge

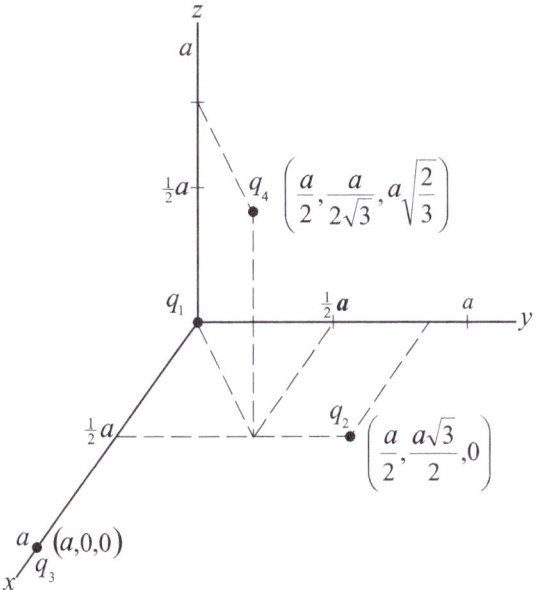

Figure 1.363

q_1:
$$\vec{F}_1 = \vec{F}_{2,1} + \vec{F}_{3,1} + \vec{F}_{4,1}$$
Find $\vec{F}_{2,1}$: $\vec{F}_{2,1} = \frac{kq_1q_2}{r_{2,1}^2}\hat{r}_{2,1} = C(-\hat{i}) = -C\hat{i}$

Find $\vec{F}_{3,1}$: $\vec{F}_{3,1} = \frac{kq_3q_1}{r_{3,1}^2}\hat{r}_{3,1} = C\frac{\left(0 - \frac{a}{2}\right)\hat{i} + \left(0 - \frac{a\sqrt{3}}{2}\right)\hat{j}}{a}$
$$= -C\left(\frac{1}{2}\hat{i} + \frac{\sqrt{3}}{2}\hat{j}\right)$$

Noting that the magnitude of point charge q_4 is three times that of the other point charges and that it is negative, express $\vec{F}_{4,1}$:
$$\vec{F}_{4,1} = 3C\hat{r}_{4,1}$$

$$= -3C\frac{\left(0 - \frac{a}{2}\right)\hat{i} + \left(0 - \frac{a}{2\sqrt{3}}\right)\hat{j} + \left(0 - \frac{a\sqrt{2}}{\sqrt{3}}\right)\hat{k}}{\sqrt{\left(\frac{a}{2}\right)^2 + \left(\frac{a}{2\sqrt{3}}\right)^2 + \left(\frac{a\sqrt{2}}{\sqrt{3}}\right)^2}}$$

$$= 3C\frac{\left(\frac{a}{2}\right)\hat{i} + \left(\frac{a}{2\sqrt{3}}\right)\hat{j} + \left(\frac{a\sqrt{2}}{\sqrt{3}}\right)\hat{k}}{a}$$

$$= 3C\left[\left(\frac{1}{2}\right)\hat{i} + \left(\frac{1}{2\sqrt{3}}\right)\hat{j} + \left(\sqrt{\frac{2}{3}}\right)\hat{k}\right]$$

Substitute in the expression for $\vec{F}_1$ to obtain:
$$\vec{F}_1 = -C\hat{i} - C\left(\frac{1}{2}\hat{i} + \frac{\sqrt{3}}{2}\hat{j}\right)$$
$$+ 3C\left[\left(\frac{1}{2}\right)\hat{i} + \left(\frac{1}{2\sqrt{3}}\right)\hat{j} + \left(\sqrt{\frac{2}{3}}\right)\hat{k}\right] = C\sqrt{6}\hat{k}$$

From symmetry considerations:
$$\vec{F}_2 = \vec{F}_3 = \vec{F}_1 = C\sqrt{6}\hat{k}$$
Express the condition that the molecule is in equilibrium:
$$\vec{F}_1 + \vec{F}_2 + \vec{F}_3 + \vec{F}_4 = 0$$
Solve for and evaluate $\vec{F}_4$:
$$\vec{F}_4 = -\left(\vec{F}_1 + \vec{F}_2 + \vec{F}_3\right) = -3\vec{F}_1 = -3C\sqrt{6}\hat{k}$$

14. **APPROACH** Let q represent the point charge at the origin and use Coulomb's law for $\vec{E}$ due to a point charge to find the electric field at $x = 6.0$ m and -10 m.

SOLUTION (a) Express the electric field at a point P located a distance x from a point charge q:
$$\vec{E}(x) = \frac{kq}{x^2}\hat{r}_{P,0}$$
Evaluate this expression for $x = 6.0$m:
$$\vec{E}(6.0\text{ m}) = \frac{\left(8.988 \times 10^9 \frac{\text{N}\cdot\text{m}^2}{\text{C}^2}\right)(4.0\mu\text{C})}{(6.0\text{ m})^2}\hat{i} = (1.0\text{kN/C})\hat{i}$$

(b) Evaluate $\vec{E}$ at $x = -10$ m:
$$\vec{E}(-10\text{ m}) = \frac{\left(8.988 \times 10^9 \frac{\text{N}\cdot\text{m}^2}{\text{C}^2}\right)(4.0\mu\text{C})}{(10\text{ m})^2}(-\hat{i})$$
$$= (-0.36\text{ kN/C})\hat{i}$$

(c) E_x vs x graph is shown in Fig.1.364.

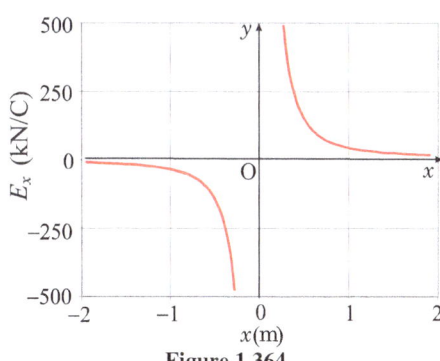

Figure 1.364

15. **APPROACH** Let q represent the point charges of $+4.0$ μC and use Coulomb's law for $\vec{E}$ due to a point charge and the principle of superposition for fields to find the electric field at the locations specified.

Noting that $q_1 = q_2$, use Coulomb's law and the principle of superposition to express the electric field due to the given charges at point P a distance x from the origin:
$$\vec{E}(x) = \vec{E}_{q_1}(x) + \vec{E}_{q_2}(x)$$

$$= \frac{kq_1}{x^2}\hat{r}_{q_1,P} + \frac{kq_2}{(8.0\,\text{m} - x)^2}\hat{r}_{q_2,P}$$

$$= kq_1\left(\frac{1}{x^2}\hat{r}_{q_1,P} + \frac{1}{(8.0\,\text{m} - x)^2}\hat{r}_{q_2,P}\right)$$

$$= \left(36\,\text{kN}\cdot\text{m}^2/\text{C}\right)\left(\frac{1}{x^2}\hat{r}_{q_1,P} + \frac{1}{(8.0\,\text{m} - x)^2}\hat{r}_{q_2,P}\right)$$

(a) Apply this equation to the point at $x = -2.0$ m:

$$\vec{E}(-2.0\,\text{m}) = \left(36\,\text{kN}\cdot\text{m}^2/\text{C}\right)\left[\frac{1}{(2.0\,\text{m})^2}(-\hat{i})\right.$$
$$\left. + \frac{1}{(10\,\text{m})^2}(-\hat{i})\right]$$
$$= (-9.4\,\text{kN/C})\hat{i}$$

(b) Evaluate $\vec{E}$ at $x = 2.0$ m:

$$\vec{E}(2.0\,\text{m}) = \left(36\,\text{kN}\cdot\text{m}^2/\text{C}\right)\left[\frac{1}{(2.0\,\text{m})^2}(\hat{i}) + \frac{1}{(6.0\,\text{m})^2}(-\hat{i})\right]$$
$$= (8.0\,\text{kN/C})\hat{i}$$

(c) Evaluate $\vec{E}$ at $x = 6.0$ m:

$$\vec{E}(6.0\,\text{m}) = \left(36\,\text{kN}\cdot\text{m}^2/\text{C}\right)\left[\frac{1}{(6.0\,\text{m})^2}(\hat{i}) + \frac{1}{(2.0\,\text{m})^2}(-\hat{i})\right]$$
$$= (-8.0\,\text{kN/C})\hat{i}$$

(d) Evaluate $\vec{E}$ at $x = 10$ m:

$$\vec{E}(10\,\text{m}) = \left(36\,\text{kN}\cdot\text{m}^2/\text{C}\right)\left[\frac{1}{(10\,\text{m})^2}(\hat{i}) + \frac{1}{(2.0\,\text{m})^2}(\hat{i})\right]$$
$$= (9.4\,\text{kN/C})\hat{i}$$

(e) From symmetry considerations:
$$\vec{E}(4.0\,\text{m}) = 0$$

(f) E_x vs x graph is shown in Fig.1.365

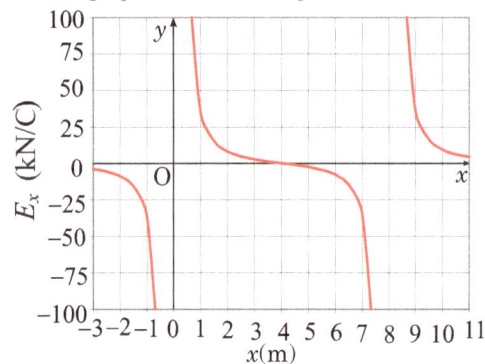

Figure 1.365

16. APPROACH If the electric field at $x = 0$ is zero, both its x and y components must be zero. The only way this condition can be satisfied with point charges of $+5.0\,\mu\text{C}$ and $-8.0\,\mu\text{C}$ on the x axis is if the point charge $+6.0\,\mu\text{C}$ is also on the x axis. Let the subscripts 5, -8, and 6 identify the point charges and their fields. We can use Coulomb's law for $\vec{E}$ due to a point charge and the principle of superposition for fields to determine where the $+6.0\,\mu\text{C}$ point charge should be located so that the electric field at $x = 0$ is zero.

SOLUTION Express the electric field at $x = 0$ in terms of the fields due to the point charges of $+5.0\,\mu\text{C}$, $-8.0\,\mu\text{C}$, and $+6.0\,\mu\text{C}$:

$$\vec{E}(0) = \vec{E}_{5\mu\text{C}} + \vec{E}_{-8\mu\text{C}} + \vec{E}_{6\mu\text{C}} = 0$$

Substitute for each of the fields to obtain:

$$\frac{kq_5}{r_5^2}\hat{r}_5 + \frac{kq_6}{r_6^2}\hat{r}_6 + \frac{kq_{-8}}{r_{-8}^2}\hat{r}_{-8} = 0$$

or

$$\frac{kq_5}{r_5^2}\hat{i} + \frac{kq_6}{r_6^2}(-\hat{i}) + \frac{kq_{-8}}{r_{-8}^2}(-\hat{i}) = 0$$

Divide out the unit vector $\hat{i}$ to obtain:

$$\frac{q_5}{r_5^2} - \frac{q_6}{r_6^2} - \frac{q_{-8}}{r_{-8}^2} = 0$$

Substitute numerical values to obtain r_6:

$$\frac{5}{(3.0\,\text{cm})^2} - \frac{6}{r_6^2} - \frac{-8}{(4.0\,\text{cm})^2} = 0 \implies r_6 = 2.4\,\text{cm}$$

17. APPROACH Fig. 1.366 shows the electric field vectors at the point of interest P due to the two point charges. We can use Coulomb's law for $\vec{E}$ due to point charges and the superposition principle for electric fields to find $\vec{E}_P$. We can apply $\vec{F} = q\vec{E}$ to find the force on an electron at $(-1.0\,\text{m}, 0)$.

SOLUTION (a) Express the electric field at $(-1.0\,\text{m}, 0)$

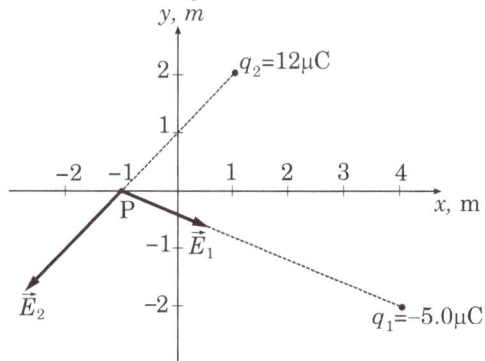

Figure 1.366

due to the point charges q_1 and q_2: $\vec{E}_P = \vec{E}_1 + \vec{E}_2$
Substitute numerical values and evaluate $\vec{E}_1$

$$\vec{E}_1 = \frac{kq_1}{r_{1,P}^2}\hat{r}_{1,P} = \frac{(8.988 \times 10^9\,\text{N}\cdot\text{m}^2/\text{C}^2)(-5.0\,\mu\text{C})}{(5.0\,\text{m})^2 + (2.0\,\text{m})^2}$$

$$\left(\frac{(-5.0\,\text{m})\hat{i} + (2.0\,\text{m})\hat{j}}{\sqrt{(5.0\,\text{m})^2 + (2.0\,\text{m})^2}}\right)$$

$$= (-1.55 \times 10^3\,\text{N/C})(-0.928\hat{i} + 0.371\hat{j})$$
$$= (1.44\,\text{kN/C})\hat{i} + (-0.575\,\text{kN/C})\hat{j}$$

Substitute numerical values and evaluate $\vec{E}_2$:

$$\vec{E}_2 = \frac{kq_2}{r_{2,P}^2}\hat{r}_{2,P}$$

$$= \frac{(8.988 \times 10^9\,\text{N}\cdot\text{m}^2/\text{C}^2)(12\,\mu\text{C})}{(2.0\,\text{m})^2 + (2.0\,\text{m})^2}$$

$$\left(\frac{(-2.0\,\text{m})\hat{i} + (-2.0\,\text{m})\hat{j}}{\sqrt{(2.0\,\text{m})^2 + (2.0\,\text{m})^2}}\right)$$

$$= \left(13.5 \times 10^3\,\text{N/C}\right)(-0.707\hat{i} - 0.707\hat{j})$$
$$= (-9.54\,\text{kN/C})\hat{i} + (-9.54\,\text{kN/C})\hat{j}$$

Substitute for $\vec{E}_1$ and $\vec{E}_2$ and simplify to find $\vec{E}_P$:

$$\vec{E}_P = (1.44\,\text{kN/C})\hat{i} + (-0.575\,\text{kN/C})\hat{j} + (-9.54\,\text{kN/C})\hat{i}$$
$$+ (-9.54\,\text{kN/C})\hat{j}$$
$$= (-8.10\,\text{kN/C})\hat{i} + (-10.1\,\text{kN/C})\hat{j}$$

The magnitude of $\vec{E}_P$ is:

$$E_P = \sqrt{(-8.10\,\text{kN/C})^2 + (-10.1\,\text{kN/C})^2} = 13\,\text{kN/C}$$

The direction of $\vec{E}_P$ is:

$$\theta_E = \tan^{-1}\left(\frac{-10.1\,\text{kN/C}}{-8.10\,\text{kN/C}}\right) = 230°$$

Note that the angle returned by your calculator for $\tan^{-1}\left(\frac{-10.1\,\text{kN/C}}{-8.10\,\text{kN/C}}\right)$ is the reference angle and must be increased by 180° to yield θ_E.

(b) Express and evaluate the force on an electron at point P:

$$\vec{F} = q\vec{E}_P = \left(-1.602 \times 10^{-19}\text{C}\right)[(-8.10\text{kN/C})\hat{i}$$
$$+(-10.1\text{kN/C})\hat{j}]$$
$$= \left(1.30 \times 10^{-15}\text{ N}\right)\hat{i} + \left(1.62 \times 10^{-15}\text{ N}\right)\hat{j}$$

Find the magnitude of $\vec{F}$:
$$F = \sqrt{\left(1.30 \times 10^{-15}\text{ N}\right)^2 + \left(1.62 \times 10^{-15}\text{ N}\right)^2}$$
$$= 2.1 \times 10^{-15}\text{ N} \text{ Find the direction of } \vec{F}:$$
$$\theta_F = \tan^{-1}\left(\frac{1.62 \times 10^{-15}\text{ N}}{1.3 \times 10^{-15}\text{ N}}\right) = 51°$$

18. **APPROACH** The electric field of the 4th charged point particle must cancel the sum of the electric fields due to the other three charged point particles[Fig.1.367]. By symmetry, the position of the 4th charged point particle must lie on the vertical center line of the triangle. Using trigonometry, one can show that the center of an equilateral triangle is a distance $L/\sqrt{3}$ from each vertex, where L is the length of the side of the triangle. Note that the x components of the fields due to the base charged particles cancel each other, so we only need concern ourselves with the y components of the fields due to the charged point particles at the vertices of the triangle. Choose a coordinate system in which the origin is at the midpoint of the base of the triangle, the $+x$ direction is to the right, and the $+y$ direction is upward.

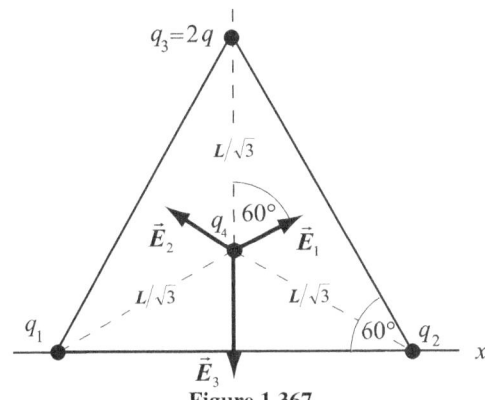

Figure 1.367

SOLUTION Express the condition that must be satisfied if the electric field at the center of the triangle is to be zero:
$$\sum_{i=1\text{to}4} \vec{E}_i = 0$$

Substituting for $\vec{E}_1, \vec{E}_2, \vec{E}_3$, and $\vec{E}_4$ yields:
$$\frac{k(q)}{\left(\frac{L}{\sqrt{3}}\right)^2}\cos 60°\hat{j} + \frac{k(q)}{\left(\frac{L}{\sqrt{3}}\right)^2}\cos 60°\hat{j}$$
$$-\frac{k(2q)}{\left(\frac{L}{\sqrt{3}}\right)^2}\hat{j} + \frac{kq}{y^2}\hat{j} = 0$$

On simplifying for y, we get-
$$y = \pm L/\sqrt{3}$$

The positive solution corresponds to the 4th point particle being a distance $L/\sqrt{3}$ above the base of the triangle, where it produces the same strength and same direction electric field caused by the three charges at the corners of the triangle. So the charged point particle must be placed a distance $L/\sqrt{3}$ below the midpoint of the triangle.

19. **APPROACH** We can determine the stability of the equilibrium in Part (a) and Part (b) by considering the forces the equal point charges q at $y = +a$ and $y = -a$ exert on the test charge when it is given a small displacement along either the x or y axis. (c) The application of Coulomb's law in Part (c) will lead to the magnitude and sign of the charge that must be placed at the origin in order that a net force of zero is experienced by each of the three point charges.

SOLUTION (a) Because E_x is in the x direction, a positive test charge that is displaced from $(0,0)$ in either the $+x$ direction or the $-x$ direction will experience a force pointing away from the origin and accelerate in the direction of the force. Consequently, the equilibrium at $(0,0)$ is unstable for a small displacement along the x axis.

If the positive test charge is displaced in the direction of increasing y (the $+y$ direction), the charge at $y = +a$ will exert a greater force than the charge at $y = -a$, and the net force is then in the $-y$ direction; i.e., it is a restoring force. Similarly, if the positive test charge is displaced in the direction of decreasing y (the $-y$ direction), the charge at $y = -a$ will exert a greater force than the charge at $y = -a$, and the net force is then in the $-y$ direction; i.e., it is a restoring force. Consequently, the equilibrium at $(0,0)$ is stable for a small displacement along the y axis*.

(b) Following the same arguments as in Part (a), one finds that, for a negative test charge, the equilibrium is stable at $(0,0)$ for displacements along the x axis and unstable for displacements along the y axis.

(c) Express the net force acting on the charge at $y = +a$:
$$\sum F_{q\text{ at }y=+a} = \frac{kqq_0}{a^2} + \frac{kq^2}{(2a)^2} = 0$$

Solve for q_0 to obtain: $q_0 = -q/4$

☞ In Part (c), we could just as well have expressed the net force acting on the charge at $y = -a$. Due to the symmetric distribution of the charges at $y = -a$ and $y = +a$, summing the forces acting on q_0 at the origin does not lead to a relationship between q_0 and q.

20. **APPROACH** The diagram1.368 shows the locations of point charges q_1 and q_2 and the point on the x axis at which we are to find $\vec{E}$. From symmetry considerations we can conclude that the y component of $\vec{E}$ at any point on the x axis is zero. We can use Coulomb's law for the electric field due to point charges and the principle of superposition of fields to find the field at any point on the x axis. We can establish the results called for in parts (b) and (c) by factoring the radicand and using the approximation $1 + \alpha \approx 1$ whenever $\alpha \ll 1$.

SOLUTION (a) Express the x-component of the electric field due to the point charges at $y = a$ and $y = -a$ as a function of the distance r from either charge to point P:
$$\vec{E}_x = 2\frac{kq}{r^2}\cos\theta\hat{i}$$

Substitute for $\cos\theta$ and r to obtain:
$$\vec{E}_x = 2\frac{kq}{r^2}\frac{x}{r}\hat{i} = \frac{2kqx}{r^3}\hat{i} = \frac{2kqx}{\left(x^2+a^2\right)^{3/2}}\hat{i} = \frac{2kqx}{\left(x^2+a^2\right)^{3/2}}\hat{i}$$

The magnitude of $\vec{E}_x$ is:
$$E_x = \frac{2kqx}{\left(x^2+a^2\right)^{3/2}}$$

(b) For $|x| \ll a$, $x^2 + a^2 \approx a^2$, so:
$$E_x \approx \frac{2kqx}{\left(a^2\right)^{3/2}} = \frac{2kqx}{a^3}$$

(c) For $x \gg a$, the charges separated by a would appear to

---
*It is important to note that the equilibrium is stable only along y axis not in all directions. So, Earnshaw's theorem (section 1.24) is not violated

be a single charge of magnitude $2q$. Its field would be given by $E_x = 2kq/x^2$.
Factorize the radicand to obtain:
$$E_x = 2kqx \left[x^2 \left(1 + \frac{a^2}{x^2} \right) \right]^{-3/2}$$
For $a \ll x$, $1 + \frac{a^2}{x^2} \approx 1$ and $E_x = 2kqx \left[x^2 \right]^{-3/2} = \frac{2kq}{x^2}$

21. **APPROACH** From problem 20 The electric field on the x axis, due to equal positive point charges located at $(0, a)$ and $(0, -a)$, is given by $E_x = 2kqx(x^2 + a^2)^{-3/2}$ We can use $T = 2\pi\sqrt{m/k'}$ to express the period of the motion of the bead in terms of the restoring constant k'.
(a) Express the force acting on the bead when its displacement from the origin is x:
$$F_x = -qE_x = -\frac{2kq^2 x}{(x^2 + a^2)^{3/2}} = -\frac{2kq^2 x}{a^3 \left(\frac{x^2}{a^2} + 1 \right)^{3/2}}$$
For $x \ll a$: $F_x = -\frac{2kq^2}{a^3} x$
That is, the bead experiences a linear restoring force.
(b) Express the period of a simple harmonic oscillator:
$$T = 2\pi \sqrt{\frac{m}{k'}}$$
Obtain k' from our result in Part (a): $k' = \frac{2kq^2}{a^3}$
Substitute for k' and simplify to obtain:
$$T = 2\pi \sqrt{\frac{m}{\frac{2kq^2}{a^3}}} = 2\pi \sqrt{\frac{ma^3}{2kq^2}}$$

22. **APPROACH** We can use Newton's second law of motion to find the acceleration of the proton in the uniform electric field and constant-acceleration equations to find the time required for it to reach a speed of $0.01c$ and the distance it travels while acquiring this speed.
SOLUTION (a) Compute e/m for a proton:
$$\frac{e}{m_p} = \frac{1.602 \times 10^{-19} \text{C}}{1.673 \times 10^{-27} \text{ kg}} = 9.58 \times 10^7 \text{C/kg}$$
Apply Newton's second law to relate the acceleration of the proton to the electric field:
$$a = \frac{F_{\text{net}}}{m_p} = \frac{eE}{m_p}$$
Substitute numerical values and evaluate a:
$$a = \frac{(1.602 \times 10^{-19} \text{C})(100 \text{ N/C})}{1.673 \times 10^{-27} \text{ kg}} = 9.576 \times 10^9 \text{ m/s}^2$$
$$= 9.58 \times 10^9 \text{ m/s}^2$$
The direction of the acceleration of a proton is in the direction of the electric field.
(b) Using the definition of acceleration, relate the time required for a proton to reach $0.01c$ to its acceleration:
$$\Delta t = \frac{v}{a} = \frac{0.01c}{a}$$
Substitute numerical values and evaluate Δt:
$$\Delta t = \frac{0.01(2.998 \times 10^8 \text{ m/s})}{9.576 \times 10^9 \text{ m/s}^2} = 0.3 \text{ ms}$$

23. **APPROACH** Because the electric field is in the $-y$ direction, the force it exerts on the electron is in the $+y$ direction. Applying Newton's second law to the electron will yield an expression for the acceleration of the electron in the y direction. We can then use a constant-acceleration equation to relate its speed to its acceleration and the distance it has travelled.

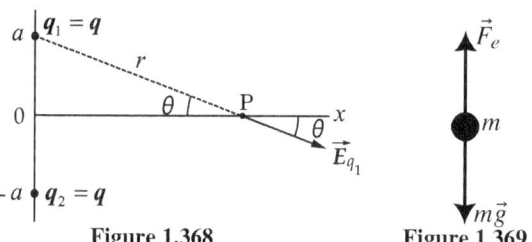

Figure 1.368   Figure 1.369

SOLUTION Applying Newton's second law: $\sum F_y = ma_y$ to the FBD of electron given in Fig.1.369, we get
$$F_e - F_g = ma_y$$
since, $F_e = eE$ and $F_g = mg$, therefore above equation gives:
$$eE - mg = ma_y \implies a_y = \frac{eE}{m} - g$$
Use a constant-acceleration equation to relate the speed of the electron to its acceleration and the distance it travels:
$$v_y^2 = v_0^2 + 2a_y \Delta y$$
Since, the electron starts at rest, i.e., $v_0 = 0$, therefore, above equation gives
$$v_y^2 = 2a_y \Delta y \implies v_y = \sqrt{2a_y \Delta y}$$
Substituting the value of for a_y in above expression for v_y, we get
$$v_y = \sqrt{2\left(\frac{eE}{m} - g \right) \Delta y}$$
Now, substituting numerical values, we get in above expression, we get:
$$v_y = \left[2 \left\{ \frac{(1.602 \times 10^{-19} \text{C})(1.50 \times 10^{-10} \text{ N/C})}{9.109 \times 10^{-31} \text{ kg}} \right. \right.$$
$$\left. \left. -9.81 \text{ m/s}^2 \right\} (1.0 \times 10^{-6} \text{ m}) \right]^{1/2} = 5.8 \text{ mm/s}$$

24. **APPROACH** We can apply the work-kinetic energy theorem to relate the change in the object's kinetic energy to the net force acting on it. We can express the net force acting on the charged body in terms of its charge and the electric field.
SOLUTION Using the work-kinetic energy theorem, express the kinetic energy of the object in terms of the net force acting on it and its displacement:
$$W = \Delta K = F_{\text{net}} \Delta x$$
Relate the net force acting on the charged particle to the electric field:
$$F_{\text{net}} = qE$$
Substitute for F_{net} to obtain:
$$\Delta K = K_f - K_i = qE\Delta x$$
Since, $K_i = 0$, therefore, $K_f = qE\Delta x \implies q = \frac{K_f}{E\Delta x}$
Substitute numerical values and evaluate q:
$$q = \frac{0.120 \text{ J}}{(300 \text{ N/C})(0.500 \text{ m})} = 800 \mu\text{C}$$

25. **APPROACH** We can use constant-acceleration equations to express the x and y coordinates of the particle in terms of the parameter t and Newton's second law to express the constant acceleration in terms of the electric field. Eliminating t will yield an equation for y as a function of x, q, and m that we can solve for E_y.
SOLUTION Express the x and y coordinates of the particle as functions of time:
$$x = (v \cos \theta)t \quad \cdots (1)$$
and
$$y = (v \sin \theta)t - \frac{1}{2} a_y t^2 \quad \cdots (2)$$
Apply Newton's second law to relate the acceleration of the

particle to the net force acting on it:
$$a_y = \frac{F_{\text{net},y}}{m} = \frac{qE_0}{m} \quad \cdots (3)$$
Substitute it in Eq.(2), to get y:
$$y = (v\sin\theta)t - \frac{qE_0}{2m}t^2 \quad \cdots (4)$$
Eliminate the parameter t between the two equations (1) and (4) to obtain:
$$y = (\tan\theta)x - \frac{qE_0}{2mv^2\cos^2\theta}x^2$$
Set $y = 0$ and solve for E_0 to obtain:
$$E_0 = \frac{mv^2 \sin 2\theta}{qx}$$
Substitute the non-particle-specific data to obtain:
$$E_0 = \frac{m(3.00 \times 10^6 \text{ m/s})^2 \sin 70°}{q(0.0150 \text{ m})}$$
$$= (5.638 \times 10^{14} \text{ m/s}^2)\frac{m}{q}$$

(a) Substitute for the mass and charge of an electron and evaluate E_0:
$$E_0 = (5.638 \times 10^{14} \text{ m/s}^2)\frac{9.109 \times 10^{-31} \text{ kg}}{-1.602 \times 10^{-19} \text{C}}$$
$$= -3.2 \text{kN/C}$$

(b) Substitute for the mass and charge of a proton and evaluate E_0:
$$E_0 = (5.64 \times 10^{14} \text{ m/s}^2)\frac{1.673 \times 10^{-27} \text{ kg}}{1.602 \times 10^{-19}\text{C}} = 5.9 \text{ MN/C}$$

26. **APPROACH** We can use the definition of dipole moment to find the dipole moment of given pair of charges.
SOLUTION (a) Apply the definition of electric dipole moment to obtain:
$$\vec{p} = q(2\vec{l})$$
and $\quad p = (2.0 \text{ pC})(4.0 \text{ } \mu\text{m}) = 8.0 \times 10^{-18}$ C.m

(b) If the dipole is oriented to the right[Fig.1.370], then $\vec{p}$ is to the right; pointing from the negative charge toward the positive charge.

Figure 1.370

27. **APPROACH** The forces the electron and the proton exert on each other constitute an action-and-reaction pair. Because the magnitudes of their charges are equal and their masses are the same, we find the speed of each particle by finding the speed of either one. We'll apply Coulomb's force law for point charges and Newton's second law to relate v to e, m, k, and their separation distance L.
SOLUTION Apply Newton's second law to the positron to obtain:
$$\frac{ke^2}{L^2} = m\frac{v^2}{\frac{1}{2}L} \implies \frac{ke^2}{L} = 2mv^2$$
Solving for v gives: $v = \sqrt{\frac{ke^2}{2mL}}$

28. **APPROACH** We can use Coulomb's force law for point particles and the condition for translational equilibrium to express the equilibrium position as a function of k, q, Q, m, and g. In Part (b) we'll need to show that the displaced point charge experiences a linear restoring force and, hence, will exhibit simple harmonic motion.
SOLUTION (a) Apply the condition for translational equilibrium to the particle:

$$\frac{kqQ}{y_0^2} - mg = 0 \implies y_0 = \sqrt{\frac{kqQ}{mg}}$$

(b) Express the restoring force that acts on the particle when it is displaced a distance Δy from its equilibrium position:
$$F = \frac{kqQ}{(y_0 + \Delta y)^2} - \frac{kqQ}{y_0^2}$$
Since, $\Delta y \ll y_0$, therefore
$$F \approx \frac{kqQ}{y_0^2 + 2y_0\Delta y} - \frac{kqQ}{y_0^2}$$
Simplify this expression further by writing it with a common denominator:
$$F = -\frac{2y_0\Delta y kqQ}{y_0^4 + 2y_0^3\Delta y} = -\frac{2y_0\Delta y kqQ}{y_0^4\left(1 + 2\frac{\Delta y}{y_0}\right)} \approx -\frac{2\Delta y kqQ}{y_0^3}$$
(Since, $\Delta y \ll y_0$.)

From the part (a) of our solution:
$$\frac{kqQ}{y_0^2} = mg$$
Substitute for $\frac{kqQ}{y_0^2}$ and simplify to obtain:
$$F = -\frac{2mg}{y_0}\Delta y$$
Apply Newton's second law to the displaced particle to obtain:
$$m\frac{d^2\Delta y}{dt^2} = -\frac{2mg}{y_0}\Delta y$$
or
$$\frac{d^2\Delta y}{dt^2} + \frac{2g}{y_0}\Delta y = 0$$
or
$$\frac{d^2\Delta y}{dt^2} + \omega^2\Delta y = 0$$
which is the differential equation of simple harmonic motion with angular frequency
$$\omega = \sqrt{2g/y_0}$$

29. **APPROACH** The free body diagram 1.371 shows the forces acting on the microsphere of mass m and having an excess charge of $q = Ne$ when the electric field is downward. Under terminal-speed conditions the sphere is in dynamic equilibrium under the influence of the electric force $\vec{F}_e$, its weight $m\vec{g}$, and the drag force $\vec{F}_d$. We can apply Newton's second law, under terminal speed conditions, to relate the number of excess charges N on the sphere to its mass and, using Stokes' law, find its terminal speed.

(a) Apply Newton's second law to the microsphere to

Figure 1.371

obtain:
$$F_e - mg - F_d = ma_y$$
since, $a_y = 0$, therefore
$$F_e - mg - F_{d,\text{ terminal}} = 0$$
Substitute for F_e, m, and $F_{d,\text{terminal}}$ to obtain:
$$qE - \rho Vg - 6\pi\eta r v_t = 0$$

since, $q = Ne$, therefore
$$NeE - \frac{4}{3}\pi r^3 \rho g - 6\pi\eta r v_t = 0$$
Solve for Ne to obtain:
$$Ne = \frac{\frac{4}{3}\pi r^3 \rho g + 6\pi\eta r v_t}{E} \qquad \cdots (1)$$
Substitute numerical values and evaluate $\frac{4}{3}\pi r^3 \rho g$:
$$\frac{4}{3}\pi r^3 \rho g = \frac{4}{3}\pi \left(5.50 \times 10^{-7} \text{ m}\right)^3 \left(1.05 \times 10^3 \text{ kg/m}^3\right)$$
$$\left(9.81 \text{ m/s}^2\right) = 7.18 \times 10^{-15} \text{ N}$$
Substitute numerical values and evaluate $6\pi\eta r v_t$:
$$6\pi\eta r v_t = 6\pi \left(1.8 \times 10^{-5} \text{ Pa}\cdot\text{s}\right)\left(5.50 \times 10^{-7} \text{ m}\right)$$
$$\left(1.16 \times 10^{-4} \text{ m/s}\right) = 2.16 \times 10^{-14} \text{ N}$$
Substitute numerical values in equation (1) and evaluate Ne:
$$Ne = \frac{7.18 \times 10^{-15} \text{ N} + 2.16 \times 10^{-14} \text{ N}}{6.00 \times 10^4 \text{ N/C}}$$
$$= 4.8 \times 10^{-19} \text{C}$$
(b) Divide the result in (a) by e to obtain:
$$N = \frac{4.80 \times 10^{-19} \text{C}}{1.602 \times 10^{-19} \text{C}} = 3$$
(c) With the field pointing upward, the electric force is downward and the application of $\sum F_y = ma_y$ to the bead yields:
$$F_{d,\text{ terminal}} - F_e - mg = 0$$
or
$$6\pi\eta r v_t - NeE - \frac{4}{3}\pi r^3 \rho g = 0$$
Solve for v_t to obtain:
$$v_t = \frac{NeE + \frac{4}{3}\pi r^3 \rho g}{6\pi\eta r}$$
Substitute numerical values and evaluate v_t:
$$v_t = \left[3\left(1.602 \times 10^{-19}\text{C}\right)\left(6.00 \times 10^4 \frac{\text{N}}{\text{C}}\right)\right.$$
$$\left. +\frac{4}{3}\pi\left(5.50 \times 10^{-7}\text{ m}\right)^3\left(1.05 \times 10^3 \frac{\text{kg}}{\text{m}^3}\right)\left(9.81\frac{\text{m}}{\text{s}^2}\right)\right]$$
$$\Big/\left[6\pi\left(1.8 \times 10^{-5}\text{ Pa.s}\right)\left(5.50 \times 10^{-7}\text{ m}\right)\right] = 0.19 \text{ mm/s}$$

30. **APPROACH** The free body diagram Fig.1.371 shows the forces acting on the microsphere of mass m and having an excess charge of $q = Ne$ when the electric field is downward. Under terminal-speed conditions the sphere is in equilibrium under the influence of the electric force $\vec{F}_e$, its weight $m\vec{g}$, and the drag force $\vec{F}_d$. We can apply Newton's second law, under terminal speed conditions (i.e., dynamic equilibrium), to relate the number of excess charges N on the sphere to its mass and, using Stokes' law, to its terminal speed.
SOLUTION **(a)** Apply Newton's second law to the microsphere when the electric field is downward:
$$F_e - mg - F_d = ma_y$$
or
$$F_e - mg - F_{d,\text{terminal}} = 0 \qquad (\because a_y = 0)$$
Substitute for F_e and $F_{d,\text{terminal}}$ to get:
$$qE - mg - 6\pi\eta r v_u = 0$$
or
$$NeE - mg - 6\pi\eta r v_u = 0 \qquad (\because q = Ne)$$
Solve for v_u to get:
$$v_u = \frac{NeE - mg}{6\pi\eta r} \qquad (1)$$
With the field pointing upward, the electric force is downward and the application of Newton's second law to the microsphere yields:
$$F_{d,\text{ terminal}} - F_e - mg = 0$$
or
$$6\pi\eta r v_d - NeE - mg = 0$$

Solve for v_d to obtain:
$$v_d = \frac{NeE + mg}{6\pi\eta r} \qquad \cdots (2)$$
Add equations (1) and (2) and simplify to obtain:
$$u = v_u + v_d = \frac{NeE - mg}{6\pi\eta r} + \frac{NeE + mg}{6\pi\eta r} = \frac{NeE}{3\pi\eta r} = \frac{qE}{3\pi\eta r}$$
Measuring both v_u and v_d has the advantage that you don't need to know the mass of the microsphere.
(b) Let Δu represents the change in the terminal speed of the microsphere due to a gain (or loss) of one electron we have:
$$\Delta u = v_{N+1} - v_N$$
Noting that Δu will be the same whether the microsphere is moving upward or downward, express its terminal speed when it is moving upward with N electronic charges on it:
$$v_N = \frac{NeE - mg}{6\pi\eta r}$$
Express its terminal speed upward when it has $N + 1$ electronic charges:
$$v_{N+1} = \frac{(N+1)eE - mg}{6\pi\eta r}$$
Substitute and simplify to obtain:
$$\Delta u = \frac{(N+1)eE - mg}{6\pi\eta r} - \frac{NeE - mg}{6\pi\eta r} = \frac{eE}{6\pi\eta r}$$
Substitute numerical values and evaluate Δu:
$$\Delta u = \frac{\left(1.602 \times 10^{-19}\text{C}\right)\left(6.00 \times 10^4 \text{ N/C}\right)}{6\pi\left(1.8 \times 10^{-5}\text{ Pa}\cdot\text{m}\right)\left(5.50 \times 10^{-7}\text{ m}\right)} = 52 \text{ }\mu\text{m/s}$$

31. **APPROACH** Choose a coordinate system in which the $+x$ direction is perpendicular to the surfaces and points to the right. Let the charge densities on the two plates be σ_1 and σ_2 and denote the three regions of interest as 1, 2, and 3 [Fig.1.372]. Choose a coordinate system in which the positive x direction is to the right. We can apply the equation for $\vec{E}$ near an infinite plane of charge and the superposition of fields to find the field in each of the three regions.

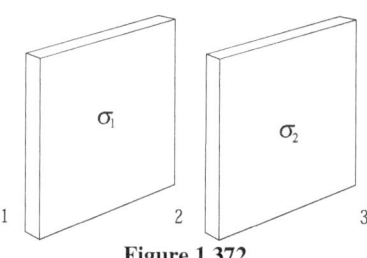

Figure 1.372

SOLUTION **(a)** Use the equation for $\vec{E}$ near an infinite plane of charge to express the field in region 1 when $\sigma_1 = \sigma_2 = +3.0 \text{ } \mu\text{C/m}^2$
$$\vec{E}_1 = \vec{E}_{\sigma_1} + \vec{E}_{\sigma_2} = -2\pi k\sigma_1 \hat{i} - 2\pi k\sigma_2 \hat{i} = -4\pi k\sigma \hat{i}$$
Substitute numerical values and evaluate $\vec{E}_1$:
$$\vec{E}_1 = -4\pi\left(8.988 \times 10^9 \text{ N}\cdot\text{m}^2/\text{C}^2\right)\left(3.0\mu\text{C/m}^2\right)\hat{i}$$
$$= -\left(3.4 \times 10^5 \text{ N/C}\right)\hat{i}$$
Proceed as above for region 2:
$$\vec{E}_2 = \vec{E}_{\sigma_1} + \vec{E}_{\sigma_2}$$
$$= 2\pi k\sigma_1 \hat{i} - 2\pi k\sigma_2 \hat{i} = 2\pi k\sigma \hat{i} - 2\pi k\sigma \hat{i} = 0$$
Proceed as above for region 3:
$$\vec{E}_3 = \vec{E}_{\sigma_1} + \vec{E}_{\sigma_2} = 2\pi k\sigma_1 \hat{i} + 2\pi k\sigma_2 \hat{i} = 4\pi k\sigma \hat{i}$$
$$= 4\pi\left(8.988 \times 10^9 \frac{\text{N}\cdot\text{m}^2}{\text{C}^2}\right)\left(3.0\mu\text{C/m}^2\right)\hat{i}$$
$$= \left(3.4 \times 10^5 \text{ N/C}\right)\hat{i}$$

(b) Use the equation for E near an infinite plane of charge to express and evaluate the field in region 1 when $\sigma_1 = +3.0\ \mu C/m^2$ and $\sigma_2 = -3.0\ \mu C/m^2$:
$$\vec{E}_1 = \vec{E}_{\sigma_1} + \vec{E}_{\sigma_2} = 2\pi k \sigma_1 \hat{i} - 2\pi k \sigma_2 \hat{i}$$
$$= 2\pi k \sigma \hat{i} - 2\pi k \sigma \hat{i} = 0$$
Proceed as above for region 2:
$$\vec{E}_2 = \vec{E}_{\sigma_1} + \vec{E}_{\sigma_2} = 2\pi k \sigma_1 \hat{i} + 2\pi k \sigma_2 \hat{i} = 4\pi k \sigma \hat{i}$$
$$= 4\pi \left(8.988 \times 10^9 \frac{N \cdot m^2}{C^2} \right) \left(3.0\ \mu C/m^2 \right) \hat{i}$$
$$= \left(3.4 \times 10^5\ N/C \right) \hat{i}$$
Proceed as above for region 3:
$$\vec{E}_3 = \vec{E}_{\sigma_1} + \vec{E}_{\sigma_2} = 2\pi k \sigma_1 \hat{i} - 2\pi k \sigma_2 \hat{i}$$
$$= 2\pi k \sigma \hat{i} - 2\pi k \sigma \hat{i} = 0$$
(c) The electric field lines for (a) and (b) are shown in Fig.1.373.

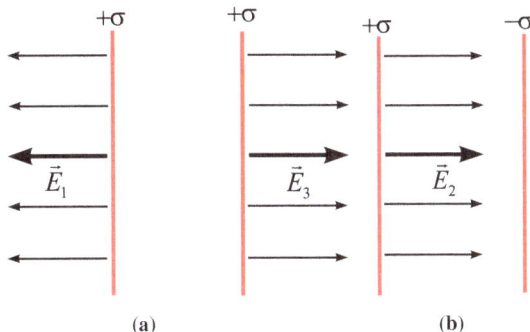

Figure 1.373

32. **APPROACH** The magnitude of the electric field at an axial distance z from a disk that has a uniform charge density σ and whose radius is R is given by
$$E_z = 2\pi k \sigma \left(1 - \frac{1}{\sqrt{1 + \frac{R^2}{z^2}}} \right)$$
We can equate this expression and $E_z = \frac{1}{2}\sigma/(2\epsilon_0)$ and solve for z.
Express the electric field on the axis of a disk charge:
$$E_z = 2\pi k \sigma \left(1 - \frac{1}{\sqrt{1 + \frac{R^2}{z^2}}} \right)$$
We're given that:
$$E_z = \frac{1}{2}\sigma/(2\epsilon_0) = \frac{\sigma}{4\epsilon_0}$$
Equating these expressions gives:
$$\frac{\sigma}{4\epsilon_0} = 2\pi k \sigma \left(1 - \frac{1}{\sqrt{1 + \frac{R^2}{z^2}}} \right)$$
Substituting for k yields:
$$\frac{\sigma}{4\epsilon_0} = 2\pi \left(\frac{1}{4\pi\epsilon_0} \right) \sigma \left(1 - \frac{1}{\sqrt{1 + \frac{R^2}{z^2}}} \right)$$
Solve for z to obtain: $z = R/\sqrt{3}$

33. The electric field at a distance z from the centre of a ring whose charge is Q and whose radius is a is given by
$$E_z = \frac{kQz}{(z^2 + a^2)^{3/2}}$$

(a) Evaluating $E_{z=0.2a}$ gives
$$E_{z=0.2a} = \frac{kQ(0.2a)}{[(0.2a)^2 + a^2]^{3/2}} = 0.189 \frac{kQ}{a^2}$$
(b) Evaluating $E_{z=0.5a}$ gives:
$$E_{z=0.5a} = \frac{kQ(0.5a)}{[(0.5a)^2 + a^2]^{3/2}} = 0.358 \frac{kQ}{a^2}$$
(c) Evaluating $E_{z=0.7a}$ gives:
$$E_{z=0.7a} = \frac{kQ(0.7a)}{[(0.7a)^2 + a^2]^{3/2}} = 0.385 \frac{kQ}{a^2}$$
(d) Evaluating $E_{z=a}$ gives:
$$E_{z=a} = \frac{kQa}{[a^2 + a^2]^{3/2}} = 0.354 \frac{kQ}{a^2}$$
(e) Evaluating $E_{z=2a}$ gives:
$$E_{z=2a} = \frac{2kQa}{[(2a)^2 + a^2]^{3/2}} = 0.179 \frac{kQ}{a^2}$$
(f) The field along the z axis is plotted in Fig.1.374. The z coordinates are in units of z/a and E is in units of kQ/a^2

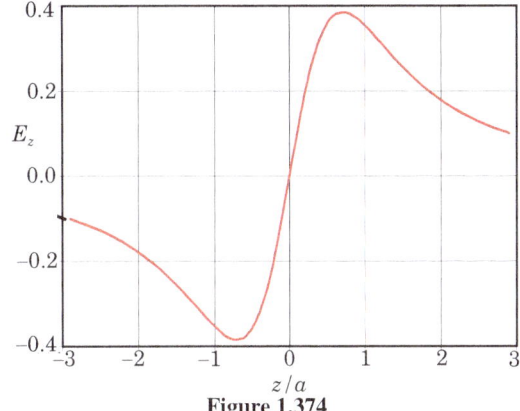

Figure 1.374

34. **APPROACH** The line charge and point $(0, y)$ are shown in the diagram 1.375. Also shown is a line element of length dx and the field $d\vec{E}$ its charge produces at $(0, y)$. We can find dE_x from $d\vec{E}$ and then integrate from $x = x_1$ to $x = x_2$.
SOLUTION Express the x component of $d\vec{E}$:

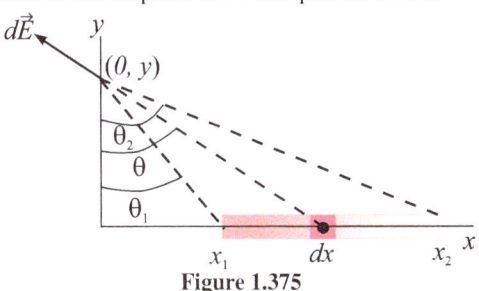

Figure 1.375

$$dE_x = -\frac{k\lambda}{x^2 + y^2} \sin\theta\, dx = -\frac{k\lambda}{x^2 + y^2} \frac{x}{\sqrt{x^2 + y^2}} dx$$
$$= -\frac{k\lambda x}{(x^2 + y^2)^{3/2}} dx$$
Integrate from $x = x_1$ to x_2 and simplify to obtain:
$$E_x = -k\lambda \int_{x_1}^{x_2} \frac{x}{(x^2 + y^2)^{3/2}} dx = -k\lambda \left[-\frac{1}{\sqrt{x^2 + y^2}} \right]_{x_1}^{x_2}$$
$$= -k\lambda \left[-\frac{1}{\sqrt{x_2^2 + y^2}} + \frac{1}{\sqrt{x_1^2 + y^2}} \right]$$

$$= -\frac{k\lambda}{y}\left[-\frac{y}{\sqrt{x_2^2+y^2}}+\frac{y}{\sqrt{x_1^2+y^2}}\right]$$

From the diagram we see that:
$$\cos\theta_2 = \frac{y}{\sqrt{x_2^2+y^2}} \text{ or } \theta_2 = \tan^{-1}\left(\frac{x_2}{y}\right)$$

and
$$\cos\theta_1 = \frac{y}{\sqrt{x_1^2+y^2}} \text{ or } \theta_1 = \tan^{-1}\left(\frac{x_1}{y}\right)$$

Substitute for $\dfrac{y}{\sqrt{x_2^2+y^2}}$ and $\dfrac{y}{\sqrt{x_1^2+y^2}}$ to Obtain:

$$E_x = -\frac{k\lambda}{y}[-\cos\theta_2+\cos\theta_1] = \frac{k\lambda}{y}[\cos\theta_2-\cos\theta_1]$$

35. APPROACH Consider the ring with its axis along the z direction shown in the Fig. 1.376. Its radius is $z = r\cos\theta$ and its width is $rd\theta$. We can use the equation for the field on the axis of a ring charge and then integrate to express the field at the center of the hemispherical shell.

SOLUTION Express the field on the axis of the ring of

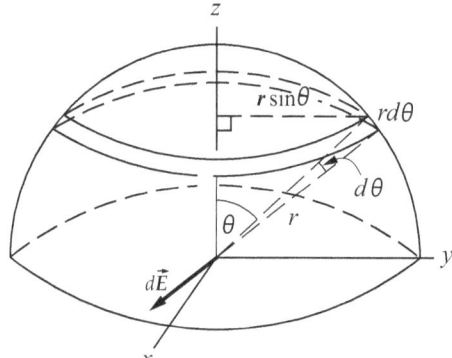

Figure 1.376

charge:
$$dE = \frac{kzdq}{\left(r^2\sin^2\theta+r^2\cos^2\theta\right)^{3/2}} = \frac{kzdq}{r^3}$$
(where $z = r\cos\theta$)

Express the charge dq on the ring:
$$dq = \sigma dA = \sigma(2\pi\cdot\sin\theta)rd\theta = 2\pi\sigma r^2\sin\theta d\theta$$

Substitute this value of dq in above expression for dE to obtain:
$$dE = \frac{k(r\cos\theta)2\pi\sigma r^2\sin\theta d\theta}{r^3} = 2\pi k\sigma\sin\theta\cos\theta d\theta$$

Integrating dE from $\theta = 0$ to $\pi/2$ yields:
$$E = 2\pi k\sigma\int_0^{\pi/2}\sin\theta\cos\theta d\theta$$
$$= 2\pi k\sigma\left[\frac{1}{2}\sin^2\theta\right]_0^{\pi/2} = \pi k\sigma$$

36. APPROACH Because the cone encloses no charge, we know, from Gauss's law, that the net flux of the electric field through the cone's surface is zero [Fig.1.377]. Thus, the number of field lines penetrating the curved surface of the cone must equal the number of field lines penetrating the base and the entering flux must equal the exiting flux.

SOLUTION The flux penetrating the base of the cone is given by:
$$\phi_{\text{entering}} = EA_{\text{base}} \quad\cdots(1)$$

The flux penetrating the curved surface of the cone is given by:

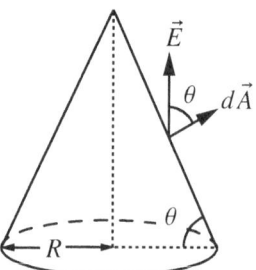

Figure 1.377

$$\phi_{\text{exiting}} = \oint_S \vec{E}\cdot d\vec{A} = \oint_S E\cos\theta dA \quad\cdots(2)$$

Equating the fluxes given by Eq.(1) and (2) and simplifying yields:
$$A_{\text{base}} = \cos\theta\oint_S dA = (\cos\theta)A_{\text{curved surface}}$$

If n represents number of field lines crossing each surface, $N_{\text{curved surface}}$ represents field line density at curved surface, and N_{base} represents field line density at base, then

The ratio of the density of field lines is:
$$\frac{N_{\text{curved surface}}}{N_{\text{base}}} = \frac{n/A_{\text{curved surface}}}{n/A_{\text{base}}} = \frac{A_{\text{base}}}{A_{\text{curved surface}}} = \cos\theta$$

37. APPROACH We can find the total charge on the sphere by expressing the charge dq in a spherical shell and integrating this expression between $r = 0$ and $r = R$. By symmetry, the electric fields must be radial. To find E_r inside the charged sphere we choose a spherical Gaussian surface of radius $r < R$. To find E_r outside the charged sphere we choose a spherical Gaussian surface of radius $r > R$. On each of these surfaces, E_r is constant. Gauss's law then relates E_r to the total charge inside the surface.

SOLUTION (a) Express the charge dq in a shell of thickness dr and volume $4\pi r^2 dr$:
$$dq = 4\pi r^2\rho dr = 4\pi r^2(Ar)dr = 4\pi Ar^3 dr$$

Integrate this expression from $r = 0$ to R to find the total charge on the sphere:
$$Q = 4\pi A\int_0^R r^3 dr = \left[\pi Ar^4\right]_0^R = \pi AR^4$$

(b) Apply Gauss's law to a spherical surface of radius $r > R$ that is concentric with the nonconducting sphere to obtain:
$$\oint_S E_r dA = \frac{1}{\epsilon_0}Q_{\text{encl}} \implies 4\pi r^2 E_r = \frac{Q_{\text{encl}}}{\epsilon_0}$$

Solving for E_r yields:
$$E_{r>R} = \frac{Q_{\text{inside}}}{4\pi\epsilon_0}\frac{1}{r^2} = \frac{kQ_{\text{inside}}}{r^2} = \frac{kA\pi R^4}{r^2} = \frac{AR^4}{4\epsilon_0 r^2}$$

Apply Gauss's law to a spherical surface of radius $r < R$ that is concentric with the non-conducting sphere to obtain:
$$\oint_S E_r dA = \frac{1}{\epsilon_0}Q_{\text{inside}} \implies 4\pi r^2 E_r = \frac{Q_{\text{inside}}}{\epsilon_0}$$

Solve for E_r and simplify to obtain:
$$E_{r<R} = \frac{Q_{\text{inside}}}{4\pi r^2\epsilon_0} = \frac{\pi Ar^4}{4\pi r^2\epsilon_0} = \frac{Ar^2}{4\epsilon_0}$$

(c) E_r versus r/R graph, with E_r in units of $A/(4\epsilon_0)$, is plotted in Fig.1.378.

Remarks: Note that the results for (a) and (b) agree at $r = R$.

38. APPROACH By symmetry, the electric fields resulting from this charge distribution must be radial. To find E_r for $r < R_1$ we choose a spherical Gaussian surface of radius $r < R_1$. To find E_r for $R_1 < r < R_2$ we choose a spherical Gaussian surface of radius $R_1 < r < R_2$. To find E_r for

Figure 1.378

$r > R_2$ we choose a spherical Gaussian surface of radius $r > R_2$. On each of these surfaces, E_r is constant. Gauss's law then relates E_r to the total charge inside the surface.

SOLUTION (a) The charge in an infinitesimal spherical shell of radius r and thickness dr is:
$$dQ = \rho dV = 4\pi \rho r^2 dr$$
Integrate dQ from $r = R_1$ to R_2 to find the total charge in the spherical shell in the interval $R_1 < r < R_2$:
$$Q_{\text{total}} = 4\pi \rho \int_{R_1}^{R_2} r^2 dr = 4\pi \rho \left[\frac{r^3}{3}\right]_{R_1}^{R_2} = \frac{4\pi \rho}{3}\left(R_2^3 - R_1^3\right)$$

(b) Apply Gauss's law to a spherical surface of radius r that is concentric with the non-conducting spherical shell to obtain:
$$\oint_S E_r dA = \frac{1}{\epsilon_0} Q_{\text{inside}} \Rightarrow 4\pi r^2 E_r = \frac{Q_{\text{inside}}}{\epsilon_0}$$

On solving for E_r, we get:
$$E_r = \frac{Q_{\text{inside}}}{4\pi \epsilon_0}\frac{1}{r^2} = \frac{kQ_{\text{inside}}}{r^2}$$

Evaluate $E_{r<R_1}$: $E_{r<R_1} = \frac{Q_{\text{inside}}}{4\pi \epsilon_0}\frac{1}{r^2} = \frac{kQ_{\text{inside}}}{r^2} = 0$

because $\rho_{r<R_1} = 0$ and, therefore, $Q_{\text{inside}} = 0$

Evaluate $E_{R_1<r<R_2}$:
$$E_{R_1<r<R_2} = \frac{kQ_{\text{inside}}}{r^2} = \frac{4\pi k \rho}{3r^2}\left(r^3 - R_1^3\right)$$
$$= \frac{\rho}{3\epsilon_0 r^2}\left(r^3 - R_1^3\right)$$

For $r > R_2$: $Q_{\text{inside}} = \frac{4\pi \rho}{3}\left(R_2^3 - R_1^3\right)$

and $E_{r>R_2} = \frac{4\pi k\rho}{3r^2}\left(R_2^3 - R_1^3\right) = \frac{\rho}{3\epsilon_0 r^2}\left(R_2^3 - R_1^3\right)$

Remark: Note that E is continuous at $r = R_2$.

39. APPROACH From symmetry, the field tangent to the surface of the cylinder must vanish. We can construct a Gaussian surface in the shape of a cylinder of radius r and length L and apply Gauss's law to find the electric field as a function of the distance from the centreline of the infinitely long non-conducting cylinder.

SOLUTION Apply Gauss's law to a cylindrical surface of radius r and length L that is concentric with the infinitely long non-conducting cylinder:
$$\oint_S E_n dA = \frac{1}{\epsilon_0} Q_{\text{inside}}$$
or
$$2\pi r L E_n = \frac{Q_{\text{inside}}}{\epsilon_0}$$
where we've neglected the end areas because there is no flux through them.

Due to symmetry, $E_n = E_r$. Solving for E_r yields:
$$E_r = \frac{Q_{\text{inside}}}{2\pi r L \epsilon_0} = \frac{2kQ_{\text{inside}}}{Lr}$$

Express Q_{inside} for $r < a$:
$$Q_{\text{inside}} = \rho(r)V = \rho_0 \left(\pi r^2 L\right)$$
Substitute to obtain:

$$E_{r<R} = \frac{2k\left(\pi \rho_0 L r^2\right)}{Lr} = \frac{\rho_0}{2\epsilon_0}r$$

since, $\lambda = \rho \pi a^2$, therefore
$$E_{r<R} = \frac{\lambda}{2\pi \epsilon_0 a^2}r$$

Express Q_{inside} for $r > a$:
$$Q_{\text{inside}} = \rho(r)V = \rho_0 \left(\pi a^2 L\right)$$
Substitute for Q_{inside} to obtain:
$$E_{r>a} = \frac{2k\left(\pi \rho_0 L a^2\right)}{Lr} = \frac{\rho_0 a^2}{2\epsilon_0 r}$$
Since, $\lambda = \rho \pi a^2$, therefore: $E_{r>a} = \frac{\lambda}{2\pi \epsilon_0 r}$

40. APPROACH From symmetry; the field tangent to the surfaces of the cylindrical shells must vanish. We can construct a Gaussian surface in the shape of a cylinder of radius r and length L and apply Gauss's law to find the electric field as a function of the distance from the centreline of the infinitely long, uniformly charged cylindrical shells.

SOLUTION (a) Apply Gauss's law to the cylindrical surface of radius r and length L that is concentric with the infinitely long, uniformly charged cylindrical shell:
$$\oint_S E_n dA = \frac{1}{\epsilon_0} Q_{\text{encl}} \implies 2\pi r L E_n = \frac{Q_{\text{encl}}}{\epsilon_0}$$
where we've neglected the end areas because there is no flux through them.

Noting that, due to symmetry, $E_n = E_r$, Solve for E_r to obtain:
$$E_r = \frac{2kQ_{\text{encl}}}{Lr} \quad \cdots (1)$$
For $r < R_1$, $Q_{\text{encl}} = 0$, so: $E_{r<R_1} = 0$
Express Q_{encl} for $a_1 < r < a_2$:
$$Q_{\text{encl}} = \sigma_1 A_1 = 2\pi \sigma_1 a_1 L$$
Substitute in equation (1) to obtain:
$$E_{a_1<r<a_2} = \frac{2k\left(2\pi \sigma_1 a_1 L\right)}{Lr} = \frac{\sigma_1 a_1}{\epsilon_0 r}$$

(b) Set $E = 0$ for $r > a_2$ to obtain:
$$\frac{\sigma_1 a_1 + \sigma_2 a_2}{\epsilon_0 r} = 0 \Rightarrow \frac{\sigma_2}{\sigma_1} = -\frac{a_1}{a_2}$$

(c) Because the electric field is determined by the enclosed charge within the Gaussian surface, the field under these conditions would be as given above:
$$E_{a_1<r<a_2} = \sigma_1 a_1 / \epsilon_0 r$$

41. APPROACH Because the penny is in an external electric field, it will have charges of opposite signs induced on its faces. The induced charge σ is related to the electric field by $E = \sigma \epsilon_0$. Once we know σ, we can use the definition of surface charge density to find the total charge on one face of the penny.

SOLUTION (a) Relate the electric field to the charge density on each face of the penny:
$$E = \frac{\sigma}{\epsilon_0} \implies \sigma = \epsilon_0 E$$
Substitute numerical values and evaluate σ:
$$\sigma = \left(8.854 \times 10^{-12} \text{C}^2/\text{N.m}^2\right)(1.60 \text{ kN/C})$$
$$= 14.17 \text{ nC/m}^2 = 14.2 \text{ nC/m}^2$$

(b) Use the definition of surface charge density to obtain:
$$\sigma = \frac{Q}{A} = \frac{Q}{\pi r^2} \implies Q = \sigma \pi r^2$$
Substitute numerical values and evaluate Q:
$$Q = \pi \left(14.17 \text{ nC/m}^2\right)(0.0100 \text{ m})^2 = 4.45 \text{ pC}$$

42. APPROACH Because the metal slab is in an external electric field, it will have charges of opposite signs induced on its faces. The induced charge σ is related to the electric field by $E = \sigma/\varepsilon_0$.

SOLUTION Relate the magnitude of the electric field to the charge density on the metal slab: $E = \dfrac{\sigma}{\epsilon_0}$

Use its definition to express σ: $\sigma = \dfrac{Q}{A} = \dfrac{Q}{L^2}$

Substitute for σ to obtain: $E = \dfrac{Q}{L^2 \varepsilon_0}$

Substitute numerical values and evaluate E:
$$E = \dfrac{1.2\,\text{nC}}{(0.12\,\text{m})^2\,(8.854\times 10^{-12}\,\text{C}^2/\text{N}\cdot\text{m}^2)} = 9.4\,\text{kN/C}$$

43. APPROACH From Gauss's law we know that the electric field at the surface of the charged sphere is given by $E = kQ/R^2$ where Q is the charge on the sphere and R is its radius. The minimum radius for dielectric breakdown corresponds to the maximum electric field at the surface of the sphere.

SOLUTION The relationship between E and R for dielectric breakdown:
$$E_{\max} = \dfrac{kQ}{R_{\min}^2} \implies R_{\min} = \sqrt{\dfrac{kQ}{E_{\max}}}$$

Substitute numerical values and evaluate $R_{\min}$:
$$R_{\min} = \sqrt{\dfrac{(8.988\times 10^9\,\text{N}\cdot\text{m}^2/\text{C}^2)(18\,\mu\text{C})}{3.0\times 10^6\,\text{N/C}}} = 23\,\text{cm}$$

44. APPROACH Because the atom is uncharged, we know that the integral of the electron's charge distribution over all of space must equal its charge q_e. Evaluation of this integral will lead to an expression for ρ_0. In (b) we can express the resultant electric field at any point as the sum of the electric fields due to the proton and the electron cloud.

SOLUTION (a) Because the atom is uncharged, the integral of the electron's charge distribution over all of space must equal its charge e:
$$e = \int_0^\infty \rho(r)\,dV = \int_0^\infty \rho(r)4\pi r^2\,dr$$

Substitute for $\rho(r)$ and simplify to obtain:
$$e = -\int_0^\infty \rho_0 e^{-2r/a} 4\pi r^2\,dr = -4\pi\rho_0 \int_0^\infty r^2 e^{-2r/a}\,dr$$

On integrating by parts, we get: $\displaystyle\int_0^\infty r^2 e^{-2r/a}\,dr = a^3/4$

On substituting for $\displaystyle\int_0^\infty r^2 e^{-2r/a}\,dr$, we get:
$$e = -4\pi\rho_0 a^3/3$$

Solving for ρ_0 yields: $e = -4\pi\rho_0\left(\dfrac{a^3}{4}\right) = -\pi a^3 \rho_0$

Solving for ρ_0 yields: $\rho_0 = \dfrac{e}{\pi a^3}$

(b) The field will be the sum of the field due to the proton and that of the electron charge cloud:
$$E = E_p + E_{\text{cloud}}$$

Express the field due to the electron cloud:
$$E_{\text{cloud}}(r) = \dfrac{kQ(r)}{r^2}$$

where $Q(r)$ is the net negative charge enclosed a distance r from the proton. Substitute for E_p and E_{cloud} to obtain:
$$E(r) = \dfrac{ke}{r^2} + \dfrac{kQ(r)}{r^2} \qquad \cdots (1)$$

$Q(r)$ is given by:
$$Q(r) = \int_0^r 4\pi r'^2 \rho(r')\,dr' = 4\pi\int_0^r r'^2 \rho_0 e^{-2r'/a}\,dr'$$

From Part (a), $\rho_0 = \dfrac{-e}{\pi a^3}$:
$$Q(r) = 4\pi\left(\dfrac{-e}{\pi a^3}\right)\int_0^r r'^2 e^{-2r'/a}\,dr'$$
$$= \dfrac{-4e}{a^3}\int_0^r r'^2 e^{-2r'/a}\,dr'$$

Using the integration by parts, we get:
$$\int_0^r x^2 e^{-2x/a}\,dx = \dfrac{1}{4}e^{-2r/a} a\left[\left(e^{-2r/a}-1\right)a^2 - 2ar - 2r^2\right]$$
$$= \dfrac{1}{4}e^{-2r/a} a^3 \left[\left(e^{-2r/a}-1\right) - 2\dfrac{r}{a} - 2\dfrac{r^2}{a^2}\right]$$
$$= \dfrac{a^3}{4}\left[\left(1 - e^{-2r/a}\right) - 2e^{-2r/a}\left(\dfrac{r}{a} + \dfrac{r^2}{a^2}\right)\right]$$

Substituting for $\displaystyle\int_0^r r'^2 e^{-2r'/a}\,dr'$ in the expression for $Q(r)$ and simplifying yields:
$$Q(r) = \dfrac{-e}{4}\left[\left(1-e^{-2r/a}\right) - 2e^{-2r/a}\left(\dfrac{r}{a} + \dfrac{r^2}{a^2}\right)\right]$$

Substitute for $Q(r)$ in equation (1) and simplify to obtain:
$$E(r) = \dfrac{ke}{r^2} - \dfrac{ke}{4r^2}\left[\left(1-e^{-2r/a}\right) - 2e^{-2r/a}\left(\dfrac{r}{a} + \dfrac{r^2}{a^2}\right)\right]$$
$$= \dfrac{ke}{r^2}\left[1 - \tfrac{1}{4}\left(1-e^{-2r/a}\right) - 2e^{-2r/a}\left(\dfrac{r}{a} + \dfrac{r^2}{a^2}\right)\right]$$

45. APPROACH See Fig.1.379. We can view the field in the hole as the sum of the field from a uniform spherical shell of charge Q plus the field due to a small patch with surface charge density equal but opposite to that of the patch cut out.

SOLUTION (a) Express the magnitude of the electric field

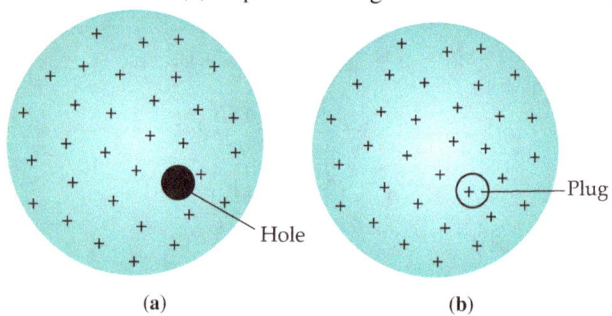

Figure 1.379

at the centre of the hole:
$$E = E_{\text{spherical shell}} + E_{\text{hole}}$$

Apply Gauss's law to a spherical gaussian surface just outside the given sphere:
$$E_{\text{spherical shell}}\left(4\pi r^2\right) = \dfrac{Q_{\text{enclosed}}}{\varepsilon_0} = \dfrac{Q}{\varepsilon_0}$$

Solve for $E_{\text{spherical shell}}$ to obtain:
$$E_{\text{spherical shell}} = \dfrac{Q}{4\pi\epsilon_0 r^2}$$

The electric field due to the small hole (small enough so that we can treat it as a plane surface) is:
$$E_{\text{hole}} = \dfrac{-\sigma}{2\varepsilon_0}$$

Substitute for $E_{\text{spherical shell}}$ and E_{hole} and shell simplify to obtain:
$$E = \dfrac{Q}{4\pi\epsilon_0 r^2} + \dfrac{-\sigma}{2\varepsilon_0} = \dfrac{Q}{4\pi\epsilon_0 r^2} - \dfrac{Q}{2\varepsilon_0(4\pi r^2)}$$
$$= \dfrac{Q}{8\pi\epsilon_0 r^2} \qquad \text{(radially outward)}$$

(b) Express the force on the patch: $F = qE$
where q is the charge on the patch.

Assuming that the patch has radius a, express the proportion between its charge and that of the spherical shell:
$$\frac{q}{\pi a^2} = \frac{Q}{4\pi r^2} \text{ or } q = \frac{a^2}{4r^2}Q$$
Substitute for q and E in the expression for F to obtain:
$$F = \left(\frac{a^2}{4r^2}Q\right)\left(\frac{Q}{8\pi\epsilon_0 r^2}\right) = \frac{Q^2 a^2}{32\pi\epsilon_0 r^4} \quad \text{(radially outward)}$$
(c) The pressure is the force exerted on the patch divided by the area of the patch:
$$P = \frac{\frac{Q^2 a^2}{32\pi\epsilon_0 r^4}}{\pi a^2} = \frac{Q^2}{32\pi^2\epsilon_0 r^4}$$

46. APPROACH (a) We can apply Newton's second law to the particle to express its speed as a function of its mass m, charge q, and the radius of its path R, and the strength of the electric field due to the infinite line charge E [Fig.1.380]. (b) The period of the particle's motion is the ratio of the circumference of the circle in which it travels divided by its orbital speed.

SOLUTION (a) Apply Newton's second law to the particle

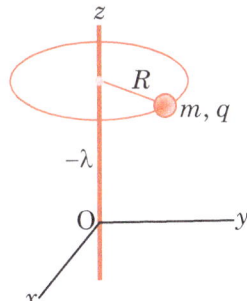

Figure 1.380

to obtain:
$$\sum F_{\text{radial}} = qE = m\frac{v^2}{R}$$
where the inward direction is positive.
Solving for v yields: $v = \sqrt{\frac{qRE}{m}}$
The strength of the electric field at a distance R from the infinite line charge is given by: $E = \frac{2k\lambda}{R}$
Substitute for E and simplify to obtain: $v = \sqrt{\frac{2kq\lambda}{m}}$
(b) The speed of the particle is equal to the circumference of its orbit divided by its period:
$$v = \frac{2\pi R}{T} \implies T = \frac{2\pi R}{v}$$
Substitute for v and simplify to obtain: $T = \pi R\sqrt{\frac{2m}{kq\lambda}}$

47. APPROACH In Part (a), you can apply Gauss's law to express $\vec{E}$ as a function of r for the uniformly charged non-conducting sphere with its center at the origin [Fig.1.381]. In Part (b), you can use the hint to express the field at a generic point $P(x, y)$ in the cavity as the sum of the fields due to equal positive and negative charge densities and then evaluate this expression at points 1 and 2.

SOLUTION (a) The electric field at a distance r from the centre of the uniformly charged non-conducting sphere is given by:
$$\vec{E}_\rho = E\hat{r} \quad \cdots (1)$$
where $\hat{r}$ is a unit vector pointing radially outward. Apply Gauss's law to a spherical surface of radius r centered at the

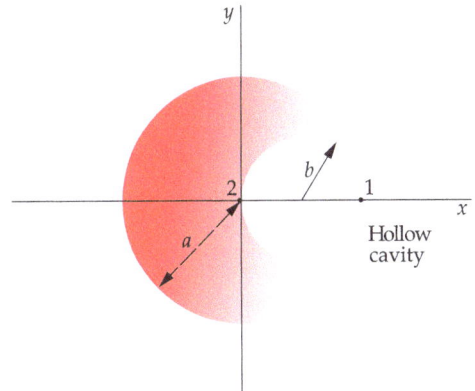

Figure 1.381

origin to obtain:
$$\oint_S E_n dA = E_\rho\left(4\pi r^2\right) = \frac{Q_{\text{inside}}}{\varepsilon_0}$$
Relate Q_{inside} to the charge density ρ:
$$\rho = \frac{Q_{\text{inside}}}{\frac{4}{3}\pi r^3} \Rightarrow Q_{\text{inside}} = \frac{4}{3}\rho\pi r^3$$
Substitute for Q_{inside}: $E_\rho\left(4\pi r^2\right) = \frac{4\rho\pi r^3}{3\epsilon_0}$
Solve for E_ρ to obtain: $E_\rho = \frac{\rho r}{3\epsilon_0}$
Substitute for E in equation (1) to obtain:
$$\vec{E}_\rho = \frac{\rho}{3\epsilon_0}r\hat{r}$$
(b) The electric field at point $P(x,y)$ is the sum of the electric fields due to the two charge distributions:
$$\vec{E} = \vec{E}_\rho + \vec{E}_{-\rho} = E_\rho\hat{r} + E_{-\rho}\hat{r}' \quad \cdots (2)$$
where $\hat{r}'$ is a unit vector normal to a spherical Gaussian surface whose center is at $x = b$.
Apply Gauss's law to a spherical surface of radius r' centred at $x = b = R/2$ to obtain:
$$\oint_S E_n dA = E_{-\rho}\left(4\pi r'^2\right) = \frac{Q_{\text{inside}}}{\varepsilon_0}$$
Relate Q_{inside} to the charge density $-\rho$:
$$-\rho = \frac{Q_{\text{inside}}}{\frac{4}{3}\pi r'^3} \Rightarrow Q_{\text{inside}} = -\frac{4}{3}\rho\pi r'^3$$
Substitute for Q_{inside} to obtain:
$$E_{-\rho}\left(4\pi r'^2\right) = -\frac{4\pi r'^3 \rho}{3\epsilon_0}$$
Solving for $E_{-\rho}$ yields: $E_{-\rho} = -\frac{\rho}{3\epsilon_0}r'$
Substitute for E_ρ and $E_{-\rho}$ in equation (2) to obtain:
$$\vec{E} = \frac{\rho}{3\epsilon_0}r\hat{r} - \frac{\rho}{3\epsilon_0}r'\hat{r}' \quad \cdots (3)$$
The vectors $\vec{r} = r\hat{r}$ and $\vec{r}' = r'\hat{r}'$ given by:
$$\vec{r} = x\hat{i} + y\hat{j} \quad \text{and} \quad \vec{r}' = (x-b)\hat{i} + y\hat{j}$$
where x and y are the coordinates of any point in the cavity.
Substitute for $r\hat{r}$ and $r'\hat{r}'$ in Eq.(3) and simplify to obtain:
$$\vec{E} = \frac{\rho}{3\epsilon_0}(x\hat{i}+y\hat{j}) - \frac{\rho}{3\epsilon_0}[(x-b)\hat{i}+y\hat{j}] = \frac{\rho b}{3\epsilon_0}\hat{i}$$
Because $\vec{E}$ is independent of x and y:
$$\vec{E}_1 = \vec{E}_2 = \frac{\rho b}{3\epsilon_0}\hat{i}$$

48. APPROACH The electric field in the cavity is the sum of the electric field due to the uniform and positive charge distribution of the sphere whose radius is a and the electric field due to any charge in the spherical cavity whose radius is b. You can use the hint given in Problem 47 to express the field at a generic point $P(x, y)$ in the cavity as the sum of the fields due to equal positive and negative charge densities to show that $\vec{E} = \dfrac{\rho}{3\varepsilon_0}b\hat{i}$.

SOLUTION The electric field at point $P(x, y)$ is the sum of the electric fields due to the two charge distributions:
$$\vec{E} = \vec{E}_\rho + \vec{E}_{-\rho} = E_\rho \hat{r} + E_{-\rho}\hat{r}' \quad \cdots (1)$$
where $\hat{r}'$ is a unit vector normal to a spherical Gaussian surface whose centre is at $x = b$.

Apply Gauss's law to a spherical surface of radius r' centred at $x = b = R/2$ to obtain:
$$\oint_S E_n dA = E_{-\rho}\left(4\pi r'^2\right) = \dfrac{Q_{\text{inside}}}{\epsilon_0}$$

Relate Q_{inside} to the charge density $-\rho$: Substitute for Q_{inside} to obtain:
$$-\rho = \dfrac{Q_{\text{inside}}}{\tfrac{4}{3}\pi r'^3} \Rightarrow Q_{\text{inside}} = -\dfrac{4}{3}\pi\rho r'^3$$

Solving for $E_{-\rho}$ yields:
$$E_{-\rho}\left(4\pi r'^2\right) = -\dfrac{4\pi r'^3 \rho}{3\epsilon_0}$$

From Problem 47: $E_{-\rho} = -\dfrac{\rho}{3\epsilon_0}r'$

Substitute for E_ρ and $E_{-\rho}$ in equation (1) to obtain:
$$E_\rho = \dfrac{\rho}{3\epsilon_0}r \quad \cdots (2)$$

The vectors $\vec{r} = r\hat{r}$ and $\vec{r}' = r'\hat{r}'$ are given by: $\vec{r} = x\hat{i} + y\hat{j}$ and $\vec{r}' = (x - b)\hat{i} + \hat{j}$, where x and y are the coordinates of any point in the cavity.

Substitute for $r\hat{r}$ and $r'\hat{r}'$ in equation (2) and simplify to obtain:
$$\vec{E} = \dfrac{\rho}{3\epsilon_0}(x\hat{i} + y\hat{j}) - \dfrac{\rho}{3\epsilon_0}[(x-b)\hat{i} + y\hat{j}] = \dfrac{\rho}{3\epsilon_0}b\hat{i}$$

49. APPROACH The electric field at a point $P(x, y)$ in the cavity is the sum of the fields due to the positive charge density and the total charge Q.

SOLUTION The electric field at point $P(x, y)$ is the sum of the electric fields due to the two charge distributions:
$$\vec{E} = \vec{E}_\rho + \vec{E}_{-\rho} + \vec{E}_Q \quad \cdots (1)$$
where $\hat{r}'$ is a unit vector normal to a spherical Gaussian surface whose center is at $x = b$.

From Problem 47: $\vec{E}_\rho + \vec{E}_{-\rho} = \dfrac{\rho b}{3\varepsilon_0}\hat{i}$

Substituting in equation (1) yields:
$$\vec{E} = \dfrac{\rho b}{3\varepsilon_0}\hat{i} + \vec{E}_Q$$

Assuming that the cavity is filled with positive charge Q:
$$\vec{E} = \dfrac{\rho b}{3\varepsilon_0}\hat{i} + \dfrac{Q}{4\pi\varepsilon_0 b^3}r'\hat{r}'$$

The vectors $\vec{r} = r\hat{r}$ and $\vec{r}' = r'\hat{r}'$ are given by:
$$\vec{r} = x\hat{i} + y\hat{j} \text{ and } \vec{r}' = (x - b)\hat{i} + y\hat{j}$$
where x and y are the coordinates of any point in the cavity.
Substitute for $r\hat{r}$ and $r'\hat{r}'$ and simplify to obtain:
$$\vec{E} = \dfrac{\rho b}{3\varepsilon_0}\hat{i} + \dfrac{Q}{4\pi\varepsilon_0 b^3}[(x-b)\hat{i} + \hat{y}]$$

At point 1, $x = 2b$ and $y = 0$:
$$\vec{E}(2b, 0) = \dfrac{\rho b}{3\varepsilon_0}\hat{i} - \dfrac{Q}{4\pi\epsilon_0 b^3}[(2b-b)\hat{i}] = \left(\dfrac{\rho b}{3\varepsilon_0} + \dfrac{Q}{4\pi\varepsilon_0 b^2}\right)\hat{i}$$

At point 2, $x = 0$ and $y = 0$:
$$\vec{E}(0, 0) = \dfrac{\rho b}{3\varepsilon_0}\hat{i} + \dfrac{Q}{4\pi\varepsilon_0 b^3}[(-b)\hat{i}] = \left(\dfrac{\rho b}{3\varepsilon_0} - \dfrac{Q}{4\pi\varepsilon_0 b^2}\right)\hat{i}$$

50. APPROACH We can find the field due to the infinitely long line charge from $E = 2k\lambda/r$ and the force that acts on the dipole using $F = pdE/dr$.

SOLUTION Express the force acting on the dipole:
$$F = p(dE/dr) \quad \cdots (1)$$
The electric field at the location of the dipole is given by:
$E = 2k\lambda/r$
Substitute for E in equation (1) to obtain:
$$F = p\dfrac{d}{dr}\left[\dfrac{2k\lambda}{r}\right] = -\dfrac{2k\lambda p}{r^2}$$
where the minus sign indicates that the dipole is attracted to the line charge.

1.38.13 Multiple Choice Assignments
1.38.13.1 Level 1

Q.No.	1	2	3	4	5	6	7	8	9
Ans.	D	D	B	A	B	A	D	A	D
Q.No.	10	11	12	13	14	15	16	17	18
Ans.	D	C	A	A	A	A	C	D	A
Q.No.	19	20	21	22	23	24	25	26	27
Ans.	C	A	A	B	B	A	C	A	B
Q.No.	28	29	30	31	32	33	34	35	36
Ans.	D	B	A	A	C	C	B	B	A
Q.No.	37	38	39	40	41	42	43	44	45
Ans.	D	A	C	A	C	B	A	B	C
Q.No.	46	47	48	49	50	51	52	53	54
Ans.	C	B	A	B	B	C	C	D	D
Q.No.	55	56	57	58	59	60	61	62	63
Ans.	C	C	A	A	C	A	C	D	B
Q.No.	64	65	66	67	68	69	70	71	72
Ans.	B	A	B	D	A	A	A	A	D
Q.No.	73	74	75	76	77	78	79	80	81
Ans.	D	C	A	C	D	C	C	B	A

1.38.13.2 Level 2

Q.No.	1	2	3	4	5	6	7	8	9
Ans.	C	D	A	D	B	B	C	B	B
Q.No.	10	11	12	13	14	15	16	17	18
Ans.	A	C	B	D	A	D	A	A	A
Q.No.	19	20	21	22	23	24	25	26	27
Ans.	A	C	A	A	C	D			

1.38.13.3 Level 3

Q.No.	1	2	3	4	5	6	7	8	9
Ans.	B	D	C	A	D	C	C	C	C
Q.No.	10	11	12	13	14	15	16	17	18
Ans.	C	C	D	B	A	B	C	C	D

1.38.13.4 Level 4
Section A

Q.No.	1	2	3	4	5	6	7	8	9
Ans.	A	B	A	D	D	B	A	D	A
Q.No.	10	11	12	13	14	15	16	17	18
Ans.	A	A	D	A	A	C	A	C	B
Q.No.	19	20	21	22	23	24	25	26	27
Ans.	D	C	B	A	B	D	D	A	B
Q.No.	28	29	30	31	32	33	34	35	36
Ans.	A	B	D	D	B	B	B	C	C
Q.No.	37	38	39	40	41	42	43	44	45
Ans.	C	C	D	B	C	B	B	C	B
Q.No.	46	47	48	49	50	51	52	53	54
Ans.	A	−48	D	A	B	C	C	B	A
Q.No.	55	56	57	58	59	60	61	62	63
Ans.	A	D	C	A	B	D	A	A	24
Q.No.	64	65	66	67	68	69	70	71	72
Ans.	C	B	B	12	640	D	2	288	A
Q.No.	73	74	75	76	77	78			
Ans.	A	A	6	B	B	C			

1.38. ANSWER KEYS AND SOLUTIONS

Section B

Q.No.	1	2	3	4	5	6	7	8
Ans.	A	B	D	B	A	C	B	D
Q.No.	9	10	11	12	13	14	15	16
Ans.	C	D	D	A	A	AD	A	D
Q.No.	17	18	19	20	21	22	23	24
Ans.	C	ACD	BD	C	C	A	2	3
Q.No.	25	26	27	28	29	30	31	32
Ans.	6	AB	B	BC	ABC	AC	6.40	3

This page intentionally left blank

Chapter 2
Electric Potential

In the previous chapter, we saw that the electric field vector $\vec{E}$ allows us to understand the exertion of electric force. In this chapter, we will build on the mechanical concepts of work and energy, and define two new scalar quantities: electric potential energy and electric potential, in terms of the work done by the electric field.

One of the advantages of using potential instead of field is that potential is a scalar quantity, while the field is a vector. This makes potential easier to handle in many cases than the electric field. If we know the electric potential of a charge distribution, we can derive the electric field due to that distribution from it. Similarly, potential energy is also a scalar quantity. By knowing the potential energy, we can find the force by taking appropriate derivatives.

2.1 Electric Potential Energy and Potential Difference in Fields

From mechanics, we know that every conservative force field is a potential field, i.e., the term potential energy can be associated with every conservative force field. Like gravitational field, the electrostatic field is also conservative force field, so we can associate the potential energy (U) with it. The electric potential energy U like the gravitational potential energy, is a scalar that depends on both the source and the test object under consideration (for example, test charge). In the case of gravity, the important property of each object is its mass. In the case of electricity, the important property is the charge. Electric potential energy depends on the charge of both the source and the object. Also, like gravity, electric potential energy depends on the configuration of the system.

The motion of a particle with positive charge q_0 in a uniform electric field is analogous to the motion of a particle of mass m in the uniform gravitational field near the earth.

When you move a particle of mass m, against the gravitational field and increase it's height from position A to B (Fig.2.1a), you have to do a positive work W_{ext}. If your external force on the particle, is just equal and opposite to the force due to the gravitational field, then it will move with a constant velocity v (say) and the kinetic energy of the particle will not change. In this case, all the external work (W_{ext}) increases only the gravitational potential energy of the earth-particle system, i.e.,

$$\boxed{W_{ext} = +\Delta U = U_f - U_i \quad (v \rightarrow \text{constant})} \quad (2.1)$$

where U_f and U_i are the final and initial potential energies respectively.

Therefore, when an external force moves a particle slowly (i.e., without acceleration), the work done by the force is equal to the change in potential energy. However, it's important to note that during accelerated motion of the particle, kinetic energy also changes. Consequently, in accelerated motion, the work done by the external force is equal to the change in mechanical energy (the sum of potential and kinetic energy), not just the change in potential energy. Therefore, Equation 2.1 is not valid for accelerated motion.

We can also write the change in potential energy in terms of work done by system's net internal conservative force acting on the par-

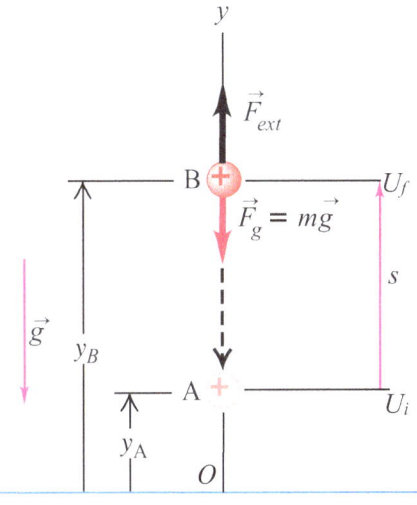

(a) Motion of a point mass in a gravitational field ($\vec{g}$).

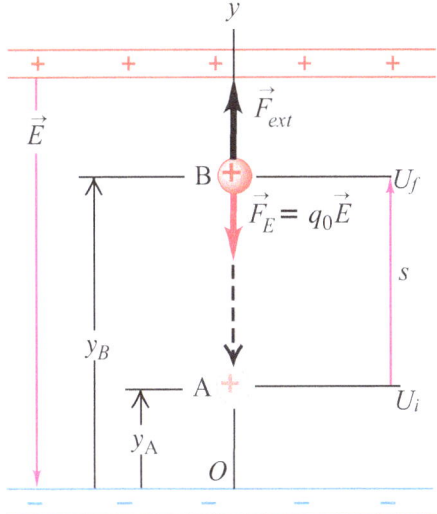

(b) Motion of a test charge in an electric field ($\vec{E}$)

Figure 2.1

ticle. If W_c is the work done by gravitational field (conservative force field), then by work kinetic energy theorem-

$$W_{net} = \underbrace{K_f - K_i}_{\text{Change in KE}}$$

Here, K_i and K_f are respectively the initial and final kinetic energies of the particle.

If W_{ext} and W_c represent work done by non conservative and conservative forces respectively, then net work (W_{net}) can be written as-

$$W_{net} = W_{ext} + W_c$$

and for constant velocity motion, $K_f = K_i$, so-

$$W_{ext} + W_c = 0 \quad \Longrightarrow \quad W_{ext} = -W_c$$

Now, if we put $W_{ext} = -W_c$ in Eq.2.1, we get-

$$\boxed{\Delta U = U_f - U_i = -W_c} \quad (2.2)$$

Equation 2.2 defines the 'potential energy difference' in terms of the physically meaningful quantity 'work' done by a conservative force (here it is the gravitational force): The change in the system's potential energy is equal to the negative of the work done by the net conservative force acting on the particle. Importantly, Eq.2.2 represents the general definition of the change in potential energy and is valid for both accelerated and non-accelerated motion. This is why we typically define the change in potential energy in terms of the work done by the system's net conservative force, rather than in terms of the work done by an external agent (Eq.2.1), which is valid only for non-accelerated, i.e., slow motion of the particle.

Clearly, potential energy of a system decreases if the net work done by all conservative forces of the system is positive, i.e., $W_c = +ve$. In other words, system does positive work at the cost of it's potential energy.

But if the net work done by all conservative forces of the system is negative, i.e., $W_c = -ve$, then potential energy of the system increases because in this case, $\Delta U = -W_c = +ve$.

Potential energy so defined is undetermined to within an additive constant. What this means is that the actual value of potential energy is not physically significant; it is only the difference of potential energy that is significant. We can always add an arbitrary constant α to potential energy at every point, since this will not change the potential energy difference:

$$\Delta U = (U_f + \alpha) - (U_i + \alpha) = U_f - U_i \quad (2.3)$$

Put it differently, there is a freedom in choosing the point where potential energy is zero. A convenient choice is to have potential energy zero at infinity. With this choice, if we take the initial position i at infinity, we get from Eq.2.3-

$$U_f - U_\infty = U_f - 0 = U_f = -W_{\infty f} \quad (2.4)$$

here, $W_{\infty f}$ represents the work done by the conservative force in moving the particle from its initial position at infinity to the given final position.

Now, if we move a point mass m from the surface of earth to height y, the work done by Earth's gravitational force (a conservative force),

$$W_c = (mg)(y)\cos 180° = -mgy$$

Therefore, from Eq.2.3, change in gravitational potential energy, $\Delta U = -W_c = mgy$.

We can obtain a function that does not depend on m by defining the gravitational potential change as the potential energy change per unit mass: $\Delta V = \Delta U/m = gy$. Clearly, ΔV increases with increase in y, i.e., ΔV increases in moving opposite to the gravitational field. The SI unit of ΔV is J/kg.

Thus, gravitational potential difference can be given as-

$$\Delta V = V_f - V_i = \frac{\Delta U}{m} = -\frac{W_c}{m} \quad (2.5)$$

Clearly, ΔV is +ve corresponding to −ve value of W_c. Thus, potential increases if work done by system's internal conservative force is −ve. If the motion of the particle is slow (i.e., non-accelerated), then $W_c = -W_{ext}$, therefore, for slow motion, Eq.2.5 can also be written as

$$\Delta V = V_f - V_i = \frac{\Delta U}{m} = -\frac{W_c}{m} = \frac{W_{ext}}{m} \quad (2.6)$$

If, we take initial position at zero potential level, i.e., $V_i = 0$, potential at final position $V_f = V$ and $m = 1$ kg, then Eq.2.5 gives-

$$\Delta V = V - 0 = -W_c \implies V = -W_c \quad (2.7)$$

If, in the above case, the motion of the particle is slow, then ($W_c = -W_{ext}$). Therefore, for particularly slow motion, the above expression can also be written in terms of external work (W_{ext})

as:

$$\Delta V = V - 0 = W_{ext} \implies V = -W_c = W_{ext} \quad (2.8)$$

From Eqs.2.7 and 2.8, we can define the gravitational potential at a given point in a gravitational field as follows:

The gravitational potential at a point is the negative of the work done by the gravitational field when we lift a unit mass from the zero level of potential to the given position, regardless of whether the motion of the mass is accelerated or non-accelerated. However, in terms of external work, it is the external work needed to lift a unit mass from the zero level of potential to the given position without a change in speed. A useful feature of the potential function is that it depends only on the source of the field (the earth) through the value of the gravitational field strength g, and not on the value of the "test mass", m.

Now, let us discuss the case of electric field instead of gravitational field. Fig.2.1b shows two points A and B in an electric field, and suppose the electric force applied by electric field on a positive test charge q_0 is $\vec{F}_E$ from point B to A. So, if we move this positive test charge infinitely slowly (without any acceleration) from A to B, we do work against the electric field $\vec{E}$.

Clearly, in the case of an electric field, the electric field $\vec{E}$ plays the role of the gravitational field $\vec{g}$, and the test charge q_0 plays the role of the point mass m.

Therefore, in electric field, the Eq.2.2 takes the form: $\Delta U = -W_c$, here W_c is the work done by electric force and similarly, Eq.2.5 takes the form-

$$\boxed{\Delta V = V_f - V_i = \frac{\Delta U}{q_0} = -\frac{W_c}{q_0}} \quad (2.9)$$

Here, ΔV is the electric potential difference between two positions in electric field, V_i is the electric potential at initial position, V_f is the electric potential at final position, W_c is the work done by electric field during this process.

Thus, when the test charge q_0 moves from one position to another in an electric field, the change in electric potential, ΔV, can be defined as the negative of work done by the external agent per unit charge or as the change in electric potential energy per unit charge.

From Eq.2.9, it is evident that potential increases corresponding to −ve value of W_c.

If an external agent moves the test charge q_0 slowly in an electric field from one position to another and performs work W_{ext} on it during this process, then $W_c = -W_{ext}$. In this case Eq.2.9 can be written as-

$$\boxed{\Delta V = V_f - V_i = \frac{\Delta U}{q_0} = \frac{W_{ext}}{q_0} \quad (v \to \text{constant})} \quad (2.10)$$

Here, it should be remembered that Eq.2.10 is valid for slow motion (i.e., for motion without acceleration) only.

From Fig. 2.1b, we observe that W_{ext} and hence ΔU are positive when q_0 moves against the electric field. Therefore, ΔV is also positive (i.e., the potential increases) when moving against the electric field. Conversely, if the test charge q_0 moves in the direction of the electric field, W_{ext} and hence ΔU are negative. Therefore, ΔV is also negative, signifying a decrease in electric potential when moving in the direction of the electric field.

The SI unit of electric potential is the volt (V), in honour of Alessandro Volta, inventor of the voltaic pile (the first primitive electric battery). Note that

$$\boxed{1\,V = 1\,J/C} \quad (2.11)$$

Definition of 1 volt: The potential difference between two points A and B is one volt if the work done by an external agent in moving one coulomb of positive charge slowly from A to B, is

2.1. ELECTRIC POTENTIAL ENERGY AND POTENTIAL DIFFERENCE IN FIELDS

one joule.
The quantity ΔV depends only on the field set up by the source charges, not on the test charge. Once the potential difference between two points is known, the external work needed to move a charge q_0, with no change in its speed, may be found from Eq.2.10 as:

$$\boxed{W_{ext} = \Delta U = q_0 \Delta V = q_0 \left(V_f - V_i \right) \quad (v \to \text{constant})} \quad (2.12)$$

The sign of this work depends on the sign of q_0 and the relative magnitudes of V_i and V_f. If $W_{ext} > 0$ [or $W_c < 0$], the positive test charge (q_0) is moving in the direction of force applied by external agent, i.e., opposite to the electric field.
Also, $W_{ext} > 0$ [or $W_c < 0$] $\implies q_0 (V_f - V_i) > 0$
Clearly, for +ve test charge (q_0), $V_f - V_i > 0 \implies V_f > V_i$.
i.e., corresponding to $W_{ext} > 0$ [or $W_c < 0$], the external force F_{ext} moves the positive test charge from lower to higher potential point.
And corresponding to $-$ve test charge, for example, $q_0 = -q_0'$ (say), where, q_0' is a positive value,

$$W_{ext} > 0 \implies q_0 (V_f - V_i) > 0 \implies -q_0' (V_f - V_i) > 0$$
$$\implies q_0' (V_f - V_i) < 0 \implies V_f - V_i < 0 \implies$$

$V_f < V_i$
i.e., corresponding to $W_{ext} > 0$, the external force F_{ext} moves the negative charge from higher to lower potential point.
If $W_{ext} < 0$ [or $W_c > 0$], the positive test charge q_0 is moving opposite to the force applied by external agent, i.e. the test charge q_0 is moving in the direction of electric field.
Also, $W_{ext} < 0$ [or $W_c > 0$] $\implies q_0 (V_f - V_i) < 0$
Therefore, corresponding to $q_0 > 0$, we have
$$V_f - V_i < 0 \implies V_f < V_i.$$
i.e., corresponding to negative external work or positive electric work, the external force moves the positive test charge from higher to lower potential point.
Corresponding to the negative test charge, as in the previous part, we can show that $V_f > V_i$, meaning the external force moves the negative test charge from a lower to a higher potential point.
From Eq.2.12, we see that only changes in potential, rather than the specific value of V_i and V_f, are significant. One can choose the reference point at which the potential is zero at some convenient point such as infinity. In electronic circuits it is convenient to choose the ground connection to earth as the zero of potential. If $V_i = 0$ and $V_f = V$, then Eq.2.9 gives-

$$V - 0 = -\frac{W_c}{q_0} \implies V = -\frac{W_c}{q_0}$$

> Thus, the electric potential at a point in an electric field is the negative of the work done by the electric field to bring a unit positive test charge from the reference point to that point.
> (Definition of Electric Potential)

If allowed to, positive charges tend to move "downhill" in potential in an electric field, just as do ordinary masses in a gravitational field. However, negative free charges tend to move "uphill" in potential. Alternatively, we can say that in an external electric field, both positive and negative charges move to decrease the electrostatic potential energy of the system.
From Eq.2.2, the potential energy difference, in terms of the work done (W_c) by the conservative force, is-
$$\Delta U = U_f - U_i = -W_c$$
The negative sign tells us that positive work done by the conservative force leads to a decrease in potential energy.
Now, consider a test charge q_0 moving in an external electric field created by a point source charge q. A and B are initial and final positions of q_0. Suppose, at any instant it's position is at

P at distance r from q and the electric field at P is $\vec{E}$. At P, the conservative force (here it is electric force) on the test charge q_0, is $F_c = F_E = q_0 E$.
If dW_E is the work done by electric field in infinitesimal displacement $d\vec{s}$ of the charged particle q_0, then
$$dW_E = \vec{F_e}.d\vec{s} = q_0 \vec{E}.d\vec{s}$$
Therefore, the infinitesimal change in electric potential energy dU, associated with an infinitesimal displacement $d\vec{s}$, is-
$$dU = -dW_E = -q_0 \vec{E} \cdot d\vec{s}$$
So, the infinitesimal change in electric potential in displacement $d\vec{s}$, is -
$$dV = \frac{dU}{q_0} = -\vec{E} \cdot d\vec{s} \quad (2.13)$$

If V_A and V_B are electric potentials at A and B, the potential change from A to B is

$$\boxed{\Delta V = V_B - V_A = -\int_A^B \vec{E} \cdot d\vec{s}} \quad (2.14)$$

If θ is the angle between $\vec{E}$ and $d\vec{s}$, then
$$\vec{E}.d\vec{s} = E ds \cos\theta = (E \cos\theta) ds = E_s ds$$
here, $E_s = E \cos\theta$ is the component of $\vec{E}$ in the direction of $d\vec{s}$.
Substituting this value in Eq.2.13, we get

$$\boxed{\Delta V = V_B - V_A = -\int_A^B \vec{E} \cdot d\vec{s} = -\int_A^B E_s ds} \quad (2.15)$$

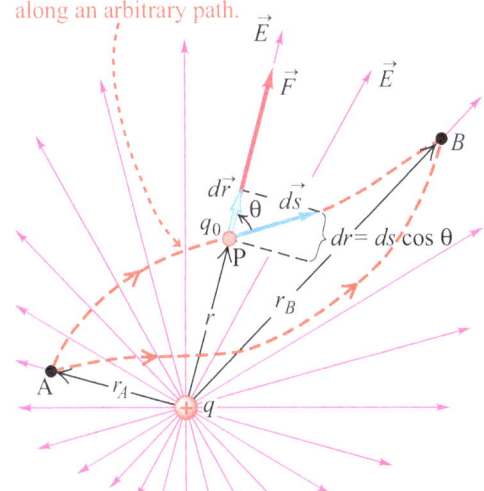

Figure 2.2: The work done on charge q_0 by the electric field of charge q does not depend on the path taken, but only on the distances r_A and r_B.

Thus, only the component of $\vec{E}$ along or opposite to the displacement is significant.
Since, the electrostatic field is a conservative field, the value of line integral $\int_A^B \vec{E} \cdot d\vec{s}$ depends only on the end points A and B, not on the path taken. The sign of the integral is determined (1) by the signs of the components of E, i.e., $E \cos\theta$ and (2) by the direction of the path taken-which is indicated by the limits.
If dr is the magnitude of projection of $d\vec{s}$ along $\vec{r}$, then from Fig. 2.2, we have-
$$ds \cos\theta = dr$$
Therefore, if $\vec{r}_A$ and $\vec{r}_B$ are the position vectors of points A and B

respectively, then Eq.(2.14) can also be written as-

$$\Delta V = V_B - V_A = -\int_A^B E(ds\cos\theta) = -\int_{r_A}^{r_B} E\,dr \quad (2.16)$$

and change in electric potential energy-

$$\Delta U = U_B - U_A = q_0(V_B - V_A) = -q_0\int_{r_A}^{r_B} E\,dr \quad (2.17)$$

From Eq.2.17, we see that if q_0 is positive, then ΔU is negative. It shows that, **a system loses it's electric potential energy when the test charge moves in the direction of the field**. This means that an electric field does work on a positive charge when the charge moves in the direction of the electric field. (This is analogous to the work done by the gravitational field on a falling object, as shown in Figure 2.1a) If a positive test charge is released from rest in this electric field, it experiences an electric force $q_0\vec{E}$ in the direction of $\vec{E}$ (downwards in Fig.2.1b). Therefore, it accelerates downward, gaining kinetic energy. **As the charged particle gains kinetic energy, the charge in electric field loses an equal amount of potential energy**. This is accordance with the conservation of energy in an isolated system.

If q_0 is negative, then ΔU in Eq.(2.17) is positive and the situation is reversed: **A negative charge in an external electric field gains electric potential energy when the charge moves in the direction of the field**. If a negative charge is released from rest in an electric field, it accelerates in a direction opposite the direction of the field. In order for the negative charge to move in the direction of the field, an external agent must apply a force and do positive work on the charge.

☞ Since, ΔV and ΔU are path independent, therefore we can apply equations 2.16 and 2.17 without bothering about actual path of motion of test charge q_0.

Remarks

- From Eq.2.16, we have $\Delta V = V_B - V_A = -\int_{r_A}^{r_B} E\,dr$. Thus, it is evident that the negative area under the E vs r graph on the r-axis corresponds to the potential difference.

- CGS unit of electric potential is esu potential. At any point within the electric field, it is equal to the work done by external agent in bringing 1 esu of charge slowly (without any acceleration) from infinity to that point.

$$1 \text{ esu potential} = \frac{1 \text{ erg}}{1 \text{ esu charge}} = \frac{10^{-7} \text{ J}}{3.335641 \times 10^{-10} \text{C}}$$
$$= 299.79245 \text{ volt} \approx 300 \text{ volt}$$

2.1.1 Potential Difference in a Uniform Field

In a uniform field, $\vec{E}$ is constant, and therefore the integral in Eq.2.14 may be written as

$$\int \vec{E}\cdot d\vec{s} = \vec{E}\cdot\int d\vec{s} = \vec{E}\cdot\Delta\vec{s}$$

The finite change in potential ΔV associated with a finite displacement $\Delta\vec{s}$ takes the form

$$\Delta V = -\vec{E}\cdot\Delta\vec{s} \quad \text{(Uniform } \vec{E}\text{)} \quad (2.18)$$

Note that Δs and ΔV depend only on the initial and final positions, not on the path taken.

Figure 2.3a shows a uniform field $\vec{E} = E\hat{i}$. Let us find the change in potential in going from point A to point B, which are separated by a distance Δx along the lines. Since, from Fig.2.3a, $\vec{E} = E\hat{i}$ and $\Delta\vec{s} = \Delta x\hat{i} + \Delta y\hat{j}$, therefore,

$$\vec{E}\cdot\Delta\vec{s} = E\hat{i}\cdot(\Delta x\hat{i} + \Delta y\hat{j}) = E\Delta x$$

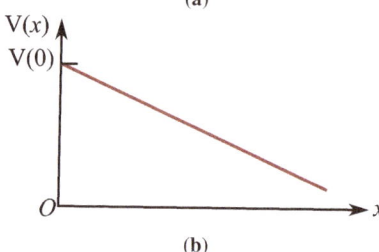

Figure 2.3

Putting this value in Eq.2.18, we get

$$\Delta V = -E\Delta x \quad \text{(Uniform } \vec{E}\text{)} \quad (2.19)$$

If we write $\Delta x = +x$ in Eq.2.19, we get

$$\Delta V = V(x) - V(0) = -Ex \quad (2.20)$$

here, V_x is the potential at B and $V(0)$ is the potential at reference point A.

Thus, the potential decreases linearly along the x axis, as depicted in the graph of Fig. 2.3b. Notice that the field lines point from high potential to low potential.

Again from Fig.2.3a,

$$\Delta\vec{s} = \overrightarrow{AB} = \overrightarrow{AC} + \overrightarrow{CB}$$

Since, $\vec{E}$ is perpendicular to the displacement along $\overrightarrow{BC}$, no work will be done by field $\vec{E}$ on a test charge along this segment. Work is done only along the segment AC parallel to the field lines. Since only the component of the displacement along, or against, the field lines is significant, Eq.2.20 is often written in the form:

$$\Delta V = \pm Ed \quad \text{(Uniform } \vec{E}\text{)} \quad (2.21)$$

where d is the magnitude of the component of the displacement along, or against, the field. The positive sign applies to a displacement opposite to the field. Thus, once again we see that V decreases in the direction of $\vec{E}$ and increases opposite to $\vec{E}$. Therefore, we can say that electric field is always directed from higher to lower potential.

From Eq.2.21 we see that an equivalent unit for electric field is V/m: $1\text{V/m} = 1\text{N/C}$

EXAMPLE 1. A point charge q moves from point P to point S along the path PQRS (Fig.2.5) in a uniform electric field $\vec{E}$ pointing parallel to the positive direction of the x-axis. The coordinates of points P, Q, R and S are $(a, b, 0)$, $(2a, 0, 0)$, $(a, -b, 0)$, $(0, 0, 0)$ respectively. Find the work done by the field in the above process.

APPROACH The work done by the conservative forces (electrostatic, gravitational, etc.) is independent of the path i.e., it depends only on the initial and final points. Thus, the work done by the field along path $P \rightarrow Q \rightarrow R \rightarrow S$ is same as the work done along the path $P \rightarrow T \rightarrow S$ as shown in Fig.2.6.

SOLUTION W_{el}(along PQRS) $= W_{el}$(along PTS)

2.1. ELECTRIC POTENTIAL ENERGY AND POTENTIAL DIFFERENCE IN FIELDS

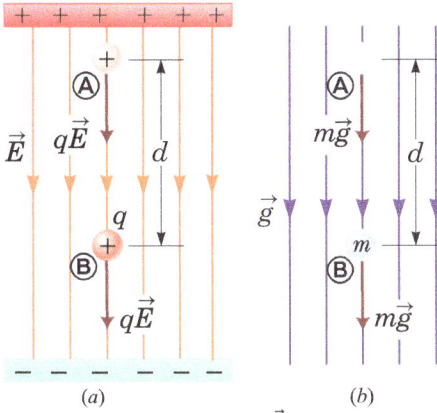

Figure 2.4: (a) When the electric field $\vec{E}$ is directed downward, point B is at a lower electric potential than point A. On a positive charge q electric force $q\vec{E}$ is acting in the direction of electric field $\vec{E}$, i.e., in the direction of lower electric potential. When a positive test charge moves from point A to point B, the charge configuration loses electric potential energy. (b) When an object of mass m moves downward in the direction of the gravitational field $\vec{g}$, the object-source system loses gravitational potential energy.

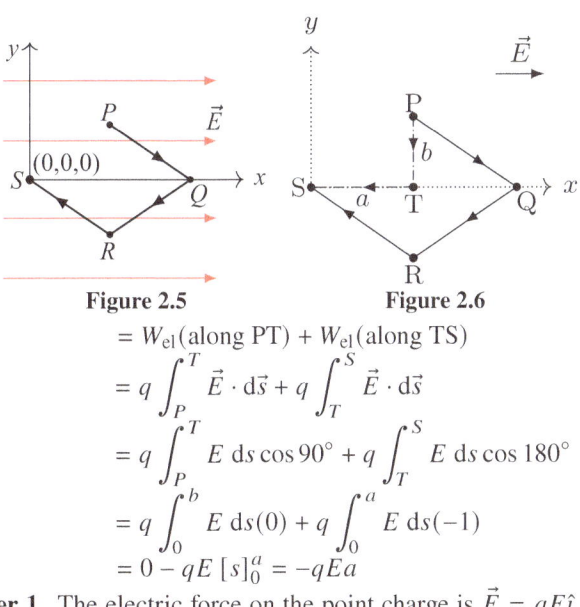

Figure 2.5   **Figure 2.6**

$$= W_{el}(\text{along PT}) + W_{el}(\text{along TS})$$
$$= q\int_P^T \vec{E}\cdot d\vec{s} + q\int_T^S \vec{E}\cdot d\vec{s}$$
$$= q\int_P^T E\,ds\cos 90° + q\int_T^S E\,ds\cos 180°$$
$$= q\int_0^b E\,ds(0) + q\int_0^a E\,ds(-1)$$
$$= 0 - qE\,[s]_0^a = -qEa$$

Aliter 1. The electric force on the point charge is $\vec{F} = qE\hat{i}$ and its displacement is
$$\vec{s} = \vec{r}_S - \vec{r}_P = 0 - (a\hat{i} + b\hat{j}) = -a\hat{i} - b\hat{j}.$$
Thus, $W_{el} = \vec{F}\cdot\vec{s} = (qE\hat{i})\cdot(-a\hat{i} - b\hat{j}) = -qEa$.

Aliter 2. By using the definition of electric potential energy: Work done by electric field is equal to negative of change of electric potential energy. i.e.,
$$W_{el} = -\Delta U = -\left[-q\int_P^S \vec{E}\cdot d\vec{s}\right] = q\int_P^S \vec{E}\cdot d\vec{s}$$
Here, $\vec{E} = E\hat{i}$, and $d\vec{s} = dx\hat{i} + dy\hat{j} + dz\hat{k}$
Therefore, $\vec{E}\cdot d\vec{s} = E\,dx$. So,
$$W_{el} = q\int_a^0 E\,dx = -qEa$$

2.1.2 Electrostatic Potential Energy Difference and Potential Energy of a System of Two Point Charges

Let us consider the field $\vec{E}$ due to a point charge q placed at the origin. Now, imagine that we bring a test charge q_0 from a point A to a point B against the repulsive force on it due to the charge q. With reference to Fig. 2.7, this will happen if q and q_0 are both positive or both negative. For definiteness, let us take $q, q_0 > 0$. Two remarks may be made here. First, we

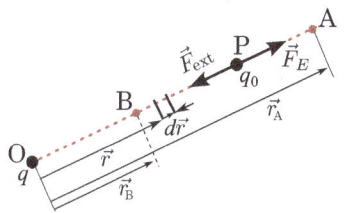

Figure 2.7

assume that the test charge q_0 is so small that it does not disturb the original configuration, namely the charge q at the origin (or else, we keep q fixed at the origin by some unspecified force). Second, in bringing the charge q_0 from A to B, we apply an external force F_{ext} just enough to counter the repulsive electric force F_E (i.e, $F_{ext} = -F_E$). This means there is no net force on or acceleration of charge q_0 when it is brought from A to B, meaning it is brought with infinitesimally slow constant speed. In this situation, the work done by the external force is the negative of the work done by the electric force, and it is fully stored as electrostatic potential energy of the charge q and q_0 system. If the external force is removed upon reaching B, the electric force will take the charge away from q. The stored energy (potential energy) for position B is used to provide kinetic energy to the test charge q_0 in such a way that the sum of the kinetic and potential energies is conserved.

☞ Electric potential energy of a system, containing the source (or sources) of electric field and the test charge q_0 at a given point within the electric field, is the work done by the external force (or −ve of work done by electric force) in bringing the test charge q_0 infinitely slowly[a] from infinity to that point.

[a] In the context of electric potential energy, "infinitely slowly" means moving charges very slowly. This is done to ensure that there is no acceleration, preventing any sudden changes in speed. The reason for this careful movement is to maintain forces in equilibrium, meaning a balance in the forces acting on the charges.

Now, from Eq.2.17, the change in potential energy when test charge q_0 moves from A to B is given as-

$$\Delta U = U_B - U_A = -q_0\int_{r_A}^{r_B} E\,dr \qquad (2.22)$$

Since, the magnitude of electric field at P, due to point charge q at O is given by-
$$E = k\frac{q}{r^2}, \quad \text{with} \quad k = \frac{1}{4\pi\epsilon_0}$$
Therefore, Eq.2.22, gives-
$$\Delta U = -q_0\int_{r_A}^{r_B} k\frac{q}{r^2}dr = -kqq_0\int_{r_A}^{r_B}\frac{dr}{r^2}$$
$$= -kqq_0\left[\frac{-1}{r}\right]_{r_A}^{r_B} = kqq_0\left(\frac{1}{r_B} - \frac{1}{r_A}\right)$$
$$\implies \Delta U = -kqq_0\left[\frac{-1}{r}\right]_{r_A}^{r_B} = kqq_0\left(\frac{1}{r_B} - \frac{1}{r_A}\right) \qquad (2.23)$$

here, $\vec{r}_A$ and $\vec{r}_B$ are the position vectors of points A and B, respectively, with respect to charge q at origin O.

Eq.2.23 shows that, the electric potential energy difference between two points A and B is path-independent and only depends on the positions of A and B. For example, consider two charges

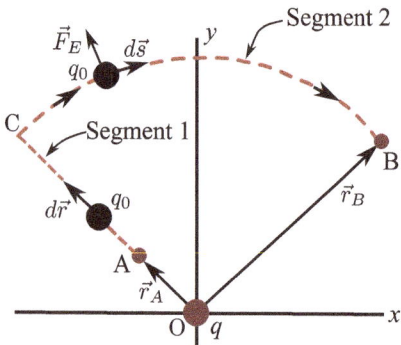

Figure 2.8: The two points A and B are not along the same radius. The path is taken to run radially outward from A to C then along circular arc from C to B. Position vectors of A and B with respect to O are $\vec{r}_A$ and $\vec{r}_B$ respectively.

q and q_0 as shown in Fig. 2.8. The source charge q is fixed at origin O and the test charge is initially at A having position vector $\overrightarrow{OA} = \vec{r}_A$. B is another point with position vector $\overrightarrow{OB} = \vec{r}_B$. The change in potential energy is still given by Eq.2.23, whether q_0 follows the dashed path or moves directly from point A to B.

Explanation: For segment 1 (Fig.2.8), which runs radially outward from A ($OA = r_A$) to C ($OC = r_B$), the result is identical to Eq.2.23. For segment 2, which follows a circumference at a distance r_B from the origin, the integral is zero because the electric force $\vec{F}_E$ is perpendicular to the path segment $d\vec{s}$ everywhere. So, the result for the change in potential energy-

$$\Delta U = \Delta U_{AC} + \Delta U_{CB} = \Delta U_{AC}$$
$$= kqq_0 \left(\frac{1}{r_B} - \frac{1}{r_A} \right)$$

It is the same result as given by Eq.2.23.

Note: Although we have shown the result for circular segment 2, the result holds for paths of any shape like one shown in Fig.2.2. Eq.2.23 shows that the change in electric potential energy is given by the difference of two functions, $U(r_B)$ and $U(r_A)$. We can therefore choose the zero of the potential energy function to be at whatever value of r we like. It is convenient and natural to choose zero potential energy to be at infinity. We can do this if we let $r_A \to \infty$ and let r_B take on a general value r in Eq.2.23:

$$\Delta U = U(r) - U(r_A)|_{r_A \to \infty} = k\frac{qq_0}{r}$$

We then say that the potential energy of a system having charge q_0 at a distance r from charge q is the difference in potential energy between that point and infinity. When we reverse the roles of q and q_0, the potential energy of q at a distance r from q_0 is again kqq_0/r. We can then say that the electric potential energy $U(r)$ for a system of two point charges q and q_0 separated by a distance r is-

$$\boxed{U(r) = k\frac{qq_0}{r}} \qquad (2.24)$$

It is indeed true that $U(r) = 0$ in the limit $r \to \infty$. Thus, the system has no potential energy when the two charges are infinitely far apart. Note that the potential energy of the two charges depends only on the distance r between them and on the magnitudes and signs of the charges.

Remarks

1. From Eq.2.24, we see that the electrical potential energy of system of two point charges is positive if the two charges are of the same type, either positive or negative, and negative if the two charges are of opposite types. This makes sense if you think of the change in the potential energy U as you bring the two charges closer or move them farther apart. Depending on the relative types of charges, you may have to work on the system or the system would do work on you, that is, your work is either positive or negative. If you have to do positive work on the system (actually push the charges closer), then the energy of the system should increase. If you bring two positive charges or two negative charges closer, you have to do positive work on the system, which raises their potential energy. Since potential energy is proportional to $1/r$, the potential energy goes up when r goes down between two positive or two negative charges.

   On the other hand, if you bring a positive and a negative charge nearer, you have to do negative work on the system (the charges are pulling you), which means that you take energy away from the system. This reduces the potential energy of the system. Since potential energy is negative in the case of a positive and a negative charge pair, the decrease in r (or increase in $1/r$) makes the potential energy more negative, which is the same as a reduction in potential energy.

2. Suppose, an electron with charge $q = e = 1.6 \times 10^{-19}$C is released in an electric field $\vec{E}$. This electric field applies an electric force $q\vec{E}$ i.e., $-e\vec{E}$ on it. Here, negative sign shows that electric force on electron is opposite to field $\vec{E}$. As a result, electron get accelerated opposite to $\vec{E}$ i.e., from lower potential point A (say) to higher potential point B (say). If V_A and V_B are potentials at A and B respectively and K_A and K_B are kinetic energies of electron at A and B respectively, then by work kinetic energy theorem, we have

   Work done by electric field = change in KE
   $$\Rightarrow \quad W_{el} = K_B - K_A = K_B - 0 = K_B$$

   Since, the work done by a conservative force field $\vec{E}$ is equal to negative of change of potential energy, therefore
   $$-\Delta U = K_B \quad \Longrightarrow \quad -(-e)(V_B - V_A) = K_B$$
   If we write $V_B - V_A = \Delta V$, then
   $$e\Delta V = K_B \quad \Longrightarrow \quad K_B = e\Delta V = 1.6 \times 10^{-19} \Delta V$$
   If we set $\Delta V = 1$ volt, then kinetic energy gained by electron
   $$K_B = e\Delta V = 1.6 \times 10^{-19} \text{J}$$
   This unit of energy is defined as 1 electron volt or 1 eV, i.e., 1 eV = 1.6×10^{-19}J.

   The units based on eV are most commonly used in atomic, nuclear and particle physics. 1 keV = 10^3eV = 1.6×10^{-16} J,
   1 MeV = 10^6eV = 1.6×10^{-13} J,
   1 GeV = 10^9eV = 1.6×10^{-10} J and
   1 TeV = 10^{12}eV = 1.6×10^{-7} J

EXAMPLE 2. $\vec{E} = (100 \text{ V/m})\hat{i} - (50 \text{ V/m})\hat{j}$. Calculate potential difference between (0,0) and (3,4).

APPROACH From Eq.2.15, electric potential difference, between two points A and B, in terms of electric field, is given by-

$$\Delta V = V_B - V_A = -\int_A^B \vec{E} \cdot d\vec{s}$$

with $d\vec{s} = dx\hat{i} + dy\hat{j} + dz\hat{k}$.
Now, substitute the given values and integrate for given limits.

SOLUTION Given that $\vec{E} = (100\text{V/m})\hat{i} - (50\text{V/m})\hat{j}$, therefore-

$$\Delta V = -\int_{(0,0)}^{(3,4)} \vec{E} \cdot d\vec{s}$$
$$= -\int_{(0,0)}^{(3,4)} (100\hat{i} - 50\hat{j}) \cdot (dx\hat{i} + dy\hat{j} + dz\hat{k})$$
$$= -\int_0^3 100 dx + \int_0^4 50 dy = -100[x]_0^3 + 50[y]_0^4$$
$$= -100[3 - 0] + 50[4 - 0] = -100 \text{ V}$$

EXAMPLE 3. $\vec{E} = (10x\hat{i} - 30y^2\hat{j})$. Calculate potential difference between $(0,0)$ and $(3,4)$.

APPROACH From Eq.2.15, electric potential difference, between two points A and B, in terms of electric field, is given by-
$$\Delta V = V_B - V_A = -\int_A^B \vec{E}.d\vec{s}$$
with $d\vec{s} = dx\hat{i} + dy\hat{j} + dz\hat{k}$.
Now, substitute the given values and integrate for given limits.

SOLUTION Given that $\vec{E} = 10x\hat{i} - 30y^2\hat{j}$, therefore-
$$\Delta V = -\int_{(0,0)}^{(3,4)} \vec{E}.d\vec{s}$$
$$= -\int_{(0,0)}^{(3,4)} (10x\hat{i} - 30y^2\hat{j}).(dx\hat{i} + dy\hat{j} + dz\hat{k})$$
$$= -\int_0^3 10x\,dx + \int_0^4 30y^2\,dy$$
$$= -10\left[\frac{x^2}{2}\right]_0^3 + 30\left[\frac{y^3}{3}\right]_0^4 = -10\left[\frac{3^2}{2}\right] + 30\left[\frac{4^3}{3}\right]$$
$$= -10\left[\frac{9}{2}\right] + 30\left[\frac{64}{3}\right] = -45 + 640 = 595\text{ V}$$

EXAMPLE 4. How much work is done by the electrical force when a point charge is brought from infinity to rest at a distance r from a fixed charge of the opposite sign? What is the meaning of the sign of your result?

SOLUTION Since, the charges are of opposite nature, therefore, the electric force between them is attractive and hence it is in the direction of displacement. So, if we bring a point charge from infinity to rest at a distance r from a fixed charge of the opposite sign with zero acceleration, the work done by electric force is positive.

Since, change in electric potential energy (ΔU) of the system is defined as the negative of work done by the electric force, so, it will be negative and hence system losses it's electric potential energy by doing positive work.

Calculation of Work done by Electric Force
Work done by electric force $W_{el} = -\Delta U$
$$\Rightarrow\quad W_{el} = -(U_r - U_\infty) = -\left(k\frac{qq_0}{r} - 0\right) = -k\frac{qq_0}{r}$$
If, fixed charge $q = -Q$, and test charge is $q_0 = +Q_0$, then
$$\Rightarrow\quad W_{el} = k\frac{QQ_0}{r}$$

EXAMPLE 5. A point charge 2×10^{-2} C is moved from P to S in a uniform electric field of 30 NC^{-1} directed along positive x-axis as shown in Fig.2.9. If coordinates of P and S are $(1\text{ m}, 2\text{ m}, 0\text{ m})$ and $(0\text{ m}, 0\text{ m}, 0\text{ m})$ respectively, find the work done by electric field during this displacement.

```
                          • $\vec{E}=30\,\hat{i}$
       •S(0, 0, 0)    P(1, 2, 0)
              Figure 2.9
```

APPROACH Work done by electric force in displacement $\overrightarrow{PS}$ can be calculated by using work formula-
$$W_{el} = \vec{F}_{el}\cdot\overrightarrow{PS} = q\vec{E}\cdot\overrightarrow{PS}$$

SOLUTION The given situation is shown in Fig.2.9. The displacement,
$$\overrightarrow{PS} = \overrightarrow{OS} - \overrightarrow{OP} = \left(0\hat{i} + 0\hat{j} + 0\hat{k}\right) - \left(\hat{i} + 2\hat{j} + 0\hat{k}\right) = -\hat{i} - 2\hat{j}$$
Given electric field vector, $\vec{E} = 30\text{ NC}^{-1}\hat{i}$
Therefore, the work done by above electric field in displacing the charge 2×10^{-2} C by displacement $\overrightarrow{PS}$, is
$$W_E = q\vec{E}\cdot\overrightarrow{PS} = 2\times10^{-2}[30\hat{i}\cdot(-\hat{i}-2\hat{j})] = 2\times10^{-2}(-30)$$
$$= -0.60\text{ J} = -600\text{ mJ}$$

☞ We encourage students to solve this problem by calculating the potential energy difference too.

2.2 Check Point 1

1. ••From the electric field $\vec{E}(r) = \left(kQ_s/r^2\right)\hat{r}$ produced by a positively charged particle, find the (usual) expression for its electric potential.

2. ••At a point (x, y, z), the electric field is given as-
$$\vec{E} = K\left[y^2\hat{i} + \left(2xy + z^2\right)\hat{j} + 2yz\hat{k}\right]$$
Here K is a constant with the appropriate units. Find the potential, using the origin as your reference point.

3. ••In some region of space, the electric field is given by $\vec{E} = Ax\hat{i} + By^2\hat{j}$. Find the electric potential difference between points whose positions are $(x_i, y_i) = (a, 0)$ and $(x_f, y_f) = (0, b)$. The constants A, B, a, and b have the appropriate SI units.

Multiple Choice Questions

4. ••The electric field in a region is given by $\vec{E} = (Ax + B)\hat{i}$, where E is in NC^{-1} and x is in metres. The values of constants are $A = 20$ SI unit and $B = 10$ SI unit. If the potential at $x = 1$ is V_1 and that at $x = -5$ is V_2, then $V_1 - V_2$ is:
   (A) 320 V   (B) –48 V   (C) 180 V   (D) –520 V

5. ••On moving a charge of 20 C slowly by 2 cm, 2 J of external work is done, then the potential difference between the points is-
   (A) 0.1 V   (B) 8 V   (C) 2 V   (D) 0.5 V

2.3 Electric Potential Energy of a System of Point Charges: Self Electric Energy

Suppose, the electric field $\vec{E}$ in which charge q_0 moves is caused by several point charges $q_1, q_2, q_3, \ldots$ at distances $r_1, r_2, r_3, \ldots$ from q_0, as shown in Fig.2.10. The total electric field at each point is the vector sum of the fields due to the individual charges, and the total work done on q_0 during any displacement is the sum of the contributions from the individual charges. Therefore, from Eq.2.24, we conclude that the potential energy associated with the test charge q_0 at point A in Fig. 2.10 is the algebraic sum:
$$U = \frac{q_0}{4\pi\epsilon_0}\left(\frac{q_1}{r_1} + \frac{q_2}{r_2} + \frac{q_3}{r_3} + \cdots\right) = \frac{q_0}{4\pi\epsilon_0}\sum_i \frac{q_i}{r_i} \quad (2.25)$$

The total electric potential energy of the system of point charges

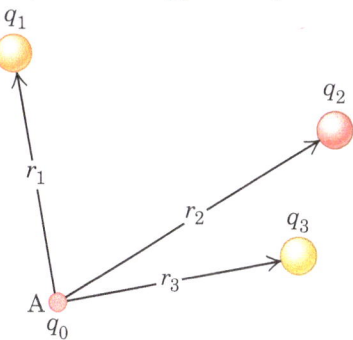

Figure 2.10: The potential energy associated with a charge q_0 at point A depends on the other charges q_1, q_2, and q_3 and on their distances r_1, r_2, and r_3 from point A.

is also referred to as the electric self-energy of the system.

In terms of the net electric field work, the net electric potential energy of the system of point charges, i.e., the self-energy of the system of charges, is defined as the negative of the net work done by the electric field in assembling the point charges from infinite separation to the given configuration.

We can also define self-energy in terms of the work done by an external agent: "The electric potential energy or self-energy of a system of point charges is the work done by an external force against the system's net electric force in assembling the charges from infinite separation to the present configuration without any change in the kinetic energy of any particle."

When q_0 is at a different point B (say), the potential energy is given by the same expression, but $r_1, r_2, \ldots$ are the respective distances from $q_1, q_2, \ldots$ to point B. The work done by electric field on test charge q_0, when it moves from A to B along any path is equal to the difference $(U_A - U_B)$ between the potential energies when q_0 is at A and at B respectively.

We can represent any charge distribution as a collection of point charges, so Eq.2.24 shows that we can always find a potential-energy function for any static electric field. It follows that for every electric field due to a static charge distribution, the force exerted by that field is conservative.

Equations 2.24 and 2.25 define U to be zero when distances $r_1, r_2, \ldots$ are infinite that is, when the test charge q_0 is very far away from all the charges that produce the field. As with any potential-energy function, the point where $U = 0$ is arbitrary. In electrostatics problems it's usually simplest to choose this point to be at infinity. When we analyse electric circuits, other choices will be more convenient.

Equation 2.25 gives the potential energy associated with the presence of the test charge q_0 in the $\vec{E}$ field produced by $q_1, q_2, q_3, \ldots$ But there is also potential energy involved in assembling these charges. If we start with charges $q_1, q_2, q_3, \ldots$ all separated from each other by infinite distances and then bring them together so that the distance between q_i and q_j is r_{ij}, the total potential energy U is the sum of the potential energies of interaction for each pair of charges. We can write this as

$$\boxed{U = \frac{1}{4\pi\epsilon_0} \sum_{i<j} \frac{q_i q_j}{r_{ij}}} \qquad (2.26)$$

This sum extends over all pairs of charges; we don't let $i = j$ (because that would be an interaction of a charge with itself), and we include only terms with $i < j$ to make sure that we count each pair only once. Thus, to account for the interaction between q_3 and q_4, we include a term with $i = 3$ and $j = 4$ but not a term with $i = 4$ and $j = 3$

Methods for Calculating U: We can find the electric potential energy of system of particles by any of the following three methods-

Method 1: You can directly apply Eq.2.26 to the given configuration.

Method 2: Initially, keep all the charges at ∞ separation from each other and then bring them one by one in present configuration and calculate the work done.
$$PE_{\text{sys}} = \Sigma W_i$$

Method 3: Find potential energies of each charge due to electric field of the other charges and then apply following Eq.2.27 to get the potential energy of the system of point charges-

$$PE_{\text{sys}} = \frac{PE_1 + PE_2 + PE_3 + \ldots}{2} \qquad (2.27)$$

where $PE_1 = PE_{12} + PE_{13} + \ldots$,
$PE_2 = P_{21} + PE_{23} + \ldots$, etc.

In the numerator of RHS of Eq.2.27, each pair of charge comes twice, that is why we divided it by 2. Eq.(2.27) is useful for symmetric charge arrangements.

EXAMPLE 6. Find the potential energy of the system shown in Fig.2.11. Each edge is of length a.

APPROACH I In this approach, we find the work done by an

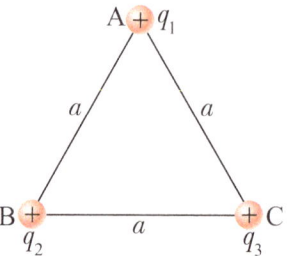

Figure 2.11

external agent in bringing each charge slowly, one by one, from infinity to shown positions. The work is done by the external agent against the electric field established by charges already reached to their position. This work is equal to change in potential energy of the system. Since, initially the charge separation is infinite, therefore, initial potential energy of the system of point charges is zero. If final potential energy is U, then
$$\Delta U = W_{\text{ext}} \implies U_f - U_i = W_{\text{ext}}$$
$$\implies U - 0 = W_{\text{ext}} \implies U = W_{\text{ext}}$$

SOLUTION Work done in bringing first charge q_1 from infinity to position A-
$$W_1 = 0$$
(in this case there was no electric field)

Now, this charge establishes an electric field in it's surroundings. So, work done in bringing second charge q_2 from infinity to position B against the electric field established by the charge q_1-
$$W_2 = \frac{kq_1q_2}{a}$$

Now, at position C, both q_1 and q_2 produce their electric fields. Therefore, in bringing, a third charge q_3 from infinity to position C, we have to do work against the electric field established by q_1 and q_2. This work done is given by-
$$W_3 = \frac{kq_1q_3}{a} + \frac{kq_2q_3}{a}$$

Net work done in forming the system ABC is stored in the form of the electric potential energy of the system. Some times we call it the self energy of the system.

So, self energy of the system ABC,
$$U = (W_{\text{ext}})_{\text{net}} = W_1 + W_2 + W_3$$
$$= 0 + \frac{kq_1q_2}{a} + \frac{kq_1q_3}{a} + \frac{kq_2q_3}{a}$$
$$\implies U = \frac{kq_1q_2}{a} + \frac{kq_1q_3}{a} + \frac{kq_2q_3}{a} \qquad (2.28)$$

So, we can say that the net potential energy of the system is equal to the sum of potential energy of each pair of charges.

APPROACH II There are $^3C_2 = 3$ pairs of charges. Find electrostatic potential energy for each pair by using Eq.2.24. Their sum will be the net electrostatic potential energy of the system.

SOLUTION Net electric potential energy of the given arrangement is-
$$U = U_{AB} + U_{AC} + U_{BC} = \frac{kq_1q_2}{a} + \frac{kq_1q_3}{a} + \frac{kq_2q_3}{a}$$
Here $k = 1/4\pi\epsilon_0$

☞ Student are advised to use approach II as a time saving trick in similar potential energy problems.

EXAMPLE 7. Determine the interaction energy of the system of point charges shown in Fig.2.12.

APPROACH There are $^4C_2 = 6$ pairs of charges. Find electrostatic potential energy for each pair by using Eq.2.24. Their sum

2.3. ELECTRIC POTENTIAL ENERGY OF A SYSTEM OF POINT CHARGES: SELF ELECTRIC ENERGY

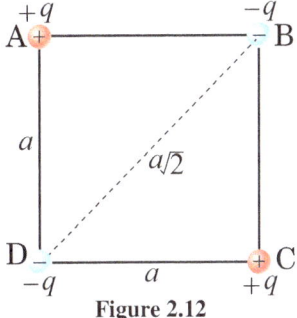

Figure 2.12

will be the net electric potential energy of the system.
SOLUTION Net electric potential energy of the arrangement is given by-

$$U = U_{AB} + U_{AC} + U_{AD} + U_{BC} + U_{BD} + U_{CD}$$
$$= -\frac{kq^2}{a} + \frac{kq^2}{(a\sqrt{2})} - \frac{kq^2}{a} - \frac{kq^2}{a} + \frac{kq^2}{(a\sqrt{2})} - \frac{kq^2}{a}$$
$$= -\frac{4kq^2}{a} + \frac{2kq^2}{a\sqrt{2}} = -\frac{kq^2}{a}\left(4 - \sqrt{2}\right)$$

EXAMPLE 8. Eight point charges are placed at the corners of a cube of edge a as shown in the Figure 2.13. Find the work done in disassembling this system of charges.

APPROACH If U_i is the initial electric potential energy of the

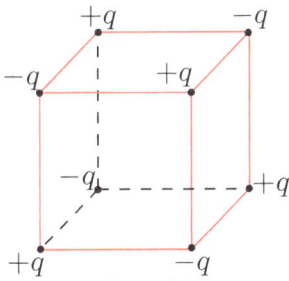

Figure 2.13

system and U_f is the final electric potential energy of the system, then work done by external agent in separating the charges slowly, is defined as-

$$W = \text{change in potential energy} = U_f - U_i \quad (1)$$

Disassembling the system means putting the charges at infinite separation. So, final electrostatic potential energy of the system $U_f = U_\infty = 0$
Now, calculate the initial potential energy (U_i) of the system by applying Eq.2.25 and substitute these values in above Eq.(1) and then solve for W.
SOLUTION In the given system, there are $^8C_2 = 28$ pairs of charges. The charge pairs on cube edges (twelve edges of length a each) have unlike charges. The charge pairs on face-diagonals (six faces, two diagonals per face, twelve face-diagonal of length $\sqrt{2}a$ each) have like charges.
The charge pairs on main-diagonals (four main-diagonal each with length $\sqrt{3}a$) have unlike charges. Thus, the initial potential energy of the system is

$$U_i = \frac{1}{4\pi\epsilon_0}\left[-12\frac{q^2}{a} + 12\frac{q^2}{\sqrt{2}a} - 4\frac{q^2}{\sqrt{3}a}\right] = -\frac{5.824}{4\pi\epsilon_0}\frac{q^2}{a}$$

Substituting these values of U_i and U_f in Eq.(1), we get-

$$W = 0 - U_i = \frac{5.824}{4\pi\epsilon_0}\frac{q^2}{a}$$

EXAMPLE 9. Three point charges q, $2q$ and $8q$ are to be placed on a 9 cm long straight line. Find the positions where the charges should be placed such that the potential energy of this system is minimum. In this situation, what is the electric field at the position of the charge q due to the other two charges?
APPROACH To obtain the minimum potential energy of the system, the charges, of larger magnitude, should be placed at the extreme positions. If the charge q is at distance x from $2q$, then corresponding to minimum potential energy, $dU/dx = 0$ and $d^2U/dx^2 = -ve$. By applying this concept, determine the value of x and then calculate the value of electric field at the position of q.
SOLUTION In Fig.2.14, the total potential energy of the given system is

Figure 2.14

$$U = \frac{1}{4\pi\epsilon_0}\left[\frac{(2q)(8q)}{9} + \frac{(2q)(q)}{x} + \frac{(q)(8q)}{9-x}\right]$$
$$= \frac{q^2}{2\pi\epsilon_0}\left[\frac{8}{9} + \frac{1}{x} + \frac{4}{9-x}\right]$$

The value of x for which potential energy is minimum is given by $dU/dx = 0$ i.e.,

$$\frac{dU}{dx} = \frac{q^2}{2\pi\epsilon_0}\left[-\frac{1}{x^2} + \frac{4}{(9-x)^2}\right] = 0$$

Simplifying, we get: $x = 3$, $x = -9$.
Now, $\dfrac{d^2U}{dx^2} = \dfrac{q^2}{2\pi\epsilon_0}\left[\dfrac{2}{x^3} - \dfrac{8}{(9-x)^3}\right]$
Corresponding to $x = 3$

$$\frac{d^2U}{dx^2} = \frac{q^2}{2\pi\epsilon_0}\left[\frac{2}{3^3} - \frac{8}{6^3}\right] = +ve$$

Since, corresponding to $x = 3$, $\frac{d^2U}{dx^2}$ is positive, therefore, potential energy is minimum at $x = 3$.
Now, corresponding to $x = -9$,

$$\frac{d^2U}{dx^2} = \frac{q^2}{2\pi\epsilon_0}\left[\frac{2}{(-9)^3} - \frac{8}{(18)^3}\right] = -ve$$

Since, corresponding to $x = -9$, $\frac{d^2U}{dx^2}$ is negative, therefore potential energy is maximum for $x = -9$.
Now, the electric field at the position of q due to the other two charges is

$$\vec{E} = \frac{1}{4\pi\epsilon_0}\left[\frac{2q}{x^2} - \frac{8q}{(9-x)^2}\right]\hat{\imath} = \frac{q}{4\pi\epsilon_0}\left[\frac{2}{(3)^2} - \frac{8}{(9-3)^2}\right]\hat{\imath} = \vec{0}$$

Note: Since, the force on the charge q is given by $\vec{F} = q\vec{E} = -dU/dx\,\hat{\imath}$. So, corresponding to minimum potential energy of the system, the charge q will be in the state of a stable equilibrium.
EXAMPLE 10. Two fixed, equal, positive charges, each of magnitude $q = 5 \times 10^{-5}$C are located at points A and B separated by a distance 6 m. An equal and opposite charge moves towards them along the line COD, the perpendicular bisector of the line AB. The moving charge, when reaches the point C at a distance of 4 m from O, has a kinetic energy of 4 J. Calculate the distance of the farthest point D which the negative charge will reach before returning towards C.
APPROACH At the farthest point, negative charge changes its direction of motion, so, the velocity of negative charge at this point will be zero. Since, there are only electrostatic forces on the charged particle, therefore, by applying the law of conservation of mechanical energy, we can find the position of farthest point.
SOLUTION If D is the farthest point at a distance r_{OD} from O, then the velocity of negative charge at D is $v_d = 0$. From Fig.2.16, we have-

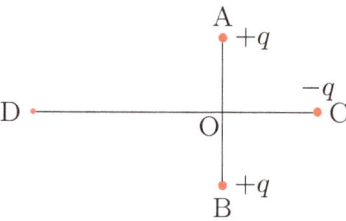

Figure 2.15

$r_{AC} = r_{BC} = \sqrt{3^2 + 4^2} = 5$ m

The electrostatic potential energies of the system for negative

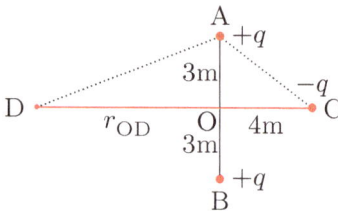

Figure 2.16

charge placed at C and D-

$$U_C = -\frac{q^2}{4\pi\epsilon_0}\left[\frac{1}{r_{AC}} + \frac{1}{r_{BC}}\right]$$
$$= -\frac{2\left(9 \times 10^9\right)\left(5 \times 10^{-5}\right)^2}{5} = -9 \text{ J}$$
$$U_D = -\frac{q^2}{4\pi\epsilon_0}\left[\frac{1}{r_{AD}} + \frac{1}{r_{BD}}\right] = -\frac{45}{r_{AD}}\text{ J}$$

The kinetic energy of the negative charge at C is $K_C = 4$ J and at D is $K_D = \frac{1}{2}mv_d^2 = 0$.

By conservation of mechanical energy, we have-
$$K_C + U_C = K_D + U_D$$

On substituting the values of K_C, U_C, K_D and U_D in above expression and simplifying for r_{AD}, we get
$$r_{AD} = 45/5 = 9 \text{ m}$$

Now, applying Pythagoras theorem in triangle AOD, we get-
$$r_{OD} = \sqrt{r_{AD}^2 - r_{AO}^2} = \sqrt{9^2 - 3^2} = \sqrt{72} = 8.48 \text{ m}$$

2.3.1 Potential Energy of a System of Two Charges in an External Field

In the previous section, we studied the potential energy of a system of point charges due to their own interactions. In this section, we will calculate the potential energy of a system of two charges q_1 and q_2 located at r_1 and r_2, respectively, in an external field. First, we calculate the work done by an external agent in bringing the charge q_1 slowly from infinity to r_1. The work done in this step is $q_1V(r_1)$. Next, we consider the work done by the external agent in bringing q_2 slowly to r_2. In this step, work is done not only against the external field $\vec{E}$ but also against the field due to q_1.

Work done on q_2 against the external field
$$= q_2V(r_2) \tag{2.29}$$

Work done on q_2 against the field due to q_1
$$= \frac{q_1q_2}{4\pi\epsilon_0 r_{12}} \tag{2.30}$$

where r_{12} is the distance between q_1 and q_2. By the superposition principle for fields, we add up the work done on q_2 against the two fields (E and that due to q_1):

Work done in bringing q_2 to r_2
$$= q_2V(r_2) + \frac{q_1q_2}{4\pi\epsilon_0 r_{12}} \tag{2.31}$$

Thus, Potential energy of the system = the total work done in assembling the configuration
$$= q_1V(r_1) + q_2V(r_2) + \frac{q_1q_2}{4\pi\epsilon_0 r_{12}} \tag{2.32}$$

EXAMPLE 11. (a) Determine the electrostatic potential energy of a system consisting of two charges 7μC and -2μC (and with no external field) placed at $(-9$ cm, 0, 0) and $(9$ cm, 0, 0) respectively. (b) How much work is required to separate the two charges infinitely away from each other? (c) Suppose that the same system of charges is now placed in an external electric field $E = A\left(1/r^2\right); A = 9 \times 10^5 \text{Cm}^{-2}$. What would the electrostatic energy of the configuration be?

SOLUTION The given charges are $q_1 = 7 \times 10^{-6}$ C and $q_2 = -2 \times 10^{-6}$ C, located at $(-9$ cm, 0, 0) and $(9$ cm, 0, 0 cm) respectively. Thus, the distance between them is
$$r = 0.18 \text{ m}.$$

(a) Electrostatic potential energy of the two-charge system:
The potential energy of a pair of point charges is given by
$$U = k\frac{q_1q_2}{r}.$$

Substituting values:
$$U = \left(9 \times 10^9\right) \cdot \frac{(7 \times 10^{-6})(-2 \times 10^{-6})}{0.18}$$
$$= \left(9 \times 10^9\right) \cdot (-7.778 \times 10^{-11}) = -0.7 \text{ J}$$

(b) Work required to separate the charges to infinity:
At infinite separation, the potential energy becomes zero. Hence, the external work required to move the charges, without acceleration, from separation r to infinity is
$$W = U_\infty - U_{\text{initial}} = 0 - (-0.7) = 0.7 \text{ J}.$$

(c) Electrostatic energy in presence of external electric field:
In this case, the mutual interaction energy of the two charges remains unchanged. In addition, there is the energy of interaction of the two charges with the external electric field.

The external field is given by $E(r) = \dfrac{A}{r^2}$, which corresponds to a potential
$$V(r) = -\int_\infty^r E(r)\,dr = \frac{A}{r}.$$

Let r_1 is the distance of $(9$ cm, 0, 0 cm) and r_2 is that of $(-9$ cm, 0, 0) from origin, then
$$r_1 = \sqrt{(-0.09 - 0)^2 + (0-0)^2 + (0-0)^2} = 0.09 \text{ m}$$
and $r_2 = \sqrt{(0.09 - 0)^2 + (0-0)^2 + (0-0)^2} = 0.09 \text{ m}$

Therefore, interaction energy with external electric field:
$$q_1V(\vec{r}_1) + q_2V(\vec{r}_2) = A\frac{7\mu C}{0.09 \text{ m}} + A\frac{-2\mu C}{0.09 \text{ m}}$$
$$= 70 \text{ J} - 20 \text{ J} = 50 \text{ J}$$

Since, the interaction energy due to interaction between charges remains the same, therefore:
$$U = -0.7 \text{ J}.$$

Therefore, from (2.32), the net electrostatic energy of the system of charges in external electric field is:
$$q_1V(\vec{r}_1) + q_2V(\vec{r}_2) + \frac{q_1q_2}{4\pi\epsilon_0 r_{12}} = 50 \text{ J} - 0.7 \text{ J} = 49.3 \text{ J}$$

2.3.2 Potential Energy of a Dipole in a Uniform Electric Field

Let us consider an electric dipole in a uniform external electric field as shown in Fig.2.17.

If τ is the magnitude of the torque applied by an external agent against the torque of the electric field for the shown position of the dipole, as depicted in Fig.2.17, then work done by external

2.3. ELECTRIC POTENTIAL ENERGY OF A SYSTEM OF POINT CHARGES: SELF ELECTRIC ENERGY

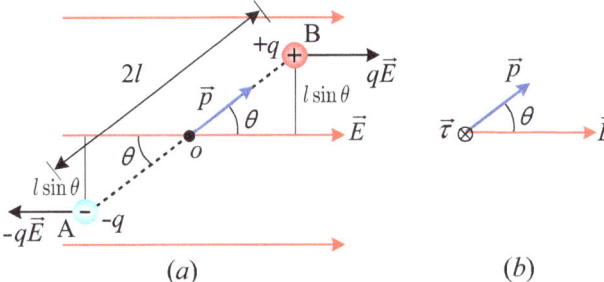

Figure 2.17: (a) An electric dipole has an electric dipole moment $\vec{p}$ in an external uniform electric field $\vec{E}$. The angle between $\vec{p}$ and $\vec{E}$ is θ. The line connecting the two charges represents their rigid connection and their centre of mass is assumed to be midway between them. (b) Representing the electric dipole moment by a vector $\vec{p}$ in the external electric field $\vec{E}$ and showing the direction of the torque $\vec{\tau}$ into the page by the symbol $\otimes$

force in rotating electric dipole anti-clock wise by small angle $d\theta$ against electric force

$$dW = \tau \, d\theta$$

Since, $\tau = pE \sin\theta$, therefore-

$$W_{\text{ext}} = \int_{\theta_1}^{\theta_2} \tau \, d\theta = \int_{\theta_1}^{\theta_2} pE \sin\theta \, d\theta$$

or
$$\boxed{W_{\text{ext}} = -pE\left(\cos\theta_2 - \cos\theta_1\right)} \quad (2.33)$$

In this case the work done by electric field

$$\boxed{W_{\text{el}} = -W_{\text{ext}} = pE\left[\cos\theta_2 - \cos\theta_1\right]} \quad (2.34)$$

If $\theta_1 = 0$, $\theta_2 = \theta$, then

$$W_{\text{ext}} = -pE\left[\cos\theta - 1\right]$$
and
$$W_{\text{el}} = pE\left[\cos\theta - 1\right]$$

The work done by external force is stored in the form of electrostatic potential energy of the system. If we take zero potential energy at $\theta_1 = 90°$, then for $\theta_2 = \theta$, we have
Change in potential energy, $\Delta U = W_{\text{ext}} = -pE\cos\theta$
i.e.,
$$\Delta U = -\vec{p} \cdot \vec{E}$$

When $\theta = 0°$, the dipole moment is in the direction of the field and the dipole is in *stable equilibrium*. If it is slightly displaced from its equilibrium position, it performs oscillations about the stable equilibrium position, which serves as the mean position. When $\theta = 180°$, the dipole moment is opposite to the direction of the field and the dipole is in *unstable equilibrium*.

2.3.3 The Electric Potential Due to a Point Charge

Let's calculate the electric potential of the simplest possible system: one point charge. The point charge q is the source of an electric potential. Consider a test charge q_0 separated by a distance r from a single point charge q. As Eq.2.24 shows, the potential energy of the system is $U(r) = q_0 q/4\pi\epsilon_0 r$, and hence $U(r)/q_0 = q/4\pi\epsilon_0 r$. We have found the electric potential of a point charge q at a distance r from the charge:

$$\boxed{V(r) = \frac{U(r)}{q_0} = \frac{1}{4\pi\epsilon_0}\frac{q}{r}} \quad (2.35)$$

In Eq.2.35, we have assumed that zero potential energy is at infinity and, as a consequence, we have taken the electric potential due to a charge q to be zero at infinity. To emphasize this point, we might say that Eq.2.35 is the potential of a single charge with respect to infinity.

As for potential energy, the only physically relevant feature of the potential is how it differs between two points. The electric potential difference due to the charge q between the points A and B at locations $\vec{r}_A$ and $\vec{r}_B$ is given by (Fig.2.8):

$$\Delta V = V_B - V_A = \frac{U_B - U_A}{q_0} = \frac{q}{4\pi\epsilon_0}\left(\frac{1}{r_B} - \frac{1}{r_A}\right) \quad (2.36)$$

Here, we have abbreviated V as a function of r_A, or $V(r_A)$, as V_A, and so forth.

EXAMPLE 12. A conducting bubble of radius a and thickness t ($t \ll a$) has potential V. Now the bubble collapses into a droplet. Find the potential of the droplet.

SOLUTION The volume of liquid in a bubble of radius a and thickness t is given by

$$V_b = \frac{4}{3}\pi(a+t)^3 - \frac{4}{3}\pi a^3 = \frac{4}{3}\pi a^3\left(\left(1+\frac{t}{a}\right)^3 - 1\right)$$

$$= \frac{4}{3}\pi a^3\left(1 + 3\frac{t}{a} - 1\right) \quad (\text{since } t \ll a)$$

$$\approx 4\pi a^2 t$$

The volume of liquid in the droplet of radius r formed by collapsing the bubble, is-

$$V_d = \frac{4}{3}\pi r^3$$

Since, mass of the material of bubble does not change during formation of drop from bubble, therefore

Mass of bubble = Mass of droplet
$$\rho V_b = \rho V_d, \quad [\rho \text{ is the density of the material of bubble}]$$
$$\implies \frac{4}{3}\pi r^3 = 4\pi a^2 t \implies r = \left(3a^2 t\right)^{1/3}$$

The potential on a spherical shell of radius a and charge q is-

$$V = \frac{q}{(4\pi\epsilon_0 a)}$$

Therefore, the charge on bubble having potential V is-
$$q = 4\pi\epsilon_0 a V$$

So, by conservation of charge, the charge on the droplet will also be, $q = 4\pi\epsilon_0 a V$.

Therefore, the electric potential on the droplet is given by

$$V_d = \frac{q}{4\pi\epsilon_0 r} = \frac{4\pi\epsilon_0 a V}{(3a^2 t)^{1/3}} = \left(\frac{a}{3t}\right)^{1/3} V$$

EXAMPLE 13. Electric potential at a point 'P' due to a point charge of 5×10^{-9} C is 50 V. Calculate the distance of 'P' from the point charge. (Assume, $1/4\pi\epsilon_0 = 9 \times 10^{+9}$ Nm^{2}C^{-2})

APPROACH Apply relation, $V = kQ/r$ and then solve for r.

SOLUTION $V_P = \dfrac{kQ}{r} \implies 50 = \dfrac{9 \times 10^9 \times 5 \times 10^{-9}}{r}$

$$\implies r = \frac{45}{50} = \frac{9}{10} = 0.9 \text{ m} = 90 \text{ cm}$$

EXAMPLE 14. Two equal point charges (each of magnitude q), are fixed at $x = -a$ and $x = +a$ on the x-axis. Another point charge Q is placed at the origin. Find the change in electrical potential energy of Q, when it is displaced by a small distance x along the x-axis.

SOLUTION Let O be the origin and O' be a point to the right

Figure 2.18

of O at a distance x.
The potentials at O and O' respectively, due to charges at $(-a, 0)$ and $(a, 0)$ are

$$V_O = \frac{q}{4\pi\epsilon_0 a} + \frac{q}{4\pi\epsilon_0 a} = \frac{q}{2\pi\epsilon_0 a}$$

$$V_{O'} = \frac{q}{4\pi\epsilon_0(a+x)} + \frac{q}{4\pi\epsilon_0(a-x)} = \frac{q}{2\pi\epsilon_0}\left(\frac{a}{a^2 - x^2}\right)$$

The potential energy associated with the charge Q, when it is placed at a point where potential is V, is QV. Thus, the change in potential energy of charge Q when it is displaced by a small distance x is

$$\Delta U = QV_{O'} - QV_O = \frac{qQ}{2\pi\epsilon_0}\left[\frac{a}{a^2-x^2} - \frac{1}{a}\right]$$
$$= \frac{qQ}{2\pi\epsilon_0}\frac{x^2}{a(a^2-x^2)} \approx \frac{qQ}{2\pi\epsilon_0}\frac{x^2}{a^3} \quad (\text{for } x \ll a)$$

EXAMPLE 15. A uniform electric field pointing in positive x direction exists in a region. Let A be the origin, B be the point on the x-axis at $x = +1$ cm, and C be the point on the y-axis at $y = +1$ cm. Then the potentials at the points A, B and C satisfy
(A) $V_A < V_B$ (B) $V_A > V_B$ (C) $V_A < V_C$ (D) $V_A > V_C$

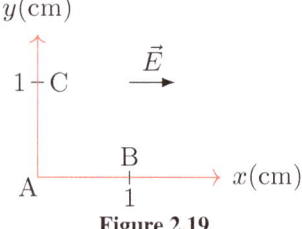

Figure 2.19

APPROACH Apply the relation, $V_B - V_A = -\int_i^f \vec{E}\cdot d\vec{r}$ along x and y directions.

SOLUTION (B) The uniform electric field in the region is $\vec{E} = E\hat{\imath}$. Let $d\vec{r}_x = dx\hat{\imath}$ and $d\vec{r}_y = dy\hat{\jmath}$ be the small displacement vectors along x- and y-axes.

The potentials at the point B and C relative to the point A are given by

$$V_B = V_A - \int \vec{E}\cdot d\vec{r}_x = V_A - \int_0^1 E\,dx = V_A - E$$
$$V_C = V_A - \int \vec{E}\cdot d\vec{r}_y = V_A \qquad (\because \vec{E}\perp d\vec{r}_y)$$

Note that the potential decreases along $\vec{E}$ but does not change in a direction perpendicular to $\vec{E}$.

2.4 Equipotential Surfaces

Any surface consisting of a continuous distribution of points having the same electric potential, is called an equipotential surface. It can be either an imaginary surface or a real, physical surface. Since, no point can be at two different potentials at the same time, so equipotential surfaces for different potentials can never touch or intersect. If a charged particle, having charge q_0, moves between two points A and B on the same equipotential surface (i.e., $V_A = V_B$), then from Eq.2.10:

$$V_B - V_A = \frac{W_{AB}}{q_0}$$

the net work done on the charged particle by an electric field:
$$W_{AB} = q_0(V_B - V_A) = q_0(0) = 0 \quad [\because \quad V_A = V_B]$$
Thus, no net work W_{AB} is done on a charged particle by an electric field when it moves between two points A and B on the same equipotential surface. Furthermore, because of the path independence of potential, this result ($W_{AB} = 0$) holds for any two points A and B on the equipotential surface, even if the path between them does not lie entirely on the equipotential surface.
As change in potential energy is equal to negative of work done by electric field, i.e., $\Delta U = -W_{AB}$, so potential energy does not change (because, $W_{AB} = 0$) as a test charge moves over an equipotential surface, the electric field can do no work on such a charge. It follows that $\vec{E}$ must be perpendicular to the surface at every point so that the electric force $q_0\vec{E}$ is always perpendicular to the displacement of a charge moving on the surface. Field lines and equipotential surfaces are always mutually perpendicular. In general, field lines are curves, and equipotentials are curved surfaces. For the special case of a uniform field, in which the field lines are straight, parallel, and equally spaced, the equipotentials are parallel planes perpendicular to the field lines.

Figure 2.20 shows a family of equipotential surfaces associated with the electric field due to some distribution of charges. The work done by the electric field on a charged particle as the particle moves from one end to the other of paths I and II is zero because each of these paths begins and ends on the same equipotential surface and thus there is no net change in potential. The work done as the charged particle moves from one end to the other of paths III and IV is not zero but has the same value for both these paths because the initial and final potentials are identical for the two paths; that is, paths III and IV connect the same pair of equipotential surfaces.

☞ In diagrams we usually show only a few representative equipotentials, often with equal potential differences between adjacent surfaces.

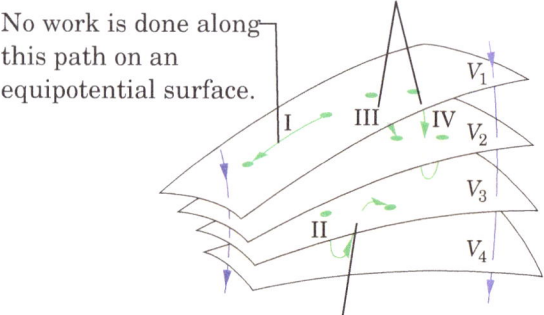

Figure 2.20: Portions of four equipotential surfaces at electric potentials V_1, V_2, V_3, and V_4 ($V_1 > V_2 > V_3 > V_4$). Four paths along which a test charge may move are shown. Two electric field lines are also indicated.

2.4.1 Various Equipotential Surfaces
2.4.1.1 Equipotential Surfaces For a Point Charge
If a point charge q is fixed at origin O $(0,0,0)$ and P (x,y,z) be a point at distance $r\left(=\sqrt{x^2+y^2+z^2}\right)$ from O, then electric potential (V) at point P is-
$$V = k\frac{q}{r^2} = k\frac{q}{\sqrt{(x^2+y^2+z^2)}} = V_0 = \text{constant}$$
$$\implies x^2 + y^2 + z^2 = \frac{k^2q^2}{V_0^2} = \text{constant} \quad (2.37)$$

Equation 2.37 is of the form of a sphere with the center at $(0,0,0)$ and a radius of kq/V_0. Thus, the surfaces are concentric spheres with the origin (the location of the charge) as the centre and a radius (R) given by
$$R = \frac{kq}{V_0}$$

2.4.1.2 Equipotential Surfaces Due to an Electric Dipole
In Fig.2.22, electric field lines and equipotential lines are plotted for a set of two charges of equal magnitude, one positive and one negative. The field lines are closest together near the charges, indicating that the electric field is strongest there. Very

2.4. EQUIPOTENTIAL SURFACES

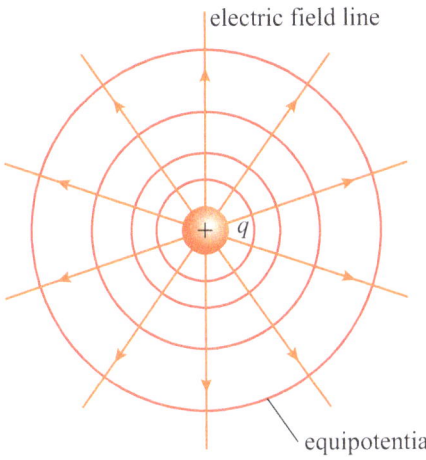

Figure 2.21

close to each charge, the effect of that charge dominates on the equipotential plots, which are not strongly different from circles, but farther away the superposition of the two electric potentials produces more noticeable effects. Note that electric field lines always cross equipotential lines perpendicularly.

In Fig.2.22, electric field lines and equipotential lines are plotted for a pair of charges with equal magnitude, one positive and one negative. The field lines are closest together near the charges, indicating that the electric field is strongest there. In close proximity to each charge, the influence of that charge dominates on the equipotential plots, which appear not significantly different from circles. However, farther away, the superposition of the two electric potentials produces more noticeable effects. It is important to note that electric field lines always cross equipotential lines perpendicularly.

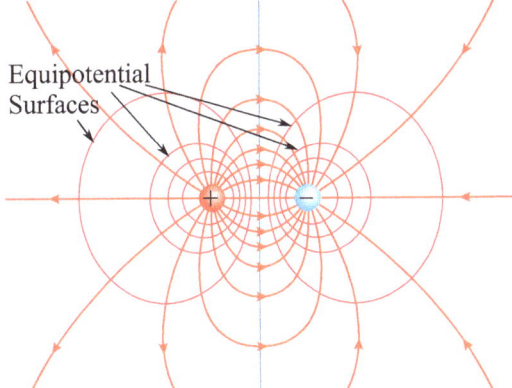

Figure 2.22: Equipotential lines and electric field lines for a pair of charges of equal magnitude but opposite sign.

2.4.1.3 Equipotential Surfaces Due to Two Identical Positive Charges:

Fig.2.23 displays equipotential surfaces and electric field lines resulting from two identical positive point charges. At first glance, it might seem that two equipotential surfaces intersect at the centre of Fig.2.23, contrary to the rule that prohibits such intersections. However, in reality, this is a single figure-eight-shaped equipotential surface.

2.4.1.4 Equipotential Surfaces of a Charged Wire of Infinite Length

Let us consider a line charge of infinite length having linear charge density λ. Suppose, it is placed along z-axis. The potential due to this line charge at a point P is given by Eq.2.60:

$$V(r) = -2k\lambda \ln r = -\frac{\lambda}{2\pi\epsilon_0} \ln r$$

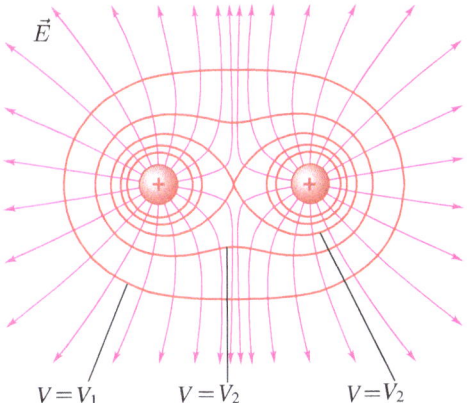

Figure 2.23: Two equal positive charges

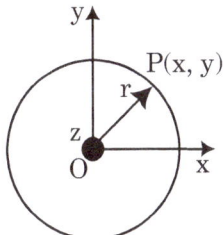

Figure 2.24: Cross-sectional view of cylindrical wire having axis along z-axis

where r is the distance of the point P from the line charge as shown in Fig.2.24.

Since, the line charge is along the z-axis, $r = \sqrt{x^2 + y^2}$, therefore,

$$V(r) = -\frac{\lambda}{4\pi\epsilon_0} \ln(x^2 + y^2)$$

Now, $V = \text{constant} = V_0$

$$\implies \ln(x^2 + y^2) = -\frac{4\pi\epsilon_0 V_0}{\lambda} \implies x^2 + y^2 = e^{-4\pi\epsilon_0 V_0/\lambda}$$

which represents cylinders with axis along the z-axis with radii $r = e^{-2\pi\epsilon_0 V_0/\lambda}$.

As V_0 increases, radius becomes smaller. Thus the cylinders are

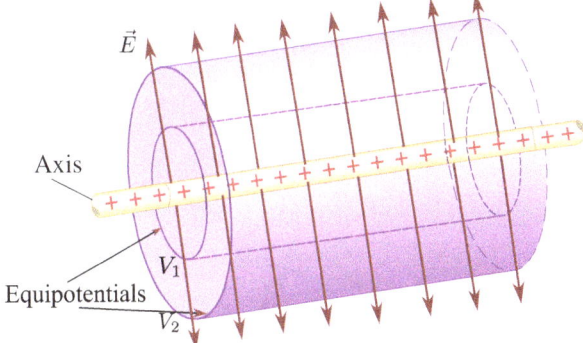

Figure 2.25

packed closer around the axis, showing that the field is stronger near the axis.

2.4.1.5 Uniformly Charged Plane Surface of Infinite Dimensions

Let us consider a plane sheet of charge, in x-y plane, having surface charge density σ. Consider a point P at distance z from this sheet. From Eq.2.62, the electric potential at P will be given by -

$$V = V_0 - \frac{\sigma}{2\epsilon_0}|z| \tag{1}$$

Here, V_0 is the potential at $z = 0$.
Equipotential surface means

$V = \text{constant} \implies V_0 - \dfrac{\sigma}{2\epsilon_0}|z| = C$

$\implies |z| = \text{constant}$

Note that, the shapes of the equipotential surfaces in above di-

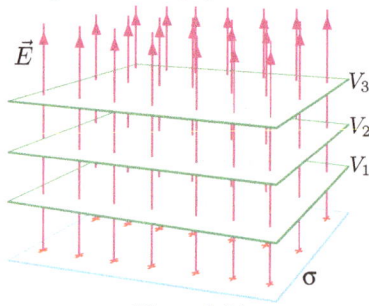

Figure 2.26

agrams will not change if the sign of each charge is reversed. If the positive charges in above diagrams, are replaced by negative charges, the equipotential surfaces will be the same but the sign of the potential will get reversed. For example, the surfaces with potentials $V = +30$ V and $V = -50$ V (say) will now have potentials $V = -30$ V and $V = +50$ V, respectively.

2.4.2 Properties of Equipotential surface

(i) The potential difference between any two points on an equipotential surface is always zero.

(ii) If a test charge q_0 moves from one point to the other on an equipotential surface, the electric potential energy $q_0 V$ remains constant.

(iii) No work is done by the electric force, when a charge moves along this surface.

(iv) Two equipotential surfaces can never intersect each other because in this scenario the point of intersection will have two potentials which is of course not acceptable.

(v) Field lines and equipotential surfaces are always mutually perpendicular.

2.5 Calculating the Field From the Potential: Potential Gradient

We can quantify the relation between potential and field by considering the potential difference dV between two nearby points A and B. Suppose they're separated by a small displacement ds in the 3-D space. Then, from Eq.2.13 we have-

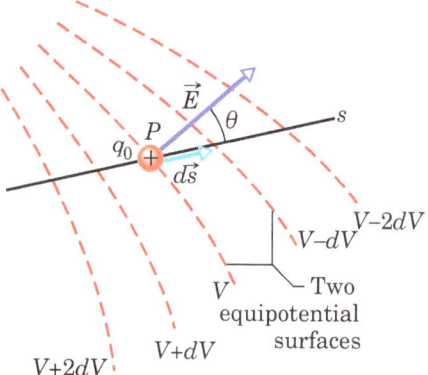

Figure 2.27: A test charge q_0 moves a distance $d\vec{s}$ from one equipotential surface to another. (The separation between the surfaces has been magnified for clarity.) The displacement $d\vec{s}$ makes an angle θ with the direction of the electric field $\vec{E}$.

$$dV = -\vec{E} \cdot d\vec{s} = -E\cos\theta \, ds \quad (2.38)$$

Here, θ is the angle between electric field vector $\vec{E}$ and displacement vector $d\vec{s}$.

Now, from Fig.2.27, we see that $E\cos\theta$ is the component of $\vec{E}$ in the direction of $d\vec{s}$. If we denote it by E_s, then above Eq.2.38 gives:

$$\boxed{E_s = E\cos\theta = -\dfrac{dV}{ds}} \quad (2.39)$$

In words, this equation states; *the negative of the rate of change of the potential with position in any direction is the component of $\vec{E}$ in that direction.* The minus sign implies that $\vec{E}$ points in the direction of decreasing V, as in Fig.2.27. It is clear from Eq.2.39 that an appropriate unit for E is the volt/meter (V/m).

According to Eq.2.39, the magnitude of $-dV/ds$ is maximum for $\theta = 0$. The quantity, E_s is also maximum for $\theta = 0$, i.e., in the direction of $\vec{E}$ which is always perpendicular to equipotential surface. Thus, corresponding to maximum magnitude of $-dV/ds$ the displacement $d\vec{s}$ should be in the direction of $\vec{E}$. Thus,

$$(E_s)_{\max} = E\cos 0 = -\left(\dfrac{dV}{ds}\right)_{\max}$$

or $\quad \boxed{E = -\left(\dfrac{dV}{ds}\right)_{\max}} \quad (2.40)$

The maximum value of dV/ds at a given point is called the potential gradient at that point. The direction of $d\vec{s}$ for which dV/ds has its maximum value is always at right angles to the equipotential surface, corresponding to the direction of $\vec{E}$ as shown in Fig.2.27.

> ⚠️ **Caution**
>
> In practical physics usage, if you simply say *potential gradient* without specifying a direction, it is generally interpreted as the rate of change of the potential with position in the direction *perpendicular* to the equipotential surfaces, and is directly related to the electric field magnitude according to Eq. (2.40):
>
> $$E = -\left(\dfrac{dV}{ds}\right)_{\max}$$
>
> If you mean any other direction, it is clearer to say "rate of change of potential in that direction"' rather than "potential gradient."

In a straightforward scenario where the electric field $\vec{E}$ is uniform, Eq.2.39 becomes

$$\boxed{E_s = -\dfrac{\Delta V}{\Delta s}} \quad (2.41)$$

where Δs is perpendicular to the equipotential surfaces. The component of the electric field is zero in any direction parallel to the equipotential surfaces because there is no change in potential along the surfaces.

If points A and B are separated by small displacement dx in the x-direction. Then similar to Eq.(2.38), we have-

$$dV = -E_x dx \quad (2.42)$$

where we again handled the dot product by considering only the component of $\vec{E}$ along the displacement.
Dividing both sides of Eq.2.42 by dx, we get-

$$E_x = -\dfrac{dV}{dx} \quad (2.43)$$

here, E_x is the electric-field component in the x-direction. Similarly, we can write the expressions for y- and z-components

2.5. CALCULATING THE FIELD FROM THE POTENTIAL: POTENTIAL GRADIENT

as given below-
$$E_y = -\frac{dV}{dy}, \quad E_z = -\frac{dV}{dz} \quad (2.44)$$

When a function depends on more than one variable, as the potential generally does, we write derivatives with the partial derivative symbol ∂ instead of d to indicate the rate of change with respect to only one variable. Thus we have $E_x = -\partial V/\partial x$, $E_y = -\partial V/\partial y$, and $E_z = -\partial V/\partial z$.

Now, the entire electric-field vector can be written as-
$$\vec{E} = E_x\hat{i} + E_y\hat{j} + E_z\hat{k} \quad (2.45)$$

On substituting the values of E_x, E_y and E_z in above equation, we get-
$$\boxed{\vec{E} = -\left(\frac{\partial V}{\partial x}\hat{i} + \frac{\partial V}{\partial y}\hat{j} + \frac{\partial V}{\partial z}\hat{k}\right)} \quad (2.46)$$

Equation 2.46 confirms that the electric field is strong where the potential changes rapidly. The negative sign reflects that the electric field always points in the direction of decreasing potential, i.e., a positive test charge naturally moves from high to low potential.

In physics, we often use a shorthand called the gradient operator, written as:
$$\textbf{grad} \text{ or } \vec{\nabla} = \frac{\partial}{\partial x}\hat{i} + \frac{\partial}{\partial y}\hat{j} + \frac{\partial}{\partial z}\hat{k}$$

It's just a compact way of writing the rate of change (i.e., derivative) of a quantity in all three directions.

Using this, we can write Eq.2.46 as-
$$\boxed{\vec{E} = -\left(\frac{\partial}{\partial x}\hat{i} + \frac{\partial}{\partial y}\hat{j} + \frac{\partial}{\partial z}\hat{k}\right)V = -\vec{\nabla}V \text{ or } -\textbf{grad } V} \quad (2.47)$$

The quantity, $\vec{\nabla}V$ is called the potential gradient.

Clearly, the unit of potential gradient is volt per meter, while electric intensity or force per unit charge is expressed in newtons per coulomb. However,
$$\frac{\text{volt}}{\text{m}} = \frac{\text{J/C}}{\text{m}} = \frac{\text{N} \cdot \text{m}}{\text{C} \cdot \text{m}} = \frac{\text{N}}{\text{C}}$$
so, the volt/meter and the newton/coulomb are equivalent units.

Note: Potential is a scalar quantity but the gradient of potential is a vector quantity. In Cartesian co-ordinates, the potential gradient is defined by Eq.2.47.

Thus if V is known for all points of space, that is, if the function $V(x, y, z)$ is known, the components of $\vec{E}$, and thus E itself, can be found by taking partial derivatives*.

We therefore have two methods for calculating E for continuous charge distributions. One is based on integrating Coulomb's law, and the other is based on differentiating V (see Eq.2.47). In practice, the second method is often less difficult.

Special Case for Radial Fields:
In certain systems, the electric potential V depends only on the radial distance r from a central point, and not on the direction. Such systems are called *spherically symmetric*. This means that all points at the same distance from the centre have the same potential, and the electric field at any point is directed purely along the radial direction.

In Eq. (2.39), $E = -\frac{dV}{ds}$, the derivative is taken along the direction of the electric field. In a spherically symmetric system,

---
*The symbol $\partial V/\partial x$ denotes a partial derivative. In taking this derivative of the function $V(x, y, z)$, the quantity x is to be viewed as a variable and y and z are to be regarded as constants. Similar considerations hold for $\partial V/\partial y$ and $\partial V/\partial z$.

the electric field at every point is purely radial, and the potential V depends only on the radial distance r. Therefore, when we move in the direction of the electric field, we are effectively moving along the radial direction itself. Hence, the displacement ds becomes identical to the radial displacement dr, and the relation simplifies to:
$$\boxed{E = -\frac{dV}{ds} = -\frac{dV}{dr}} \quad (2.48)$$

This form is valid only in spherically symmetric cases, where the direction of the field and the radial direction coincide.

Although $\vec{r}$ is a three-dimensional vector, the potential depends only on its magnitude r, and not on the angular coordinates. Hence, the variation of potential occurs only along the radial direction, and the electric field magnitude can be calculated using a single-variable derivative with respect to r.

This is the **scalar form** of the relation between electric field and potential in a radial field. Here, E represents the magnitude of the electric field, and $\frac{dV}{dr}$ is the rate at which the potential changes with distance from the centre.

The negative sign is important: it reflects the physical fact that the electric field always points in the direction of decreasing potential. Even though Eq. (2.48) is a scalar equation, the sign tells us about the relative direction between field and potential. For example, if V decreases with increasing r, then $\frac{dV}{dr} < 0$, making $E > 0$, which means the field points outward. On the other hand, if V increases with r, then $E < 0$, indicating an inward-pointing field. The negative sign, therefore, preserves the correct physical relationship between the electric field and the potential.

As an example, in Eq.2.35 we have shown that the potential at a point at the radial distance r from a source point charge q is
$$V = \frac{1}{4\pi\epsilon_0}\frac{q}{r}$$

To get E at the position of V, we consider the displacement $d\vec{s}$ in the direction of $\vec{E}$, i.e., radially away from point charge q.

Therefore, $E = -\frac{d}{dr}\left(\frac{1}{4\pi\epsilon_0}\frac{q}{r}\right) = \frac{1}{4\pi\epsilon_0}\frac{q}{r^2}$

which is in agreement with Coulomb's law.

Key Point

1. The component of the electric field $\vec{E}$ along any direction is related to the rate of change of potential with position by
$$E_s = -\frac{dV}{ds},$$
where ds is the infinitesimal displacement along that direction and dV is the corresponding change in potential.
If θ is the angle between $\vec{E}$ and $d\vec{s}$, then this component can also be written as
$$E_s = E\cos\theta.$$

2. For a spherically symmetric field, V depends only on the radial distance r from the source. In this case,
$$E = -\frac{dV}{dr},$$
where dr is the infinitesimal change in radial distance. The negative sign indicates that $\vec{E}$ points in the direction of decreasing potential.

EXAMPLE 16. Figure 2.28 shows lines of constant potential in a region in which an electric field is present. The values of the potential of each line is also shown. Of the points A, B and C, which one has the maximum magnitude of electric field $\vec{E}$

APPROACH From relation $E = -\partial V/\partial x$, we can say that, the

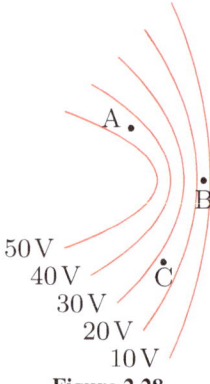

Figure 2.28

magnitude of $\vec{E}$ will be greater for greater value $\partial V/\partial x$.

SOLUTION The potential difference between all the successive lines of constant potential is $\Delta V = 10$ V. The perpendicular distances between successive lines at the point A and at the point C are almost equal but it is smaller at the point B i.e., $\Delta x_A = \Delta x_C > \Delta x_B$. Hence, $|E_B| = \Delta V/\Delta x_B > \Delta V/\Delta x_A = |E_A| = |E_C|$.

EXAMPLE 17. If $V = 3x^2y + y^2 + yz$, then find an expression for electric field $\vec{E}$

APPROACH From Eq.2.47, the electric field in terms of electric potential is defined as-
$$\vec{E} = -\left(\frac{\partial V}{\partial x}\hat{i} + \frac{\partial V}{\partial y}\hat{j} + \frac{\partial V}{\partial z}\hat{k}\right) \quad \cdots (1)$$

SOLUTION Given that, $V = 3x^2y + y^2 + yz$ therefore-
$$\frac{\partial V}{\partial x} = \frac{\partial}{\partial x}\left(3x^2y + y^2 + yz\right) = \frac{\partial}{\partial x}\left(3x^2y\right) = 3y\frac{d}{dx}\left(x^2\right) = 6xy$$

Similarly, $\frac{\partial V}{\partial y} = \frac{\partial}{\partial y}\left(3x^2y + y^2 + yz\right) = 3x^2 + 2y + z$,

and $\frac{\partial V}{\partial z} = \frac{\partial}{\partial z}\left(3x^2y + y^2 + yz\right) = \frac{\partial}{\partial z}yz = y$

Substituting these values in Eq.(1), we get-
$$\vec{E} = -6xy\hat{i} - (3x^2 + 2y + z)\hat{j} - y\hat{k}$$

EXAMPLE 18. If $V = -100/r$, then calculate intensity of electric field at point (3, 4).

APPROACH Since, electric potential is given in terms of position r in 2-D, therefore, we use Eq.(2.48): $\boxed{E = -\frac{dV}{dr}}$ to get an expression for E. Here, $\vec{r} = x\hat{i} + y\hat{j}$ and $r = \sqrt{x^2 + y^2}$

SOLUTION $E = -\frac{dV}{dr} = -\frac{d}{dr}\left(-\frac{100}{r}\right) = -\frac{100}{r^2}$

The negative sign in the definition $E = -dV/dr$ indicates that the electric field vector points in the direction in which the potential decreases most rapidly.

In this problem, the potential is given by $V = -100/r$. Because of the negative sign in V, the potential *increases* with increasing r. Hence, $\vec{E}$ points opposite to the outward radial direction $\hat{r}$, i.e., towards the origin.

The magnitude of electric field is: $E = \frac{100}{r^2} > 0$

At (3, 4), $r = \sqrt{3^2 + 4^2} = 5$, therefore, the magnitude of electric field:
$$E = \frac{100}{5^2} = 4 \text{ V/m}$$

In vector form the electric field can be written as
$$\vec{E} = -|E|\hat{r} = -4\hat{r} \quad (1)$$

where, $\vec{r} = 3\hat{i} + 4\hat{j}$, therefore, $\hat{r} = \frac{\vec{r}}{r} = \frac{3}{5}\hat{i} + \frac{4}{5}\hat{j}$
Substituting this value of $\hat{r}$ in Eq. (1), we get
$$\vec{E} = -\frac{12}{5}\hat{i} - \frac{16}{5}\hat{j} \text{ V/m}.$$

EXAMPLE 19. $V = 3x + 4y + 5z$. Find electric field intensity.

APPROACH Since, electric potential is given in terms of Cartesian coordinates, therefore, apply Eq. 2.46.

SOLUTION Given that- $V = 3x + 4y + 5z$, therefore-
$$\frac{\partial V}{\partial x} = 3, \frac{\partial V}{\partial y} = 4, \text{ and } \frac{\partial V}{\partial z} = 5$$

Substituting these values in Eq.2.46:
$$\vec{E} = -\left(\frac{\partial V}{\partial x}\hat{i} + \frac{\partial V}{\partial y}\hat{j} + \frac{\partial V}{\partial z}\hat{k}\right),$$

we get-
$$\vec{E} = -\left(\frac{\partial V}{\partial x}\hat{i} + \frac{\partial V}{\partial y}\hat{j} + \frac{\partial V}{\partial z}\hat{k}\right) = -\left(3\hat{i} + 4\hat{j} + 5\hat{k}\right) \text{ V/m}$$

Therefore, the magnitude of electric field is-
$$E = \sqrt{(-3)^2 + (-4)^2 + (5)^2} = \sqrt{50} = 5\sqrt{2} \text{ V/m}$$

EXAMPLE 20. $V = x^2y + y^2z + z^2x$. Find electric field intensity at (1, 2, 3).

APPROACH Here, electric potential is given in terms of Cartesian coordinates, therefore, apply Eq.2.46.

SOLUTION Given that- $V = x^2y + y^2z + z^2x$, therefore-
$$\frac{\partial V}{\partial x} = 2xy + z^2, \frac{\partial V}{\partial y} = x^2 + 2zy, \text{ and } \frac{\partial V}{\partial z} = y^2 + 2zx$$

Substituting these values in Eq.2.46, we get-
$$\vec{E} = -\left(\frac{\partial V}{\partial x}\hat{i} + \frac{\partial V}{\partial y}\hat{j} + \frac{\partial V}{\partial z}\hat{k}\right)$$
$$= -\left((2xy + z^2)\hat{i} + (x^2 + 2zy)\hat{j} + (y^2 + 2zx)\hat{k}\right) \text{ V/m}$$

Therefore, at point (1, 2, 3)
$$\vec{E} = -((2(1)(2) + 3^2)\hat{i} + (1^2 + 2(3)(2))\hat{j} + (3^2 + 2(3)(1))\hat{k})\text{V/m}$$
$$= -\left(13\hat{i} + 13\hat{j} + 10\hat{k}\right) \text{V/m}$$

and $|\vec{E}| = \sqrt{13^2 + 13^2 + 10^2} = \sqrt{438}$ V/m

EXAMPLE 21. If the electric potential at any point (x, y, z) m space is given by $V = 3x^2$ volt, find the electric field at the point (1, 0, 3) m

SOLUTION Given: $V = 3x^2$, therefore, x, y and z components of electric field are-
$$E_x = -\frac{\partial V}{\partial x} = -6x, E_y = -\frac{\partial V}{\partial y} = 0, E_z = -\frac{\partial V}{\partial z} = 0,$$

respectively.

Now, net electric field,
$$\vec{E} = E_x\hat{i} + E_y\hat{j} + E_z\hat{k} = -6x\,\hat{i} + 0\hat{j} + 0\hat{k} = -6x\,\hat{i}$$

Therefore, at position (1 m, 0 m, 3 m), $\vec{E} = -6(1) = -6$ V/m

2.6 Electric Potential For a System of Charges

For a total of N point charges, the potential V at any point P can be derived from the principle of superposition.

Recall that potential due to q_1 at point P:
$$V_1 = \frac{1}{4\pi\epsilon_0}\frac{q_1}{r_1}$$

Therefore, the total potential at point P due to all N point charges:
$$V = V_1 + V_2 + V_3 + \cdots + V_N \quad \text{(principle of superposition)}$$

2.6. ELECTRIC POTENTIAL FOR A SYSTEM OF CHARGES

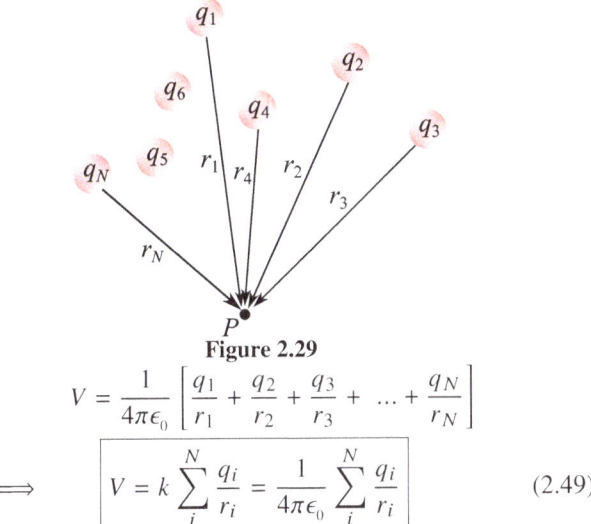

Figure 2.29

$$V = \frac{1}{4\pi\epsilon_0}\left[\frac{q_1}{r_1} + \frac{q_2}{r_2} + \frac{q_3}{r_3} + \ldots + \frac{q_N}{r_N}\right]$$

$$\Rightarrow \boxed{V = k\sum_i^N \frac{q_i}{r_i} = \frac{1}{4\pi\epsilon_0}\sum_i^N \frac{q_i}{r_i}} \quad (2.49)$$

When we have a continuous distribution of charge along a line, over a surface, or through a volume, we divide the charge into elements dq, and the sum in Eq.2.49 becomes an integral:

$$\boxed{V = \frac{1}{4\pi\epsilon_0}\int \frac{dq}{r}} \quad (2.50)$$

Note: Remember that there does not have to be a charge at a given point for a potential V to exist at that point. (In the same way, an electric field can exist at a given point even if there's no charge there to respond to it.)

EXAMPLE 22. Positive and negative point charges of equal magnitude are kept at $(0,0,a/2)$ and $(0,0,-a/2)$, respectively. Find the work done by the electric field when another positive point charge is moved from $(-a,0,0)$ to $(0,a,0)$.

APPROACH Work done by electric field, when a point charge Q moves from potential V_A to potential V_B, is:
$$W = -Q(V_B - V_B)$$

SOLUTION The charge configuration is shown in the Fig.2.30. The point $A(-a,0,0)$ is at a distance $r_A = \sqrt{5}a/2$ from both the charges. Also, the point $B(0,a,0)$ is at a distance $r_B = \sqrt{5}a/2$ from both the charges.

The potentials at the point A and B are given by

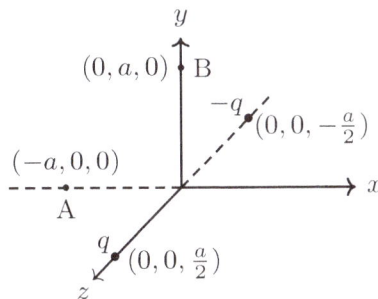

Figure 2.30

$$V_A = \frac{1}{4\pi\epsilon_0}\frac{q}{r_A} - \frac{1}{4\pi\epsilon_0}\frac{q}{r_A} = 0,$$
$$V_B = \frac{1}{4\pi\epsilon_0}\frac{q}{r_B} - \frac{1}{4\pi\epsilon_0}\frac{q}{r_B} = 0$$

Since $V_A = V_B$, therefore the work done in taking a point charge from A to B is
$$W = Q(V_B - V_A) = 0.$$

EXAMPLE 23. A charge $+q$ is fixed at each of the points $x = x_0, x = 3x_0, x = 5x_0, \cdots, \infty$ on the x-axis and a charge $-q$ is fixed at each of the points $x = 2x_0, x = 4x_0, x = 6x_0, \cdots, \infty$.

Here x_0 is a positive constant. Take the electric potential at a point due to a charge Q at a distance r from it to be $Q/(4\pi\epsilon_0 r)$. Find the potential at the origin due to the above system of charges.

APPROACH Due to several charge, the electric potential at any point is given by Eq.2.49, so apply it and solve for V at origin-

SOLUTION Using Eq.2.49, the electric potential at the origin due to the given system of charges is

$$V = \frac{q}{4\pi\epsilon_0 x_0}\left[\frac{1}{1} - \frac{1}{2} + \frac{1}{3} - \frac{1}{4} + \ldots\right] = \frac{q}{4\pi\epsilon_0 x_0}\ln 2$$

EXAMPLE 24. In Figure 2.31, six point charges are kept at the vertices of a regular hexagon of side L and centre O. Given that $k = \frac{1}{4\pi\epsilon_0}\frac{q}{L^2}$, which of the following statement(s) is (are) correct?

(A) The electric field at O is $6k$ along OD.
(B) The potential at O is zero.
(C) The potential at all points on the line PR is same.
(D) The potential at all points on the line ST is same.

APPROACH Apply the principle of superposition of electric

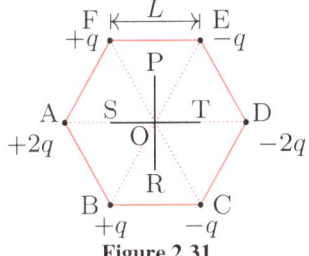

Figure 2.31

field as given in last chapter and principle of super position of electric potentials as given by Eq.2.49.

SOLUTION (A) (B) (C) The electric field at O due to the charges at A and D is $4K$ along OD, due to the charges at B and E is $2K$ along OE and due to the charges at C and F is $2K$ along OC. For the given geometry, resultant of these fields is $6K$ along OD.

The potential at O is
$$V_O = \sum \frac{1}{4\pi\epsilon_0}\frac{q_i}{L} = \frac{1}{4\pi\epsilon_0 L}\sum q_i = 0$$

For any point on PR, we have pairs of equal and opposite charges at the same distance making the potential at any point on PR zero. It may be seen that potential at points on OS is positive and that on OT is negative.

Comment: Students are advised to show that the potential on ST (at a distance x from O, taken positive towards the right) is

$$V(x) = \frac{q}{4\pi\epsilon_0}\left[\frac{2}{\sqrt{L^2 + x^2 + xL}} - \frac{2}{\sqrt{L^2 + x^2 - xL}} - \frac{4x}{L^2 - x^2}\right]$$

EXAMPLE 25. Six charges are placed around a regular hexagon of side length a as shown in the Fig.2.32. Five of them have charge q, and the remaining one has charge x. The perpendicular from each charge to the nearest hexagon side passes through the centre O of the hexagon and is bisected by the side. Which of the following statement(s) is (are) correct in SI units?

(A) When $x = q$, the magnitude of the electric field at O is zero.
(B) When $x = -q$, the magnitude of the electric field at O is $\frac{q}{6\pi\epsilon_0 a^2}$.
(C) When $x = 2q$, the potential at O is $\frac{7q}{4\sqrt{3}\pi\epsilon_0 a}$.
(D) When $x = -3q$, the potential at O is $\frac{3q}{4\sqrt{3}\pi\epsilon_0 a}$.

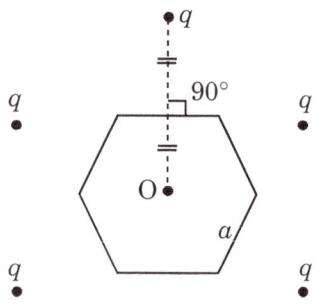

Figure 2.32

SOLUTION (A) When $x = q$, due to symmetry $\vec{E}_O = 0$

Figure 2.33

(B) When $x = -q$,
$$E_O = \frac{kq}{(2d)^2} \times 2 = \frac{2q \times 4}{4\pi\epsilon_0 \cdot 4 \cdot 3a^2} = \frac{q}{6\pi\epsilon_0 a^2}$$

(C) When $x = 2q$, the potential at O is
$$V_O = \frac{7kq}{2d} = \frac{7q}{4\pi\epsilon_0 \cdot \sqrt{3}a} = \frac{7q}{4\sqrt{3}\pi\epsilon_0 q}$$

(D) When $x = -3q$, the potential at O is
$$V_O = \frac{2kq}{2d} = \frac{2q}{4\pi\epsilon_0 \cdot \sqrt{3}a} = \frac{q}{2\sqrt{3}\pi\epsilon_0 q}$$

From above we see that the correct choices are- A, B and C.

EXAMPLE 26. In Fig.2.34a, four point charges $+8\ \mu C$, $-1\ \mu C$, $-1\ \mu C$, and $+8\ \mu C$ are fixed at the points $-\sqrt{27/2}$ m, $-\sqrt{3/2}$ m, $+\sqrt{3/2}$ m and $+\sqrt{27/2}$ m respectively on the y-axis. A particle of mass 6×10^{-4} kg and charge $+0.1\mu C$ moves along the x direction. Its speed at $x = +\infty$ is v_0. Find the least value of v_0 for which the particle will cross the origin. Also find the kinetic energy of the particle at the origin. Assume that space is gravity free. $\left[\frac{1}{4\pi\epsilon_0} = 9 \times 10^9\ \text{Nm}^2/\text{C}^2\right]$

APPROACH To cross the origin, the initial kinetic energy of the particle should be enough to cross the potential barrier applied by given four charges for any position $x = x_0$ (say) of the particle. Since, electric force is conservative, therefore, we can always apply the law of conservation of mechanical energy. If electric potential at any point on the x axis is V, then the position of potential barrier i.e., maximum potential, can be obtained by applying the relation $\partial V/\partial x = 0$.

Once you get the value of x_0, put $x = x_0$ in the expression for V and find net potential energy of the system for this position of moving charged particle. By conservation of energy, this energy will be equal to net initial kinetic energy of the particle for $x = \infty$. To get the kinetic energy of the particle at origin, again apply the conservation of energy principle and solve for KE.

SOLUTION Given $q = 1\mu C = 10^{-6}$C, $Q = 8\mu C = 8 \times 10^{-6}$C $= 8q$, $q_0 = 0.1\mu C = 10^{-7}$C, $m = 6 \times 10^{-4}$ kg and $a = \sqrt{3/2}$ m. Consider a point P at a distance x from the origin.
The potential at P due to given charge distribution is

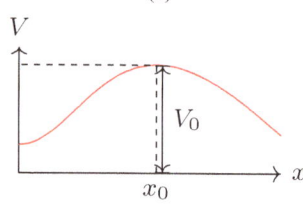

Figure 2.34

$$V(x) = \frac{1}{4\pi\epsilon_0}\left[\frac{2Q}{\sqrt{x^2 + 9a^2}} - \frac{2q}{\sqrt{x^2 + a^2}}\right]$$

The potential varies with x and attains its maximum at x_0 (Fig. 2.34b). For maximum value of $V(x)$, we have-
$$\frac{dV(x)}{dx} = -\frac{2x}{4\pi\epsilon_0}\left[\frac{Q}{(x^2 + 9a^2)^{3/2}} - \frac{q}{(x^2 + a^2)^{3/2}}\right] = 0 \quad \cdots (1)$$

On substituting, $Q = 8q$ in equation (1) and solving for x_0, we get $x_0 = a\sqrt{5/3} = \sqrt{5/2}$ m. The potential at x_0 is $V_0 = V(x_0) = 2.7 \times 10^4$ V.

Now, by conservation of mechanical energy, we can write-
$$\frac{1}{2}mv_0^2 = q_0 V_0, \implies v_0 = \sqrt{2q_0 V_0/m} = 3\text{ m/s}$$

The potential energy of the system for q_0 at the origin is
$$U = \frac{1}{4\pi\epsilon_0}\left[\frac{2Qq_0}{3a} - \frac{2qq_0}{a}\right] = 2.4 \times 10^{-3}\text{ J}$$

If K is the kinetic energy of q_0 at the origin, then the conservation of mechanical energy, $\frac{1}{2}mv_0^2 = K + U$, gives
$$K = \frac{1}{2}mv_0^2 - U = 3 \times 10^{-4}\text{ J}$$

EXAMPLE 27. Two fixed charges $-2Q$ and Q are located at the points with coordinates $(-3a, 0)$ and $(+3a, 0)$ respectively in the x-y plane.
(a) Show that all points in the $x - y$ plane where the electric potential due to the two charges is zero, lie on a circle. Find its radius and the location of its centre.
(b) Give the expression $V(x)$ at a general point on the x-axis and sketch the function $V(x)$ on the whole x-axis.
(c) If a particle of charge $+q$ starts from rest at the centre of the circle, show by a short quantitative argument that the particle eventually crosses the circle. Find its speed when it does so.

APPROACH For part (a), find the electric potential at any point $P(x, y)$ due to the charge $-2Q$ located at $(-3a, 0)$ and the charge Q located at $(3a, 0)$. For zero potential points, equate it to zero and find a relation between the coordinates x and y. The relation between x and y will give you the equation of path of zero potential points. For part (b), just find the electric potential at $(x, 0, 0)$ due to both the charges and apply the principle of superposition. To

2.6. ELECTRIC POTENTIAL FOR A SYSTEM OF CHARGES

get speed of particle at the center of circle, apply the principle of conservation of energy.

SOLUTION (a) The net electric potential at the point $P(x, y)$ due to the charge $-2Q$ located at $(-3a, 0)$ and the charge Q located at $(3a, 0)$ is given by

$$V = \frac{Q}{4\pi\epsilon_0}\left[\frac{1}{\sqrt{(x-3a)^2+y^2}} - \frac{2}{\sqrt{(x+3a)^2+y^2}}\right]$$

For zero potential points: $V = 0$, therefore above expression gives-

$$\implies \frac{1}{\sqrt{(x-3a)^2+y^2}} = \frac{2}{\sqrt{(x+3a)^2+y^2}}$$

$$\implies (x-5a)^2 + y^2 = (4a)^2.$$

which is an equation of circle of radius $4a$ and centre $(5a, 0)$. **(b)**

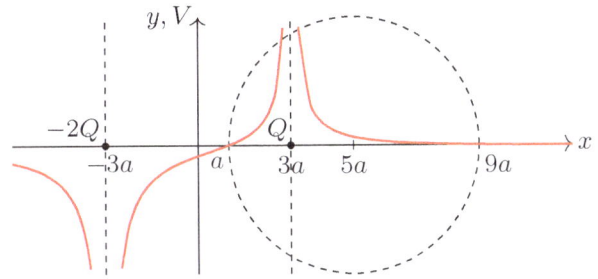

Figure 2.35

The potential on the x axis is given by

$$V(x) = \frac{Q}{4\pi\epsilon_0}\left(\frac{-2}{|x+3a|} + \frac{1}{|x-3a|}\right) \quad (2.51)$$

By definition of modulus function, we have-

$$|x| = \begin{cases} x, & \text{if } x \geq 0; \\ -x & \text{if } x < 0 \end{cases} \quad (2.52)$$

Using definition of modulus function (Eq.2.52), we can write Eq. 2.51 as -

$$V(x) = \begin{cases} \frac{Q}{4\pi\epsilon_0}\left(\frac{2}{3a+x} + \frac{1}{3a-x}\right), & \text{if } x \leq -3a; \\ \frac{Q}{4\pi\epsilon_0}\left(\frac{-2}{x+3a} + \frac{1}{3a-x}\right), & \text{if } -3a < x \leq 3a; \\ \frac{Q}{4\pi\epsilon_0}\left(\frac{-2}{x+3a} + \frac{1}{x-3a}\right), & \text{if } x > 3a. \end{cases} \quad (2.53)$$

From Eq.2.53, it is clear that- $V \to -\infty$ as $x \to -3a$ and $V \to \infty$ as $x \to 3a$. The potential is zero at $x = a$ and at $x = 9a$. V vs x graph is shown in Fig.2.35.
The potential at the centre of circle ($x = 5a$) is

$$V = \frac{Q}{4\pi\epsilon_0}\left(\frac{-2}{8a} + \frac{1}{2a}\right) = \frac{Q}{16\pi\epsilon_0 a}$$

which has a positive value. The potential at the circumference of the circle is zero.
(c) A positive charge moves from a higher potential to a lower potential. By conservation of energy, decrease in the potential energy is equal to increase in kinetic energy i.e.,

$$\frac{1}{2}mv^2 = \frac{qQ}{16\pi\epsilon_0 a}, \quad \text{which gives} \quad v = \sqrt{\frac{Qq}{8\pi\epsilon_0 ma}}$$

2.6.1 Electric Potential at a Point $P(r, \theta)$ Due to an Electric Dipole

Let an electric dipole AB, of length $2l$, is placed along x-axis. The centre of the dipole is at origin O and charges $-q$ and $+q$ are situated at points A and B respectively (Fig.2.36a). $P(r, \theta)$ is a point where electric potential due to this electric dipole is required. If $r \gg l$, then we can assume $AP||OP||BP$. It is also shown in a separate Fig.2.36b.
In this case, $\angle PAB = \angle POB = \theta$.

Now, draw perpendiculars OC and BD from points O and B on the lines AP and BP respectively. From Fig. 2.36a, we have

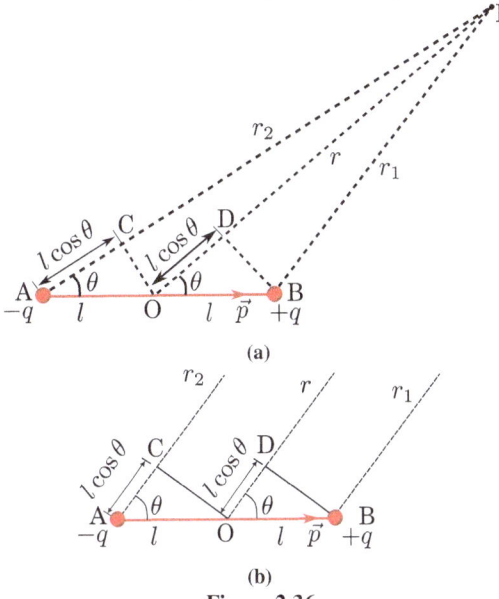

Figure 2.36

$AC = OD = l\cos\theta$ and $CP \approx OP$, $BP \approx DP$.
Therefore, $AP = AC + CP \approx AC + OP = l\cos\theta + r$
i.e., $r_2 \approx r + l\cos\theta$ and $BP \approx DP = OP - OD$
i.e., $r_1 \approx r - l\cos\theta$
Electric potential at point $P(r, \theta)$, due to charge $-q$ at A is-

$$V_{(-)} = \frac{kq}{r_2} \approx \frac{-kq}{r + l\cos\theta} \qquad [\because \quad r_2 \approx r + l\cos\theta]$$

And, the electric potential at point $P(r, \theta)$, due to charge $+q$ at B is -

$$V_{(+)} = \frac{kq}{r_1} \approx \frac{kq}{r - l\cos\theta} \qquad [\because \quad r_1 \approx r - l\cos\theta]$$

Therefore, net electric potential at P due to both the charges of the electric dipole,

$$V = V_- + V_+ = kq\left[\frac{1}{r - l\cos\theta} - \frac{1}{r + l\cos\theta}\right]$$

$$= kq\left[\frac{r + l\cos\theta - r + l\cos\theta}{r^2 - l^2\cos^2\theta}\right] = \frac{kp\cos\theta}{[r^2 - l^2\cos^2\theta]}$$

For $r \gg l$

$$\boxed{V \approx \frac{kp\cos\theta}{r^2}} \quad (2.54)$$

here angle θ is measured from the dipole moment $\vec{p}$ to the position vector $\vec{r}$ that extends from the centre of the dipole to the point of interest. The sign of the electric potential depends on the angle θ (Fig. 2.37).
Special Cases
(i) End on Position: For End on positions, $\theta = 0$ or $\theta = \pi$ and then $\cos\theta = \pm 1$, therefore-

$$\boxed{V \approx \frac{kp\cos\theta}{r^2} = \pm\frac{kp}{r^2}} \quad (2.55)$$

(ii) Broadside on Position: For broadside on positions, $\theta = \pi/2$ and then $\cos\theta = 0$, therefore -

$$\boxed{V \approx \frac{kp\cos\theta}{r^2} = 0} \quad (2.56)$$

(iii) For $0 < \theta < \pi/2$, we have $\cos\theta > 0$. So-

$$\boxed{V \approx \frac{kp\cos\theta}{r^2} > 0} \quad (2.57)$$

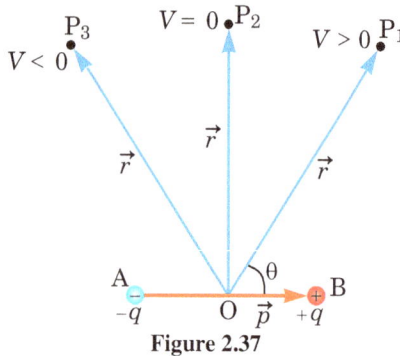

Figure 2.37

For $\pi/2 < \theta < \pi$, we have $\cos\theta < 0$. So -

$$\boxed{V \approx \frac{kp\cos\theta}{r^2} < 0} \qquad (2.58)$$

Figure 2.38 shows equipotential surfaces for the dipole in Figure 2.37. Most of the equipotential surfaces look like flattened spherical shells. Solid lines indicate positive electric potentials, and dashed lines indicate negative electric potentials. The solid black line in the centre of the Fig.2.38 represents the zero-potential surface, a flat plane. For a dipole, the electric potential is zero both at infinity and at any point on the plane shown in black. Notice that the sharpest change in the potential is between the two charges, near the plane. **EXAMPLE 28.** An electric dipole is formed

Equipotential surfaces

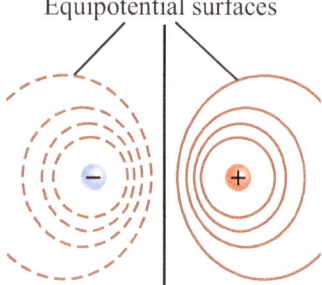

Figure 2.38: Equipotential surfaces for an electric dipole

by two charges $+q$ and $-q$ located in xy-plane at $(0, 2)$ mm and $(0, -2)$ mm, respectively, as shown in the Fig.2.39. The electric potential at point $P(100, 100)$ mm due to the dipole is V_0. The charges $+q$ and $-q$ are then moved to the points $(-1, 2)$ mm and $(1, -2)$ mm, respectively. What is the value of electric potential at P due to the new dipole?
(A) $V_0/4$ (B) $V_0/2$ (C) $V_0/\sqrt{2}$ (D) $3V_0/4$

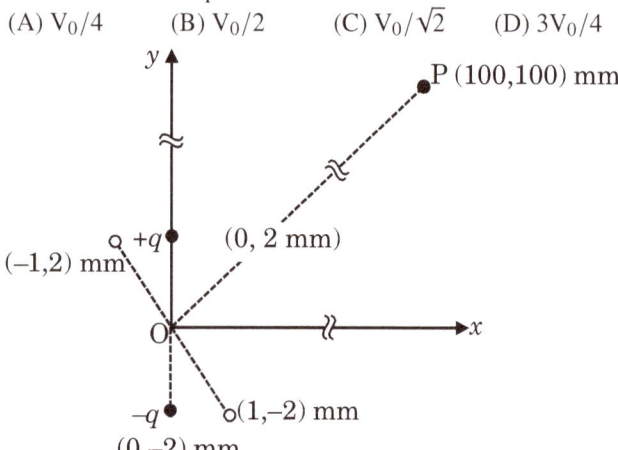

Figure 2.39

SOLUTION Initial separation between charges (Fig.2.40a):
$$2l = \sqrt{(0-0)^2 + (-2-2)^2} = 4 \text{ mm} = 4 \times 10^{-3} \text{ m}$$
Therefore, initial dipole moment,

$$\vec{p}_1 = p_1 \hat{j} = q(2l)\hat{j} = q\left(4 \times 10^{-3}\right)\hat{j},$$

Since, the centre of dipole divides it in the ratio of 1 : 1, therefore, the coordinates of centre of dipole will be $\left(\frac{0+0}{2}, \frac{2+(-2)}{2}\right)$, i.e., $(0, 0)$.

Position vector of point P (100, 100) with respect to centre of the dipole-
$$\vec{r} = 100\hat{i} + 100\hat{j} = 100\left(\hat{i} + \hat{j}\right).$$
Therefore, magnitude of $\vec{r}$, i.e., $r = \sqrt{100^2 + 100^2} = 100\sqrt{2}$ m
Initial voltage at point P,
$$V_0 = \frac{k\vec{p}_1 \cdot \vec{r}}{r^3} = \frac{k(100\, p_1)}{(100\sqrt{2})^3} \qquad \cdots (1)$$
When the charges $+q$ and $-q$ are moved to the points $(-1, 2)$ mm

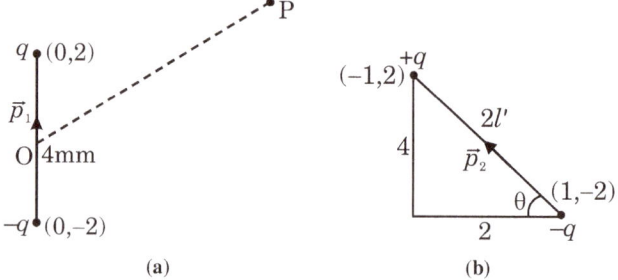

(a) (b)
Figure 2.40

and $(1, -2)$ mm, respectively (Fig.2.40b), coordinates of new centre of dipole becomes $\left(\frac{-1+1}{2}, \frac{2+(-2)}{2}\right)$, i.e., $(0, 0)$.

New length of dipole is, $2l' = \sqrt{(1+1)^2 + (-2-2)^2} = 2\sqrt{5}$ m
Therefore, magnitude of new dipole moment
$$p_2 = q(2l') = q\left(2\sqrt{5}\right) \times 10^{-3} \text{ C.m}$$
New dipole moment vector,
$$\vec{p}_2 = -p_2\cos\theta\hat{i} + p_2\sin\theta\hat{j}$$
New position vector of point P with respect to centre of dipole
$$\vec{r}' = (100-0)\hat{i} + (100-0)\hat{j} = 100\hat{i} + 100\hat{j} = \vec{r} \text{ and } r' = r$$
New potential at point P
$$V = \frac{k\vec{p}_2 \cdot \vec{r}'}{r'^3} = k\frac{-100p_2\cos\theta + 100p_2\sin\theta}{\left(100\sqrt{2}\right)^3}$$
$$\implies V = k\frac{100}{\left(100\sqrt{2}\right)^3}(-p_2\cos\theta + p_2\sin\theta) \qquad \cdots (2)$$

Again from Fig.2.40b, $\cos\theta = \frac{2}{2\sqrt{5}} = \frac{1}{\sqrt{5}}$,
$$d\sin\theta = \frac{4}{2\sqrt{5}} = \frac{2}{\sqrt{5}}.$$
Therefore, eq.(2) gives
$$V = k\frac{100}{\left(100\sqrt{2}\right)^3}p_2\left(-\frac{1}{\sqrt{5}} + \frac{2}{\sqrt{5}}\right) = k\frac{100}{\left(100\sqrt{2}\right)^3}\frac{p_2}{\sqrt{5}}$$
Now, from Eq.(1) and (2), we get
$$\frac{V}{V_0} = \frac{p_2}{p_1\sqrt{5}} = \frac{q\left(2\sqrt{5}\right) \times 10^{-3}}{q\left(4 \times 10^{-3}\right)\sqrt{5}} = \frac{1}{2} \implies V = \frac{V_0}{2}$$
Thus, the correct choice is B.

2.7 Check Point 2

1. ●●If $V = -5x + 3y + \sqrt{15}z$ then find magnitude of electric field at point (x, y, z).

2.7. CHECK POINT 2

2. ••Fig. 2.41 is a graph of the electric potential in a region of space where $\vec{E}$ is parallel to the x-axis. Draw a graph of E_x versus x.

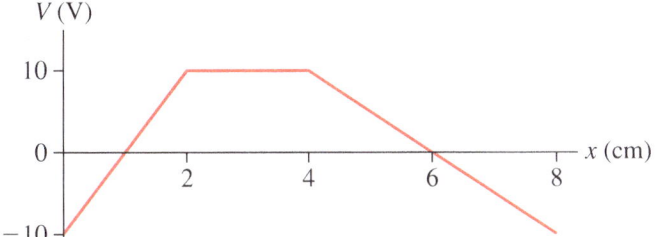

Figure 2.41

3. ••In Fig.2.42, a 1 cm × 1 cm grid is superimposed on a contour map of the potential. Estimate the strength and direction of the electric field at points 1, 2, and 3. Show your results graphically by drawing the electric field vectors on the contour map.

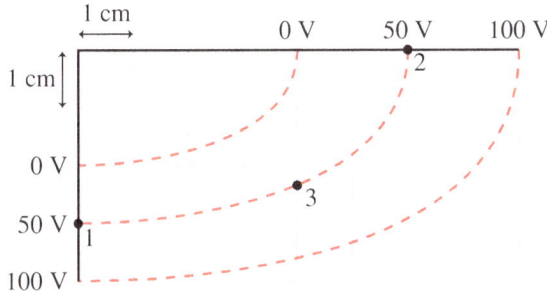

Figure 2.42: Equipotential lines

4. ••In a certain region of space, the electric potential is $V(x, y, z) = Axy - Bx^2 + Cy$, where A, B, and C are positive constants. (a) Calculate the x–, y–, and z-components of the electric field. (b) At which points is the electric field equal to zero?

5. ••In a certain region of space the electric potential is given by $V = +Ax^2y - Bxy^2$, where $A = 5.00$ V/m^3 and $B = 8.00$ V/m^3. Calculate the magnitude and direction of the electric field at the point in the region that has coordinates $x = 2.00$ m, $y = 0.400$ m, and $z = 0$

6. ••The electric potential V at any point (x, y, z) (all in metre) in space is given by $V = 4x^2$ volt. Find the electric field at the point (1 m, 0 m, 2 m).

7. ••A metal sphere with radius r_a is supported on an insulating stand at the center of a hollow, metal spherical shell with radius r_b. There is charge $+q$ on the inner sphere and charge $-q$ on the outer spherical shell.
(a) Calculate the potential $V(r)$ for (i) $r < r_a$; (ii) $r_a < r < r_b$; (iii) $r > r_b$. (**Hint:** The net potential is the sum of the potentials due to the individual spheres.) Take V to be zero when r is infinite.
(b) Show that the potential of the inner sphere with respect to the outer is
$$V_{ab} = \frac{q}{4\pi\epsilon_0}\left(\frac{1}{r_a} - \frac{1}{r_b}\right)$$
(c) Use relation $E = -\partial V/\partial r$ and the result from part (a) to show that the electric field at any point between the spheres has magnitude
$$E(r) = \frac{V_{ab}}{(1/r_a - 1/r_b)}\frac{1}{r^2}$$
(d) Use Eq. $E = -\partial V/\partial r$ and the result from part (a) to find the electric field at a point outside the larger sphere at a distance r from the centre, where $r > r_b$.

(e) Suppose the charge on the outer sphere is not $-q$ but a negative charge of different magnitude, say $-Q$. Show that the answers for parts (b) and (c) are the same as before but the answer for part (d) is different.

8. ••If $\vec{E}$ is zero throughout a certain region of space, is the potential necessarily also zero in this region? Why or why not? If not, what can be said about the potential?

Multiple Choice Questions

9. ••In Figure 2.43, two points A and B are located within a region in which there is an electric field. The potential difference $\Delta V = V_B - V_A$ is-
(A) +ve (B) –ve (C) 0 (D) insufficient data

10. ••In Fig.2.44, a point charge $+q$ is placed at the origin O. Work done in taking another point charge $-Q$ from the point A [coordinates $(0, a)$] to another point B [coordinates $(a, 0)$] along the straight path AB, is-
(A) zero
(B) $\left(\frac{qQ}{4\pi\epsilon_0}\frac{1}{a^2}\right) \cdot \sqrt{2}a$
(C) $\left(\frac{-qQ}{4\pi\epsilon_0}\frac{1}{a^2}\right) \cdot \sqrt{2}a$
(D) $\left(\frac{qQ}{4\pi\epsilon_0}\frac{1}{a^2}\right) \cdot \frac{a}{\sqrt{2}}$

Figure 2.43

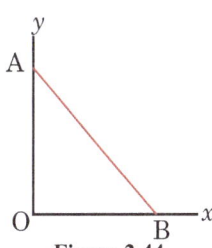
Figure 2.44

11. ••The electric potential at a point (x, y, z) is given by $V = -x^2y - xz^3 + 4$. The electric field at that point is
(A) $\vec{E} = \hat{i}2xy + \hat{j}(x^2 + y^2) + \hat{k}(3xz - y^2)$
(B) $\vec{E} = \hat{i}z^3 + \hat{j}xyz + \hat{k}z^2$
(C) $\vec{E} = \hat{i}(2xy - z^3) + \hat{j}xy^2 + \hat{k}3z^2x$
(D) $\vec{E} = \hat{i}(2xy + z^3) + \hat{j}x^2 + \hat{k}3xz^2$

12. ••If potential (in volts) in a region is expressed as $V(x, y, z) = 6xy - y + 2yz$, the electric field (in N/C) at point (1, 1, 0) is
(A) $-(2\hat{i} + 3\hat{j} + \hat{k})$ (B) $-(6\hat{i} + 9\hat{j} + \hat{k})$
(C) $-(3\hat{i} + 5\hat{j} + 3\hat{k})$ (D) $-(6\hat{i} + 5\hat{j} + 2\hat{k})$

13. ••In a region, the potential is represented by $V(x, y, z) = 6x - 8xy - 8y + 6yz$, where V is in volts and x, y, z are in metres. The electric force experienced by a charge of 2 coulomb situated at point (1, 1, 1) is
(A) $6\sqrt{5}$ N (B) 30 N
(C) 24 N (D) $4\sqrt{35}$ N

14. ••An alpha particle of energy 5 MeV is scattered through 180° by a fixed uranium nucleus. The distance of closest approach is of the order of
(A) 10^{-9} cm (B) 10^{-10} cm
(C) 10^{-12} cm (D) 10^{-15} cm

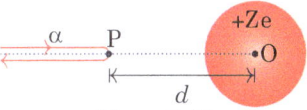
Figure 2.45

15. ••A dipole of dipole moment $\vec{p}$ is kept along an electric field $\vec{E}$ such that $\vec{E}$ and $\vec{p}$ are in the same direction. Find the

work done in rotating the dipole by an angle π.
(A) $W = 2Ep$ (B) $W = -2Ep$
(C) $W = Ep$ (D) $W = -Ep$

16. An electric dipole, made up of a positive and a negative charge, each of magnitude 1 μC and placed at a distance 2 cm apart, is placed in an electric field 10^5 N/C. Compute the maximum torque which the field can exert on the dipole, and the work that must be done to turn the dipole from a position $\theta = 0°$ to $\theta = 180°$
    (A) 6×10^{-3} N.m and 4×10^6 joule
    (B) 3×10^{-3} N.m and 4×10^9 joule
    (C) 4×10^{-3} N.m and $4 \times 10^{+6}$ joule
    (D) 2×10^{-3} N.m and 4×10^3 joule

2.7.1 Electric Potential due to a Line charge of Finite Length

EXAMPLE 29. Find the electric potential of a uniformly charged, non-conducting wire with linear density λ (in coulomb/meter) and length L at a point that lies on a line that divides the wire into two equal parts.

APPROACH We choose Cartesian coordinates in such a way as to exploit the symmetry in the problem as much as possible. We place the origin at the centre of the wire and orient the y-axis along the wire so that the ends of the wire are at $y = \pm L/2$. The field point P is in the xy-plane and since the choice of axes is up to us, we choose the x-axis to pass through the field point P, as shown in Figure 2.46.

SOLUTION Consider a small wire element dy of the charge

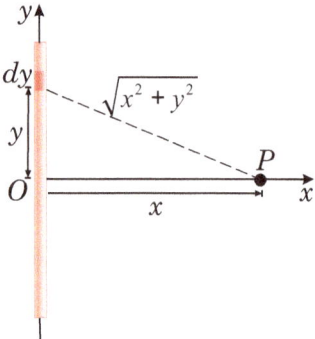

Figure 2.46: Electric potential due to a line of charge

distribution at distance y from O [Fig. 2.46]. The charge on the element is $dq = \lambda\, dy$ and the distance from the charge element to the field point P is $\sqrt{x^2 + y^2}$.
Therefore, the potential becomes

$$V = k \int \frac{dq}{r} = k \int_{-L/2}^{L/2} \frac{\lambda\, dy}{\sqrt{x^2 + y^2}}$$

$$= k\lambda \left[\ln\left(y + \sqrt{y^2 + x^2}\right) \right]_{-L/2}^{L/2}$$

$$= k\lambda \left[\ln\left(\left(\tfrac{L}{2}\right) + \sqrt{\left(\tfrac{L}{2}\right)^2 + x^2}\right) \right.$$
$$\left. - \ln\left(\left(-\tfrac{L}{2}\right) + \sqrt{\left(-\tfrac{L}{2}\right)^2 + x^2}\right) \right]$$

$$\implies \boxed{V = k\lambda \ln\left[\frac{L + \sqrt{L^2 + 4x^2}}{-L + \sqrt{L^2 + 4x^2}}\right]} \qquad (2.59)$$

2.7.2 Potential Due to an Infinite Charged Wire

In Example 29, we have already calculated the potential at any point due to a wire of finite length 'L.' Now, in Example 30, we will be considering a special case where the length of the wire is infinite.

EXAMPLE 30. Find the electric potential due to an infinitely long uniformly charged wire.

APPROACH We have already worked out the potential of a finite wire of length L in EXAMPLE 29. Now, taking $L \to \infty$ in Eq.2.59, we get

$$V = \lim_{L \to \infty} k\lambda \ln\left(\frac{L + \sqrt{L^2 + 4x^2}}{-L + \sqrt{L^2 + 4x^2}}\right)$$

$$= \lim_{L \to \infty} k\lambda \ln\left(\frac{1 + \sqrt{1 + \frac{4x^2}{L^2}}}{-1 + \sqrt{1 + \frac{4x^2}{L^2}}}\right) = k\lambda \ln\left(\frac{2}{0}\right)$$

Since, $\ln\left(\frac{2}{0}\right)$ is not defined, therefore this way of finding V of an infinite wire does not work. The reason for this problem may be traced to the fact that the charges are not localized in some space but continue to infinity in the direction of the wire. Hence, our (unspoken) assumption that zero potential must be an infinite distance from the wire is no longer valid.

To avoid this difficulty in calculating limits, we use the definition

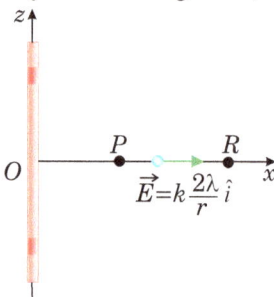

Figure 2.47: Electric potential of an infinite line of charge

of potential by integrating over the electric field, and the value of the electric field from this charge configuration from the previous chapter.

SOLUTION We use the integral

$$V_P - V_R = -\int_R^P \vec{E} \cdot d\vec{s}$$

where R is a finite distance from the line of charge, as shown in Figure 2.47.

With this setup, we use $\vec{E}_P = 2k\lambda \frac{1}{r}\hat{i}$ and $d\vec{s} = dr$ to obtain

$$V_P - V_R = -\int_{r_R}^{r_P} 2k\lambda \frac{1}{r} dr = -2k\lambda \ln \frac{r_P}{r_R}$$

Now, if we define the reference potential $V_R = 0$ at $r_R = 1$m, this simplifies to

$$\boxed{V_P = -2k\lambda \ln r_P} \qquad (2.60)$$

Note that this form of the potential is quite usable; it is 0 at 1m and is undefined at infinity, which is why we could not use the latter as a reference.

2.7.3 Electric Potential due to a Charged Thin Plane Sheet

Let's assume we have a charged thin plane sheet with surface charge density σ (charge per unit area) and it extends infinitely in both directions. The electric field due to an infinite plane sheet of charge is given by:

$$E = \frac{\sigma}{2\epsilon_0} \qquad (2.61)$$

where ϵ_0 is the electric permittivity of free space.
Now, to find the potential at a point P at distance $|z|$ from the sheet, we integrate this electric field expression with respect to distance (from a reference point at $r = \infty$ to '$|z|$' from the sheet):

$$V = -\int_\infty^{|z|} E\,dr$$

Substitute the the value of E from Eq.2.61, we get

$$V = -\int_\infty^{|z|} \frac{\sigma}{2\epsilon_0} dr = -\frac{\sigma}{2\epsilon_0} \int_\infty^{|z|} dr$$

Integrating $\int_\infty^{|z|} dr$ is problematic as it gives undefined result. Instead, we should choose a reference point at a finite distance. Let's set the reference point at $r = 0$ (the location of the charged sheet). The potential at this reference point is V_0. In this scenario, the electric potential can be calculates as follows:

$$\int_{V_0}^V dV = -\int_0^{|z|} \frac{\sigma}{2\epsilon_0} dr \implies V - V_0 = -\frac{\sigma}{2\epsilon_0}|z|$$

$$\implies \boxed{V = V_0 - \frac{\sigma}{2\epsilon_0}|z|} \qquad (2.62)$$

This is the expression for the electric potential due to a charged thin plane sheet.

2.7.4 Electric Potential Due to a Charged Ring

In Fig.2.48, a ring of radius a is shown. A charge Q is uniformly distributed over the circumference of of this ring. P is a point on the axis of the ring at a distance x from the centre of the ring, where electric potential is to be calculated.
Notice that no vector considerations are necessary here because electric potential is a scalar.

☞ Because the ring consists of a continuous distribution of charge rather than a set of discrete charges, we must use the integration technique to calculate electric potential.

We divide the ring into infinitesimal segments and use Eq.2.50 to find V. All parts of the ring (and therefore all elements of the charge distribution) are at the same distance from P.
The electric potential at P due to the charge element dq of the ring is given by

$$dV = \frac{1}{4\pi\epsilon_0} \frac{dq}{r} = \frac{1}{4\pi\epsilon_0} \frac{dq}{(a^2 + x^2)^{1/2}}$$

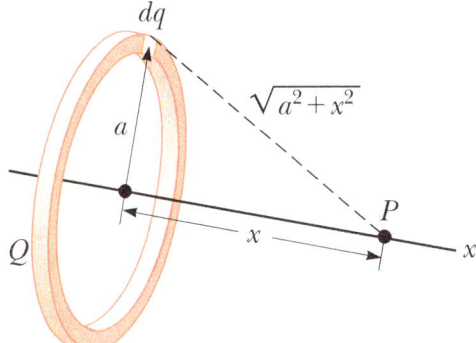

Figure 2.48

Hence, the electric potential at P due to the uniformly charged ring is given by-

$$V = \int_0^Q \frac{1}{4\pi\epsilon_0} \frac{dq}{(a^2+x^2)^{1/2}}$$

Figure 2.48 shows that the distance from each charge element dq to P is $r = \sqrt{x^2 + a^2}$. Therefore, on taking the factor $1/\sqrt{x^2 + a^2}$ outside the integral in above equation, we get-

$$V = \frac{1}{4\pi\epsilon_0} \frac{1}{(a^2+x^2)^{1/2}} \int_0^Q dq$$

or $\qquad \boxed{V = \frac{1}{4\pi\epsilon_0} \frac{Q}{\sqrt{(a^2+x^2)}}} \qquad (2.63)$

☞ Eq.2.63 is also valid for a ring segment having radius a and charge Q.
Note: When x is much larger than a, our expression for V becomes approximately $V = Q/4\pi\epsilon_0 x$, which is the potential at a distance x from a point charge Q. Very far from a charged ring, its electric potential looks like that of a point charge.

EXAMPLE 31. What is the potential on the axis of a non-uniform ring of charge, where the charge density is $\lambda(\theta) = \lambda\cos\theta$?

SOLUTION Since, $\cos\theta$ is positive in first and fourth quadrants whereas it is negative in second and third quadrants, therefore for all points on the axis, there are equal and opposite charges equidistant from the point of interest. So, electric potential at all points on the axis will be zero. Note that this distribution will, in fact, have a dipole moment.

EXAMPLE 32. A circular ring of radius R with uniform positive charge density λ per unit length is located in the y-z plane with its centre at the origin O. A particle of mass m and positive charge q is projected from the point $P(R\sqrt{3}, 0, 0)$ on the positive x-axis directly towards O, with an initial speed v. Find the smallest (non-zero) value of the speed v such that the particle does not return to P.

(a)

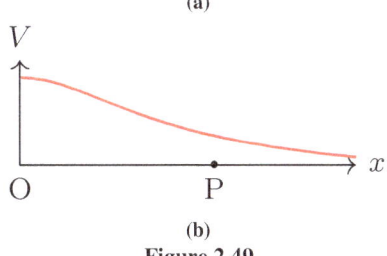

(b)
Figure 2.49

APPROACH From Eq.2.63, the electric potential at any point $P(x, 0, 0)$ on the axis of charged ring, is given by-

or $\qquad \boxed{V = \frac{1}{4\pi\epsilon_0} \frac{Q}{\sqrt{(a^2+x^2)}}} \qquad$ (i)

The potential decreases monotonically from centre O $(0, 0, 0)$ to $P(x, 0, 0)$ and it is maximum for $x = 0$, i.e., at the centre O of the ring. So, the particle cannot come back to P if it just crosses O. Since, the force acting on the particle is only electrostatic force, which is conservative in nature, therefore, the total mechanical energy of the ring-particle system will remain conserved. Now, apply conservation of mechanical energy to get the velocity of particle at point P.

SOLUTION The charge on the ring is $Q = 2\pi R\lambda$. Therefore, from Eq.(i), the electric potential at a point $P(R\sqrt{3}, 0, 0)$-

$$V_P = \frac{1}{4\pi\epsilon_0} \frac{Q}{2R} \qquad (ii)$$

The electric potentials at the point $O(0, 0, 0)$-

$$V_O = \frac{1}{4\pi\epsilon_0} \frac{Q}{R}, \qquad (iii)$$

Net mechanical energy of the particle-ring system at point P $(R\sqrt{3}, 0, 0)$-
$$E_P = K_P + U_{el} = \frac{1}{2}mv^2 + qV_P \qquad (iv)$$
here, v is the speed of the particle at point P.
If the charged particle just crosses the centre O of the ring, then it's velocity and hence it's kinetic energy at O will be zero. So, net mechanical energy of the particle ring system at point O (0,0,0)-
$$E_O = qV_O \qquad (v)$$
Applying, the principle of conservation of mechanical energy for points P and O, we get-
$$E_P = E_O$$
Substituting the values from Eq.v and iv in above, we get-
$$\frac{1}{2}mv^2 + qV_P = qV_O$$
or $\quad \frac{1}{2}mv^2 = q(V_O - V_P) \qquad (vi)$
Substituting the values of V_P and V_O, from Eq.ii and iii in Eq.vi, we get
$$\frac{1}{2}mv^2 = q\left(\frac{1}{4\pi\epsilon_0}\frac{Q}{R} - \frac{1}{4\pi\epsilon_0}\frac{Q}{2R}\right) = \frac{1}{4\pi\epsilon_0}\frac{qQ}{2R}$$
Since, $Q = \lambda(2\pi R)$, therefore-
$$\frac{1}{2}mv^2 = \frac{1}{4\pi\epsilon_0}\frac{q\lambda(2\pi R)}{2R} = \frac{q\lambda}{4\epsilon_0}$$
Solving for v, we get
$$v = \sqrt{q\lambda/(2\epsilon_0 m)}$$

EXAMPLE 33. Two identical thin rings, each of radius R, are co-axially placed a distance R apart. If Q_1 and Q_2 are respectively the charges uniformly spread on the two rings, the work done in moving a charge q slowly from the centre of one ring to that of the other is:

(A) zero (B) $q(Q_1 - Q_2)\left(\dfrac{\sqrt{2}-1}{\sqrt{2}\pi\epsilon_0 R}\right)$

(C) $q\sqrt{2}\left(\dfrac{Q_1 + Q_2}{4\pi R}\right)$ (D) $q(Q_1/Q_2)\left(\dfrac{\sqrt{2}+1}{\sqrt{2}\pi\epsilon_0 R}\right)$

APPROACH To find the work done in moving a charge q from the centre of one ring to that of the other, apply relation-
$$W = \Delta U = U_B - U_A$$
here U_A is the electric potential energy of the system when the charged particle was at any point on ring 1 and U_B is the potential energy of the system for the position of the charged particle on ring 2.

SOLUTION (B) The potential at A due to the charge Q_1 on the ring 1 is given as:
$$V_{A1} = k\frac{Q_1}{R}$$
The potential at A due to the charge Q_2 on the ring 2 is given as:
$$V_{A2} = k\frac{Q_2}{\sqrt{R^2 + R^2}} = k\frac{Q_2}{R\sqrt{2}}$$
Total potential at A is
$$V_A = V_{A1} + V_{A2} = k\frac{1}{R}\left(Q_1 + \frac{Q_2}{\sqrt{2}}\right)$$
The potential energy of charge q at A is
$$U_A = qV_A = k\frac{q}{R}\left(Q_1 + \frac{Q_2}{\sqrt{2}}\right)$$
Similarly, the potential energy of charge q at B is
$$U_B = qV_B = k\frac{q}{R}\left(Q_2 + \frac{Q_1}{\sqrt{2}}\right)$$
The work done in moving a charge q from point A to B is:

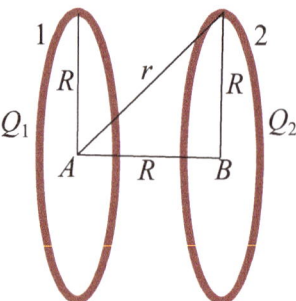

Figure 2.50

$$W = \Delta U = U_B - U_A = k\frac{q}{R}\left(Q_2 + \frac{Q_1}{\sqrt{2}} - Q_1 - \frac{Q_2}{\sqrt{2}}\right)$$
$$= \frac{1}{4\pi\epsilon_0 R}(Q_2 - Q_1)\frac{\sqrt{2}-1}{\sqrt{2}}$$

EXAMPLE 34. Calculate the electric potential at the centre of two concentric half rings of radii R_1 and R_2, having same linear charge density λ.

SOLUTION The situation is shown in Fig.2.51. The net potential (V_{centre}) at the centre will be equal to algebraic sum of potentials produced due to both semicircular concentric half rings.
$$V_{\text{centre}} = \frac{kQ_1}{R_1} + \frac{kQ_2}{R_2} \qquad \cdots (1)$$

Figure 2.51

here, Q_1 and Q_2 are charges on semicircular rings of radii R_1 and R_2 respectively.
Since, charge density on each semicircular ring is λ, therefore
$$Q_1 = \lambda \cdot \pi R_1 \text{ and } Q_2 = \lambda \cdot \pi R_2$$
Substituting these values in Eq.(1), we get-
$$V_{\text{centre}} = \frac{(\lambda \cdot \pi R_1)}{4\pi\epsilon_0 R_1} + \frac{(\lambda \cdot \pi R_2)}{4\pi\epsilon_0 R_2} = \frac{\lambda}{2\epsilon_0}$$

2.7.5 Electric Potential Due to a Charged Disc at a Point on it's Geometric Axis

A non-conducting disc of radius R has a uniform surface charge density σ C/m^2. Let us calculate the potential at a point on the axis of the disc at a distance x from its centre. The symmetry of the disc tells us that the appropriate choice of element is a ring of radius x and thickness dx. All points on this ring are at the same distance $z = \sqrt{x^2 + r^2}$, from the point P. The charge on the ring is $dq = \sigma dA = \sigma(2\pi r dr)$ and so the potential due to the ring is
$$dV = \frac{1}{4\pi\epsilon_0}\frac{dq}{z} = \frac{1}{4\pi\epsilon_0}\frac{(\sigma 2\pi r dr)}{\sqrt{r^2 + x^2}}$$
Since potential is scalar, the potential due to the whole disc is given by
$$V = \frac{\sigma}{2\epsilon_0}\int_0^R \frac{r dr}{\sqrt{r^2 + x^2}}$$
$$= \frac{\sigma}{2\epsilon_0}[(r^2 + x^2)^{1/2}]_0^R = \frac{\sigma}{2\epsilon_0}[(R^2 + x^2)^{1/2} - x]$$
or $\quad \boxed{V = \dfrac{\sigma}{2\epsilon_0}[(R^2 + x^2)^{1/2} - x]} \qquad (2.64)$

Let us see this expression at large distance when $x \gg R$. In this case, $R^2 + x^2 \approx x^2$, therefore, corresponding to $x \gg R$, we have-

2.7. CHECK POINT 2

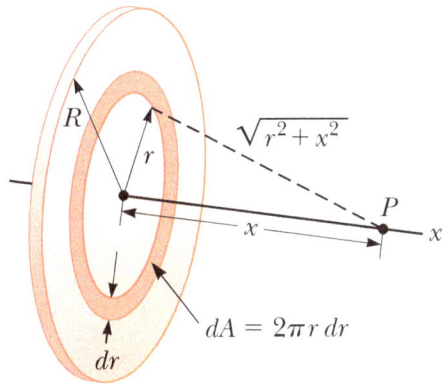

Figure 2.52

$$V = \frac{1}{4\pi\epsilon_0}\frac{Q}{x}$$

where $Q = \pi R^2 \sigma$ is the total charge on the disc.
Thus, we conclude that at large distance, the potential due to the disc is the same as that of a point charge Q.
At the centre of the disc $x = 0$, therefore

$$V = \frac{\sigma}{2\epsilon_0} R$$

EXAMPLE 35. A non-conducting disc of radius a and uniform positive surface charge density σ is placed on the ground with its axis vertical. A particle of mass m and positive charge q is dropped, along the axis of the disc from a height H with zero initial velocity. The particle has $q/m = 4\epsilon_0 g/\sigma$
(a) Find the value of H if the particle just reaches the disc.
(b) Sketch the potential energy of the particle as a function of its height and find its equilibrium position.

(a) APPROACH There are two forces on the charged particle:
1. Downward gravitational force
2. Upward electrostatic repulsive force due to electric field produced by the disc.

Both forces are conservative in nature, therefore, total mechanical energy of the earth-disc-particle system, will always remain conserved.
To find the height H, from where the particle is released, we apply the principle of conservation of mechanical energy.
Electric potential due to disc of radius R at distance x from it's centre is given by Eq.2.64-

$$V = \frac{\sigma}{2\epsilon_0}[(R^2 + x^2)^{1/2} - x] \quad \text{(i)}$$

In this problem, radius, $R = a$ and $x = H$, therefore V at height H-

$$V_P = \frac{\sigma}{2\epsilon_0}[(a^2 + H^2)^{1/2} - H]$$

Considering the position of disc as reference level, the gravitational potential energy of the earth-particle system for the particle at height H is-

$$[U_{\text{grav}}]_P = mgH$$

The electric potential energy of the disc particle system for the position of particle at height H-

$$[U_{\text{el}}]_P = qV_P = \frac{q\sigma}{2\epsilon_0}[(a^2 + H^2)^{1/2} - H]$$

Since, the particle is released from rest, therefore it's kinetic energy at height H-

$$K_P = 0$$

So, net mechanical energy of the disc-particle-earth system for the position of particle at height H, is given by-

$$E_P = [U_{\text{grav}}]_i + [U_{\text{el}}]_i + K_i$$
$$= mgH + \frac{q\sigma}{2\epsilon_0}[(a^2 + H^2)^{1/2} - H] + 0$$

$$E_P = mgH + \frac{q\sigma}{2\epsilon_0}[(a^2 + H^2)^{1/2} - H] \quad \text{(iii)}$$

Electric potential at the centre O of disc is-

$$V_O = \frac{\sigma a}{2\epsilon_0}$$

Therefore, electrostatic potential energy of the disc-particle system for particle at O is

$$[U_{\text{el}}]_O = qV_O = q\frac{\sigma a}{2\epsilon_0}$$

Since, disc surface is selected as reference level for gravitational potential energy, therefore at O-

$$[U_{\text{grav}}]_O = 0$$

Since, the particle just reaches the disc, so its kinetic energy at O will be zero.

$$K_O = 0$$

So, net mechanical energy of the earth-disc-particle system for position of particle at O, is given by-

$$E_O = [U_{\text{grav}}]_f + [U_{\text{el}}]_O + [K]_O$$
$$\implies E_O = q\frac{\sigma a}{2\epsilon_0} + 0 + 0 = q\frac{\sigma a}{2\epsilon_0} \quad \text{(iv)}$$

Now, apply conservation of mechanical energy and solve for H

SOLUTION By conservation of mechanical energy, we have

$$E_P = E_O \quad \text{(v)}$$

Substituting the values of E_P and E_O in Eq.(v), we get

$$mgH + \frac{q\sigma}{2\epsilon_0}[(a^2 + H^2)^{1/2} - H] = q\frac{\sigma a}{2\epsilon_0}$$

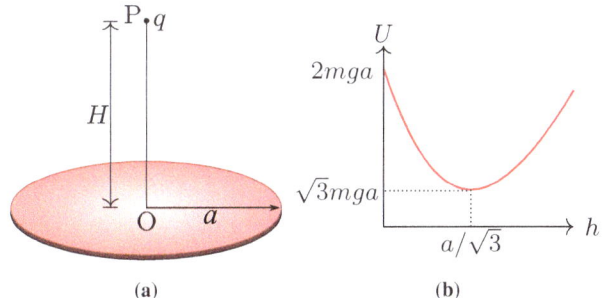

Figure 2.53

$$\implies mgH = \frac{q\sigma}{2\epsilon_0}\left[a + H - \sqrt{a^2 + H^2}\right]$$

$$\implies H = \frac{q\sigma}{2\epsilon_0 mg}\left[a + H - \sqrt{a^2 + H^2}\right]$$

On substitute $q/m = 4\epsilon_0 g/\sigma$, we get

$$H = 2\left[a + H - \sqrt{a^2 + H^2}\right]$$

$$\implies \sqrt{a^2 + H^2} = a + H/2 \implies H = 4a/3$$

(b) APPROACH Net conservative force acting on the particle, is equal to $-ve$ of potential energy gradient, i.e.,

$$F = -\frac{\partial U}{\partial H} \quad \text{(vi)}$$

At equilibrium position, $F = 0$, so from Eq.(vi), we have-

$$\frac{\partial U}{\partial H} = 0$$

Therefore, the potential energy attains extremum at the equilibrium position.

SOLUTION Total potential energy (U) of the particle at point P is the sum of its gravitational and electrostatic potential energies i.e.,

$$U = mgH + \frac{q\sigma}{2\epsilon_0}\left[\sqrt{a^2 + H^2} - H\right]$$
$$= mg\left[2\sqrt{a^2 + H^2} - H\right]$$

At equilibrium position, we have

$$\frac{dU}{dH} = mg\left[\frac{2H}{\sqrt{a^2+H^2}} - 1\right] = 0$$

which gives $H_{min} = a/\sqrt{3}$ and $U_{min} = \sqrt{3}mga$. The Fig. 2.53 shows the variation of U with height H. Note that the equilibrium is stable (i.e., $d^2U/dH^2 > 0$).

EXAMPLE 36. A disc of radius R with uniform positive charge density σ is placed on the xy plane with its centre at the origin. The Coulomb potential along the z-axis is

$$V(z) = \frac{\sigma}{2\epsilon_0}\left(\sqrt{R^2+z^2} - z\right)$$

A particle of positive charge q is placed initially at rest at a point on the z axis with $z = z_0$ and $z_0 > 0$. In addition to the Coulomb force, the particle experiences a vertical force $\vec{F} = -c\hat{k}$ with $c > 0$. Let $\beta = \frac{2c\epsilon_0}{q\sigma}$. Which of the following statement(s) is (are) correct?

(A) For $\beta = 1/4$ and $z_0 = 25R/7$, the particle reaches the origin.
(B) For $\beta = 1/4$ and $z_0 = 3R/7$, the particle reaches the origin.
(C) For $\beta = 1/4$ and $z_0 = R/\sqrt{3}$, the particle returns back to $z = z_0$.
(D) For $\beta > 1$ and $z_0 > 0$, the particle always reaches the origin.

SOLUTION By work–kinetic energy theorem, we have-
$$W_{el} + W_{ext} = K_f - K_i$$
$$\implies qV_i - qV_f + W_{ext} = K_f - K_i$$

Here, $V_i = \frac{\sigma}{2\epsilon_0}\left(\sqrt{R^2 + z_0^2} - z_0\right)$,

$V_f = \frac{\sigma}{2\epsilon_0}\left(\sqrt{R^2+0^2} - 0\right) = \frac{\sigma R}{2\epsilon_0}$,

$W_{ext} = cz = \frac{q\sigma\beta z}{2\epsilon_0}\left(\because \beta = \frac{2c\epsilon_0}{q\sigma}\right)$ and $K_i = 0$

Substituting these values and solving for K_f, we get
$$\implies \frac{q\sigma}{2\epsilon_0}\left[\sqrt{R^2+z^2} - z\right] - \frac{q\sigma R}{2\epsilon_0} + \frac{q\sigma\beta z}{2\epsilon_0} = K_f - 0$$
$$\implies K_f = \frac{q\sigma}{2\epsilon_0}\left[\sqrt{R^2+z^2} - z\right] - \frac{q\sigma R}{2\epsilon_0} + \frac{q\sigma\beta z}{2\epsilon_0}$$

Now, substitute the values of β and z_0 from given options and solve for final kinetic energy K_f.
If kinetic energy is positive, then particle will reach at origin
If kinetic energy is negative, then particle will not reach at origin.
Thus correct choices are A, C and D.

2.7.6 Electric Potential Due to a Charged Conducting Shell or Sphere

Suppose, charge Q is uniformly distributed over the surface of a conducting shell of radius R. We will calculate the electric potential at a point-
(a) outside the shell; $(r > R)$
(b) on the surface of the shell $(r = R)$
(c) inside the shell $(r < R)$.
(a) At points outside a uniform spherical distribution, the electric field is
$$\vec{E} = \frac{1}{4\pi\epsilon_0}\frac{Q}{r^2}\hat{r}$$
Since, $\vec{E}$ is radially outward, therefore-
$$\vec{E}.d\vec{s} = E\,dr\cos 0 = E\,dr$$
Since, $V(\infty) = 0$, therefore we have-
$$V(r) - V(\infty) = -\int_\infty^r \vec{E}\cdot d\vec{r}$$
$$\implies V - 0 = -\int_\infty^r \frac{Q}{4\pi\epsilon_0 r^2}dr$$

Integrating right hand side, we get-
$$\boxed{V = \frac{1}{4\pi\epsilon_0}\frac{Q}{r}} \qquad (2.65)$$

From Eq.2.65, we see that the potential due to a uniformly charged shell is the same as that due to a point charge Q at the center of the shell.
(b) For a point on the surface of the shell, $r = R$, therefore Eq.2.65, we have-
$$\boxed{V = \frac{1}{4\pi\epsilon_0}\frac{Q}{R}} \qquad (2.66)$$
(c) At points inside the shell, $E = 0$. So, the work done in bringing a unit positive charge from a point on the surface to any point inside the shell is zero. Thus, the potential has a fixed value at all points within the spherical shell and is equal to the potential at the surface.
$$\boxed{V = \frac{1}{4\pi\epsilon_0}\frac{Q}{R}} \qquad (2.67)$$

Variation of electric potential with the distance from the centre (r) is given in Fig.2.54.

☞*Note that, all above results also hold for a conducting sphere*

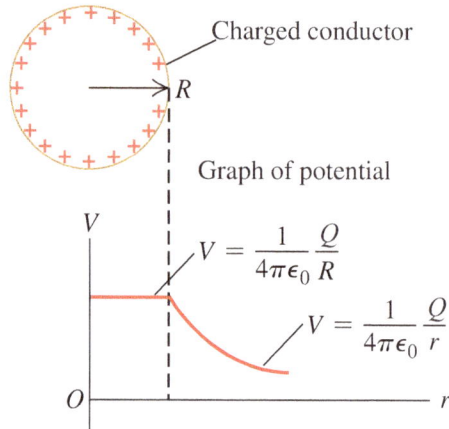

Figure 2.54: Electric potential V at points inside and outside a positively charged spherical shell/spherical conductor

whose charge lies entirely on the outer surface.

2.7.7 Electric Potential Due to a Non-Conducting Charged Sphere

Suppose a charge Q is uniformly distributed throughout a non-conducting spherical volume of radius R. Let the volume charge density of this non-conducting sphere is ρ. We have to find expressions for the potential at-
(a) an external point $(r > R)$; (b) on the surface $(r = R)$;
(c) an internal point $(r < R)$
where r is the distance of the point from the centre of the sphere.
(a) Electric Potential at an External Point of Uniformly Charged Sphere: Let O be the centre of a non-conducting sphere of radius R, have a charge Q distributed uniformly over its entire volume.
For $r \geq R$, the electric field E is radial and has a magnitude:
$$E_r = k\frac{Q}{r^2} \quad (r \geq R) \qquad (2.68)$$
This is the same as the electric field due to a point charge, and hence the electric potential at any point of radius r in this region is given by:
$$V_r = -\int_\infty^r E_r\,dr = -\int_\infty^r \frac{kQ}{r^2}dr$$

Integrating right hand side, we get

$$V_r = k\frac{Q}{r} \quad (r \geq R) \quad (2.69)$$

(b) Electric Potential at a Point on the Surface of the Uniformly Charged Non-conducting Sphere: For a point on the surface of the non-conducting sphere, $r = R$, therefore Eq.(2.69) gives-

$$V_R = k\frac{Q}{R} \quad (2.70)$$

(c) Electric Potential at an internal Point of Uniformly Charged Sphere ($r < R$): To find the potential at some point r inside the sphere, we use Eq.2.16,

$$V_B - V_A = -\int_{r_A}^{r_B} E\,dr$$

along a radial path that begins at some point where the potential is known. We choose a point on the surface of the sphere, and integrate the electric field from that point (at $r_A = R$) to a point inside where we want to find the potential (at $r_B = r$). From Eq.1.145, the electric field at any point inside the non conducting sphere of radius R at $r = r$, is given by $E_r = k\frac{Qr}{R^3}$ and electric potential at $r = R$ is $V_R = k\frac{Q}{R}$, therefore-

$$V_r - V_R = -\int_R^r E_r\,dr = -k\frac{Q}{R^3}\int_R^r r\,dr$$

$$= -k\frac{Q}{R^3}\left|\frac{r^2}{2}\right|_{r=R}^{r=r} = k\frac{Q}{2R^3}\left(R^2 - r^2\right)$$

$$\implies V_r - V_R = k\frac{Q}{2R^3}\left(R^2 - r^2\right) \quad (2.71)$$

Using $V_R = kQ/R$ in the last result we reach to the following relation:

$$V_r = k\frac{Q}{2R}\left(3 - \frac{r^2}{R^2}\right) \quad (0 \leq r \leq R) \quad (2.72)$$

At $r = 0$, we have $V_0 = 3kQ/2R$, and at $r = R$, we get $V_R = kQ/R$ as expected. Figure 2.55 shows the electric potential in the two regions $0 \leq r \leq R$ and $r \geq R$.

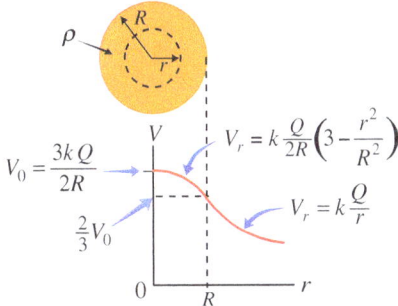

Figure 2.55: A sketch of the electric potential $V(r)$ as a function of r in the two regions $0 \leq r \leq R$ and $r \geq R$. The curve for the region $0 \leq r \leq R$ is parabolic and joins smoothly with the curve for the region $r \geq R$, which is hyperbola

☞ A point charge q can be considered to be the limiting case of a small spherical conductor whose radius tends to zero and potential to infinity.

EXAMPLE 37. In the Fig.2.56, the inner (shaded) region A represents a sphere of radius $r_A = 1$, within which the electrostatic charge density varies with the radial distance r from the centre as $\rho_A = kr$, where k is positive. In the spherical shell B of outer radius r_B, the electrostatic charge density varies as $\rho_B = 2k/r$. Assume that dimensions are taken care of. All physical quantities are in their SI units.

Which of the following statement(s) is(are) correct?

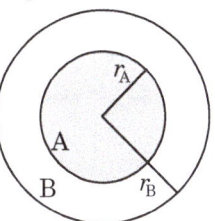

Figure 2.56

(A) If $r_B = \sqrt{3/2}$, then the electric field is zero everywhere outside B.
(B) If $r_B = 3/2$, then the electric potential just outside B is k/ϵ_0.
(C) If $r_B = 2$, then the total charge of the configuration is $15\pi k$.
(D) If $r_B = 5/2$, then the magnitude of the electric field just outside B is $13\pi k/\epsilon_0$.

SOLUTION $q_1 = \int_0^1 kr 4\pi r^2 dr = \frac{4\pi k}{4} = \pi k$,

$$q_2 = \int_1^r \frac{2k}{r} 4\pi r^2 dr = \frac{8\pi k\left(r^2 - 1^2\right)}{2},$$

$$= 4\pi k\left[r^2 - 1\right] = 4\pi k r^2 - 4\pi k$$

Therefore, $q_{\text{net}} = q_1 + q_2$
$$= 4\pi k r r^2 - 3\pi k = \pi k\left[4r^2 - 3\right]$$

(A) $E_{\text{net}} = 0 \implies q_{\text{net}} = 0 \implies r = \frac{\sqrt{3}}{2}$

(B) $V = \frac{kQ_{\text{net}}}{r} = \frac{1}{4\pi\epsilon_0}\frac{\pi k\left(4r^2 - 3\right)}{r} = \frac{k}{4\epsilon_0}\left[4r - \frac{3}{r}\right]$

$$= \frac{k}{4\epsilon_0}\left[4 \times \frac{3}{2} - \frac{3 \times 2}{3}\right] = \frac{k}{\epsilon_0}$$

(C) $q_{\text{sat}} = \pi k\left[4(2)^2 - 3\right] = 13\pi k$

(D) $E_2 = \frac{kQ}{r^2} = \frac{1}{4\pi\epsilon_0}\frac{\pi k\left(4r^2 - 3\right)}{r^2} = \frac{k}{4\epsilon_0}\left[\frac{4\left(\frac{5}{2}\right)^2 - 3}{(5/2)^2}\right]$

$$= \frac{k}{25\epsilon_0}[25 - 3] = \frac{22}{25}\frac{k}{\epsilon_0}$$

From above it is evident that the correct choice is B.

2.7.8 Potential on the Edge of a Uniformly Charged Disc:

To calculate the potential at point O on the circumference of the disc, let us divide the disc in large number of rings with O as centre. ABD and EFG are the arcs of two such concentric rings having common centre O. The potential due to one segment between r and $r + dr$ is given by-

$$dV = \frac{1}{4\pi\epsilon_0}\cdot\frac{dq}{r}$$

Here, $dq = \sigma$ (Area of ring) $= \sigma$ (arc $ABD \times dr$)
$= \sigma\left(r \times 2\theta \times dr\right) = \sigma\left(2r\theta\right)dr$

$$\therefore \quad dV = \frac{1}{4\pi\epsilon_0}\frac{\sigma\left(2r\theta\right)dr}{r} = \frac{\sigma}{2\pi\epsilon_0}\theta\,dr$$

Further, $r = OH\cos\theta = 2R\cos\theta$
$\therefore \quad dr = -2R\sin\theta\,d\theta$
Hence, $dV = -\frac{\sigma}{2\pi\epsilon_0}2R\theta\sin\theta\,d\theta$

$$\therefore \quad V = \int_{\pi/2}^0 dV = \frac{\sigma R}{\pi\epsilon_0}\int_0^{\pi/2}\theta\sin\theta\,d\theta$$

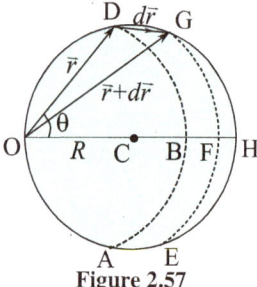

Figure 2.57

Simplifying it, we get -

$$V = \frac{\sigma R}{\pi \epsilon_0} \qquad (2.73)$$

EXAMPLE 38. A point charge q_1 is at the origin, and a second point charge q_2 is on the x axis at $x = a$. Find an expression for the electric potential everywhere on the x axis as a function of x.
APPROACH Find electric potentials due to each charge sepa-

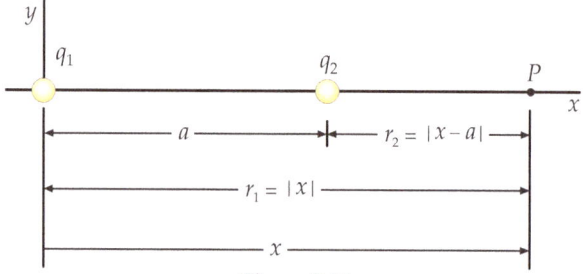

Figure 2.58

rately at a point on the x-axis and add them to find the net electric potential at that point.
SOLUTION Fig.2.58 shows the x axis and the two q_1, q_2 charges on it. Let r_1 be the distance from q_1 to an arbitrary field point P at position x on the x axis, that is, $r_1 = |x|$. Let r_2 be the distance from q_2 to P, that is, $r_2 = |x - a|$.
The potential as a function of the distances to the two charges, is given as-

$$V = \frac{kq_1}{r_1} + \frac{kq_2}{r_2} = \frac{kq_1}{|x|} + \frac{kq_2}{|x-a|} \qquad (x \neq 0, x \neq a)$$

Note that, $V \to \infty$ both as $x \to 0$ and as $x \to a$, and $V \to 0$ both

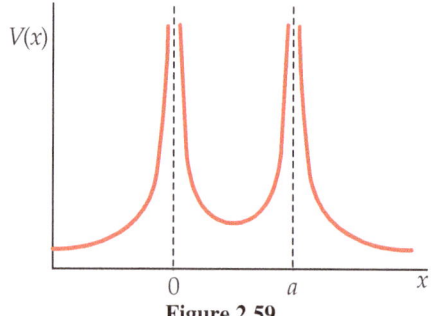

Figure 2.59

as $x \to -\infty$ and as $x \to \infty$ as one would expect.
Figure 2.59 shows V versus x on the x axis for $q_1 = q_2 > 0$.
EXAMPLE 39. 64 identical drops each charged up to potential of 10 mV are combined to form a bigger drop. Find the potential of the bigger drop.

SOLUTION If q is the charge on each drop and r is it's radius, then electric potential at the surface of the drop-

$$V = k\frac{q}{r} = 10\,\text{mV} \qquad \cdots (1)$$

If the radius of above 64 combined drops is R, then

$$64 \times (\text{volume})_{\text{smaller drop}} = (\text{volume})_{\text{bigger drop}}$$
$$\implies 64 \times \frac{4}{3}\pi r^3 = \frac{4}{3}\pi R^3 \implies R = 4r$$

And by charge conservation, the net charge on bigger drop, $Q = 64q$. Therefore, surface potential of the bigger drop-

$$V_{\text{bigger drop}} = \frac{kQ}{R} = \frac{k(64q)}{4r} = 16\frac{kq}{r} = 16 \times 10\,\text{mV}$$
$$= 160\,\text{mV} \quad (\text{since, from Eq.(1)}, kq/r = 10\,\text{mV})$$

2.8 Equipotentials and Conductors

☞ When all charges are at rest, the surface of a conductor is always an equipotential surface.

The electric field $\vec{E}$ is always perpendicular to an equipotential

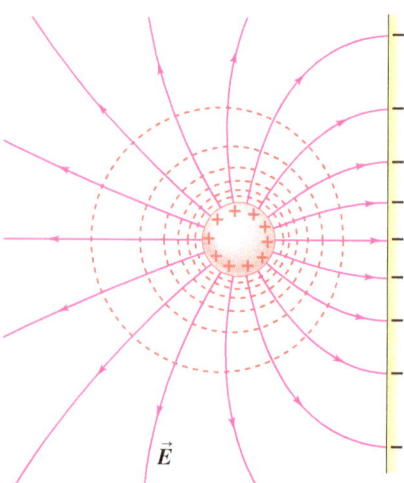

---- Cross sections of equipotential surfaces
→ Electric field lines

Figure 2.60: When charges are at rest, a conducting surface is always an equipotential surface. Field lines are perpendicular to a conducting surface.

surface.
To prove it by conservative nature of electric field, let us transport a test charge q_0 around the loop $abcda$ as shown in Fig. 2.61a. The segments bc and da (and the associated work) may be made arbitrarily small (because the length is very small. Think $\vec{F}.\vec{s}$ i.e., work), and no work is done in the segment cd because, as already shown, the field is zero everywhere inside the conductor. If the field just outside the conductor has a component $E_\parallel$, parallel to the surface, this component does work equal to $q_0 E_\parallel$ that is the parallel component of $\vec{E}$ vector at the surface (if any) does work on the test charge q_0 equal to q_0 times l (length of ab) times parallel component of $\vec{E}$ at surface, i.e., $E_\parallel$.
So, the net work done in transporting the test charge q_0 along a closed path $abcda$,

$$W_{abcda} = W_{ab} + W_{bc} + W_{cd} + W_{da} \neq 0$$

i.e., the net work done is non zero. It is impossible, because, it shows that the electric force field is non-conservative. To avoid this contradiction, we must conclude that there cannot be a component of $\vec{E}$ parallel to the surface, and that $\vec{E}$ is therefore perpendicular to the surface.
Thus, $\vec{E}$ is perpendicular to the surface at each point, proving our statement.
It also follows that when all charges are at rest, the entire solid volume of a conductor is at the same potential. Eq.2.14: $\left[V_B - V_A = \frac{W_{AB}}{q_0} = -\int_A^B \vec{E}\cdot d\vec{s}\right]$, states that the potential

2.8. EQUIPOTENTIALS AND CONDUCTORS

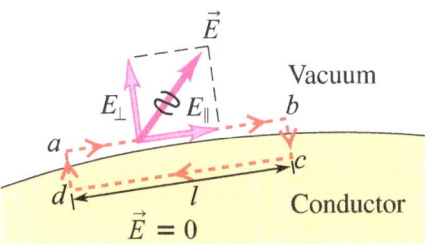

(a) Electric field at the surface of conductor.

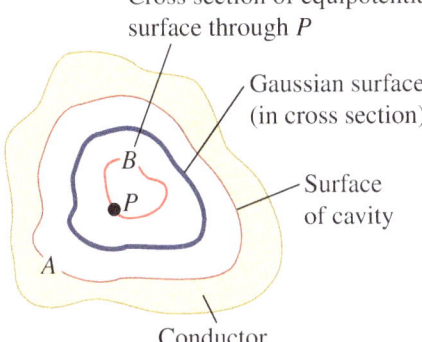

(b) Cavity in a conductor.

Figure 2.61: (a) At all points on a conductor's surface, the electric field must be perpendicular to the surface. If $\vec{E}$ had a tangential component, a net amount of work would be done on a test charge by moving it around a loop as shown-which is impossible because the electric force is conservative. (b) If the cavity contains no charge, every point in the cavity is at the same potential, the electric field is zero everywhere in the cavity, and there is no charge anywhere on the surface of the cavity.

difference between two points A and B within the conductor's solid volume, $V_B - V_A$, is equal to the line integral $\int_A^B \vec{E} \cdot d\vec{s}$ of the electric field from A to B. Since $\vec{E} = 0$ everywhere inside the conductor, the integral is guaranteed to be zero for any two such points A and B. Hence the potential is the same for any two points within the solid volume of the conductor. We describe this by saying that the solid volume of the conductor is an equipotential volume.

Theorem 1: *In an electrostatic situation, if a conductor contains a cavity and if no charge is present inside the cavity, then there can be no net charge anywhere on the surface of the cavity.*

Proof: To prove this theorem, we first prove that every point in the cavity is at the same potential. In Fig.2.61b, the conducting surface A of the cavity is an equipotential surface. Suppose point P in the cavity is at a different potential; then we can construct a different equipotential surface B including point P.
Now consider a Gaussian surface, shown in Fig.2.61b, between the two equipotential surfaces. Because of the relationship between $\vec{E}$ and the equipotentials, we know that the field at every point between the equipotentials is from A towards B, or conversely, at every point, it is from B towards A, depending on which equipotential surface is at higher potential.
In either case the flux through this Gaussian surface is certainly not zero. But then Gauss's law says that the charge enclosed by the Gaussian surface cannot be zero. This contradicts our initial assumption that there is no charge in the cavity. So the potential at P cannot be different from that at the cavity wall.
The entire region of the cavity must therefore be at the same potential. But for this to be true, the electric field inside the cavity must be zero everywhere. Finally, Gauss's law shows that the electric field at any point on the surface of a conductor is proportional to the surface charge density σ at that point. We conclude that the surface charge density on the wall of the cavity is zero at every point.

☞ Don't confuse equipotential surfaces with the Gaussian surfaces. Gaussian surfaces have relevance only when we are using Gauss's law, and we can choose any Gaussian surface that's convenient. We cannot choose equipotential surfaces; the shape is determined by the charge distribution.

EXAMPLE 40. A solid conducting sphere, having a charge Q, is surrounded by an uncharged conducting hollow spherical shell. Let the potential difference between the surface of the solid sphere and that of the outer surface of the hollow shell be V. If the shell is now given a charge of $-4Q$, calculate the new potential difference between the same two surfaces.

APPROACH First indicate the charge distribution and then find the potential difference by applying the formula of potential at any point due to charged sphere/spherical shell.

SOLUTION Given situation is shown in Fig.2.62. Since, the

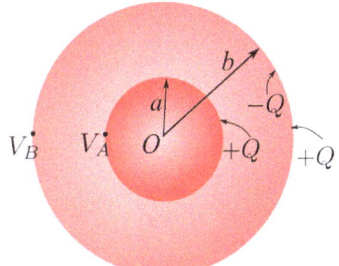

Figure 2.62

uncharged shell initially encloses charge Q, therefore, due to induction, the charge on inner surface of the shell will be $-Q$. By charge conservation, the charge on the outer surface of the shell will be $+Q$.
The electric potential at any point A on the inner surface of shell is given by-

$$V_A = \frac{kQ}{a} + \frac{k(-Q)}{b} + \frac{kQ}{b} \qquad (2.74)$$

where, $k = \dfrac{1}{4\pi\epsilon_0} = 9.0 \times 10^9 \text{N.m}^2/\text{C}^2$

Electric potential on surface of outer shell is

$$V_B = \frac{kQ}{b} + \frac{k(-Q)}{b} + \frac{kQ}{b} \qquad (2.75)$$

Therefore, the potential difference is

$$\Delta V_{AB} = V_A - V_B = kQ\left(\frac{1}{a} - \frac{1}{b}\right)$$

Given, that $\Delta V_{AB} = V$, so-

$$kQ\left(\frac{1}{a} - \frac{1}{b}\right) = V \qquad (2.76)$$

Now, if the shell is given an extra charge $-4Q$, it will uniformly get spread on the outer surface of the shell. In this case, the new potential difference will be

$$\Delta V_{AB} = V_A - V_B$$
$$= \left(\frac{kQ}{a} + \frac{k(-Q)}{b}\right) - \left(\frac{kQ}{b} + \frac{k(-4Q)}{b}\right)$$
$$= kQ\left(\frac{1}{a} - \frac{1}{b}\right) = V \quad [\text{ from Eq.(2.76)}]$$

Remark: It is a very important result which shows that the potential difference between the outer surface of the spherical shell and inner sphere is independent on the charge of the shell.

EXAMPLE 41. Two conducting spheres 1 and 2, having radii a and b charged to q_1 and q_2 respectively. Find the potential difference between 1 and 2.

SOLUTION The potential on the surface of the sphere 1 is given

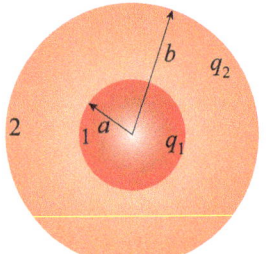

Figure 2.63

by -
$$V_1 = \frac{1}{4\pi\epsilon_0}\frac{q_1}{a} + \frac{1}{4\pi\epsilon_0}\frac{q_2}{b} \qquad (2.77)$$

The potential on the surface of the sphere 2 is given by,
$$V_2 = \frac{1}{4\pi\epsilon_0}\frac{q_1}{b} + \frac{1}{4\pi\epsilon_0}\frac{q_2}{b} \qquad (2.78)$$

$$\Rightarrow \quad V_1 - V_2 = \frac{1}{4\pi\epsilon_0}\frac{q_1}{a} - \frac{1}{4\pi\epsilon_0}\frac{q_1}{b}$$
$$= \frac{q_1}{4\pi\epsilon_0}\left(\frac{1}{a} - \frac{1}{b}\right)$$

EXAMPLE 42. A point charge q is kept at a distance l from a charged conducting sphere of radius R ($l > R$), having charge Q over its surface. Find electric potential due to induced charges at any point P inside the sphere.

SOLUTION The given situation is shown in Fig.2.64. Potential due to induced charges is zero as centre C is equidistant from all induced charges. Since, every point inside this conducting sphere will be at equal potential, therefore-
$$V_C = V_P = \frac{kq}{l} + \frac{kQ}{R}$$
Also at point P, we can write

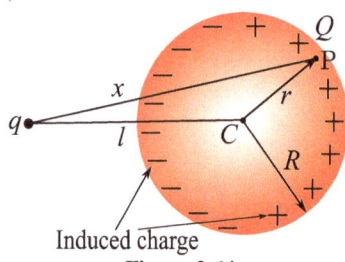

Figure 2.64

$V_P = V_{\text{due to } q} + V_{\text{due to induced charges}} + V_{\text{due to } Q}$

Comparing above two values of V_P, we get-
$$\frac{kq}{l} + \frac{kQ}{R} = \frac{kq}{x} + \frac{kQ}{R} + V_{\text{due to induced charges}}$$
$$\Rightarrow \quad V_{\text{due to induced charges}} = \frac{kq}{l} - \frac{kq}{x}$$

Also, $\vec{E}_{\text{external}} + \vec{E}_{\text{conductor}} + \vec{E}_{\text{induced}} = 0$
Resultant electric field inside material is zero

EXAMPLE 43. A metal sphere of radius R, carrying charge q, is surrounded by a thick concentric metal shell (inner radius a, outer radius b, as in Fig.2.65). The shell carries no net charge.
(a) Find the surface charge density σ at R, at a, and at b.
(b) Find the potential at the center, using infinity as the reference point.
(c) Now the outer surface is touched to a grounding wire, which lowers its potential to zero (same as at infinity). How do your answers to (a) and (b) change?

SOLUTION (a) $\sigma_R = \frac{q}{4\pi R^2}$; $\sigma_a = \frac{-q}{4\pi a^2}$; $\sigma_b = \frac{-q}{4\pi b^2}$

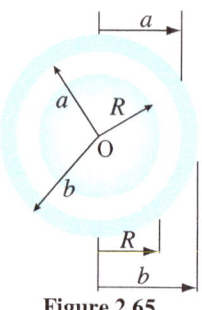

Figure 2.65

(b) $V(0) = -\int_\infty^0 E\, dr$

To simplify right hand side of above expression, we break the limit ∞ to 0 into pieces: $(\infty \to b) + (b \to a) + (a \to R) + (R \to 0)$ and then apply corresponding E values:

$$V(0) = -\int_\infty^b \left(\frac{1}{4\pi\epsilon_0}\frac{q}{r^2}\right) dr - \int_b^a (0) dr$$
$$-\int_a^R \left(\frac{1}{4\pi\epsilon_0}\frac{q}{r^2}\right) dr - \int_R^0 (0) dr$$
$$= \frac{1}{4\pi\epsilon_0}\left(\frac{q}{b} + \frac{q}{R} - \frac{q}{a}\right)$$

(c) $\sigma_b \to 0$ (the charge "drains off");
$$V(0) = -\int_\infty^a (0) dr - \int_a^R \left(\frac{1}{4\pi\epsilon_0}\frac{q}{r^2}\right) - \int_R^0 (0) dr$$
$$= \frac{1}{4\pi\epsilon_0}\left(\frac{q}{R} - \frac{q}{a}\right)$$

EXAMPLE 44. A total charge q is spread uniformly over the inner surface of a non-conducting hemispherical cup of inner radius a. Calculate (a) the electric field and (b) the electric potential at the centre of the hemisphere (Consider the cup as a stack of rings).

SOLUTION Consider a circular strip symmetric about z-axis of radius r and width $ad\theta$ [Fig. 2.66]. The charge on the strip is
$$dq = q\,\frac{2\pi r\, ad\theta}{2\pi a^2} = \frac{qr\, d\theta}{a} = q\sin\theta\, d\theta$$

(a) At the centre of the hemisphere, the x-component of the field

Figure 2.66

will be cancelled for reasons of symmetry. The entire field will be contributed by the z-component alone.
$$dE = dE_z = \frac{q\sin\theta\, d\theta\cos\theta}{4\pi\epsilon_0 a^2}$$
$$\therefore \quad E = \int dE_z = \frac{q}{4\pi\epsilon_0 a^2}\int_0^{\pi/2} \sin\theta\cos\theta\, d\theta = \frac{q}{8\pi\epsilon_0 a^2}$$

(b) $dV = \frac{q\sin\theta\, d\theta}{4\pi\epsilon_0 a}$;
$$\Rightarrow \quad V = \int dV = \frac{q}{4\pi\epsilon_0 a}\int_0^{\pi/2} \sin\theta\, d\theta = \frac{q}{4\pi\epsilon_0 a}$$

EXAMPLE 45. An alpha particle (two protons, two neutrons) moves into a stationary gold atom (79 protons, 118 neutrons), passing through the electron region that surrounds the gold nucleus like a shell and headed directly toward the nucleus (Fig.

2.9. CONNECTED CONDUCTING SPHERES

2.67). The alpha particle slows until it momentarily stops when its centre is at radial distance $r = 9.23$ fm from the nuclear centre. Then it moves back along its incoming path. (Because the gold nucleus is much more massive than the alpha particle, we can assume the gold nucleus does not move.) What was the kinetic energy K_i of the alpha particle when it was initially far away (hence external to the gold atom)? Assume that the only force acting between the alpha particle and the gold nucleus is the (electrostatic) Coulomb force and treat each as a single charged particle.

APPROACH During the entire process, the mechanical energy

Figure 2.67

of the *alpha particle + gold atom* system is conserved. When the alpha particle is outside the atom, the system's initial electric potential energy U_i is zero because the atom has an equal number of electrons and protons, which produce a *net* electric field of zero. However, once the alpha particle passes through the electron region surrounding the nucleus on its way to the nucleus, the electric field due to the electrons goes to zero. The reason is that the electrons act like a closed spherical shell of uniform negative charge and such a shell produces zero electric field in the space it encloses. The alpha particle still experiences the electric field of the protons in the nucleus, which produces a repulsive force on the protons within the alpha particle.

As the incoming alpha particle is slowed by this repulsive force, its kinetic energy is transferred to electric potential energy of the system. The transfer is complete when the alpha particle momentarily stops and the kinetic energy is zero, i.e., $K_f = 0$.

Since, there is only electrostatic force between α particle and gold nucleus which is conservative, therefore, we can apply the principle of conservation of mechanical energy.

SOLUTION According to principle of conservation of mechanical energy, we have-
$$K_i + U_i = K_f + U_f$$
Here, $U_i = 0$ and $K_f = 0$, therefore-
$$K_i = k\frac{(2e)(79e)}{9.23 \times 10^{-15}} = 3.94 \times 10^{-12} \text{J} = 24.6 \text{ MeV}$$

EXAMPLE 46. Two particles of mass m and $2m$ carry a charge q each. Initially the heavier particle is at rest on a smooth horizontal plane and the other is projected along the plane directly towards the first from a distance d with speed u. Find the closest distance of approach.

APPROACH As the mass $2m$ is not fixed, it will also move away from m due to repulsion. The distance between the particles is minimum when their relative velocity is zero i.e., when they have equal velocities.

If v_1 and v_2 are the velocities of m and $2m$ respectively at closest distance of approach, then $v_1 = v_2$.

Additionally, the electrostatic force is the only force acting between the masses, therefore you can apply the principle of conservation of mechanical energy.

SOLUTION By conservation of linear momentum, we have-
$$mu = mv_1 + 2mv_2 \implies u = v_1 + 2v_2$$
Since, $v_1 = v_2$, therefore,
$$u = v_1 + 2v_1 \implies v_1 = u/3$$
$$\implies v_2 = v_1 = u/3$$

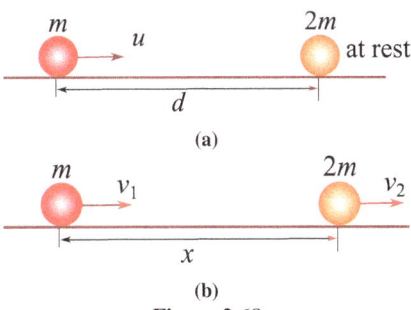

Figure 2.68

By conservation of mechanical energy, we have-
Loss in KE = gain in PE
$$\frac{1}{2}mu^2 - \left(\frac{1}{2}mv_1^2 + \frac{1}{2}2mv_2^2\right) = \frac{q^2}{4\pi\epsilon_0}\left(\frac{1}{x} - \frac{1}{d}\right)$$
$$\implies \frac{1}{2}mu^2 - \frac{1}{2}m\frac{u^2}{9}(1+2) = \frac{q^2}{4\pi\epsilon_0}\left(\frac{1}{x} - \frac{1}{d}\right)$$
$$\implies \frac{1}{3}mu^2 = \frac{q^2}{4\pi\epsilon_0}\left(\frac{1}{x} - \frac{1}{d}\right)$$
$$\implies \frac{1}{x} = \frac{1}{d} + \frac{4\pi\epsilon_0 mu^2}{3q^2} \implies x = \frac{3q^2 d}{3q^2 + 4\pi\epsilon_0 mu^2 d}$$

2.9 Connected Conducting Spheres

Now consider two charged conducting spheres with different radii connected with a long conducting wire, as shown in Figure 2.69. We assume that the spheres are far enough apart that the charge distribution on one does not directly affect the other. We already know that the electric field is stronger where the electric field lines are closer together and weaker where the electric field lines are widely separated. Electric charge is placed on one of the spheres. Some of that charge will then flow through the conducting wire so both spheres are charged. In electric equilibrium, the charge must be distributed so the electric potential is the same on the two spheres (since otherwise charges would flow in the conducting wire and the system would not be in electrical equilibrium as assumed). Two conductors connected can be seen as

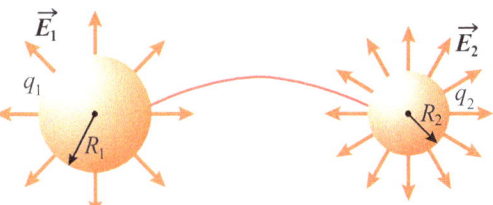

Figure 2.69

a single conductor. Therefore, electric potential will be identical everywhere on the connected spheres.

Potential at any point on the surface of charged sphere having charge q_1 and radius R_1 is-
$$V_1 = \frac{q_1}{4\pi\epsilon_0 R_1}$$
Potential at any point on the surface of charged sphere having charge q_2 and radius R_2 is-
$$V_2 = \frac{q_2}{4\pi\epsilon_0 R_2}$$
Therefore, in electrostatic equilibrium:
$$V_1 = V_2 \implies \frac{q_1}{R_1} = \frac{q_2}{R_2} \implies \frac{q_1}{q_2} = \frac{R_1}{R_2}$$
Surface charge density of the sphere of radius R_1, is-
$$\sigma_1 = \frac{q_1}{4\pi R_1^2}$$
Surface charge density of the sphere of radius R_2, is

$$\sigma_2 = \frac{q_2}{4\pi R_2^2}$$

$$\therefore \boxed{\frac{\sigma_1}{\sigma_2} = \frac{q_1}{q_2} \cdot \frac{R_2^2}{R_1^2} = \frac{R_1}{R_2} \cdot \frac{R_2^2}{R_1^2} = \frac{R_2}{R_1}} \quad (2.79)$$

So, if $R_1 < R_2$, then $\sigma_1 > \sigma_2$ and the surface electric field $E_1 > E_2$.

☞ Charge distribution on a conductor not necessarily have to be uniform.

EXAMPLE 47. Two isolated metallic solid spheres of radii R and $2R$ are charged such that both have same charge density σ. The spheres are then connected by a thin conducting wire. If the new charge density of the bigger sphere is σ', find the ratio σ'/σ
SOLUTION The initial situation is shown in Fig.2.70. Charge on sphere of radius R is

$$Q_1 = \sigma\left(4\pi R^2\right) = 4\pi R^2\sigma$$

and charge on sphere of radius $2R$ is

$$Q_2 = \sigma\left(4\pi(2R)^2\right) = 16\pi R^2\sigma$$

Therefore, net charge on both spheres

Figure 2.70

$$Q_1 + Q_2 = 4\pi R^2\sigma + 16\pi R^2\sigma = 20\pi R^2\sigma \quad \cdots (1)$$

When both spheres are connected, the redistribution of charge starts until they both reach a common potential. Let their common potential be 'V,' and the charge on the sphere with a radius of R be Q_1' and the charge on the sphere with a radius of $2R$ be Q_2'. This situation is shown in Fig.2.71.

The common potential, of both the spheres, is given as

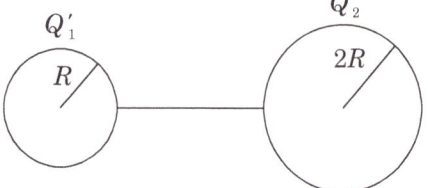

Figure 2.71

$$V = \frac{Q_1'}{4\pi\epsilon_0 R} = \frac{Q_2'}{4\pi\epsilon_0(2R)} = \frac{Q_1' + Q_2'}{4\pi\epsilon_0 R + 4\pi\epsilon_0(2R)}$$

By charge conservation,

$$Q_1' + Q_2' = Q_1 + Q_2 = 20\pi R^2\sigma \text{ (By Eq.(1))}$$

Therefore, $V = \dfrac{Q_1'}{4\pi\epsilon_0 R} = \dfrac{Q_2'}{4\pi\epsilon_0(2R)} = \dfrac{20\pi R^2\sigma}{12\pi\epsilon_0 R} = \dfrac{5R\sigma}{3\epsilon_0}$

$$\implies Q_1' = \frac{5R\sigma}{3\epsilon_0} \times 4\pi\epsilon_0 R = \frac{20}{3}\pi\sigma R^2$$

and $Q_2' = \dfrac{5R\sigma}{3\epsilon_0} \times 8\pi\epsilon_0 R = \dfrac{40}{3}\pi\sigma R^2$

Clearly, $Q_2' = 2Q_1'$
If new charge density of bigger sphere is σ', then

$$\sigma' = \frac{Q_2'}{4\pi(2R)^2} = \frac{40\pi\sigma R^2}{3\left(16\pi R^2\right)} = \frac{5}{6}\sigma \implies \frac{\sigma'}{\sigma} = \frac{5}{6}$$

2.10 Earthing

Earth is very big conducting sphere. We assume potential of earth is zero because its size is very large.

When a conducting body is joined to earth through a conducting wire. Potential of the body also becomes zero.

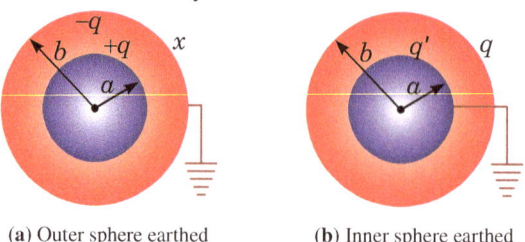

(a) Outer sphere earthed (b) Inner sphere earthed
Figure 2.72

In Fig.2.72a, charge $+q$ is given to inner sphere of radius a placed inside a larger concentric sphere of radius $b(> a)$. By using Gausses' law, we can show that the induced charge produced at inner surface of bigger sphere is $-q$. Let the charge on the outer surface of larger sphere is x.

Since, the outer sphere is earthed, therefore, the net electric potential at any point on the surface of outer sphere must be zero, i.e.,-

$$V_{\text{outer}} = 0 \implies k\frac{q}{b} + k\frac{(-q+x)}{b} = 0 \implies x = 0$$

So, there will not be any charge on the outer surface of the larger sphere, only $-q$ charge will be induced on the inner surface of the outer sphere.

In Fig. 2.72b the inner sphere is earthed and the net given charge on the outer sphere is q. If induced charge on outer surface of inner sphere is q', then -

$$V_{\text{inner}} = 0 \implies \frac{kq'}{a} + \frac{kq}{b} = 0 \implies \boxed{q' = -q\left(\frac{a}{b}\right)}$$

2.11 Corona Discharge

Let us consider an arbitrarily shaped charged conductor, as shown in Fig.2.73. Now, if we fit circles at different positions on the surface of the conductor, then the radii of these circles will be the radii of curvatures at these positions, respectively. Since the radius of curvature is smallest at the sharpest end of the conductor, therefore, from Eq.2.79, it is clear that the charge density is greatest for the sharpest regions. As a result of this, the electric field ($\propto$ charge density) is much greater in sharply pointed regions, and it may become large enough for the air or other gas to break down and conduct a current because we have ionized material near those points. *Corona discharge is an electric discharge due to the ionization of material in a fluid (such as air) surrounding a conductor. This discharge happens near pointed regions on conductors.* The minimum electric field needed for breakdown is called the breakdown field and is typically 3MV/m for fairly dry air. Corona discharge is used in ozone generators, in electrical precipitators that remove particulates in smokestacks, and in forced-air heating systems. It is also important in many other applications, including lightning protection rods.

EXAMPLE 48. Electric charge can accumulate on an airplane in flight. You may have observed needle-shaped metal extensions on the wing tips and tail of an airplane. Their purpose is to allow charge to leak off before much of it accumulates. The electric field around the needle is much larger than the field around the body of the airplane, and can become large enough to produce dielectric breakdown of the air, discharging the airplane. To model this process, assume that two charged spherical conductors are connected by a long conducting wire, and a charge of $1.20 \mu C$ is placed on the combination. One sphere, representing the body

2.11. CORONA DISCHARGE

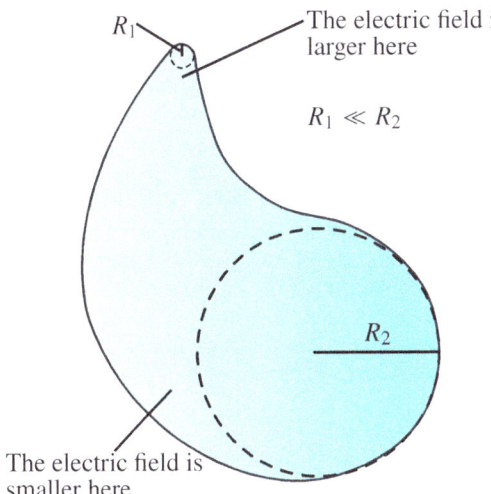

Figure 2.73

of the airplane, has a radius of 6.00 cm, and the other, representing the tip of the needle, has a radius of 2.00 cm. (a) What is the electric potential of each sphere? (b) What is the electric field at the surface of each sphere?

SOLUTION (a) Let q_1 and q_2 are charges on larger and smaller spheres. Radii of these spheres are $r_1 = 6.00$ cm and $r_2 = 2.00$ cm respectively.

Both spheres are connected with each other with metal body, therefore they must be at the same potential, i.e.,

$$\frac{kq_1}{r_1} = \frac{kq_2}{r_2} \implies q_1 = \frac{q_2 r_1}{r_2}$$

where also the total charge is-

$$q_1 + q_2 = 1.20 \times 10^{-6} \text{C}$$

Substituting the value of q_1 in above expression, we get

$$\frac{q_2 r_1}{r_2} + q_2 = 1.20 \times 10^{-6} \text{C}$$

$$\implies q_2 \left(\frac{r_1}{r_2} + 1\right) = 1.20 \times 10^{-6} \text{C} \implies q_2 = \frac{1.20 \times 10^{-6} \text{C}}{\frac{r_1}{r_2} + 1}$$

Now, substituting the values of r_1, r_2 in above expression and solving for q_2, we get

$$q_2 = \frac{1.20 \times 10^{-6} \text{C}}{1 + 6 \text{ cm}/2 \text{ cm}} = 0.300 \times 10^{-6} \text{C}$$

Thus, charge on smaller sphere is 0.300×10^{-6}C.

Since, $q_1 + q_2 = 1.20 \times 10^{-6}$C, therefore charge on larger sphere,

$$q_1 = 1.20 \times 10^{-6}\text{C} - q_2 = 1.20 \times 10^{-6}\text{C} - 0.300 \times 10^{-6} \text{ C}$$
$$= 0.900 \times 10^{-6}\text{C}$$

Now, electric potential

$$V = \frac{kq_1}{r_1} = \frac{(8.99 \times 10^9 \text{ N} \cdot \text{m}^2/\text{C}^2)(0.900 \times 10^{-6}\text{C})}{6 \times 10^{-2} \text{ m}}$$
$$= 1.35 \times 10^5 \text{ V}$$

(b) Outside the larger sphere,

$$E_1 = \frac{kq_1}{r_1^2}\hat{r} = \frac{V_1}{r_1}\hat{r} = \frac{1.35 \times 10^5 \text{ V}}{0.06 \text{ m}}\hat{r}$$
$$= 2.25 \times 10^6 \text{ V/m away}$$

Outside the smaller sphere,

$$E_2 = \frac{1.35 \times 10^5 \text{ V}}{0.02 \text{ m}}\hat{r} = 6.74 \times 10^6 \text{ V/m away}$$

The smaller sphere carries less charge but creates a much stronger electric field than the larger sphere.

2.11.1 Electric Potential Energy for Continuous Charge System

Electric potential energy of a continuously charged object is also called self-energy* or internal energy of the object.

2.11.1.1 Self (or Internal) Energy of a Uniformly Charged Conducting Spherical Shell or Conducting Sphere

For a conducting spherical shell or a sphere, charges are stored only on the outer surface, so the calculation of potential energy for the conducting sphere will be identical to that of a conducting spherical shell. Here, we will focus on discussing the conducting spherical shell.

Take an uncharged shell. Now bring charges one by one from infinity to the surface of the shell. The work required in this process will be stored as potential Energy.

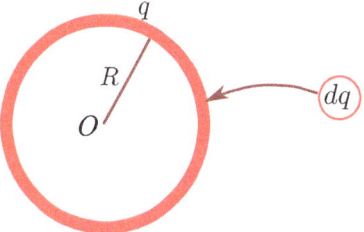

Figure 2.74

Suppose, at any instant the given charge on the conducting spherical shell is q and now we are giving extra charge dq to it. The work required to bring this extra charge dq from infinity to the shell is -

$$dW = (dq)(V_f - V_i) = dq\left(\frac{kq}{R} - 0\right) = \frac{kq}{R}dq$$

Therefore, net work done in charging the shell up to charge Q, is given by-

$$W = \int_0^Q \frac{kq}{R}dq = \frac{kQ^2}{2R}$$

This work will be stored in the form of electrostatic potential energy (self energy). So, the electric potential energy of the charged spherical shell-

$$\boxed{U = \frac{kQ^2}{2R}} \qquad (2.80)$$

Eq.2.80 gives the self energy (or electric potential energy) of a uniformly charged conducting shell (or sphere).

2.11.1.2 Self Energy of a Uniformly Charged Non Conducting Solid Sphere

To make a non conducting charged solid sphere, we have to assemble charged particles (each of infinitely small step-sized charge dq) one over the other until a sphere of required charge and radius is formed. For it, we bring the charged particles one by one from infinity to the sphere so that the size of the sphere increases. Suppose, at any instant when the collected charge is q,

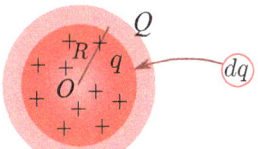

Figure 2.75

then the radius of sphere is r. Now, let us giving extra charge dq to it, which increases it's radius by dr.

---
*For a system of stationary or moving discrete charges, the net electrostatic potential energy can be calculated as discussed in section 2.3 but for system of objects having a continuous charge distribution, we must also consider the self-energy of each object in the system.

∴ Work required to bring charge dq from infinity to the sphere

$$dW = dq\,(V_f - V_i) = dq\left(\frac{kq}{r} - 0\right) = \frac{kq\,dq}{r}$$

here, $q = \rho\left(\frac{4}{3}\pi r^3\right) \implies dq = \rho(4\pi r^2 dr)$

∴ Total work required to give charge Q to the sphere

$$W = \int_0^R k\,\frac{\rho\left(\frac{4}{3}\pi r^3\right)\rho(4\pi r^2 dr)}{r}$$

On simplifying it, we get:

$$W = \frac{3}{5}\frac{kQ^2}{R} = U_{\text{self}}$$

(for a non conducting solid sphere)

◊ The total electrical energy of a system of objects is always equal to the sum of self energies of each object and the electrical potential energy due to mutual interaction between the objects. If the objects are charged particles (i.e., point charges), then the self energy term will be zero.

EXAMPLE 49. A uniformly charged spherical shell of radius R and charge q, is expanded to a radius $2R$. Find the work performed by the external agent against electric forces and work done by electric forces, in this process.

APPROACH The work done by electric forces is always equal to the –ve of change in electric potential energy of the system, i.e.,

$$W_{\text{el}} = -\Delta U = -\left(U_f - U_i\right)$$

Here, U_i and U_f are respectively the initial and final electric potential energies of the spherical shell.

SOLUTION $W_{\text{el}} = -(U_f - U_i) = -\left[\dfrac{q^2}{16\pi\epsilon_0 R} - \dfrac{q^2}{8\pi\epsilon_0 R}\right]$

$$= \frac{q^2}{16\pi\epsilon_0 R}$$

Now, work done by an external agent against electric forces-

$$W_{\text{ext}} = -W_{\text{el}} = -\frac{q^2}{16\pi\epsilon_0 R}$$

☞ +ve work of electric field or negative work of an external agent shows the loss of potential energy of the system.

EXAMPLE 50. Two non-conducting hollow uniformly charged spheres of radii R_1 and R_2 with charge Q_1 and Q_2 respectively are placed at a distance r. Find out total energy of the system.

APPROACH To find the total energy of the system, add the self energies of both spheres and the potential energy due to mutual interaction.

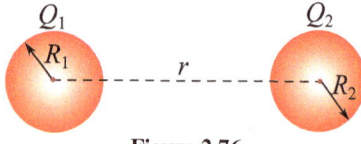

Figure 2.76

SOLUTION $U_{\text{total}} = U_{\text{self}} + U_{\text{interaction}}$

$$= \frac{Q_1^2}{8\pi\epsilon_0 R_1} + \frac{Q_2^2}{8\pi\epsilon_0 R_2} + \frac{Q_1 Q_2}{4\pi\epsilon_0 r}$$

EXAMPLE 51. Two uniformly charged concentric spherical shells of radii R_1 and R_2 ($R_2 > R_1$) have charges Q_1 and Q_2 respectively. Find out total energy of the system.

Figure 2.77

APPROACH To find the total energy of the system, add the self energies of both spherical shells and the potential energy due to mutual interaction.

SOLUTION $U_{\text{total}} = U_{\text{self1}} + U_{\text{self2}} + U_{\text{interaction}}$

$$= \frac{Q_1^2}{8\pi\epsilon_0 R_1} + \frac{Q_2^2}{8\pi\epsilon_0 R_2} + \frac{Q_1 Q_2}{4\pi\epsilon_0 R_2}$$

2.12 The Van de Graaff Generator

This is a machine that can build up high voltages of the order of a few million volts and electric field close to breakdown field of air which is about 3×10^6 V/m. The resulting large electric fields are used to accelerate charged particles (electrons, protons, ions) to high energies needed for experiments to probe the small scale structure of matter. The principle underlying the machine is as follows:

Principle: Suppose we have a large spherical conducting shell of radius R, on which we place a charge Q. This charge spreads itself uniformly over the entire surface of the shell. The field outside the sphere is like that of a point charge Q at the centre; while inside the sphere, the field vanishes. The potential outside is that of a point charge; whereas inside, it remains constant and equal to the value of potential at the surface of the conducting shell.

Thus, the potential inside conducting spherical shell of radius R carrying charge Q,

$$V_{\substack{\text{due to charge }Q\\ \text{on spherical shell}}} = \frac{1}{4\pi\epsilon_0}\frac{Q}{R} \qquad (2.81)$$

Now, as shown in Fig. 2.78, let us introduce a small sphere of radius r, carrying some charge q, into the large one, and place it at the centre.

Now, the potential due to small sphere of radius r carrying charge q:

1. At surface of small sphere $= \dfrac{1}{4\pi\epsilon_0}\dfrac{q}{r}$

2. At large shell of radius $R = \dfrac{1}{4\pi\epsilon_0}\dfrac{q}{R}$

Taking both charges q and Q into account, the net potential at any point on the surface of conducting shell:

$$V(R) = \frac{1}{4\pi\epsilon_0}\left(\frac{Q}{R} + \frac{q}{R}\right) \qquad (2.82)$$

And net potential at any point on the surface of inner sphere:

$$V(r) = \frac{1}{4\pi\epsilon_0}\left(\frac{Q}{R} + \frac{q}{r}\right) \qquad (2.83)$$

Therefore, the potential difference between the surface of inner sphere and spherical shell:

$$V(r) - V(R) = \frac{q}{4\pi\epsilon_0}\left(\frac{1}{r} - \frac{1}{R}\right) \qquad (2.84)$$

Because $r < R$, therefore, if q is positive then $V(r) - V(R)$ will also be positive.

From Eq.2.84, we see that, $V(r) - V(R)$ is independent of charge Q that may have accumulated on the larger sphere and even if it is positive, the inner sphere is always at a higher potential for +ve value of q (because, the difference $V(r) - V(R)$ is positive). The potential due to Q is constant up to radius R and so cancels out in the difference.

This means that if we now connect the smaller and larger sphere by a wire, the charge q on the smaller will immediately flow onto the larger, even though the charge Q may be quite large. The natural tendency is for positive charge to move from higher to lower potential. Thus, provided we are somehow able to introduce the small charged sphere into the larger one, we can in this way keep piling up larger and larger amount of charge on the latter. The

2.13 The Millikan Oil-Drop Experiment

Robert Millikan conducted a series of experiments during 1909-1913 in which he measured e, the magnitude of the elementary charge on an electron, and demonstrated the quantized nature of this charge. His apparatus, diagrammed in Figure 2.80, contains two parallel metallic plates separated by distance d. Oil droplets from an atomizer are allowed to pass through a small hole in the upper plate. Millikan used x-rays to ionize the air in the chamber so that freed electrons would adhere to the oil drops, giving them a negative charge. A horizontally directed light beam is used to illuminate the oil droplets, which are viewed through a telescope whose long axis is perpendicular to the light beam. When viewed in this manner, the droplets appear as shining stars against a dark background and the rate at which individual drops fall can be determined.

Figure 2.78

Figure 2.79

Figure 2.80

Let's assume a single drop having a mass m and carrying a charge q is being viewed and its charge is negative. If no electric field is present between the plates (Fig.2.81a), the two forces acting on the charge are-
1. Gravitational force $m\vec{g}$ (acting downwards)
2. Viscous drag force $\vec{F}_D$ (acting upwards)

Note that there is also a buoyant force on the oil drop due to the surrounding air. This force can be taken as a correction term in the gravitational force $m\vec{g}$ on the drop. For now, we will not consider it in our analysis.

The drag force is proportional to the drop's speed. When the drop reaches its terminal speed (v_T) the two forces balance each other, i.e.

$$mg = F_D$$

At terminal speed of the drop, the viscous drag force is given by-

$$F_D = 6\pi r \eta v_T$$

where r is the radius of the oil drop, η is the viscosity of the air, and v_T is the terminal velocity of the falling drop.

From above two equations, we can write-

$$mg = 6\pi r \eta v_T$$
$$\implies r = \frac{mg}{6\pi \eta v_T} \qquad (2.85)$$

Now suppose a battery connected to the plates sets up an electric field between the plates such that the upper plate is at the higher electric potential. In this case, a third force $q\vec{E}$ acts on the charged drop. Since, q is negative and $\vec{E}$ is directed downward, this electric force is directed upward as shown in Figure 2.81b. If this upward force is strong enough, the drop moves upward and the drag force $\vec{F}'_D$ acts downward. When the upward electric force $q\vec{E}$ balances the sum of the gravitational force and the downward drag force $\vec{F}'_D$, the drop reaches a new terminal speed v'_T in the

potential (Eq.2.82) at the outer sphere would also keep rising, at least until we reach the breakdown field of air.

This is the principle of the van de Graaff generator.

Construction and Working: A schematic diagram of the van de Graaff generator is given in Fig.2.79. A large spherical conducting shell (of few meters radius) is supported at a height several meters above the ground on an insulating column. A long narrow endless belt insulating material, like rubber or silk, is wound around two pulleys – one at ground level, one at the center of the shell. This belt is kept continuously moving by a motor driving the lower pulley. It continuously carries positive charge, sprayed on to it by a brush at ground level, to the top. There it transfers its positive charge to another conducting brush connected to the large shell. Thus positive charge is transferred to the shell, where it spreads out uniformly on the outer surface. In this way, voltage differences of as much as 6 or 8 million volts (with respect to ground) can be built up.

Note: It is possible to increase the potential of the dome until electrical ionization occurs in the air. Since the ionization breakdown of air occurs at an electric field of about 3×10^6 V/m, a sphere of 1m can be raised to maximum of $V_{\max} = ER = (3 \times 10^6 \text{ V/m})(1 \text{ m}) = 3 \times 10^6$ V. The dome's electric potential can be increased further by placing the dome in vacuum and by increasing the radius of the sphere.

(a) With the electric field off, the droplet falls at terminal velocity $\vec{v}_T$ under the influence of the gravitational and drag forces.

(b) When the electric field is turned on, the droplet moves upward at terminal velocity $\vec{v}'_T$ under the influence of the electric, gravitational, and drag forces.

Figure 2.81: The forces acting on a negatively charged oil droplet in the Millikan experiment.

upward direction. In this situation-
$$qE = mg + F_D$$
$$\Rightarrow \quad qE = mg + 6\pi r \eta v'_T \quad (2.86)$$

Substituting the value of r from Eq.2.85 to Eq.2.86, we get-
$$qE = mg + 6\pi \left(\frac{mg}{6\pi \eta v_T}\right) \eta v'_T = mg + \frac{mg}{v_T} v'_T$$

$$\therefore \quad \boxed{q = \frac{mg}{E}\left(1 + \frac{v'_T}{v_T}\right)} \quad (2.87)$$

Since the electrodes are parallel plates, the magnitude of the electric field can be readily calculated from the potential difference (V) between the plates and the plate separation (d). Thus, by measuring the terminal velocities both with the field off and with the field on, it is possible to calculate the charge on a drop with the help of Eq.2.87.

Millikan made the necessary measurements on a number of drops and calculated the charge on each drop. He showed that all the drops had charges that are integer multiples of a fundamental unit of charge (e), i.e.,
$$q = ne \quad n = 0, \pm 1, \pm 2, \pm 3, \ldots$$

In his initial results, published in 1910, he reported a value of 1.63×10^{-19} C for the fundamental charge. After modifying his equipment and measuring a larger number of drops, he published a revised value of 1.59×10^{-19} C in 1913. The currently accepted value for the fundamental (or elementary) charge, e, is approximately 1.602×10^{-19} C.

2.14 Check Point 3

1. ••The labelled points in Figure 2.82 are on a series of equipotential surfaces associated with an electric field. Rank (from greatest to least) the work done by the electric field on a positively charged particle that moves from A to B; from B to C; from C to D; from D to E.

Figure 2.82

Multiple Choice Questions

2. ••Equipotential surfaces associated with an electric field which is increasing in magnitude along the X-direction are-
   (A) planes parallel to YZ-plane
   (B) planes parallel to XY-plane
   (C) planes parallel to XZ-plane
   (D) coaxial cylinders of increasing radii around the x-axis

3. ••For the equipotential surfaces in Figure 2.82, what is the approximate direction of the electric field?
   (A) Out of the page
   (B) Into the page
   (C) Toward the top of the page
   (D) Toward the bottom of the page.

4. ••A metallic solid sphere is placed in a uniform electric field. The lines of force follow the path(s) shown in the Fig.2.83

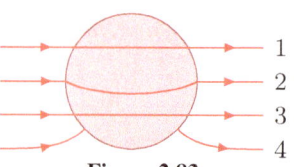

Figure 2.83

   (A) 1   (B) 2   (C) 3   (D) 4

5. ••A uniformly charged non-conducting solid sphere of radius R has potential V_0 (measured with respect to ∞) on its surface. For this sphere, the equipotential surfaces with potentials $3V_0/2$, $5V_0/4$, $3V_0/4$ and $V_0/4$ have radii $R_1, R_2, R_3,$ and R_4 respectively. Then,
   (A) $R_1 = 0$ and $R_2 > (R_4 - R_3)$
   (B) $R_1 \neq 0$ and $(R_2 - R_1) > (R_4 - R_3)$
   (C) $R_1 = 0$ and $R_2 < (R_4 - R_3)$
   (D) $2R < R_4$

6. ••A hollow metal sphere of radius 5 cm is charged such that the potential on its surface is 10 V. The potential at the centre of the sphere is
   (A) zero
   (B) 10 V
   (C) same as at a point 5 cm away from the surface
   (D) same as at a point 25 cm away from the surface.

7. ••The arc AB with the centre C and the infinitely long wire having linear charge density λ are lying in the same plane as shown in Fig.2.84. The minimum amount of work to be expended to move a point charge q_0 from point A to B through a circular path AB of radius a is equal to:
   (A) $\frac{q_0}{2\pi \epsilon_0} \ln \frac{2}{3}$ (B) $\frac{q_0 \lambda}{2\pi \epsilon_0} \ln \frac{3}{2}$
   (C) $\frac{q_0 \lambda}{2\pi \epsilon_0} \ln \frac{2}{3}$ (D) $\frac{q_0 \lambda}{\sqrt{2}\pi \epsilon_0}$

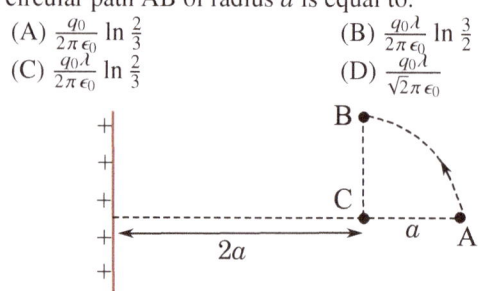

Figure 2.84

8. •••Consider a system of three charges $q/3$, $q/3$ and $-2q/3$ placed at points A, B and C, respectively, as shown in the Fig.2.85. Take O to be the centre of the circle of radius R and angle CAB = 60°.

2.15. QUESTIONS AND EXERCISES

(A) The electric field at point O is $\frac{q}{8\pi\epsilon_0 R^2}$ directed along the negative x-axis
(B) The potential energy of the system is zero.
(C) The magnitude of the force between the charge C and B is $\frac{q^2}{54\pi\epsilon_0 R^2}$
(D) The potential at point O is $\frac{q}{12\pi\epsilon_0 R}$.

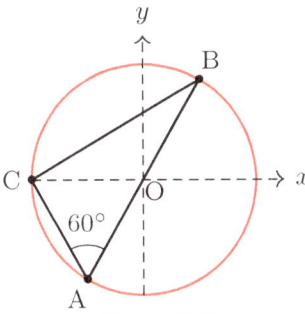

Figure 2.85

2.15 Questions and Exercises
2.15.1 Conceptual Questions

1. A negative charge moves in the direction of a uniform electric field. Does the potential energy of the charge–field system increase or decrease? Does the charge move to a position of higher or lower potential?
2. A proton is moved to the left in a uniform electric field that points to the right. Is the proton moving in the direction of increasing or decreasing electric potential? Is the electrostatic potential energy of the proton, in presence of field, increasing or decreasing?
3. An electron is moved to the left in a uniform electric field that points to the right. Is the electron moving in the direction of increasing or decreasing electric potential? Is the electrostatic potential energy of the electron, in presence of field, is increasing or decreasing?
4. If the electric potential is uniform throughout a region of space, what can be said about the electric field in that region?
5. Can there be a potential difference between two conductors that carry like charges of the same magnitude?
6. If V is known at only a single point in space, can $\vec{E}$ be found at that point? Explain your answer.
7. If two points are at the same potential, does this mean that no work is done in moving a test charge from one point to the other? Does this imply that no force must be exerted? Explain.
8. If a negative charge is initially at rest in an electric field, will it move toward a region of higher potential or lower potential? What about a positive charge? How does the potential energy of the charge change in each instance?
9. Figure 2.86 shows a point particle that has a positive charge $+Q$ and a metal sphere that has a charge $-Q$. Sketch the electric field lines and equipotential surfaces for this system of charges.

Figure 2.86

10. Can a free particle ever move from a region of low electric potential to one of high potential and yet have its electric potential energy decrease? Explain.
11. Figure 2.87 shows a point particle and metal sphere. Both have equal charge $+Q$. Sketch the electric field lines and equipotential surfaces for this system of charges.

Figure 2.87

12. Two equal positive point charges are separated by a finite distance. Sketch the electric field lines and the equipotential surfaces for this system.
13. Two point charges are fixed on the x-axis. (a) Each has a positive charge q. One is at $x = -a$ and the other is at $x = +a$. At the origin, which of the following is true?
    (A) $\vec{E} = 0$ and $V = 0$,
    (B) $\vec{E} = 0$ and $V = 2kq/a$,
    (C) $\vec{E} = (2kq/a^2)\hat{i}$ and $V = 0$,
    (D) $\vec{E} = (2kq/a^2)\hat{i}$ and $V = 2kq/a$,
    (b) One point charge has a positive charge $+q$ and the other has a negative charge $-q$. The positive point charge is at $x = -a$ and the negative point charge is at $x = +a$. At the origin, which of the following is true?
    (A) $\vec{E} = 0$ and $V = 0$,
    (B) $\vec{E} = 0$ and $V = 2kq/a$,
    (C) $\vec{E} = (2kq/a^2)\hat{i}$ and $V = 0$,
    (D) $\vec{E} = (2kq/a^2)\hat{i}$ and $V = 2kq/a$.
14. A uniform electric field is parallel to the x axis. In what direction can a charge be displaced in this field without any external work being done on the charge?
15. The electric potential (in volts) is given by $V(x, y, z) = 4.00|x| + V_0$, where V_0 is a constant, and x is in meters. (a) Sketch the electric field for this potential. (b) Which of the following charge distributions is most likely responsible for this potential:
    (A) A negatively charged flat sheet in the $x = 0$ plane,
    (B) a point charge at the origin,
    (C) a positively charged flat sheet in the $x = 0$ plane,
    (D) a uniformly charged sphere centered at the origin?
    Explain your answer.
16. The electric potential is the same everywhere on the surface of a conductor. Does this mean that the surface charge density is also the same everywhere on the surface? Explain your answer.
17. If $V = 0$ at a point in space, must $\vec{E} = 0$? If $\vec{E} = 0$ at some point, must $V = 0$ at that point? Explain. Give examples for each.
18. When dealing with practical devices, we often take the ground (the Earth) to be 0 V. (a) If instead we said the ground was −10 V, how would this affect V and E at other points? (b) Does the fact that the Earth carries a net charge affect the choice of V at its surface?
19. Explain why electric field lines are always perpendicular to equipotential surfaces.
20. Can two equipotential lines cross? Explain.
21. A satellite orbits the Earth along a gravitational equipotential line. What shape must the orbit be?
22. Suppose a charged ring is non-uniformly charged, so that the

density of charge was twice as great near the top as near the bottom. Assuming the total charge on the ring is Q. Would this affect the potential at point P on the axis (Fig.2.88) as obtained for a uniformly charged ring? Would it affect the value of $\vec{E}$ at that point obtained for uniformly charged ring? Is there a discrepancy here? Explain.

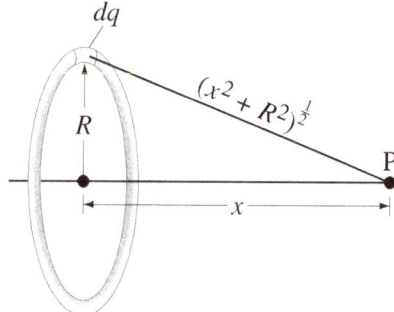

Figure 2.88

23. Consider a metal conductor in the shape of a football. If it carries a total charge Q, where would you expect the charge density σ to be greatest, at the ends or along the flatter sides? Explain. [Hint: Near the surface of a conductor, $E = \sigma/\epsilon_0$.]

24. If you know V at a point in space, can you calculate $\vec{E}$ at that point? If you know $\vec{E}$ at a point can you calculate V at that point? If not, what else must be known in each case?

25. A conducting sphere carries a charge Q and a second identical conducting sphere is neutral. The two are initially isolated, but then they are placed in contact. (a) What can you say about the potential of each when they are in contact? (b) Will charge flow from one to the other? If so, how much? (c) If the spheres do not have the same radius, how are your answers to parts (a) and (b) altered?

26. At a particular location, the electric field points due north. In what direction(s) will the rate of change of potential be (a) greatest, (b) least, and (c) zero?

27. Why is it important, when soldering connectors onto a piece of electronic circuitry, to leave no pointy protrusions from the solder joints?

28. Equipotential lines are spaced 1.00 V apart. Does the distance between the lines in different regions of space tell you anything about the relative strengths of $\vec{E}$ in those regions? If so, what?

29. If the electric field $\vec{E}$ is uniform in a region, what can you infer about the electric potential V? If V is uniform in a region of space, what can you infer about $\vec{E}$?

30. Is the electric potential energy of two unlike charges positive or negative? What about two like charges? What is the significance of the sign of the potential energy in each case?

31. Fig.2.89 shows the x-component of $\vec{E}$ as a function of x. Draw a graph of V versus x in this same region of space. Let $V = 0$ V at $x = 0$ m and include an appropriate vertical scale.

32. Fig.2.90 shows the electric potential as a function of x. Draw a graph of E_x versus x in this same region of space.

33. For each contour map in Fig.2.91, estimate the electric fields $\vec{E}_1$ and $\vec{E}_2$ at points 1 and 2.

34. An electron is released from rest at $x = 2$ m in the potential shown in Fig.2.92. Does it move? If so, to the left or to the right? Explain.

35. Fig.2.93 shows an electric field diagram. Dashed lines 1 and 2 are two surfaces in space, not physical objects.
(a) Is the electric potential at point a higher than, lower than,

Figure 2.89

Figure 2.90

Figure 2.91

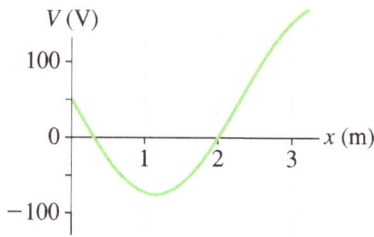

Figure 2.92

or equal to the electric potential at point b ? Explain.
(b) Rank in order, from largest to smallest, the magnitudes of the potential differences ΔV_{ab}, ΔV_{cd}, and ΔV_{ef}.
(c) Is surface 1 an equipotential surface? What about surface 2? Explain why or why not.

36. Fig.2.94 shows a negatively charged electroscope. The gold leaf stands away from the rigid metal post. Is the electric potential of the leaf higher than, lower than, or equal to the

2.15. QUESTIONS AND EXERCISES

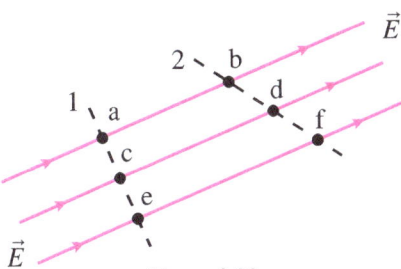

Figure 2.93

potential of the post? Explain.

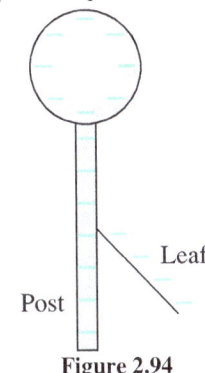

Figure 2.94

37. The two metal spheres in Fig.2.95 are connected by a metal wire with a switch in the middle. Initially the switch is open. Sphere 1, with the larger radius, is given a positive charge. Sphere 2, with the smaller radius, is neutral. Then the switch is closed. Afterward, sphere 1 has charge Q_1, is at potential V_1, and the electric field strength at its surface is E_1. The values for sphere 2 are Q_2, V_2, and E_2

    (a) Is V_1 larger than, smaller than, or equal to V_2? Explain.

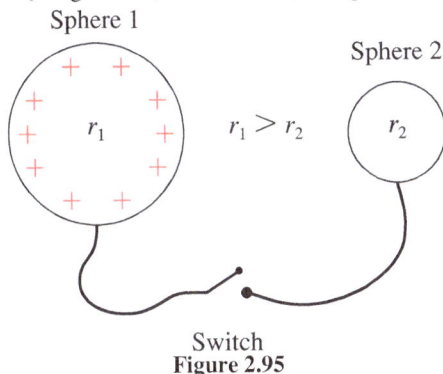

Figure 2.95

(b) Is Q_1 larger than, smaller than, or equal to Q_2? Explain.
(c) Is E_1 larger than, smaller than, or equal to E_2? Explain.

38. Fig.2.96 shows a graph of E_x vs x. What is the potential difference between $x_i = 1.0$ m and $x_f = 3.0$ m?

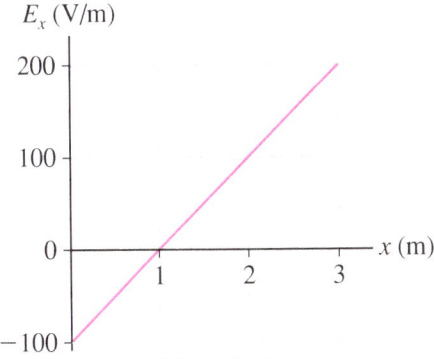

Figure 2.96

39. Fig.2.97 is a graph of E_x vs x. The potential at the origin is -50 V. What is the potential at $x = 3.0$ m?

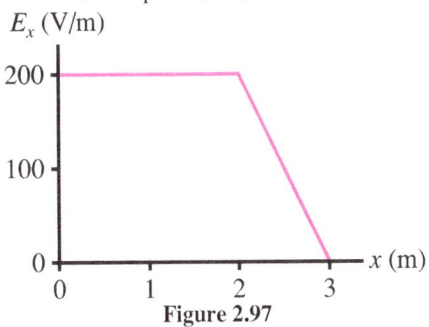

Figure 2.97

40. What are the magnitude and direction of the electric field at the dot in Fig.2.98?

Figure 2.98

41. What are the magnitude and direction of the electric field at the dot in Fig.2.99?

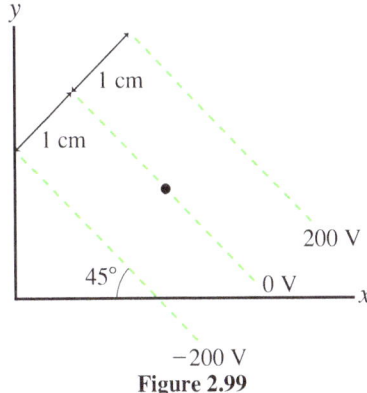

Figure 2.99

42. Fig.2.100 shows V versus x graphs. Draw the corresponding graphs of E_x versus x.

43. The electric potential in a region of uniform electric field is -1000 V at $x = -1.0$ m and $+1000$ V at $x = +1.0$ m. What is E_x?

44. The electric potential along the x-axis is $V = 100x^2$ V, where x is in meters. What is E_x at (a) $x = 0$ m and (b) $x = 1$ m?

45. The electric potential along the x-axis is $V = 100\,e^{-2x}$ V, where x is in meters. What is E_x at (a) $x = 1.0$ m and (b) $x = 2.0$ m?

2.15.2 Problems

In all the problems in this section, assume that the electric potential is zero at distances far from all charges unless otherwise stated.

Potential Differences in a Uniform Electric Field

1. ●●When an electron moves from A to B along an electric field line in Fig.2.101, the electric field does 3.94×10^{-19} J of work on it. What are the electric potential differences (a) $V_B - V_A$ (b) $V_C - V_A$, and (c) $V_C - V_B$?

2. ●●A graph of the x component of the electric field as a function of x in a region of space is shown in Fig. 2.102.

Figure 2.100

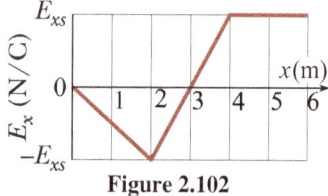

Figure 2.101

The scale of the vertical axis is set by $E_{xs} = 20.0$ N/C. The y and z components of the electric field are zero in this region. If the electric potential at the origin is 10 V, (a) what is the electric potential at $x = 2.0$ m, (b) what is the greatest positive value of the electric potential for points on the x axis for which $0 \leq x \leq 6.0$ m, and (c) for what value of x is the electric potential zero?

Figure 2.102

3. ••What are (a) the charge and (b) the charge density on the surface of a conducting sphere of radius 0.15 m whose potential is 200 V (with $V = 0$ at infinity)?

4. ••The difference in potential between the accelerating plates in the electron gun of a TV picture tube is about 25000 V. If the distance between these plates is 1.50 cm, what is the magnitude of the uniform electric field in this region?

5. ••A uniform electric field of magnitude 325 V/m is directed in the negative y direction in Figure 2.103. The coordinates of point A are $(-0.200$ m, -0.300 m$)$, and those of point B are $(0.400$ m, 0.500 m$)$. Calculate the potential difference $V_B - V_A$, using the path ACB.

6. ••A block having mass m and charge $+Q$ is connected to a spring having spring constant k. The block lies on a frictionless horizontal track, and the system is immersed in a uniform electric field of magnitude E, directed as shown in Fig.2.104. If the block is released from rest when the

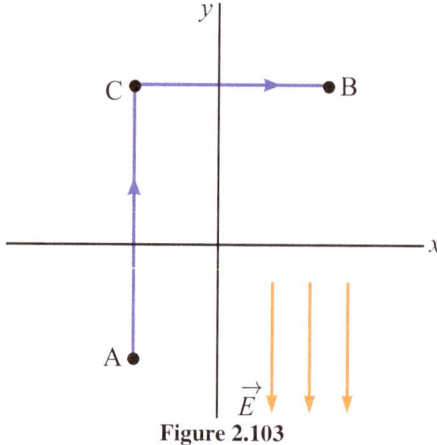

Figure 2.103

spring is unstretched (at $x = 0$), (a) by what maximum amount does the spring expand? (b) What is the equilibrium position of the block? (c) Show that the block's motion is simple harmonic, and determine its period. (d) What If? Repeat part (a) if the coefficient of kinetic friction between block and surface is μ_k.

Figure 2.104

7. ••An insulating rod having linear charge density $\lambda = 40.0$ μC/m and linear mass density $\mu = 0.100$ kg/m is released from rest in a uniform electric field $E = 100$ V/m directed perpendicular to the rod (Fig. 2.105). (a) Determine the speed of the rod after it has travelled 2.00 m. (b) How does your answer to part (a) change if the electric field is not perpendicular to the rod? Explain.

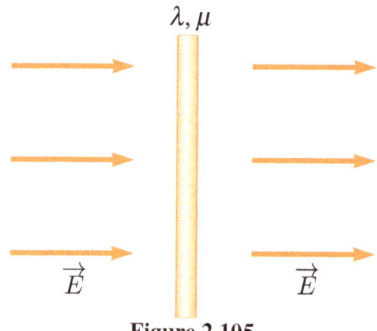

Figure 2.105

8. ••A particle having charge $q = +2.00$ μC and mass $m = 0.0100$ kg is connected to a string that is $L = 1.50$ m long and is tied to the pivot point P in Figure 2.106. The particle, string and pivot point all lie on a frictionless horizontal table. The particle is released from rest when the string makes an angle $\theta = 60.0°$ with a uniform electric field of magnitude $E = 300$ V/m. Determine the speed of the particle when the string is parallel to the electric field (point a in Fig.2.106).

Electric Potential and Potential Energy Due to Point Charges

9. •(a) Find the potential at a distance of 1.00 cm from a proton. (b) What is the potential difference between two points that are 1.00 cm and 2.00 cm from a proton? (c) What If? Repeat parts (a) and (b) for an electron.

10. ••Consider a particle with charge $q = 1.0$ μC, point A at

2.15. QUESTIONS AND EXERCISES

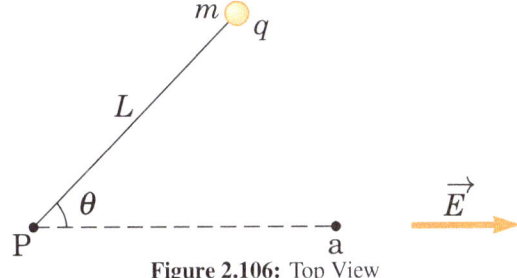

Figure 2.106: Top View

distance $d_1 = 2.0$ m from q, and point B at distance $d_2 = 1.0$ m. (a) If A and B are diametrically opposite to each other, as in Fig.2.107a, what is the electric potential difference $V_A - V_B$? (b) What is that electric potential difference if A and B are located as in Fig. 2.107b?

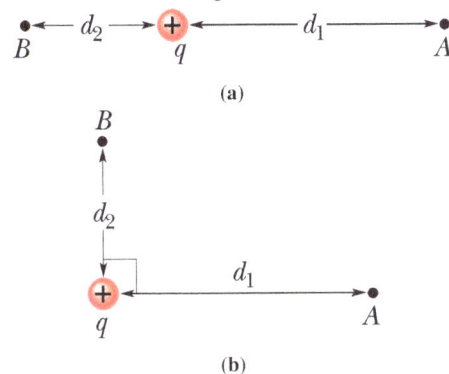

Figure 2.107

11. ●●A spherical drop of water carrying a charge of 30 pC has a potential of 500 V at its surface (with $V = 0$ at infinity). (a) What is the radius of the drop? (b) If two such drops of the same charge and radius combine to form a single spherical drop, what is the potential at the surface of the new drop?

12. ●●Figure 2.108 shows a rectangular array of charged particles fixed in place, with distance $a = 39.0$ cm and the charges shown as integer multiples of $q_1 = 3.40$ pC and $q_2 = 6.00$ pC. With $V = 0$ at infinity, what is the net electric potential at the rectangle's centre? (Hint: Thoughtful examination of the arrangement can reduce the calculation.)

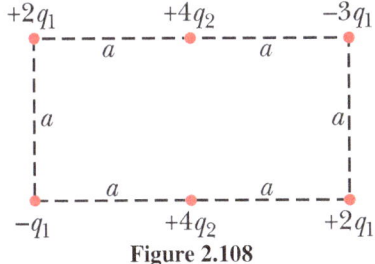

Figure 2.108

13. ●●In Fig.2.109, what is the net electric potential at point P due to the four particles if $V = 0$ at infinity, $q = 5.00$ fC, and $d = 4.00$ cm?

14. ●●In Fig.2.110, particles with the charges $q_1 = +5e$ and $q_2 = -15e$ are fixed in place with a separation of $d = 24.0$ cm. With electric potential defined to be $V = 0$ at infinity, what are the finite (a) positive and (b) negative values of x at which the net electric potential on the x axis is zero?

15. ●●Two particles, of charges q_1 and q_2, are separated by distance d in Fig. 2.110. The net electric field due to the particles is zero at $x = d/4$. With $V = 0$ at infinity, locate (in terms of d) any point on the x axis (other than at infinity) at which the electric potential due to the two particles is zero.

Figure 2.109

Figure 2.110

16. ●●Given two 2.00 μC charges, as shown in Figure 2.111, and a positive test charge $q = 1.28 \times 10^{-18}$C at the origin, (a) what is the net force exerted by the two 2.00 μC charges on the test charge q? (b) What is the electric field at the origin due to the two 2.00 μC charges? (c) What is the electric potential at the origin due to the two 2.00 μC charges?

Figure 2.111

17. ●●A charge $+q$ is at the origin. A charge $-2q$ is at $x = 2.00$ m on the x axis. For what finite value(s) of x is (a) the electric field zero? (b) the electric potential zero?

18. ●●The three charges in Figure 2.112 are at the vertices of an isosceles triangle. Calculate the electric potential at the midpoint of the base, taking $q = 7.00$ μC.

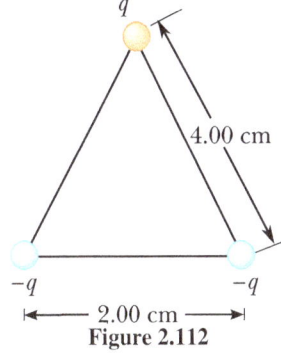

Figure 2.112

19. ●●Two point charges, $Q_1 = +5.00$ nC and $Q_2 = -3.00$ nC, are separated by 35.0 cm. (a) What is the potential energy of the pair? What is the significance of the algebraic sign of your answer? (b) What is the electric potential at a point midway between the charges?

20. ●●Show that the amount of work required to assemble four identical point charges of magnitude Q at the corners of a square of side s is $5.41kQ^2/s$

21. ●●●Two insulating spheres have radii 0.300 cm and 0.500 cm, masses 0.100 kg and 0.700 kg, and uniformly distributed charges of -2.00 μC and 3.00 μC. They are released from rest when their centres are separated by 1.00 m. (a) How fast will each be moving when they collide? (b) What If? If the spheres were conductors, would the speeds be

greater or less than those calculated in part (a)? Explain.

22. •••Two insulating spheres have radii r_1 and r_2, masses m_1 and m_2, and uniformly distributed charges $-q_1$ and q_2. They are released from rest when their centres are separated by a distance d. (a) How fast is each moving when they collide? (Suggestion: consider conservation of energy and conservation of linear momentum.) (b) If the spheres were conductors, would their speeds be greater or less than those calculated in part (a)? Explain.

23. ••Two particles, with charges of 20.0 nC and −20.0 nC, are placed at the points with coordinates (0, 4.00 cm) and (0, −4.00 cm), as shown in Figure 2.113. A particle with charge 10.0 nC is located at the origin. (a) Find the electric potential energy of the configuration of the three fixed charges. (b) A fourth particle, with a mass of 2.00×10^{-13} kg and a charge of 40.0 nC, is released from rest at the point (3.00 cm, 0). Find its speed after it has moved freely to a very large distance away.

24. ••A light unstressed spring has length d. Two identical particles, each with charge q, are connected to the opposite ends of the spring. The particles are held stationary a distance d apart and then released at the same time. The system then oscillates on a horizontal frictionless table. The spring has a bit of internal kinetic friction, so the oscillation is damped. The particles eventually stop vibrating when the distance between them is $3d$. Find the increase in internal energy that appears in the spring during the oscillations. Assume that the system of the spring and two charges is isolated.

25. ••Two point charges of equal magnitude are located along the y axis at equal distances above and below the x axis, as shown in Figure 2.114. (a) Plot a graph of the potential at points along the x axis over the interval $-3a < x < 3a$. (**Hint:** You should plot the potential in units of kQ/a.) (b) If the charge located at $-a$ be negative then plot the potential along the y axis over the interval $-4a < y < 4a$.

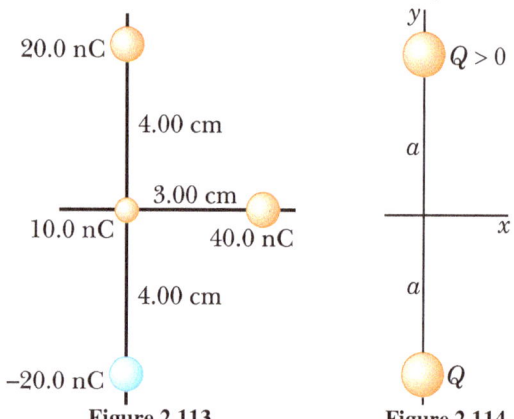

Figure 2.113   Figure 2.114

26. •••An alpha particle, has charge $+2e$ and mass 6.64×10^{-27} kg. Assume an alpha particle, initially very far from a gold nucleus, is fired with a velocity of 2.00×10^7 m/s directly toward the nucleus (charge $+79\,e$). How close does the alpha particle get to the nucleus before turning around? Assume the gold nucleus remains stationary.

27. ••An electron starts from rest 3.00 cm from the centre of a uniformly charged insulating sphere of radius 2.00 cm and total charge 1.00 nC. What is the speed of the electron when it reaches the surface of the sphere?

28. ••Calculate the energy required to assemble the array of charges shown in Fig.2.115, where $a = 0.200$ m, $b = 0.400$ m, and $q = 6.00\mu C$.

29. ••Four identical particles each have charge q and mass m.

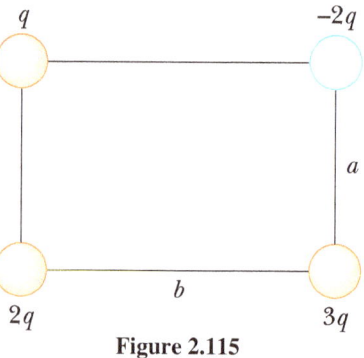

Figure 2.115

They are released from rest at the vertices of a square of side L. How fast is each charge moving when their distance from the centre of the square doubles?

30. ••How much work is required to assemble eight identical point charges, each of magnitude q, at the corners of a cube of side s?

31. ••A positive point charge $+Q$ is located on the x axis at $x = -a$. (a) How much work is required to bring an identical point charge from infinity to the point on the x axis at $x = +a$? (b) With the two identical point charges in place at $x = -a$ and $x = +a$, how much work is required to bring a third point charge $-Q$ from infinity to the origin? (c) How much work is required to move the charge $-Q$ from the origin to the point on the x axis at $x = 2a$ along the semicircular path shown in Fig.2.116?

Figure 2.116

Obtaining the Value of the Electric Field from the Electric Potential

32. ••An ion accelerated through a potential difference of 115 V experiences an increase in kinetic energy of 7.37×10^{-17} J. Calculate the charge on the ion.

33. ••(a) Calculate the speed of a proton that is accelerated from rest through a potential difference of 120 V. (b) Calculate the speed of an electron that is accelerated through the same potential difference.

34. ••What potential difference is needed to stop an electron having an initial speed of 4.20×10^5 m/s?

35. ••The potential in a region between $x = 0$ and $x = 6.00$ m is $V = a + bx$, where $a = 10.0$ V and $b = -7.00$ V/m. Determine (a) the potential at $x = 0$, 3.00 m, and 6.00 m, and (b) the magnitude and direction of the electric field at $x = 0$, 3.00 m, and 6.00 m.

36. ••The electric potential inside a charged spherical conductor of radius R is given by $V = kQ/R$, and the potential outside is given by $V = kQ/r$. Using $E_r = -dV/dr$, derive the electric field (a) inside and (b) outside this charge distribution.

37. ••Over a certain region of space, the electric potential is $V = 5x - 3x^2y + 2yz^2$. Find the expressions for the x, y and z components of the electric field over this region. What is the magnitude of the field at the point P that has coordinates $(1, 0, -2)$ m?

38. ••Fig.2.117 shows several equipotential lines each labelled by its potential in volts. The distance between the lines of the square grid represents 1.00 cm. (a) Is the magnitude of the field larger at A or at B? Why? (b) What is E at B? (c) Represent what the field looks like by drawing at least eight

2.15. QUESTIONS AND EXERCISES

field lines.

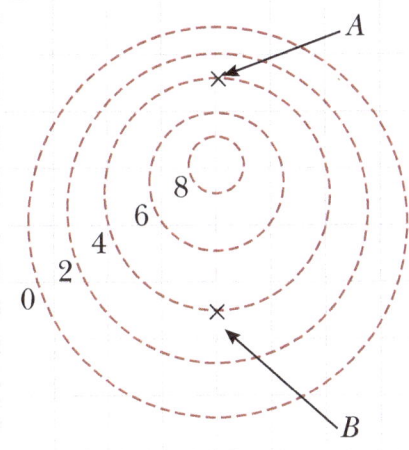

Figure 2.117

Electric Potential Due to Continuous Charge Distributions

39. ••Consider a ring of radius R with the total charge Q spread uniformly over its perimeter. What is the potential difference between the point at the centre of the ring and a point on its axis a distance $2R$ from the centre?

40. ••In Fig.2.118, a plastic rod having a uniformly distributed charge $Q = -25.6$ pC has been bent into a circular arc of radius $R = 3.71$ cm and central angle $\phi = 120°$. With $V = 0$ at infinity, what is the electric potential at P, the centre of curvature of the rod?

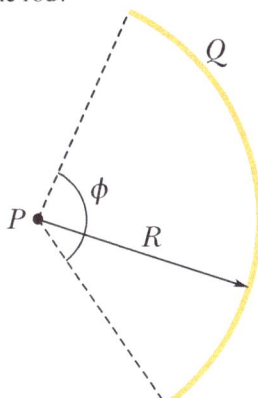

Figure 2.118

41. ••A plastic rod has been bent into a circle of radius $R = 8.20$ cm. It has a charge $Q_1 = +4.20$ pC uniformly distributed along one-quarter of its circumference and a charge $Q_2 = -6Q_1$ uniformly distributed along the rest of the circumference (Fig. 2.119). With $V = 0$ at infinity, what is the electric potential at (a) the center C of the circle and (b) point P, on the central axis of the circle at distance $D = 6.71$ cm from the center?

42. ••In Fig.2.120, three thin plastic rods form quarter-circles with a common center of curvature at the origin. The uniform charges on the three rods are $Q_1 = +30$ nC, $Q_2 = +3.0Q_1$, and $Q_3 = -8.0Q_1$. What is the net electric potential at the origin due to the rods?

43. ••A rod of length L (Fig.2.121) lies along the x axis with its left end at the origin. It has a non-uniform charge density $\lambda = \alpha x$, where α is a positive constant. (a) What are the units of α? (b) Calculate the electric potential at A.

44. ••For the arrangement described in the previous problem, calculate the electric potential at point B, which lies on the perpendicular bisector of the rod a distance b above the x axis.

45. ••Calculate the electric potential at point P on the axis of the annulus shown in Fig.2.122, which has a uniform charge density σ.

Figure 2.119

Figure 2.120

Figure 2.121

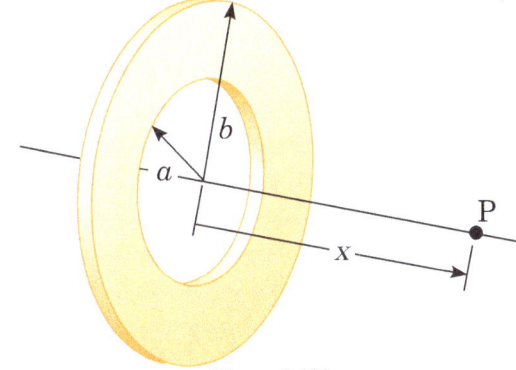

Figure 2.122

46. ••A wire having a uniform linear charge density λ is bent into the shape shown in Figure 2.123. Find the electric potential at point O.

Figure 2.123

Electric Potential Due to a Charged Conductor

47. ••How many electrons should be removed from an initially

uncharged spherical conductor of radius 0.300 m to produce a potential of 7.50 kV at the surface?

48. ••A spherical conductor has a radius of 14.0 cm and charge of 26.0 μC. Calculate the electric field and the electric potential at (a) $r = 10.0$ cm, (b) $r = 20.0$ cm, and (c) $r = 14.0$ cm from the center.

Applications of Electrostatics

49. ••An electron is released from rest on the axis of a uniform positively charged ring, 0.100 m from the ring's centre. If the linear charge density of the ring is +0.100 μC/m and the radius of the ring is 0.200 m, how fast will the electron be moving when it reaches the centre of the ring?

50. ••Four balls, each with mass m, are connected by four non-conducting strings to form a square with side a, as shown in Figure 2.124. The assembly is placed on a horizontal non-conducting frictionless surface. Balls 1 and 2 each have charge q, and balls 3 and 4 are uncharged. Find the maximum speed of balls 1 and 2 after the string connecting them is cut.

Figure 2.124

51. ••A disc of radius R (Fig.2.125) has a non-uniform surface charge density $\sigma = Cr$, where C is a constant and r is measured from the centre of the disk. Find (by direct integration) the potential at P.

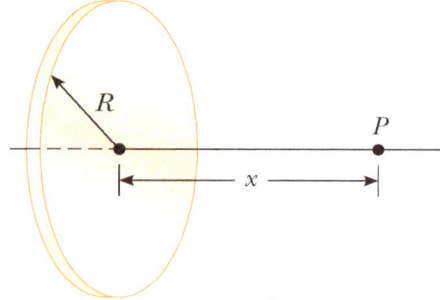

Figure 2.125

52. ••Suppose, an alpha particle is shot at a gold foil from a large distance with kinetic energy of 5.0 MeV. If the alpha particle is aimed directly at the gold nucleus, and the only force acting on it is the electric force of repulsion exerted on it by the gold nucleus, how close will it approach the gold nucleus before turning back?

2.15.3 Multiple Choice Questions
Level 1
Potential and Potential Difference

1. When charge of 3 coulomb is placed in a Uniform electric field, it experiences a force of 3000 newton, within this field, potential difference between two points separated by a distance of 1 cm is-
   (A) 10 V (B) 90 V (C) 1000 V (D) 3000 V

2. A uniform electric field having a magnitude E_0 and direction along positive x-axis exists. If the electric potential (V) is zero at $x = 0$ then its value at $x = +x$ will be-
   (A) $V_x = xE_0$ (B) $V_x = -x \cdot E_0$
   (C) $V_x = x^2 E_0$ (D) $V_x = x^2 E_0$

3. The dimensions of potential difference are -
   (A) $[ML^2 T^{-2} Q^{-1}]$ (B) $[MLT^{-2} Q^{-1}]$
   (C) $[MT^{-2} Q^{-2}]$ (D) $[ML^2 T^{-1} Q^{-1}]$

4. Three equal charges are placed at the three corners of an isosceles triangle as shown in the Figure 2.126. The statement which is true for electric potential V and the field intensity E at the centre of the triangle is-
   (A) $V = 0, E = 0$ (B) $V = 0, E \neq 0$
   (C) $V \neq 0, E = 0$ (D) $V \neq 0, E \neq 0$

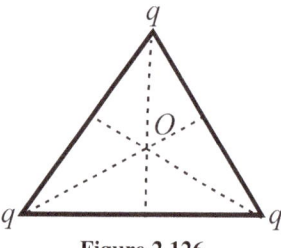

Figure 2.126

5. 1 esu of potential is equal to-
   (A) 1/300 volt (B) 8×10^{10} volt
   (C) 300 volt (D) 3 volt

6. The earth's surface is considered to be at -
   (A) Zero potential (B) Negative Potential
   (C) Infinite Potential (D) Positive Potential

7. Electric potential is a -
   (A) vector quantity (B) scalar quantity
   (C) fictious quantity (D) none of above

8. The electric potential V at any point (x, y, z) in space is given by $V = 4x^2$ volt. The electric field E in V/m at the point $(1, 0, 2)$ is
   (A) +8 in x direction (B) 8 in $-x$ direction
   (C) 16 in +x direction (D) 16 in $-x$ direction

9. ABC is equilateral triangle of side 1 m. Charges are placed at its corners as shown in Fig.2.127. O is the mid point of side BC. The potential at point O, is-
   (A) 2.7×10^3 V (B) 1.52×10^5 V
   (C) 1.3×10^3 V (D) -1.52×10^5 V

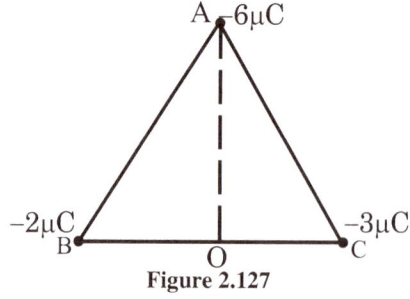

Figure 2.127

10. In a region where $E = 0$, the potential (V) varies with distance r as -
    (A) $V \propto 1/r$ (B) $V \propto r$
    (C) $V \propto 1/r^2$ (D) V is independent on r

11. Charges of $+\left(\frac{10}{3}\right) \times 10^{-9}$ are placed at each of the four corners of a square of side 8 cm. The potential at the intersection of the diagonals is
    (A) $150\sqrt{2}$ V (B) $1500\sqrt{2}$ V
    (C) $900\sqrt{2}$ V (D) 900 V

2.15. QUESTIONS AND EXERCISES

12. An equipotential surface is that surface-
    (A) On which each and every point has the same potential
    (B) Which has negative potential
    (C) Which has positive potential
    (D) Which has zero potential

13. The surface of a conductor -
    (A) is a non-equipotential surface
    (B) has all the points at the same potential
    (C) has different points at different potential
    (D) has at least two points at the same potential

14. The electron potential (V) as a function of distance (x) [in meters] is given by $V = (5x^2 + 10x - 9)$ volt. The value of electric field at x = 1 m would be-
    (A) 20 volt/m   (B) 6 volt/m
    (C) 11 volt/m   (D) –23 volt/m

15. Some equipotential lines are as shown is Fig.2.128 E_1, E_2 and E_3 are the electric fields at points 1, 2 and 3 then-
    (A) $E_1 = E_2 = E_3$ (B) $E_1 > E_2 > E_3$
    (C) $E_1 > E_2, E_2 < E_3$ (D) $E_1 < E_2 < E_3$

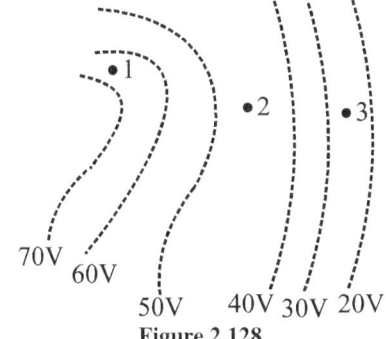

Figure 2.128

16. Three charges $2q$, $-q$, $-q$ are located at the vertices of an equilateral triangle. At the circumcenter of the triangle-
    (A) the field is zero but potential is not zero.
    (B) the field is non-zero but the potential is zero.
    (C) both, field and potential are zero.
    (D) both, field and potential are non- zero

Electric Potential Energy and Work Done

17. A positive point charge of Q' units is moved round another point positive charge of Q units in circular path. If the radius of the circle is r, then the work done on the charge Q' in making one complete revolution is-
    (A) $\dfrac{Q}{4\pi \epsilon_0 r}$ (B) $\dfrac{QQ'}{4\pi \epsilon_0 r}$ (C) $\dfrac{Q'}{4\pi \epsilon_0 r}$ (D) 0

18. A proton is projected with velocity 7.45×10^5 m/s towards an other proton which is at rest. The minimum distance of approach is-
    (A) 10^{-12} m   (B) 10^{-14} m
    (C) 10^{-10} m   (D) 10^{-8} m

19. Three charges are placed as shown in Fig.2.129. If the electric potential energy of system is zero, then $Q : q$ -
    (A) $Q/q = -2/1$ (B) $Q/q = 2/1$
    (C) $Q/q = -1/2$ (D) $Q/q = 1/4$

Figure 2.129

20. If a unit charge is taken from one point to another over an equipotential surface then-
    (A) work is done on the charge
    (B) work is done by the charge
    (C) work on the charge is constant
    (D) no work is done

21. In an electric field the work done in moving a unit positive charge between two points is the measures of -
    (A) Resistance
    (B) Potential difference
    (C) Intensity of electric field
    (D) Capacitance

22. State which one of the following is correct?
    (A) joule = coulomb × volt
    (B) joule = coulomb / volt
    (C) joule = volt / ampere
    (D) joule = volt × ampere

23. One electron volt (eV) of energy is equal to -
    (A) 1.6×10^{-12} ergs.   (B) 4.8×10^{-10} ergs.
    (C) 9×10^{11} ergs.   (D) 3×10^{9} ergs.

24. The kinetic energy in electron volt gained by an $\alpha-$ particle when it moves from rest at point where its potential is 70 volt to a point where potential is 50 volts, is -
    (A) 20 eV   (B) 20 MeV
    (C) 40 eV   (D) 40 MeV

25. An α - particle moves towards a nucleus at rest, if kinetic energy of α-particle is 10 MeV and atomic number of nucleus is 50 . The distance of closest approach will be -
    (A) 1.44×10^{-14} m   (B) 2.88×10^{-14} m
    (C) 1.44×10^{-10} m   (D) 2.88×10^{-10} m

26. A point charge q moves form point P to point S along the path PQRS as shown in Fig.2.130 in a uniform electric field $\vec{E}$, pointing co-parallel to the positive direction of the x-axis. The coordinates of the points P, Q, R and S are $(a, b, 0)$, $(2a, 0, 0)$, $(a, -b, 0)$ and $(0, 0, 0)$ respectively. The work done by the field in the above process is given by the expression
    (A) qEq (B) $-qEa$
    (C) $q\,E\,a\sqrt{2}$ (D) $qE\sqrt{(2a)^2 + b^2}$

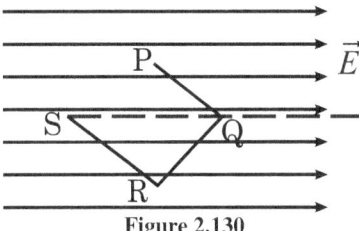

Figure 2.130

27. Two identical thin rings, each of radius R metres, are coaxially placed at a distance (R) metres apart. Q_1 and Q_2 respectively are the charges on two rings. The charge distribution on both the rings is uniform. The work done in moving a charge q from the centre of one ring to that of other is-
    (A) zero   (B) $\dfrac{q(Q_1 - Q_2)(\sqrt{2}-1)}{(4\sqrt{2}\epsilon_0 \pi R)}$
    (C) $\dfrac{q\sqrt{2}(Q_1+Q_2)}{(4\epsilon_0 \pi R)}$ (D) $\dfrac{q(Q_1+Q_2)(\sqrt{2}+1)}{(4\sqrt{2}\epsilon_0 \pi R)}$

Applications of Gauss's Law

28. A solid conducting sphere having a charge Q is surrounded by an uncharged concentric conducting hollow spherical shell. Let the potential difference between the surface of the solid sphere and that of the outer surface of the hollow shell be V. If the shell is now given a charge of $3Q$ the new potential difference between the same two surfaces is

(A) V (B) 2 V (C) 4 V (D) √2 V

29. The dependence of electric potential V on the distance 'r' from the centre of a charged spherical shell is shown by which of the graph in Fig.??-

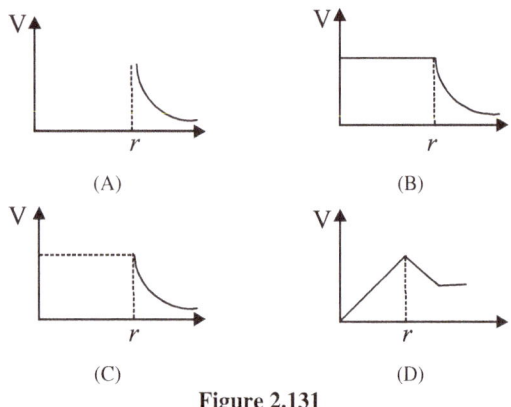

Figure 2.131

Electric dipole

30. The electric potential at a point due to an electric dipole will be-
    (A) $k\dfrac{\vec{p}\cdot\vec{r}}{r^3}$
    (B) $k\dfrac{\vec{p}\cdot\vec{r}}{r^2}$
    (C) $\dfrac{k(\vec{p}\times\vec{r})}{r}$
    (D) $\dfrac{k(\vec{p}\times\vec{r})}{r^2}$

31. An electric dipole of dipole moment p is aligned parallel to a uniform electric field E. The energy required to rotate the dipole by 90° is-
    (A) p^2E (B) pE (C) ∞ (D) pE^2

32. An electric dipole of moment p is placed in an electric field of intensity E. The dipole acquires a position such that the axis of the dipole makes an angle θ with direction of the field. Assuming that the potential energy of the dipole to be zero when $\theta = 90°$, the torque and the potential energy of the dipole will respectively be-
    (A) $pE\sin\theta, -pE\cos\theta$
    (B) $pE\sin\theta, -2pE\cos\theta$
    (C) $pE\sin\theta, 2pE\cos\theta$
    (D) $pE\cos\theta, -pE\sin\theta$

Miscellaneous

33. A proton is first placed at A and then at B between the two plates of a parallel plate capacitor charged to a potential difference of V volt as shown in Fig.2.132. The force on proton at A is-
    (A) more than at B
    (B) less than at B
    (C) equal to that at B
    (D) nothing can be said

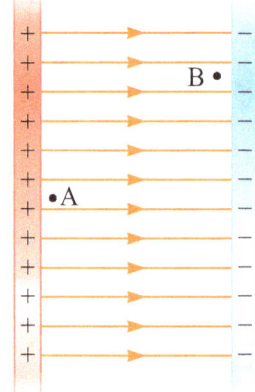

Figure 2.132

34. In an electric field, a 6.75μC charge experiences 2.5 N force, when placed at distance of 5 m from the origin. Then potential gradient at this point will be- (in MKS)
    (A) 5.71×10^5 (B) 3.71×10^5
    (C) 18.81×10^5 (D) 1.881×10^5

35. The electric flux coming out of the equipotential surface is-
    (A) perpendicular to the surface
    (B) parallel to the surface
    (C) in all directions
    (D) zero

36. A charge Q is distributed over two concentric hollow spheres of radii r and $R(>r)$ such that the surface densities are equal. Find the potential at the common centre-
    (A) $\dfrac{Q}{4\pi\epsilon_0}\times\dfrac{(r+R)}{(R+r)^2}$
    (B) $\dfrac{Q(R^2+r)^2}{4\pi\epsilon_0(r+R)}$
    (C) $\dfrac{Q(r+R)}{4\pi\epsilon_0(R^2+r^2)}$
    (D) none of these

37. The earth had a net charge equivalent to 1 electron per square metre of surface area of radius 6.4×10^6 m. Its potential would be-
    (A) $+0.12\ V$ (B) $-0.12\ V$ (C) $+1.2\ V$ (D) $-1.2\ V$

38. The electric potential at the surface of an atomic nucleus ($Z = 50$) of radius 9×10^{-15} m is -
    (A) 80 V (B) 8×10^6 V
    (C) 8×10^4 V (D) 8×10^2 V

39. A hollow metal sphere of radius 5 cm is charged such that the potential on its surface is 10 V. The potential at the centre of the sphere is -
    (A) 0 V
    (B) 10 V
    (C) same as at point 5 cm away from the surface
    (D) same as at point 25 cm away from the surface

2.15.3.1 Level 2

1. Electric potential in an electric field is given as $V = K/r$, (K being constant), if position vector $\vec{r} = 2\hat{i} + 3\hat{j} + 6\hat{k}$ then electric field will be
    (A) $\dfrac{(2\hat{i}+3\hat{j}+6\hat{k})K}{243}$
    (B) $\dfrac{(2\hat{i}+3\hat{j}+6\hat{k})K}{343}$
    (C) $\dfrac{(3\hat{i}+2\hat{j}+6\hat{k})K}{243}$
    (D) $\dfrac{(6\hat{i}+2\hat{j}+3\hat{k})K}{343}$

2. At any point $(x, 0, 0)$ the electric potential V is $\left(\dfrac{1000}{x} + \dfrac{1500}{x^2} + \dfrac{500}{x^3}\right)$ volt, then electric field at $x = 1$ m
    (A) $5500(\hat{j}+\hat{k})V/m$
    (B) $5500\hat{i}V/m$
    (C) $\dfrac{5500}{\sqrt{2}}(\hat{j}+\hat{k})V/m$
    (D) $\dfrac{5500}{\sqrt{2}}(\hat{i}+\hat{k})V/m$

3. Potential difference between centre and the surface of a sphere of radius R with uniform charge density σ with in it will be
    (A) $\sigma R^2/6\epsilon_0$
    (B) $\sigma R^2/4\epsilon_0$
    (C) zero
    (D) $\sigma R^2/2\epsilon_0$

4. Two conducting spheres of radii r_1 and r_2 are equally charged. The ratio of their potential is-
    (A) r_1^2/r_2^2 (B) r_2^2/r_1^2 (C) r_1/r_2 (D) r_2/r_1

5. Two similar rings P and Q (radius = 0.1 m) are placed co-axially at a distance 0.5 m apart. The charge on P and Q is 2 μC and 4 μC respectively. Work done in moving a 5 μC charge from center of P to the center of Q is -
    (A) 1.28 J (B) 0.72 J (C) 0.144 J (D) 1.44 J

6. Three point charge $-q, +q$ and $-q$ are placed along a straight line at equl distances (say r meter). Electric potential energy of this system of charges will be if $+q$ charge is in the middle-
    (A) $\dfrac{-3q^2}{4\pi\epsilon_0 r}$
    (B) $\dfrac{-8q^2}{3\pi\epsilon_0 r}$
    (C) $\dfrac{-3q^2}{8\pi\epsilon_0 r}$
    (D) $\dfrac{-q^2}{8\pi\epsilon_0 r}$

2.15. QUESTIONS AND EXERCISES

7. Four equal charges of charge q are placed at corner of a square of side a. Potential energy of the whole system is-
   (A) $\frac{4kq^2}{a}$
   (B) $\frac{4kq^2}{a}\left(1+\frac{1}{2\sqrt{2}}\right)$
   (C) $\frac{1}{2\sqrt{2}}\frac{kq^2}{a}$
   (D) $\frac{kq^2}{a}\left(4+\frac{1}{2\sqrt{2}}\right)$

8. The potential of a charged drop is v. This is divided into n smaller drops, then each drop will have the potential as;
   (A) $n^{-1}v$ (B) $n^{2/3}v$ (C) $n^{3/2}v$ (D) $n^{-2/3}v$

9. 8 small droplets of water of same size and same charge form a large spherical drop. The potential of the large drop, in comparison to potential of a small drop will be-
   (A) 2 times   (B) 4 times
   (C) 8 times   (D) same

10. In Millikan's oil drop experiment an oil drop carrying a charge Q is held stationary by a potential difference 2400 V between the plates. To keep a drop of half the radius stationary the potential difference had to be made 600 V. What is the charge on the second drop-
    (A) Q/4   (B) Q/2   (C) Q   (D) 3Q/2

11. There is an electric field E in X-direction. If the work done on moving a charge 0.2 C through a distance of 2 m along a line making an angle 60° with the X-axis, is 4.0 J, what is the value of E
    (A) $\sqrt{3}$ N/C   (B) 4 N/C
    (C) 5 N/C   (D) None of these

12. A ball of mass 1 g and charge 10^{-3}C moves from a point A. where potential is 800 volt to the point B where potential is zero. Velocity of the ball at the point B is 20 cm/s. The velocity of the ball at the point A will be-
    (A) 22.8 cm/s   (B) 228 cm/s
    (C) 16.8 cm/s   (D) 168 m/s

13. An electric dipole is placed along the x-axis at the origin O. A point P is at a distance of 20 cm from this origin such that OP makes an angle $\pi/3$ with the x-axis. If the electric field at P makes an angle θ with the x-axis. the value of θ would be
    (A) $\pi/3$
    (B) $\frac{\pi}{3}+\tan^{-1}\left(\frac{\sqrt{3}}{2}\right)$
    (C) $2\pi/3$
    (D) $\tan^{-1}\left(\frac{\sqrt{3}}{2}\right)$

14. An electric dipole of moment $\vec{p}$ is placed normal to the lines of force of electric intensity $\vec{E}$, then the work done in deflecting it through an angle of 180° is
    (A) pE (B) $+2pE$ (C) $-2pE$ (D) Zero

15. An electric dipole of moment $\vec{p}$ placed in a uniform electric field $\vec{E}$ has minimum potential energy when the angle between $\vec{p}$ and $\vec{E}$ is-
    (A) Zero   (B) $\pi/2$ (C) π (D) $3\pi/2$

16. An electric dipole has the magnitude of its charge as q and its dipole moment is p. It is placed in a uniform electric field $\vec{E}$. If its dipole moment is along the direction of the field, the force on it and its potential energy are respectively-
    (A) $2qE$, minimum   (B) qE, pE
    (C) 0, minimum   (D) qE, maximum

2.15.3.2 Level 3

1. A conducting sphere of radius R is charged to a potential of V volt. Then the electric field at a distance $r(>R)$ from the centre of the sphere would be -
   (A) RV/r^2 (B) rV/R^2 (C) V/r (D) R^2V/r^3

2. The variation of electric potential with distance from a fixed point is shown in Fig.2.133. The value of electric field at $x=2$ m -
   (A) 0   (B) 6/2   (C) 6/1   (D) 6/3

3. A positive point charge q is carried from a point B to a

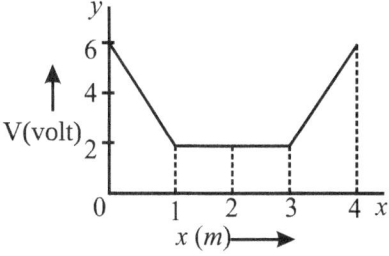

Figure 2.133

point A in the electric field of a point charge $+Q$ at origin O. If the permittivity of free space is ϵ_0, the work done in the process is given by (where $a = OA$ and $b = OB$)
   (A) $\frac{qQ}{4\pi\epsilon_0}\left(\frac{1}{a}+\frac{1}{b}\right)$
   (B) $\frac{qQ}{4\pi\epsilon_0}\left(\frac{1}{a}-\frac{1}{b}\right)$
   (C) $\frac{qQ}{4\pi\epsilon_0}\left(\frac{1}{a^2}-\frac{1}{b^2}\right)$
   (D) $\frac{qQ}{4\pi\epsilon_0}\left(\frac{1}{a^2}+\frac{1}{b^2}\right)$

4. Two conducting spheres each of radius R carry charge q. They are placed at a distance r from each other, where $r > 2R$. The neutral point lies at a distance $r/2$ from either sphere. If the electric field at the neutral point due to either sphere be E, then the total electric potential at that point will be -
   (A) $rE/2$ (B) rE (C) $RE/2$ (D) RE

5. A ring of radius R carries a charge $+q$. A test charge $-q_0$ is released on its axis at a distance $\sqrt{3}R$ from its center. How much kinetic energy will be acquired by the test charge when it reaches the centre of the ring-
   (A) $\frac{1}{4\pi\epsilon_0}\frac{qq_0}{R}$
   (B) $\frac{1}{4\pi\epsilon_0}\frac{qq_0}{2R}$
   (C) $\frac{1}{4\pi\epsilon_0}\frac{qq_0}{\sqrt{3}R}$
   (D) $\frac{1}{4\pi\epsilon_0}\frac{qq_0}{3R}$

6. Two spheres of radii r_1 and r_2 are at the same potentials. If their surface densities of charges be σ_1 and σ_2 respectively, then $\sigma_1/\sigma_2 -$
   (A) r_1/r_2 (B) r_2/r_1 (C) $(r_1/r_2)^2$ (D) $(r_2/r_1)^2$

7. A proton and an electron are released from an infinite distance apart and are attracted to each other. Which of the following statements about their kinetic energy is true?
   (A) Kinetic energy of electron is more than that of proton
   (B) Kinetic energy of electron is less than that of proton
   (C) Kinetic energy of electron = kinetic energy of proton
   (D) None of the above is true as it depending on the distance between the particles

8. Two conducting spheres of radii r_1 and r_2 are charged such that they have the same electric field on their surfaces. The ratio of the electric potential at their centres is -
   (A) $\sqrt{r_1/r_2}$ (B) r_1/r_2
   (C) r_1^2/r_2^2 (D) None of the above

9. A charge $+Q$ at A (Figure 2.134) produces electric field E and electric potential V at D. If we now put charges $-2Q$ and $+Q$ at B and C respectively, then the electric field and potential at D will be -
   (A) E and 0   (B) 0 and V
   (C) $\sqrt{2}E$ and $\frac{V}{\sqrt{2}}$ (D) $\frac{E}{\sqrt{2}}$ and $\frac{V}{\sqrt{2}}$

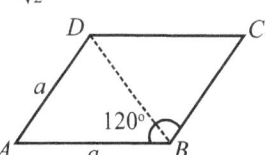

Figure 2.134

10. If a positive charge is shifted from a low-potential region to a high-potential region, the electric potential energy-

(A) increases
(B) decreases
(C) remains the same
(D) may increase or decrease

11. A particle of mass 0.002 kg and a charge 1 μC is held at rest on a frictionless horizontal surface at a distance of 1 m from a fixed charge of 1 mC. If the particle is released, it will be repelled. The speed of the particle when it is at a distance of 10 m from the fixed charge is-
    (A) 60 ms^{-1} (B) 75 ms^{-1} (C) 90 ms^{-1} (D) 100 ms^{-1}

12. Electric potential is given by: $V = 6x - 8xy^2 - 8y + 6yz - 4z^2$
    Electric field at the origin is-
    (A) $-6\hat{i} + 8\hat{j}$ (B) $6\hat{i} - 8\hat{j}$ (C) $\hat{i} + \hat{j}$ (D) Zero

13. A hollow conducting sphere of radius R has charge $(+Q)$ on its surface. The electric potential within the sphere at a distance $r = R/3$ from the centre is -
    (A) 0 (B) $\frac{1}{4\pi\epsilon_0}\frac{Q}{r}$ (C) $\frac{1}{4\pi\epsilon_0}\frac{Q}{R}$ (D) $\frac{1}{4\pi\epsilon_0}\frac{Q}{r^2}$

14. A particle has a mass 400 times than that of the electron and charge is double than that of a electron. It is accelerated by 5 V of potential difference. Initially the particle was at rest. Then its final kinetic energy will be-
    (A) 5 eV (B) 10 eV (C) 100 eV (D) 2000 eV

15. Two equal positive charges are kept at points A and B. The electric potential at the points between A and B (excluding these points) is studied while moving from A to B. The potential-
    (A) continuously increases
    (B) continuously decreases
    (C) increases then decreases
    (D) decreases then increases

Passage Type Questions (Q.16 and 18)
In the diagram shown in Fig.2.135, the broken lines represent the paths followed by particles W, X, Y and Z respectively through the constant field E. The numbers below the field represents meters.

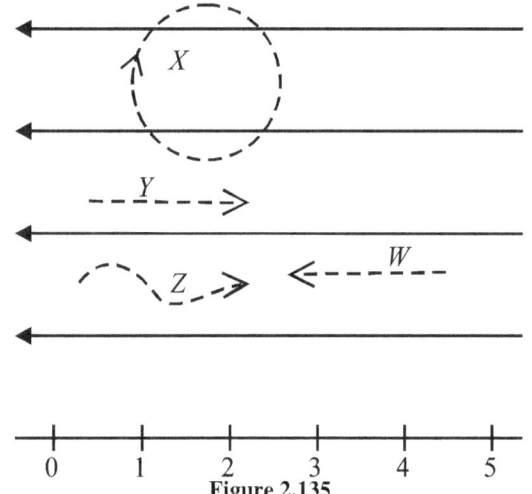

Figure 2.135

16. If the particles begin and end at rest, and all are positively charged, the same amount of work was done on which particles?
    (A) W and Z   (B) W, Y and Z
    (C) Y and Z   (D) W, X, Y and Z

17. If the particles started from rest and all are positively charged which particles must have been acted upon by a force other than that produced by the electric field.
    (A) W and Y   (B) X and Z
    (C) X, Y and Z   (D) W, X, Y and Z

18. If the particles are positively charged, which particles increased their electrical potential energy -
    (A) X and Z
    (B) Y and Z
    (C) W, X, Y and Z
    (D) Since the electric field is constant none of the particles increased their electrical potential energy.

19. Five equal and similar charges are placed at the corners of a regular hexagon as shown in the Fig.2.136. What is the electric field and potential at the centre of the hexagon -
    (A) $\frac{5}{4\pi\epsilon_0}\frac{q}{l}, \frac{5}{4\pi\epsilon_0}\frac{q}{l}$ (B) $\frac{1}{4\pi\epsilon_0}\frac{q}{l}, \frac{5}{4\pi\epsilon_0}\frac{q}{l}$
    (C) $\frac{1}{4\pi\epsilon_0}\frac{q}{l^2}, \frac{5}{4\pi\epsilon_0}\frac{q}{l}$ (D) $\frac{1}{4\pi\epsilon_0}\frac{q}{l}, \frac{1}{4\pi\epsilon_0}\frac{q}{l^2}$

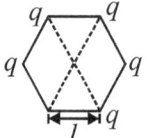

Figure 2.136

Statements Type Question:-
Each of the questions given below consist of Statement-I and Statement- II. Use the following Key to choose the appropriate answer.
(A) If both Statement-I and Statement-II are true, and Statement-II is the correct explanation of Statement-I.
(B) If both Statement-I and Statement-II are true but Statement-II is not the correct explanation of Statement $-I$.
(C) If Statement-I is true but Statement-II is false.
(D) If Statement-I is false but Statement-II is true.

20. **Statement I:** If a point charge q is placed in front of an infinite grounded conducting plane surface, the point charge will experience a force.
    Statement II: The force is due to the induced charge on the conducting surface which is at zero potential.

21. **Statement I:** Electrons move away from a region of lower potential to a region of higher potential.
    Statement II: An electron has negative charge.

22. **Statement I:** Work done in moving a charge between any two points in an electric field is independent of the path followed by the charge, between these points.
    Statement II: Electrostatic forces are non conservative.

2.15.3.3 Level 4 (Previous Years JEE Main and Advanced Questions)

Section A: JEE Main

1. On moving a charge of 20 coulombs by 2 cm, 2J of work is done, then the potential difference between the points is –
    [AIEEE-2002]
    (A) 0.1 V   (B) 8 V   (C) 2 V   (D) 0.5 V

2. A thin spherical conducting shell of radius R has a charge q. Another charge Q is placed at the centre of the shell. The electrostatic potential at a point P at a distance $R/2$ from the centre of the shell is- [AIEEE 2003]
    (A) $\frac{2Q}{4\pi\epsilon_0 R} - \frac{2q}{4\pi\epsilon_0 R}$ (B) $\frac{2Q}{4\pi\epsilon_0 R} + \frac{q}{4\pi\epsilon_0 R}$
    (C) $\frac{(q+Q)}{4\pi\epsilon_0}\frac{2}{R}$ (D) $\frac{2Q}{4\pi\epsilon_0 R}$

3. A charged particle 'q' is shot towards another charged particle 'Q', which is fixed, with a speed v. It approaches Q upto a closest distance r and then returns. If q were given a speed of $2v$, the closest distances of approach would be-
    [AIEEE-2004]
    (A) r (B) $2r$ (C) $r/2$ (D) $r/4$

4. Two thin wire rings each having a radius R are placed at a distance d apart with their axes coinciding. The charges

on the two rings are $+q$ and $-q$. The potential difference between the centres of the two rings is- [AIEEE-2005]
(A) $QR/4\pi\epsilon_0 d^2$
(B) $\dfrac{Q}{2\pi\epsilon_0}\left[\dfrac{1}{R}-\dfrac{1}{\sqrt{R^2+d^2}}\right]$
(C) zero
(D) $\dfrac{Q}{4\pi\epsilon_0}\left[\dfrac{1}{R}-\dfrac{1}{\sqrt{R^2+d^2}}\right]$

5. Two insulating plates are both uniformly charged, as shown in Fig.2.137, in such a way that the potential difference between them is $V_2 - V_1 = 20$ V (i.e. plate 2 is at a higher potential). The plates are separated by $d = 0.1$ m and can be treated as infinitely large. An electron is released from rest on the inner surface of plate 1. What is its speed when it hits plate 2? $(e = 1.6 \times 10^{-19}$ C, $m_e = 9.11 \times 10^{-31}$ kg) – [AIEEE 2006]
   (A) 1.87×10^6 m/s
   (B) 32×10^{-19} m/s
   (C) 2.65×10^6 m/s
   (D) 7.02×10^{12} m/s

Figure 2.137

6. Two spherical conductors A and B of radii 1 mm and 2 mm are separated by a distance of 5 cm and are uniformly charged. If the sphere are connected by a conducting wire then in equilibrium condition, the ratio of the magnitude of the electric fields at the surfaces of spheres A and B is - [AIEEE 2007]
   (A) 2 : 1 (B) 1 : 4 (C) 4 : 1 (D) 1 : 2

7. An electric charge $10^{-3}\mu$C is placed at the origin $(0,0)$ of X-Y co-ordinate system. Two points A and B are situated at $(\sqrt{2},\sqrt{2})$ and $(2,0)$ respectively. The potential difference between the points A and B will be- [2006]
   (A) 9 V (B) 0 V (C) 2 V (D) 4.5 V

8. Charges are placed on the vertices of a square as shown. Let $\vec{E}$ be the electric field and V the potential at the centre. If the charges on A and B are interchanged with those on D and C respectively, then [AIEEE 2007]
   (A) $\vec{E}$ remains unchanged, V changes
   (B) Both $\vec{E}$ and V change
   (C) $\vec{E}$ and V remain unchanged
   (D) $\vec{E}$ changes, V remains unchanged

Figure 2.138

9. The potential at a point x (measured in μm) due to some changes situated on the x-axis is given by $V(x) = 20/(x^2 - 4)$ volts. The electric field E at $x = 4$ μm is given by [2007]
   (A) $5/3$ volt $/\mu$m and in the $-$ve x direction
   (B) $5/3$ volt $/\mu$m and in the $+$ve x direction
   (C) $10/9$ volt μm and in the $-$ve x direction
   (D) $10/9$ volt μm and in the $+$ve x direction

10. Two points P and Q are maintained at the potentials of 10 V and -4 V, respectively. The work done in moving 100 electrons from P to Q is - [AIEEE-2009]
    (A) -9.60×10^{-17} J
    (B) 9.60×10^{-17} J
    (C) -2.24×10^{-16} J
    (D) 2.24×10^{-16} J

11. The electrostatic potential inside a charged spherical ball is given by $\phi = ar^2 + b$ where r is the distance from the centre; a, b are constants. Then the charge density inside ball is [AIEEE 2011]
    (A) $-6a\epsilon_0 r$
    (B) $-24\pi a\epsilon_0 r$
    (C) $-6a\epsilon_0$
    (D) $-24\pi a\epsilon_0 r$

12. This question has statement 1 and statement 2. Of the four choices given after the statements, choose the one that best describes the two statements.

    An insulating solid sphere of radius R has a uniformly positive charge density ρ. As a result of this uniform charge distribution there is a finite value of electric potential at the centre of the sphere, at the surface of the sphere and also at a point outside the sphere. The electric potential at infinity is zero. [2012]

    Statement 1: When a charge q is taken from the centre to the surface of the sphere, its potential energy changes by $q\rho/(3\epsilon_0)$

    Statement 2: The electric field at a distance $r (r<R)$ from the centre of the sphere is $\rho r/3\epsilon_0$

    (A) Statement 1 is true, Statement 2 is true, Statement 2 is not the correct explanation for statement 1.
    (B) Statement 1 is true, Statement 2 is false
    (C) Statement 1 is false, Statement 2 is true
    (D) Statement 1 is true, Statement 2 is the correct explanation for statement 1

13. A charge Q is uniformly distributed over a long rod AB of length L as shown in the Fig.2.139 The electric potential at the point O lying at distance L from the end A is: [2013]
    (A) $\dfrac{Q}{8\pi\epsilon_0 L}$
    (B) $\dfrac{3Q}{4\pi\epsilon_0 L}$
    (C) $\dfrac{Q}{4\pi\epsilon_0 L \ln 2}$
    (D) $\dfrac{Q\ln 2}{4\pi\epsilon_0 L}$

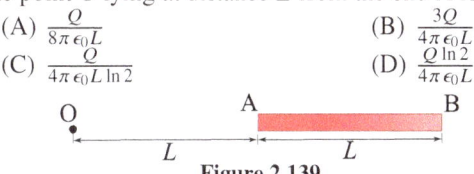

Figure 2.139

14. Assume that an electric field $\vec{E} = 30x^2\hat{i}$ exists in space. Then the potential difference $V_A - V_0$, where V_0 is the potential at the origin and V_A the potential at $x = 2$ m is: [2014]
    (A) -80 J (B) 80 J (C) 120 J (D) -120 J

15. Three concentric metal shells A, B and C of respective radii a, b and c $(a<b<c)$ have surface charge densities $+\sigma, -\sigma$ and $+\sigma$ respectively. The potential of shell B is [2018]
    (A) $\dfrac{\sigma}{\epsilon_0}\left[\dfrac{a^2-b^2}{a}+c\right]$
    (B) $\dfrac{\sigma}{\epsilon_0}\left[\dfrac{a^2-b^2}{b}+c\right]$
    (C) $\dfrac{\sigma}{\epsilon_0}\left[\dfrac{b^2-c^2}{b}+a\right]$
    (D) $\dfrac{\sigma}{\epsilon_0}\left[\dfrac{b^2-c^2}{c}+a\right]$

16. A charge Q is distributed over two concentric conducting thin spherical shells of radii r and $R (R>r)$ (Fig.2.140). If the surface charge densities on the two shells are equal, the electric potential at the common centre is: [2019]
    (A) $\dfrac{1}{4\pi\epsilon_0}\dfrac{2(R+r)}{(R^2+r^2)}Q$
    (B) $\dfrac{1}{4\pi\epsilon_0}\dfrac{(R+r)}{2(R^2+r^2)}Q$
    (C) $\dfrac{1}{4\pi\epsilon_0}\dfrac{(R+2r)Q}{2(R^2+r^2)}$
    (D) $\dfrac{1}{4\pi\epsilon_0}\dfrac{(R+r)}{2(R^2+r^2)}Q$

17. A solid conducting sphere, having a charge Q, is surrounded by an uncharged conducting hollow spherical shell. Let the potential difference between the surface of the solid sphere and that of the outer surface of the hollow shell be V. If

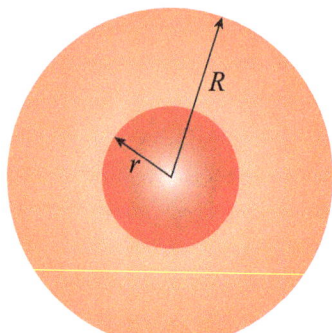

Figure 2.140

the shell is now given a charge of $-4Q$, the new potential difference between the same two surfaces is: [2019]
(A) $-2V$ (B) $2V$ (C) $4V$ (D) V

18. A positive point charge is released from rest at a distance r_o from a positive line charge with uniform density. The speed (v) of the point charge, as a function of instantaneous distance r from line charge, is proportional to: [2019]

(A) $v \propto e^{+r/r_o}$ (B) $v \propto \sqrt{\ln\left(\frac{r}{r_o}\right)}$
(C) $v \propto \ln\left(\frac{r}{r_o}\right)$ (D) $v \propto \left(\frac{r}{r_o}\right)$

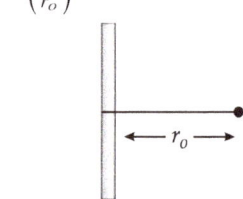

Figure 2.141

19. The electric field in a region is given by $\vec{E} = (Ax + B)\hat{i}$, where E is in NC^{-1} and x is in metres. The values of constants are $A = 20$ SI unit and $B = 10$ SI unit. If the potential at $x = 1$ is V_1 and that at $x = -5$ is V_2, then $V_1 - V_2$ is: [2019]
(A) 320 V (B) -48 V (C) 180 V (D) -520 V

20. A system of three charges are placed as shown in the Fig.2.142. If $D \gg d$, the potential energy of the system is best given by: [2019]

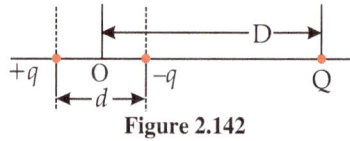

Figure 2.142

(A) $\frac{1}{4\pi\epsilon_o}\left[-\frac{q^2}{d} - \frac{qQd}{2D^2}\right]$ (B) $\frac{1}{4\pi\epsilon_o}\left[-\frac{q^2}{d} + \frac{2qQd}{D^2}\right]$
(C) $\frac{1}{4\pi\epsilon_o}\left[+\frac{q^2}{d} + \frac{qQd}{D^2}\right]$ (D) $\frac{1}{4\pi\epsilon_o}\left[-\frac{q^2}{d} - \frac{qQd}{D^2}\right]$

21. In free space, a particle A of charge $1\,\mu$C is held fixed at a point P. Another particle B of the same charge and mass $4\,\mu g$ is kept at a distance of 1 mm from P. If B is released, then its velocity at a distance of 9 mm from P is (Take $\frac{1}{4\pi\epsilon_o} = 9 \times 10^9$ Nm^{2}C^{-2}): [2019]

(A) 1.0 m/s (B) 3.0×10^4 m/s
(C) 2.0×10^3 m/s (D) 1.5×10^2 m/s

22. A point dipole $\vec{p} = -p_0\hat{x}$ is kept at the origin. The potential and electric field due to this dipole on the y-axis at a distance d are, respectively: (Take $V = 0$ at infinity) [2019]

(A) $\frac{|\vec{p}|}{4\pi\epsilon_0 d^2}, \frac{\vec{p}}{4\pi\epsilon_0 d^3}$ (B) $0, \frac{-\vec{p}}{4\pi\epsilon_0 d^3}$
(C) $0, \frac{\vec{p}}{4\pi\epsilon_0 d^3}$ (D) $\frac{|\vec{p}|}{4\pi\epsilon_0 d^2}, \frac{-\vec{p}}{4\pi\epsilon_0 d^3}$

23. A charge Q is distributed over three concentric spherical shells of radii a, b, c ($a < b < c$) such that their surface charge densities are equal to one another. The total potential at a point at distance r from their common centre, where $r < a$, would be: [2019]

(A) $\frac{Q}{12\pi\epsilon_0}\frac{ab+bc+ca}{abc}$ (B) $\frac{Q(a^2+b^2+c^2)}{4\pi\epsilon_0(a^3+b^3+c^3)}$
(C) $\frac{Q}{4\pi\epsilon_0(a+b+c)}$ (D) $\frac{Q(a+b+c)}{4\pi\epsilon_0(a^2+b^2+c^2)}$

24. Two electric dipoles, A, B with respective dipole moments $\vec{d}_A = -4qa\,\hat{i}$ and $\vec{d}_B = -2qa\,\hat{i}$ are placed on the x-axis with a separation R, as shown in the Fig.2.143 The distance from A at which both of them produce the same potential is: [2019]
(A) $\frac{R}{\sqrt{2}+1}$ (B) $\frac{\sqrt{2}R}{\sqrt{2}+1}$ (C) $\frac{R}{\sqrt{2}-1}$ (D) $\frac{\sqrt{2}R}{\sqrt{2}-1}$

```
|←——R——→|
A         B    → x
```
Figure 2.143

25. Four equal point charges Q each are placed in the xy plane at $(0,2)$, $(4,2)$, $(4,-2)$ and $(0,-2)$. The work required to put a fifth charge Q at the origin of the coordinate system will be: [2019]

(A) $\frac{Q^2}{4\pi\epsilon_0}\left(1 + \frac{1}{\sqrt{3}}\right)$ (B) $\frac{Q^2}{4\pi\epsilon_0}\left(1 + \frac{1}{\sqrt{5}}\right)$
(C) $\frac{Q^2}{2\sqrt{2}\pi\epsilon_0}$ (D) $\frac{Q^2}{4\pi\epsilon_0}$

26. Three charges Q, $+q$ and $+q$ are placed at the vertices of a right-angle isosceles triangle as shown in Fig.2.144. The net electrostatic energy of the configuration is zero, if the value of Q is: [2019]

(A) $\frac{-q}{1+\sqrt{2}}$ (B) $+q$ (C) $-2q$ (D) $\frac{-\sqrt{2}q}{\sqrt{2}+1}$

Figure 2.144

27. The given graph in Fig.2.145, shows variation (with distance r from centre) of: [2019]
(A) Electric field of a uniformly charged sphere
(B) Potential of a uniformly charged spherical shell
(C) Potential of a uniformly charged sphere
(D) Electric field of a uniformly charged spherical shell

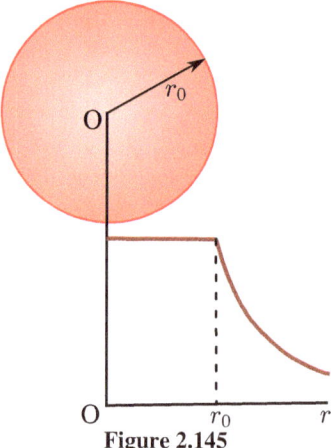

Figure 2.145

28. An electric field of 1000 V/m is applied to an electric

dipole at angle of 45°. The value of electric dipole moment is 10^{-29} C.m. What is the potential energy of the electric dipole? [2019]
(A) -20×10^{-18} J (B) -7×10^{-27} J
(C) -10×10^{-29} J (D) -9×10^{-20} J

29. Determine the electric dipole moment of the system of three charges, placed on the vertices of an equilateral triangle, as shown in the Fig.2.146 [2019]

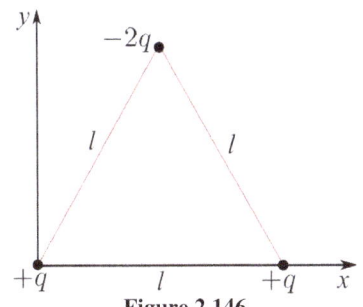

Figure 2.146

(A) $\sqrt{3}ql \left(\frac{\hat{j}-\hat{i}}{\sqrt{2}}\right)$ (B) $(ql)\left(\frac{\hat{i}+\hat{j}}{\sqrt{2}}\right)$
(C) $2ql\hat{j}$ (D) $-\sqrt{3}ql\hat{j}$

30. There is a uniform spherically symmetric surface charge density at a distance R_0 from the origin. The charge distribution is initially at rest and starts expanding because of mutual repulsion. The choice of Fig.2.147 that represents best the speed $v(R(t))$ of the distribution as a function of its instantaneous radius $R(t)$ is: [2019]

(A)   (B)

(C)   (D)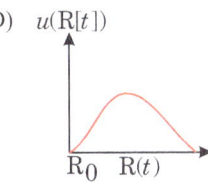

Figure 2.147

31. An electric field of 1000 V/m is applied to an electric dipole at angle of 45°. The value of electric dipole moment is 10^{-29} C·m. What is the potential energy of the electric dipole? [11 Jan 2019, (II)]
(A) -20×10^{-18} J (B) -10×10^{-29} J
(C) -7×10^{-27} J (D) -9×10^{-20} J

32. Two isolated conducting spheres S_1 and S_2 of radius $2R/3R$ and $R/3$ have 12 μC and −3 μC charges, respectively, and are at a large distance from each other. They are now connected by a conducting wire. A long time after this is done the charges on S_1 and S_2 are respectively: [2020]
(A) 6 μC and 3 μC (B) 4.5 μC on both
(C) +4.5 μC and −4.5 μC (D) 3 μC and 6 μC

33. Concentric metallic hollow spheres of radii R and $4R$ hold charges Q_1 and Q_2 respectively. Given that surface charge densities of the concentric spheres are equal, the potential difference $V(9R) - V(4R)$ is: [2020]

(A) $\frac{Q_2}{4\pi\epsilon_0 R}$ (B) $\frac{3Q_2}{4\pi\epsilon_0 R}$
(C) $\frac{3Q_1}{16\pi\epsilon_0 R}$ (D) $\frac{3Q_1}{4\pi\epsilon_0 R}$

34. Two point charges $4q$ and $-q$ are fixed on the x-axis at $x = -d/2$ and $x = d/2$, respectively. If the third point charge 'q' is taken from the origin to $x = d$ along the semicircle as shown in the Fig.2.148, the energy of the charge will: [2020]
(A) increase by $\frac{2q^2}{3\pi\epsilon_0 d}$ (B) increase by $\frac{2q^2}{4\pi\epsilon_0 d}$
(C) decrease by $\frac{q^2}{4\pi\epsilon_0 d}$ (D) decrease by $\frac{4q^2}{3\pi\epsilon_0 d}$

Figure 2.148

35. A particle of charge q and mass m is subjected to an electric field $E = E_0\left(1 - ax^2\right)$ in the x-direction, where a and E_0 are constants. Initially the particle was at rest at $x = 0$. Other than the initial position the kinetic energy of the particle becomes zero when the distance of the particle from the origin is: [2020]
(A) a (B) $\sqrt{1/a}$ (C) $\sqrt{3/a}$ (D) $\sqrt{2/a}$

36. A solid sphere of radius R carries a charge $Q + q$ distributed uniformly over its volume. A very small point like piece of it of mass m gets detached from the bottom of the sphere and falls down vertically under gravity. This piece carries charge q. If it acquires a speed v when it has fallen through a vertical height y (Fig.2.149), then (assume the remaining portion to be spherical) [2020]

(A) $v^2 = 2y\left[\frac{QqR}{4\pi\epsilon_0(R+y)^3 m} + g\right]$

(B) $v^2 = y\left[\frac{qQ}{4\pi\epsilon_0 R^2 ym} + g\right]$

(C) $v^2 = y\left[\frac{qQ}{4\pi\epsilon_0 R(R+y)m} + g\right]$

(D) $v^2 = 2y\left[\frac{qQ}{4\pi\epsilon_0 R(R+y)m} + g\right]$

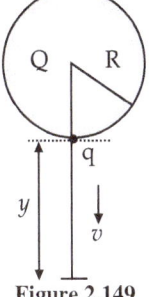

Figure 2.149

37. Ten charges are placed on the circumference of a circle of radius R with constant angular separation between successive charges. Alternate charges 1, 3, 5, 7, 9 have charge $(+q)$ each, while 2, 4, 6, 8, 10 have charge $(-q)$ each. The potential V and the electric field E at the centre of the circle are respectively: [Take $V = 0$ at infinity] [2020]
(A) $V = 0; E = 0$ (B) $V = \frac{10q}{4\pi\epsilon_0 R}; E = 0$
(C) $V = \frac{10q}{4\pi\epsilon_0 R}; E = \frac{10q}{4\pi\epsilon_0 R^2}$ (D) $V = 0, E = \frac{10q}{4\pi\epsilon_0 R^2}$

38. Two identical electrical point dipoles have dipole moments $\vec{p}_1 = p\hat{i}$ and $\vec{p}_2 = -p\hat{i}$ and held on the x-axis at distance 'a' from each other. When released, they move along the x-axis with the direction of their dipole

moments remaining unchanged. If the mass of each dipole is 'm', their speed when they are infinitely far apart is: [2020]

(A) $\frac{p}{a}\sqrt{\frac{3}{2\pi\epsilon_0 ma}}$ (B) $\frac{pP}{a}\sqrt{\frac{1}{2\pi\epsilon_0 ma}}$
(C) $\frac{p}{a}\sqrt{\frac{1}{\pi\epsilon_0 ma}}$ (D) $\frac{p}{a}\sqrt{\frac{2}{\pi\epsilon_0 ma}}$

39. Consider two charged metallic spheres S_1 and S_2 of radii R_1 and R_2, respectively. The electric fields E_1 (on S_1) and E_2 (on S_2) on their surfaces are such that $\frac{E_1}{E_2} = \frac{R_1}{R_2}$. Then the ratio $\frac{V_1(\text{on }S_1)}{V_2(\text{on }S_2)}$ of the electrostatic potentials on each sphere is: [2020]

(A) $\left(\frac{R_1}{R_2}\right)^3$ (B) $\left(\frac{R_2}{R_1}\right)$ (C) $\frac{R_1}{R_2}$ (D) $\left(\frac{R_1}{R_2}\right)^2$

40. The two thin coaxial rings, each of radius 'a' and having charges $+Q$ and $-Q$ respectively are separated by a distance of 's'. The potential difference between the centres of the two ring is: [2021]

(A) $\frac{Q}{4\pi\epsilon_0}\left[\frac{1}{a} - \frac{1}{\sqrt{s^2+a^2}}\right]$ (B) $\frac{Q}{2\pi\epsilon_0}\left[\frac{1}{a} - \frac{1}{\sqrt{s^2+a^2}}\right]$
(C) $\frac{Q}{4\pi\epsilon_0}\left[\frac{1}{a} + \frac{1}{\sqrt{s^2+a^2}}\right]$ (D) $\frac{Q}{2\pi\epsilon_0}\left[\frac{1}{a} + \frac{1}{\sqrt{s^2+a^2}}\right]$

41. If the electric potential at any point (x, y, z) m space is given by $V = 3x^2$ volt. The electric field at the point $(1, 0, 3)$ m will be: [29 June 22 Shift II]

(A) 3 Vm^{-1}, directed along positive x-axis.
(B) 3 Vm^{-1}, directed along negative x-axis.
(C) 6 Vm^{-1}, directed along positive x-axis.
(D) 6 Vm^{-1}, directed along negative x-axis.

42. The electric potential at the centre of two concentric half rings of radii R_1 and R_2, having same linear charge density λ is- [24 Jan 2023 Shift II]

(A) $\frac{2\lambda}{\epsilon_0}$ (B) $\frac{\lambda}{2\epsilon_0}$ (C) $\frac{\lambda}{4\epsilon_0}$ (D) $\frac{\lambda}{\epsilon_0}$

43. Two isolated metallic solid spheres of radii R and $2R$ are charged such that both have same charge density σ. The spheres are then connected by a thin conducting wire. If the new charge density of the bigger sphere is σ'. The ratio σ'/σ is: [30 Jan 2023 Shift I]

(A) 9/4 (B) 4/3 (C) 5/3 (D) 5/6

44. Which of the graphs shown in Fig.2.150 correctly represents the variation of electric potential (V) of a charged spherical conductor of radius (R) with radial distance (r) from the centre? [31 Jan 2023 Shift I]

(A)

(B)

(C)
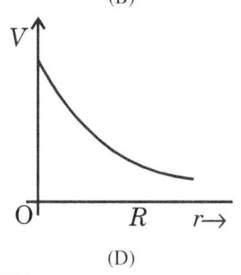
(D)

Figure 2.150

45. Which of the graphs shown in Fig.2.152 represents the electric potential (V) radially away from the centre 'O' of a uniformly charged thin spherical shell as shown in Fig.2.151?

[06 April 23 Shift I]

Figure 2.151

(A) (B)

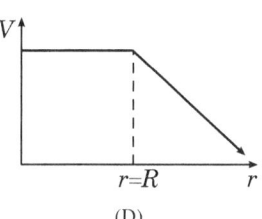

(C) (D)

Figure 2.152

46. Electric potential at a point 'P' due to a point charge of 5×10^{-9} C is 50 V. The distance of 'P' from the point charge is: [08 April 23 Shift II]
(Assume, $1/4\pi\epsilon_0 = 9 \times 10^{+9}$ Nm^{2}C^{-2})
(A) 3 cm (B) 9 cm (C) 90 cm (D) 0.9 cm

47. Three concentric spherical metallic shells X, Y and Z of radius a, b and c respectively [$a < b < c$] (Fig.2.153) have surface charge densities σ, $-\sigma$ and σ, respectively. The shells X and Z are at same potential. If the radii of X & Y are 2 cm and 3 cm, respectively. The radius of shell Z is ____ cm. [10 April 23 Shift I]

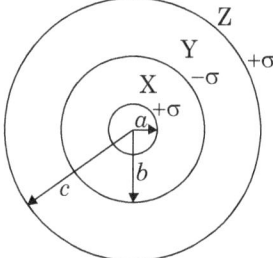

Figure 2.153

48. 64 identical drops each charged up to potential of 10 mV are combined to form a bigger drop. The potential of the bigger drop will be ____ mV. [12 April 23 Shift I]

Section - B (JEE Advanced)

1. Two identical thin rings, each of radius R, are coaxially placed a distance R apart. If Q_1 and Q_2 are respectively the charges uniformly spread on the two rings, the work done in moving a charge q from the centre of one ring to that of the other is [1992]

(A) zero (B) $\frac{q(Q_1-Q_2)(\sqrt{2}-1)}{\sqrt{2}4\pi\epsilon_0 R}$
(C) $\frac{q\sqrt{2}(Q_1+Q_2)}{4\pi\epsilon_0 R}$ (D) $\frac{q(Q_1/Q_2)(\sqrt{2}+1)}{\sqrt{2}4\pi\epsilon_0 R}$

2. The electric potential V at any point x, y, z in space is given by $V = 4x^2$ V/m^2. The electric field at the point

(1 m, 0, 2 m) is [1992]
(A) 8 V/m (B) 4 V/m (C) 16 V/m (D) $\frac{4}{3}$ V/m

3. A non-conducting ring of radius 0.5 m carries a total charge of 1.11×10^{-10} C distributed non-uniformly on its circumference producing an electric field E everywhere in space. The value of the line integral $\int_{l=\infty}^{l=0} -\vec{E} \cdot d\vec{l}$ ($l = 0$ being centre of the ring) in volts is [1994]
(A) +2 (B) −1 (C) −2 (D) 0

4. A charge $+q$ is fixed at each of the point $x = x_0$, $x = 3x_0$, $x = 5x_0$, ... ad inf. on the x-axis, and charges $-q$ is fixed at each of the point $x = 2x_0$, $x = 4x_0$, $x = 6x_0$, ... ad inf. Here x_0 is a positive constant. Take the electric potential at a point due to a charge Q at a distance r from it to be $Q/(4\pi\epsilon_0 r)$. Then the potential at the origin due to the above system of charges is [1998]
(A) −1 (B) $\frac{q}{8\pi\epsilon_0 x_0/\ln 2}$
(C) ∞ (D) $\frac{q \ln 2}{4\pi\epsilon_0 x_0}$

5. An ellipsoidal cavity is carved within a perfect conductor. A positive charge q is placed at the centre of the cavity. The points A and B are on the cavity surface as shown in the Fig.2.154. Then [1999]
(A) Electric field near A in the cavity = electric field near B in the cavity
(B) Charge density at A = charge density at B
(C) Potential at $A \neq$ potential at B
(D) Total electric field flux through the surface of the cavity is q/ϵ_0.

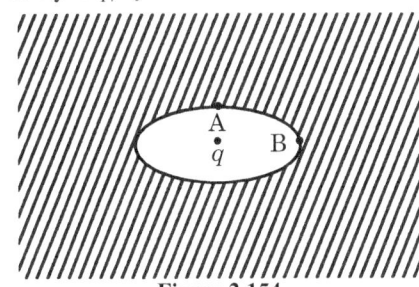

Figure 2.154

6. Three charges Q, $+q$ and $+q$ are placed at the vertices of a right-angled isosceles triangle as shown in fig. The net electrostatics energy of the configuration is zero if Q is equal to [2000]
(A) $\frac{-q}{1+\sqrt{2}}$ (B) $\frac{-2q}{2+\sqrt{2}}$ (C) $-2q$ (D) $+q$

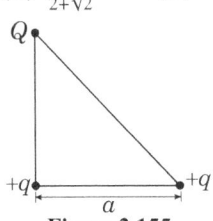

Figure 2.155

7. A uniform electric field pointing in positive x-direction exists in a region. Let A be the origin, B be the point on the x-axis at $x = +1$ cm and C be the point on the y axis at $y = +1$ cm. Then the potentials at the points A, B, and C satisfy. [2001]
(A) $V_A < V_B$ (B) $V_A > V_B$
(C) $V_A < V_C$ (D) $V_A > V_C$

8. Two equal point charges are fixed at $x = -a$ and $x = +a$ on the x-axis. Another point charge Q is placed at the origin. The change in the electrical potential energy of Q, when it is displaced by a small distance x along the x-axis, is

approximately proportional to - [2002]
(A) x (B) x^2 (C) x^3 (D) $1/x$

9. Positive and negative point charges of equal magnitude are kept at $(0, 0, a/2)$ and $(0, 0, -a/2)$, respectively. The work done by the electric field when another positive point charge is moved from $(-a, 0, 0)$ to $(0, a, 0)$ is [2007]
(A) positive
(B) negative
(C) zero
(D) depends on the path connecting the initial and final positions

10. Consider a system of three charges $q/3$, $q/3$ and $-2q/3$ placed at point A, B and C, respectively, as shown in the Fig.2.156. Take O to be the centre of the circle of radius R and angle CAB = 60°. [2008]
(A) The electric field at point O is $\frac{q}{8\pi\epsilon_0 R^2}$ direction along the negative x-axis.
(B) The potential energy of the system is zero.
(C) The magnitude of the force between the charges at C and B is $\frac{q^2}{54\pi\epsilon_0 R^2}$.
(D) The potential at point O is $\frac{q}{12\pi\epsilon_0 R}$.

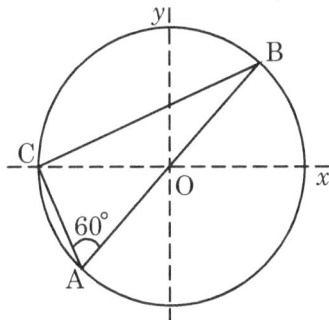

Figure 2.156

11. **STATEMENT 1:** For practical purposes the earth is used as a reference at zero potential in electrical circuits.
STATEMENT 2: The electrical potential of a sphere of radius R with charge Q uniformly distributed on the surface is given by $\frac{Q}{4\pi\epsilon_0 R}$.
(A) STATEMENT-1 is True, STATEMENT-2 is True; STATEMENT-2 is a correct explanation for STATEMENT-1
(B) STATEMENT-1 is True, STATEMENT-2 is True; STATEMENT-2 is NOT a correct explanation for STATEMENT-1
(C) STATEMENT-1 is True, STATEMENT-2 is False
(D) STATEMENT-1 is False, STATEMENT-2 is True

12. Three concentric metallic spherical shells of radii R, $2R$, $3R$ are given charges Q_1, Q_2, Q_3 respectively. It is found that the surface charge densities on the outer surfaces of the shells are equal. Then, the ratio of the charges given to the shells, $Q_1 : Q_2 : Q_3$ is- [2009]
(A) 1 : 2 : 3 (B) 1 : 3 : 5
(C) 1 : 4 : 9 (D) 1 : 8 : 18

13. A spherical metal shell A of radius R_A and a solid metal sphere B of radius R_B ($< R_A$) are kept far apart and each is given charge $+Q$. Now they are connected by a thin metal wire. Then [2011]
(A) $E_A^{\text{inside}} = 0$ (B) $Q_A > Q_B$
(C) $\sigma_A/\sigma_B = R_B/R_A$ (D) $E_A^{\text{surface}} < E_B^{\text{surface}}$

Paragraph 14-15
A dense collection of equal number of electrons and positive ions is called neutral plasma. Certain solids containing fixed

positive ions surrounded by free electrons can be treated as neutral plasma. Let 'N' be the number density of free electrons, each of mass 'm'. When the electrons are subjected to an electric field, they are displaced relatively away from the heavy positive ions. If the electric field becomes zero, the electrons begin to oscillate about the positive ions with a natural angular frequency 'ω_p' which is called the plasma frequency. To sustain the oscillations, a time varying electric field needs to be applied that has an angular frequency ω, where a part of the energy is absorbed and a part of it is reflected. As ω approaches ω_p all the free electrons are set to resonance together and all the energy is reflected. This is the explanation of high reflectivity of metals.

14. Taking the electronic charge as 'e' and the permittivity as 'ϵ_0'. Use dimensional analysis to determine the correct expression for ω_p. [2011]
    (A) $\sqrt{Ne/me_e}$
    (B) $\sqrt{me_0/Ne}$
    (C) $\sqrt{Ne^2/m\epsilon_0}$
    (D) $\sqrt{m\epsilon_0/Ne^2}$

15. Estimate the wavelength at which plasma reflection will occur for a metal having the density of electrons $N \approx 4 \times 10^{27}$ m^{-3}. Taking $e_0 = 10^{-11}$ and $m \approx 10^{-30}$, where these quantities are in proper SI units. [2011]
    (A) 800 nm (B) 600 nm (C) 300 nm (D) 200 nm

16. Two large vertical and parallel metal plates having a separation of 1 cm are connected to a DC voltage source of potential difference X. A proton is released at rest midway between the two plates. It is found to move at 45° to the vertical just after release. Then X is nearly. [2012]
    (A) 1×10^{-5} V
    (B) 1×10^{-7} V
    (C) 1×10^{-9} V
    (D) 1×10^{-10} V

17. Consider a thin spherical shell of radius R with centre at the origin, carrying uniform positive surface charge density. The variation of the magnitude of the electric field $|\vec{E}(r)|$ and the electric potential $V(r)$ with the distance r from the centre, is best represented by which of the graph shown in Fig.2.157? [2012]

(A)

(B)

(C)

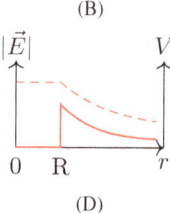
(D)

Figure 2.157

18. Two non-conducting spheres of radii R_1 and R_2 and carrying uniform volume charge densities $+\rho$ and $-\rho$, respectively, are placed such that they partially overlap, as shown in the Fig.2.158. At all points in the overlapping region, [2013]
    (A) The electrostatic field is zero
    (B) The electrostatic potential is constant
    (C) The electrostatic field is constant in magnitude
    (D) The electrostatic field has same direction

19. A point charge $+Q$ is placed just outside an imaginary hemispherical surface of radius R as shown in the Fig.2.159. Which of the following statements is/are correct? [2017]

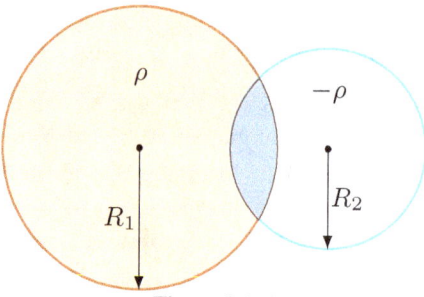

Figure 2.158

(A) The circumference of the flat surface is an equipotential
(B) The electric flux passing through the curved surface of the hemisphere is $-\frac{Q}{2\epsilon_0}\left(1 - \frac{1}{\sqrt{2}}\right)$
(C) Total flux through the curved and the flat surfaces is Q/ϵ_0
(D) The component of the electric field normal to the flat surface is constant over the surface.

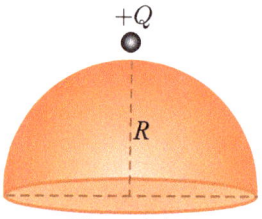

Figure 2.159

20. A particle, of mass 10^{-3}kg and charge 1.0 C, is initially at rest. At time $t = 0$, the particle comes under the influence of an electric field $\vec{E}(t) = E_0 \sin \omega t \hat{i}$, where $E_0 = 1.0$ NC^{-1} and $\omega = 10^3$ rads^{-1}. Consider the effect of only the electrical force on the particle. Then the maximum speed, in ms^{-1}, attained by the particle at subsequent times is ____. [2018]

21. Two large circular discs separated by a distance of 0.01 m are connected to a battery via a switch as shown in the Fig.2.160. Charged oil drops of density 900 kg m^{-3} are released through a tiny hole at the centre of the top disc. Once some oil drops achieve terminal velocity, the switch is closed to apply a voltage of 200 V across the discs. As a result, an oil drop of radius 8×10^{-7}m stops moving vertically and floats between the discs. The number of electrons present in this oil drop is ____ (neglect the buoyancy force, take acceleration due to gravity = 10 ms^{-2} and charge on an electron (e) = 1.6×10^{-19}C) [2020]

Figure 2.160

22. A point charge q of mass m is suspended vertically by a string of length l. A point dipole of dipole moment $\vec{p}$ is now brought towards q from infinity so that the charge moves

away. The final equilibrium position of the system including the direction of the dipole, the angles and distances is shown in Fig.2.161. If the work done in bringing the dipole to this position is $N \times (mgh)$, where g is the acceleration due to gravity, then find the value of N.
(Note that for three coplanar forces keeping a point mass in equilibrium, $F/\sin\theta$ is the same for all forces, where F is any one of the forces and θ is the angle between the other two forces) [2020]

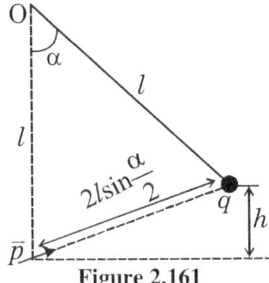

Figure 2.161

Paragraph for Q.23-24
Two point charges $-Q$ and $+Q/\sqrt{3}$ are placed in the xy-plane at the origin $(0,0)$ and a point $(2,0)$, respectively, as shown in the Fig.2.162. This results in an equipotential circle of radius R and potential $V = 0$ in the xy-plane with its centre at $(b,0)$. All lengths are measured in meters. [2021]

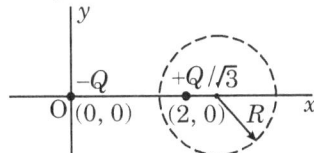

Figure 2.162

23. The value of R is ____ meter.
24. The value of b is ____ meter.
25. Six charges are placed around a regular hexagon of side length a as shown in the Fig.2.163. Five of them have charge q, and the remaining one has charge x. The perpendicular from each charge to the nearest hexagon side passes through the centre O of the hexagon and is bisected by the side. Which of the following statement(s) is (are) correct in SI units? [2022]
    (A) When $x = q$, the magnitude of the electric field at O is zero.
    (B) When $x = -q$, the magnitude of the electric field at O is $\dfrac{q}{6\pi\epsilon_0 a^2}$.
    (C) When $x = 2q$, the potential at O is $\dfrac{7q}{4\sqrt{3}\pi\epsilon_0 a}$.
    (D) When $x = -3q$, the potential at O is $\dfrac{3q}{4\sqrt{3}\pi\epsilon_0 a}$.
26. A disc of radius R with uniform positive charge density σ is placed on the xy plane with its centre at the origin. The Coulomb potential along the z-axis is
$$V(z) = \dfrac{\sigma}{2\epsilon_0}\left(\sqrt{R^2 + z^2} - z\right)$$
A particle of positive charge q is placed initially at rest at a point on the z axis with $z = z_0$ and $z_0 > 0$. In addition to the Coulomb force, the particle experiences a vertical force $\vec{F} = -c\hat{k}$ with $c > 0$. Let $\beta = \dfrac{2c\epsilon_0}{q\sigma}$. Which of the following statement(s) is(are) correct? [2022]

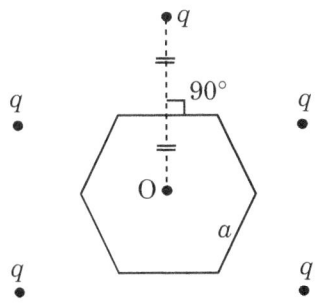

Figure 2.163

(A) For $\beta = 1/4$ and $z_0 = 25R/7$, the particle reaches the origin.
(B) For $\beta = 1/4$ and $z_0 = 3R/7$, the particle reaches the origin.
(C) For $\beta = 1/4$ and $z_0 = R/\sqrt{3}$, the particle returns back to $z = z_0$.
(D) For $\beta > 1$ and $z_0 > 0$, the particle always reaches the origin.

27. An electric dipole is formed by two charges $+q$ and $-q$ located in xy-plane at $(0,2)$ mm and $(0,-2)$ mm, respectively, as shown in the Fig.2.164. The electric potential at point P $(100, 100)$ mm due to the dipole is V_0. The charges $+q$ and $-q$ are then moved to the points $(-1,2)$ mm and $(1,-2)$ mm, respectively. What is the value of electric potential at P due to the new dipole? [2023]
(A) $V_0/4$ (B) $V_0/2$ (C) $V_0/\sqrt{2}$ (D) $3V_0/4$

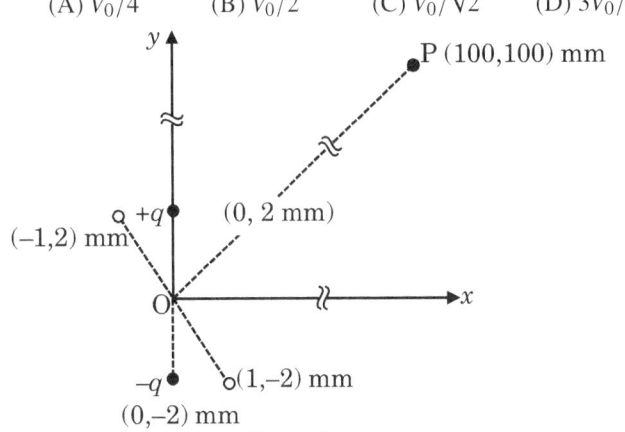

Figure 2.164

2.16 Answer Keys and Solutions
2.16.1 Check Point 1
1. **APPROACH** Sketch the electric field and include a path from r_i to r_f (Fig. 2.165).
To get potential difference, apply Eq. 2.16:
$$\Delta V = -\int_{r_A}^{r_B} E\, dr \qquad \cdots (1)$$
from limit r_i to r_f.
SOLUTION From Eq.(1), we can write-
$$\Delta V = -\int_{r_i}^{r_f} E\, dr = -\int_{r_i}^{r_f}\left(k\dfrac{Q}{r^2}\right) dr$$
$$= -kQ\int_{r_i}^{r_f}\left(\dfrac{1}{r^2}\right) dr = -kQ\left[-\dfrac{1}{r}\right]_{r_i}^{r_f} = kQ\left[\dfrac{1}{r}\right]_{r_i}^{r_f}$$
Evaluate between limits and reduce.
$$\Delta V = \dfrac{kQ}{r_f} - \dfrac{kQ}{r_i}$$

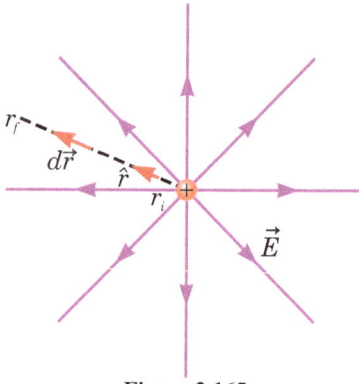

Figure 2.165

By convention, we can take the electrostatic potential zero at infinity, therefore, the expression for the electric potential at a distance r from the source will become-

$$\Delta V = V(r) - V(\infty) = V(r) - 0 = V$$

or

$$\boxed{V = \frac{kQ}{r}}$$

2. **APPROACH** The potential difference between any two points in an electric field is given by Eq.2.14:

$$\Delta V = V_B - V_A = -\int_A^B \vec{E} \cdot d\vec{s} \quad (1)$$

here, $d\vec{s} = dx\hat{i} + dy\hat{j} + dz\hat{k}$

You must select a specific path to integrate along. It doesn't matter what path you choose, since the answer is path-independent, but you simply cannot integrate unless you have a definite path in mind.

SOLUTION Let's go by the indicated path:

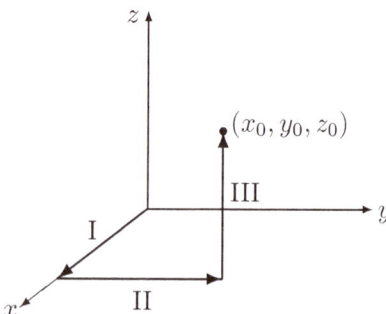

Figure 2.166

$$\vec{E} \cdot d\vec{s} = \left(y^2 dx + \left(2xy + z^2\right) dy + 2yz\, dz\right) K$$

Along Path I: $y = z = 0$; $dy = dz = 0$. $\vec{E} \cdot d\vec{s} = Ky^2 dx = 0$.

$\therefore \int_I \vec{E} \cdot d\vec{s} = 0$.

Along Path II: $x = x_0$, $y : 0 \to y_0$, $z = 0$. $dx = dz = 0$.

$$\vec{E} \cdot d\vec{s} = K\left(2xy + z^2\right) dy = 2Kx_0 y\, dy.$$

$\therefore \int_{II} \vec{E} \cdot d\vec{s} = 2Kx_0 \int_0^{y_0} y\, dy = Kx_0 y_0^2.$

Along Path III: $x = x_0$, $y = y_0$, $z : 0 \to z_0$; $dx = dy = 0$.

$$\vec{E} \cdot d\vec{s} = 2Kyz\, dz = 2Ky_0 z\, dz.$$

$\therefore \int_{III} \vec{E} \cdot d\vec{s} = 2y_0 K \int_0^{z_0} z\, dz = K y_0 z_0^2.$

Therefore, potential at point (x_0, y_0, z_0)-

$$V(x_0, y_0, z_0) = -\int_I \vec{E} \cdot d\vec{s} - \int_{II} \vec{E} \cdot d\vec{s} - \int_{III} \vec{E} \cdot d\vec{s}$$

$$= -K\left(x_0 y_0^2 + y_0 z_0^2\right)$$

Generalizing the result, we get

$$V(x, y, z) = -K\left(xy^2 + yz^2\right)$$

3. $(Aa^2/2) - (Bb^3/3)$

4. From Eq.2.14, electric potential difference, between two points A and B, in terms of electric field, is given by-

$$\Delta V = V_B - V_A = -\int_A^B \vec{E} \cdot d\vec{s} \quad (1)$$

with $d\vec{s} = dx\hat{i} + dy\hat{j} + dz\hat{k}$.

Now, substitute the given values and integrate within the given limits.

SOLUTION (C) Given that- $\vec{E} = (Ax + B)\hat{i}$, with $A = 20$, $B = 10$, therefore:

$$\vec{E} \cdot d\vec{s} = (Ax + B)\hat{i} \cdot (dx\hat{i} + dy\hat{j} + dz\hat{k}) = (Ax + B)dx$$

potential at $x = 1$ is V_1, potential at $x = -5$ is V_2, all the values are in SI units. Therefore, equation 1, gives-

$$V_2 - V_1 = -\int_{x=1}^{x=-5} (Ax + B) dx$$

Put values of A and B.

$$V_2 - V_1 = -\frac{20}{2}[x^2]_{x=1}^{x=-5} - 10[x]_{x=1}^{x=-5}$$

$$= -10(24) - 10(-6) = -240 + 60 = -180\text{ V}$$

So, $V_1 - V_2 = 180$ V

5. **APPROACH** From Eq.2.9, the potential difference between two points and the work done by external force W_{ext}, in moving a point charge q_0 slowly, from one point to other are related as-

$$\Delta V = \frac{W_{ext}}{q_0}$$

Substitute the given values in above equation and solve for ΔV.

SOLUTION (A) Put the given values of W_{ext} and q_0 and solve for $V_B - V_A$

SOLUTION Given: $W_{ext} = 2$ J, $q_0 = 20$ C

So, $$\Delta V = \frac{W_{ext}}{q_0} = \frac{2}{20} = 0.1\text{ V}$$

2.16.2 Check Point 2

1. The electric field is the negative of the slope of the potential graph.

   There are three regions of different slope:

   $0 < x < 2$ cm
   $\begin{cases} \Delta V/\Delta x = (20\text{ V})/(0.020\text{ m}) = 1000\text{ V/m} \\ E_x = -1000\text{ V/m} \end{cases}$

   $2 < x < 4$ cm
   $\begin{cases} \Delta V/\Delta x = 0\text{ V/m} \\ E_x = 0\text{ V/m} \end{cases}$

   $4 < x < 8$ cm
   $\begin{cases} \Delta V/\Delta x = (-20\text{ V})/(0.040\text{ m}) = -500\text{ V/m} \\ E_x = 500\text{ V/m} \end{cases}$

   The results are shown in Fig.2.167.

2. **APPROACH** The electric field is perpendicular to the equipotential lines, points "downhill", and depends on the slope of the potential hill.

   The potential is highest on the bottom and the right. An elevation graph of the potential would look like the lower-right quarter of a bowl or a football stadium.

   SOLUTION Some distant but unseen source charges have created an electric field and potential. We do not need to see the source charges to relate the field to the potential. Because $E \approx -\Delta V/\Delta s$, the electric field is stronger where the equipotential lines are closer together and weaker where they are farther apart. If Figure 2.42 were a topographic

2.16. ANSWER KEYS AND SOLUTIONS

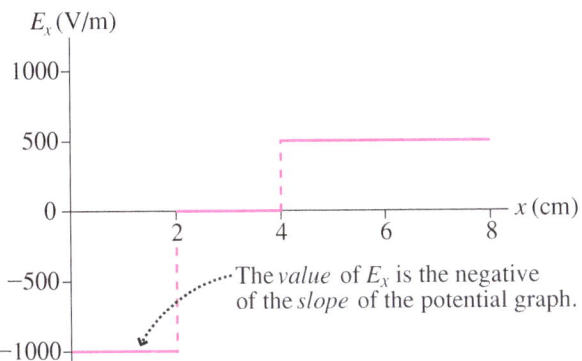

Figure 2.167

map, you would interpret the closely spaced contour lines at the bottom of the figure as a steep slope.

Fig.2.168 shows how measurements of Δs from the grid are combined with values of ΔV to determine $\vec{E}$. Point 3 requires an estimate of the spacing between the 0 V and the 100 V surfaces. Notice that we're using the 0 V and 100 V equipotential surfaces to determine $\vec{E}$ at a point on the 50 V equipotential.

Note: The directions of $\vec{E}$ are found by drawing downhill

Figure 2.168

vectors perpendicular to the equipotentials. The distances between the equipotential surfaces are needed to determine the field strengths.

3. **APPROACH** Components of $\vec{E}$, are given by-
$$E_x = -\frac{\partial V}{\partial x}, E_y = -\frac{\partial V}{\partial y} \text{ and } E_z = -\frac{\partial V}{\partial z}$$
SOLUTION $V = Axy - Bx^2 + Cy$
(a) $E_x = -\frac{\partial V}{\partial x} = -Ay + 2Bx, \quad E_y = -\frac{\partial V}{\partial y} = -Ax - C,$
$E_z = \frac{\partial V}{\partial z} = 0$
(b) $E = 0$ requires that $E_x = E_y = E_z = 0$. $E_z = 0$ everywhere. $E_y = 0$ at $x = -C/A$. And E_x is also equal to zero for this x, any value of z and $y = 2Bx/A = (2B/A)(-C/A) = -2BC/A^2$.
Since, V doesn't depend on z, so $E_z = 0$ everywhere.

4. **APPROACH** Components of $\vec{E}$, are given by-
$$E_x = -\frac{\partial V}{\partial x}, E_y = -\frac{\partial V}{\partial y} \text{ and } E_z = -\frac{\partial V}{\partial z}$$
SOLUTION The electric field at distance r due to a point charge q is given by:
$$\vec{E} = \frac{1}{4\pi\epsilon_0}\frac{q}{r^2}\hat{r}$$
(a) $E_x = -\frac{\partial V}{\partial x} = -\frac{\partial}{\partial x}\left(\frac{kQ}{\sqrt{x^2+y^2+z^2}}\right)$
$= \frac{kQx}{(x^2+y^2+z^2)^{3/2}} = \frac{kQx}{r^3}$
Similarly, $E_y = \frac{kQy}{r^3}$ and $E_z = \frac{kQz}{r^3}$.

(b) From part (a), $\vec{E} = \frac{kQ}{r^2}\left(\frac{x\hat{i}}{r} + \frac{y\hat{j}}{r} + \frac{z\hat{k}}{r}\right) = \frac{kQ}{r^2}\hat{r}$

Note: V is a scalar. $\vec{E}$ is a vector and has components.

5. The electric field is given by
$$\vec{E} = -\frac{\partial V}{\partial x}\hat{i} - \frac{\partial V}{\partial y}\hat{j} - \frac{\partial V}{\partial z}\hat{k} = -8x\hat{i}$$
Therefore, $\vec{E}_{(1m, 0m, 2m)} = -8\hat{i}$.

6. **APPROACH** For a solid metal sphere or for a spherical shell, $V = kq/r$ outside the sphere and $V = kq/R$ at all points inside the sphere, where R is the radius of the sphere. When the electric field is radial, $E = -\frac{\partial V}{\partial r}$

SOLUTION (a) (i) $r < r_a$: This region is inside both spheres.
$$V = \frac{kq}{r_a} - \frac{kq}{r_b} = kq\left(\frac{1}{r_a} - \frac{1}{r_b}\right)$$
(ii) $r_a < r < r_b$: This region is outside the inner shell and inside the outer shell.
$$V = \frac{kq}{r} - \frac{kq}{r_b} = kq\left(\frac{1}{r} - \frac{1}{r_b}\right)$$
(iii) $r > r_b$: This region is outside both spheres and $V = 0$ since outside a sphere the potential is the same as for a point charge. Therefore the potential is the same as for two oppositely charged point charges at the same location. These potentials cancel.

(b) $V_a = k\left(\frac{q}{r_a} - \frac{q}{r_b}\right)$ and $V_b = 0$,
so, $V_{ab} = kq\left(\frac{1}{r_a} - \frac{1}{r_b}\right)$

(c) Between the spheres $r_a < r < r_b$ and $V = kq\left(\frac{1}{r} - \frac{1}{r_b}\right)$
So, $E = -\frac{\partial V}{\partial r} = -kq\frac{\partial}{\partial r}\left(\frac{1}{r} - \frac{1}{r_b}\right)$
$= +k\frac{q}{r^2} = \frac{V_{ab}}{\left(\frac{1}{r_a} - \frac{1}{r_b}\right)}\frac{1}{r^2}$
$\left[\text{Since, from part (b) } kq = \frac{V_{ab}}{\left(\frac{1}{a} - \frac{1}{b}\right)}\right]$

(d) Since, V is constant, therefore $E = -\partial V/\partial r = 0$, outside the spheres.

(e) If the outer charge is different, then outside the outer sphere the potential is no longer zero but is $V = k\frac{q}{r} - k\frac{Q}{r} = k\frac{(q-Q)}{r}$. All potentials inside the outer shell are just shifted by an amount $V = kQ/r_b$. Therefore relative potentials within the shells are not affected. Thus (b) and (c) do not change. However, now that the potential does vary outside the spheres, there is an electric field there:

$$E = -\frac{\partial V}{\partial r} = -\frac{\partial}{\partial r}\left(\frac{kq}{r} + \frac{-kQ}{r}\right) = \frac{kq}{r^2}\left(1 - \frac{Q}{q}\right)$$
$$= \frac{k}{r^2}(q - Q)$$

Note: In part (a) the potential is greater than zero for all $r < r_b$.

7. Since, $E = -\partial V/\partial r$, therefore, $\vec{E} = 0$ everywhere implies a constant potential and not necessarily zero potential.

8. (B). When moving straight from A to B, $\vec{E}$ and $d\vec{s}$ both point toward the right. Thus, the dot product $\vec{E} \cdot d\vec{s}$ in equation $\Delta V = V_B - V_A = -\vec{E} \cdot d\vec{s}$ is positive and so ΔV is negative.

9. (A) Work done is equal to zero because the potential of A and B are the same $= \dfrac{1}{4\pi\epsilon_0}\dfrac{q}{a}$

No work is done if a particle does not change its potential energy. i.e. initial potential energy = final potential energy.

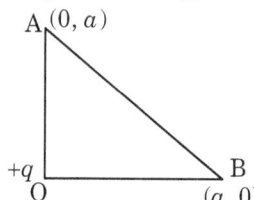

Figure 2.169

10. $\vec{E} = -\left[\dfrac{\partial V}{\partial x}\hat{i} + \dfrac{\partial V}{\partial y}\hat{j} + \dfrac{\partial V}{\partial z}\hat{k}\right] = -(-5\hat{i} + 3\hat{j} + \sqrt{15}\hat{k})$

[since, $V = -5x + 3y + \sqrt{15}z$]

$\Rightarrow |\vec{E}| = \sqrt{25 + 9 + 15} = \sqrt{49} = 7$ unit

11. (D): The electric potential at a point,
$$V = -x^2y - xz^3 + 4$$
The field $\vec{E} = -\vec{\nabla}V = -\left(\dfrac{\partial V}{\partial x}\hat{i} + \dfrac{\partial V}{\partial y}\hat{j} + \dfrac{\partial V}{\partial z}\hat{k}\right)$

$\therefore \vec{E} = \hat{i}\left(2xy + z^3\right) + \hat{j}x^2 + \hat{k}\left(3xz^2\right)$

12. (D): The electric field $\vec{E}$ and potential V in a region are related as $\vec{E} = -\left[\dfrac{\partial V}{\partial x}\hat{i} + \dfrac{\partial V}{\partial y}\hat{j} + \dfrac{\partial V}{\partial z}\hat{k}\right]$

Here, $V(x, y, z) = 6xy - y + 2yz$

$\therefore \vec{E} = -\left[\dfrac{\partial}{\partial x}(6xy - y + 2yz)\hat{i} + \dfrac{\partial}{\partial y}(6xy - y + 2yz)\hat{j} + \dfrac{\partial}{\partial z}(6xy - y + 2yz)\hat{k}\right]$

$= -\left[(6y)\hat{i} + (6x - 1 + 2z)\hat{j} + (2y)\hat{k}\right]$

At point (1, 1, 0),
$\vec{E} = -[(6(1))\hat{i} + (6(1) - 1 + 2(0))\hat{j} + (2(1))\hat{k}]$
$= -(6\hat{i} + 5\hat{j} + 2\hat{k})$

13. (D): Here, $V(x, y, z) = 6x - 8xy - 8y + 6yz$
The x, y and z components of electric field are
$$E_x = -\dfrac{\partial V}{\partial x} = -\dfrac{\partial}{\partial x}(6x - 8xy - 8y + 6yz)$$
$= -(6 - 8y) = -6 + 8y,$
$$E_y = -\dfrac{\partial V}{\partial y} = -\dfrac{\partial}{\partial y}(6x - 8xy - 8y + 6yz)$$
$= -(-8x - 8 + 6z) = 8x + 8 - 6z,$
$$E_z = -\dfrac{\partial V}{\partial z} = -\dfrac{\partial}{\partial z}(6x - 8xy - 8y + 6yz) = -6y$$
So, $\vec{E} = E_x\hat{i} + E_y\hat{j} + E_z\hat{k}$
$= (-6 + 8y)\hat{i} + (8x + 8 - 6z)\hat{j} - 6y\hat{k}$

At point (1, 1, 1)
$\vec{E} = (-6 + 8)\hat{i} + (8 + 8 - 6)\hat{j} - 6\hat{k} = 2\hat{i} + 10\hat{j} - 6\hat{k}$
The magnitude of electric field $\vec{E}$ is
$\vec{E} = \sqrt{E_x^2 + E_y^2 + E_z^2} = \sqrt{(2)^2 + (10)^2 + (-6)^2}$
$= \sqrt{140} = 2\sqrt{35}$ NC^{-1}

Electric force experienced by the charge
$F = qE = 2C \times 2\sqrt{35}$ NC$^{-1} = 4\sqrt{35}$ N

14. **APPROACH** If we assume the space as gravity free, then the force acting on the alpha particle is electrostatic in nature, which is conservative. So, we can apply the principle of conservation of mechanical energy. According to this principle,
$$K_i + U_i = K_f + U_f$$
Here letters have their usual meanings.

SOLUTION (C) Initially, kinetic energy of the alpha particle is $K_i = 5$ MeV and its potential energy is $U_i = 0$ (because it is far away from the nucleus). The charge of the uranium nucleus is $Q = Ze = 92e$ and charge on the alpha particle is $q = 2e$. Let O be the centre of the uranium nucleus. The alpha particle starts moving towards O and is scattered by an angle 180° at P.

The distance of the closest approach is $OP = d$. The kinetic energy of the alpha particle at P is $K_f = 0$ (since its velocity is zero) and its potential energy is
$$U_f = 2Ze^2/(4\pi\epsilon_0 d)$$
Apply conservation of energy, $K_i + U_i = K_f + U_f$, to get
$$d = \dfrac{2Ze^2}{4\pi\epsilon_0 K_i} = \dfrac{(9 \times 10^9)(2)(92)(1.6 \times 10^{-19})^2}{(5 \times 10^6)(1.6 \times 10^{-19})}$$
$= 5.3 \times 10^{-14}$ m

15. (A) From Eq.2.33,
$W_{ext} = -pE(\cos\theta_2 - \cos\theta_1) = -pE(\cos\pi - \cos 0)$
$= -pE(-1 - 1) = 2pE$

16. (D) The torque exerted by an electric field E on a dipole of moment p is given by-
$$\tau = pE\sin\theta \quad \cdots (1)$$
where θ is the angle which the dipole is making with the electric-field.
Corresponding to maximum torque, $\theta = 90°$, therefore from Eq.(1), we get
$$\tau_{max} = pE$$
Here, $p = q(2l) = 1 \times 10^{-6} \times 0.02$ C/m and $E = 10^5$ N/C.
Therefore, $\tau_{max} = 1 \times 10^{-6} \times 0.02 \times 10^5 = 2 \times 10^{-3}$ N.m
The work done in rotating the dipole from an angle $\theta = 0°$ to 180° is given by-
$$W = \int_{\theta_o}^{\theta} pE\sin\theta\, d\theta = pE(\cos\theta_0 - \cos\theta)$$
Here, $\theta_0 = 0°$ and $\theta = 180°$
Therefore, $W = pE(\cos 0° - \cos 180°)$
$= 2pE = 4 \times 10^{-3}$ joule

2.16.3 Check Point 3

1. B → C, C → D, A → B, D → E Moving from B to C decreases the electric potential by 2 V, so the electric field performs 2 J of work on each coulomb of positive charge that moves. Moving from C to D decreases the electric potential by 1 V, so 1 J of work is done by the field. It takes no work to move the charge from A to B because the electric potential does not change. Moving from D to E increases the electric potential by 1 V, and thus the field does −1 J of work per unit of positive charge that moves.

2. (A) Any surface which has same electrostatic potential at every point is called an equipotential surface. Electric field is always perpendicular to an equipotential surface. Therefore, X-direction is perpendicular only to YZ-plane.
3. (D) The electric field points in the direction of decreasing electric potential.
4. (D) In electrostatics (i.e., when charges are stationary and charge density does not vary with the time), electric field inside a conductor remains zero and the electric field lines remain normal to the surface and never enter inside the conductor.
5. **APPROACH** Electric potential at a distance r from the centre of a uniformly charged non-conducting sphere of radius R, as shown in Fig.2.170, is given by-

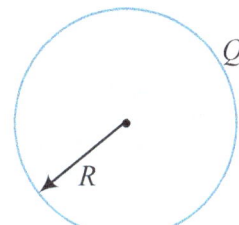

Figure 2.170

$$V_r = k\frac{Q}{r} \quad (r \geq R) \quad \text{[see Eq.(2.69)]} \quad (1)$$

At the surface of non-conducting uniformly charged sphere, electric potential is given as:

$$V_R = k\frac{Q}{R} \quad \text{[see Eq.(2.70)]} \quad (2)$$

And inside the non-conducting uniformly charged sphere, the electric potential is given by:

$$V_r = k\frac{Q}{2R}\left(3 - \frac{r^2}{R^2}\right) \quad (0 \leq r \leq R) \quad \text{[see Eq.(2.52)]} \quad (3)$$

Substitute the given values of V in above appropriate equations and simplify for r.

SOLUTION (C, D) Given that the electric potential of the sphere with respect to ∞, is V_0, therefore, from Eq.(2)-

$$V_0 = \frac{kQ}{R} \quad \Longrightarrow \quad \frac{kQ}{R} = V_0 \quad (4)$$

At the centre of sphere, $r = 0$, therefore, from Eq.(3), we have-

$$V_C = k\frac{Q}{2R}\left(3 - \frac{0^2}{R^2}\right) = \frac{3kQ}{2R} = \frac{3}{2}V_0$$

$$\text{or} \quad V_C = \frac{3}{2}V_0 \quad (5)$$

Calculation of distances from the centre of the sphere for given potentials:
(i) For potential $V_{R_1} = 3V_0/2$
From Eq.(5), it is clear that-
$$V_{R_1} = V_C \quad \Longrightarrow \quad R_1 = 0$$
(ii) For potential $V_{R_2} = 5V_0/4$
For a non-conducting sphere of radius R, potential $V < V_0$ for $r > R$ and the point for which $V_{R_2} = 5V_0/4 > V_0$ lies inside the sphere. Therefore, we use Eq.(3) for it.

$$V_{R_2} = \frac{5V_0}{4} = k\frac{Q}{2R}\left(3 - \frac{R_2^2}{R^2}\right)$$

$$V_{R_2} = \frac{5V_0}{4} = \frac{kQ}{2R^3}(3R^2 - R_2^2)$$

$$\text{or} \quad \frac{5V_0}{4} = \frac{V_0}{2R^2}(3R^2 - R_2^2) \quad \Longrightarrow \quad \frac{5}{2} = 3 - \left(\frac{R_2}{R}\right)^2$$

$$\Longrightarrow \quad \left(\frac{R_2}{R}\right)^2 = 3 - \frac{5}{2} = \frac{1}{2} \quad \Longrightarrow \quad R_2 = \frac{R}{\sqrt{2}}$$

Similarly, (iii) For potential $V_{R_3} = 3V_0/4$

$$V_{R_3} = \frac{3V_0}{4} \quad \Longrightarrow \quad \frac{kQ}{R_3} = \frac{3}{4} \times \frac{kQ}{R} \quad \Longrightarrow \quad R_3 = 4R/3$$

(iv) Now, for potential $V = V_0/4$

$$V_{R_4} = \frac{kQ}{R_4} = \frac{V_0}{4} \quad \Longrightarrow \quad \frac{kQ}{R_4} = \frac{1}{4} \times \frac{kQ}{R} \quad \Longrightarrow \quad R_4 = 4R$$

6. (B) Note that the potential inside a hollow conducting sphere is always constant and its value is equal to the potential at it's surface. So, the potential at the center will be 10 V. Note that the electric field inside the hollow conducting sphere is zero.

7. (B) $E = \dfrac{\lambda}{2\pi\epsilon_0 x}$

$$\int_{V_A}^{V_B} dV = -E\,dx = -\frac{\lambda}{2\pi\epsilon_0}\int_{3a}^{2a}\frac{dx}{x}$$

$$\Longrightarrow \quad V_B - V_A = \frac{\lambda}{2\pi\epsilon_0}\ln\frac{3}{2}$$

Therefore, the work done by external agent $= \dfrac{q_0\lambda}{2\pi\epsilon_0}\ln\dfrac{3}{2}$

8. (C) The charges at $A, B,$ and C are $q_A = q/3, q_B = q/3,$ and $q_C = -2q/3$. The electric fields at O due to q_A and q_B are equal in magnitude but opposite in direction. Thus, the resultant electric field at O is only due to charge q_C and is given by

$$\vec{E}_O = -\frac{q}{6\pi\epsilon_0 R^2}\hat{\imath}$$

The triangle ABC is right-angled with $\angle A = 60°, \angle C = 90°$, and $r_{AB} = 2R$. Thus, $r_{AC} = R$ and $r_{BC} = \sqrt{3}R$. The potential energy for the given charge distribution is

$$U = \frac{1}{4\pi\epsilon_0}\left[\frac{q_A q_B}{r_{AB}} + \frac{q_A q_C}{r_{AC}} + \frac{q_B q_C}{r_{BC}}\right]$$

$$= \frac{1}{4\pi\epsilon_0}\left[\frac{q^2}{18R} - \frac{2q^2}{9R} - \frac{2q^2}{9\sqrt{3}R}\right] \neq 0$$

The magnitude of force between q_C and q_B is

$$F_{BC} = \frac{1}{4\pi\epsilon_0}\frac{q_B q_C}{r_{BC}^2} = \frac{q^2}{54\pi\epsilon_0 R^2}$$

The potential at O is

$$V = \frac{1}{4\pi\epsilon_0}(q_A/R + q_B/R + q_C/R) = 0$$

2.16.4 Conceptual Questions

1. The potential energy increases. When an outside agent makes it a negative charge to move in the direction of the field, the charge moves to a region of lower electric potential. The product of its negative charge with a lower number of volts gives a higher number of joules. Keep in mind that a negative charge feels an electric force in the opposite direction to the field, while the potential is the work done on the charge to move it in a field per unit charge.

2. Given that the electric field points to the right and the proton is moving to the left, it means the proton is moving in the opposite direction of the electric field. Since an electric field always directs from higher potential to lower potential, we can conclude that the proton is moving toward a region of higher potential. We have already studied that if a charged particle is free to move in an electric field, it moves in a direction that minimizes its potential energy. Now, if the proton is free to move, then to minimize its electric potential energy in the electric field, it would move in the direction of the electric field, i.e., towards a lower potential point. So, it

is clear that there must be some external agent to transport the proton from a lower potential point to a higher one. As it moves toward the higher potential point, the electric potential energy of the proton in the electric field increases. This increase occurs at the cost of work done by the external agent against the electric force.

3. Since the electron is moving toward a region of higher electric potential, therefore, its electric potential energy in the external electric field is decreasing.

4. If V is constant, its gradient is zero; consequently the electric field is zero throughout the region.

5. Yes.

6. No. $\vec{E}$ cannot be determined without knowing V at a continuum of points.

7. Not necessarily. If two points are at the same potential, then no net work is done in moving a charge from one point to the other, but work (both positive and negative) could be done at different parts of the path. No. It is possible that positive work was done over one part of the path (for example accelerated motion), and negative work done over another part of the path (decelerated or retarded motion), so that these two contributions to the net work sum to zero. In this case, a non-zero force would have to be exerted over both parts of the path.

8. The negative charge will move toward a region of higher potential and the positive charge will move toward a region of lower potential. In both cases, the potential energy of the charge will decrease.

9. The electric field lines, shown as solid lines, and the equipotential surfaces (intersecting the plane of the paper), shown as dashed lines, are sketched in the Figure 2.171. The point charge $+Q$ is the point at the right, and the metal sphere with charge $-Q$ is at the left. Near the two charges the equipotential surfaces are spheres, and the field lines are normal to the metal sphere at the sphere's surface,

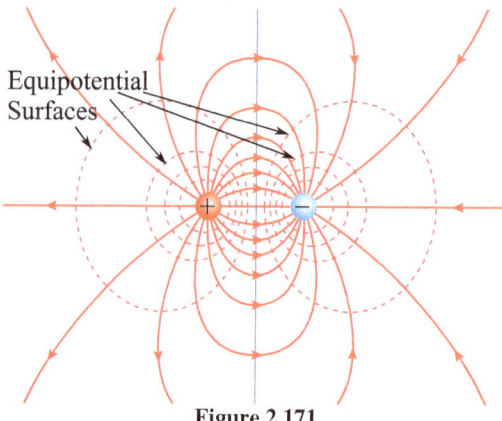

Figure 2.171

10. Yes. If the charge on the particle is negative then under the electric force only, it moves from a region of low electric potential to a region of high electric potential, its electric potential energy will decrease.

11. The electric field lines, shown as solid lines, and the equipotential surfaces (intersecting the plane of the paper), shown as dashed lines, are sketched in the adjacent Figure 2.172. The point charge $+Q$ is the point at the right, and the metal sphere with charge $+Q$ is at the left. Near the two charges the equipotential surfaces are spheres, and the field lines are normal to the metal sphere at the sphere's surface. Very far from both charges, the equipotential surfaces and field lines approach those of a point charge $2Q$ located at the midpoint.

12. The equipotential surfaces are shown with dashed lines, the

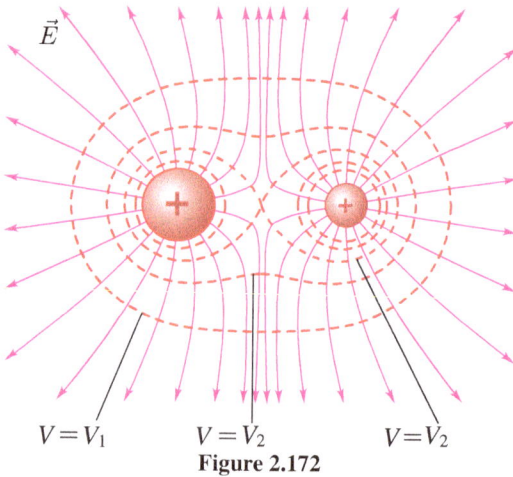

Figure 2.172

electric field lines are shown with solid lines. Near each charge, the equipotential surfaces are spheres centered on each charge; far from the charges, the equipotential surface is a sphere centered at the midpoint between the charges. The electric field lines are perpendicular to the equipotential surfaces.

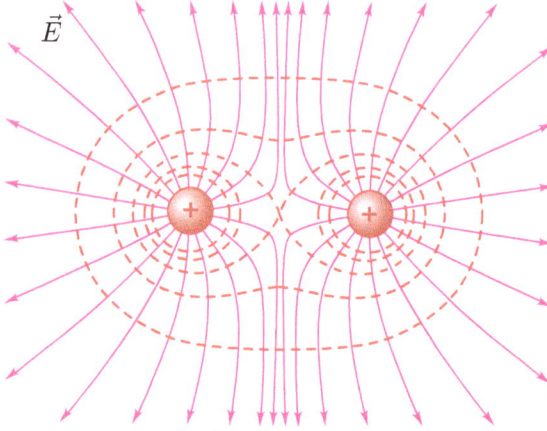

Figure 2.173

13. **APPROACH** We can use Coulomb's law and the superposition of fields to find $\vec{E}$ at the origin and the definition of the electric potential due to a point charge to find V at the origin.

SOLUTION (a) Apply Coulomb's law and the superposition of fields to find the electric field $\vec{E}$ at the origin:

$$\vec{E} = \vec{E}_{+q \text{ at } -a} + \vec{E}_{+q \text{ at } +a} = \frac{kq}{a^2}\hat{i} - \frac{kq}{a^2}\hat{i} = 0$$

The potential V at the origin is given by:

$$V = V_{+q \text{ at } -a} + V_{+q \text{ at } +a} = \frac{kq}{a} + \frac{kq}{a} = \frac{2kq}{a}$$

and (B) is correct.

(b) Apply Coulomb's law and the superposition of fields to find the electric field $\vec{E}$ at the origin:

$$\vec{E} = \vec{E}_{+q \text{ at } -a} + \vec{E}_{-q \text{ at } +a} = \left(\frac{kq}{a^2}\right)\hat{i} + \left(\frac{kq}{a^2}\right)\hat{i} = \left(\frac{2kq}{a^2}\right)\hat{i}$$

The potential V at the origin is given by:

$$V = V_{+q \text{ at } -a} + V_{-q \text{ at } +a} = \frac{kq}{a} + \frac{k(-q)}{a} = 0$$

and (C) is correct.

14. The charge can be moved along any path parallel to the y-z plane, namely perpendicular to the field.

15. **APPROACH** Use $\vec{E} = -\frac{\partial V}{\partial x}\hat{i}$ to find the electric field corresponding to the given potential and then compare its

form to those produced by the four alternatives listed.
SOLUTION (a) Find the electric field corresponding to the given potential function: $V = 4.00|x| + V_0$

$$\vec{E} = -\frac{\partial V}{\partial x}\hat{i} = -\frac{\partial}{\partial x}[4.00|x| + V_0]\hat{i} = -4.00\frac{\partial}{\partial x}[|x|]\hat{i}$$

Note that $\frac{d}{dx}|x| = \begin{cases} 1 & \text{if } x > 0 \\ -1 & \text{if } x < 0 \end{cases}$

If $x > 0$, then $\frac{\partial}{\partial x}|x| = 1$ therefore-

$$\vec{E}_{x>0} = \left(-4.00\frac{\text{V}}{\text{m}}\right)\hat{i}$$

If $x < 0$, then $\frac{\partial}{\partial x}|x| = -1$ and hence-

$$\vec{E}_{x<0} = \left(4.00\frac{\text{V}}{\text{m}}\right)\hat{i}$$

A sketch of the electric field in this region is shown in Fig. 2.174.
(b) (1) is correct because field lines end on negative

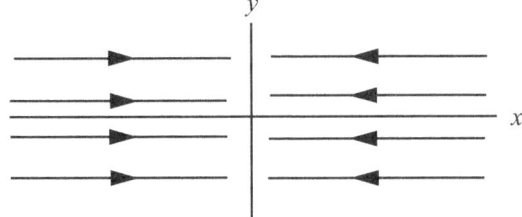

Figure 2.174

charges.

16. No. The local surface charge density is proportional to the normal component of the electric field ($E = \sigma/\epsilon_0 \implies \sigma = \epsilon_0 E$), not the potential on the surface.

17. No. Electric potential is the potential energy per unit charge at a point in space and electric field is the electric force per unit charge at a point in space. If one of these quantities is zero, the other is not necessarily zero. For example, the point exactly between two charges with equal magnitudes and opposite signs will have a zero electric potential because the contributions from the two charges will be equal in magnitude and opposite in sign. (Net electric potential is a scalar sum.) This point will not have a zero electric field, however, because the electric field contributions will be in the same direction (towards the negative and away from the positive) and so will add. (Net electric field is a vector sum.) As another example, consider the point exactly between two equal positive point charges. The electric potential will be positive since it is the sum of two positive numbers, but the electric field will be zero since the field contributions from the two charges will be equal in magnitude but opposite in direction.

18. (a) V at other points would be lower by 10 V. E would be unaffected, since E is the negative gradient of V, and a change in V by a constant value will not change the value of the gradient.
(b) If V represents an absolute potential, then yes, the fact that the Earth carries a net charge would affect the value of V at the surface. If V represents a potential difference, then no, the net charge on the Earth would not affect the choice of V.

19. If the field lines are not perpendicular to the surface, then there is a component of electric field parallel to the surface. If there is an electric field component parallel to the surface, then there must be a potential difference along the surface too.

But the surface is an equipotential surface, so, there is no potential difference along it. This gives a contradiction which means the initial assumption of field lines being not perpendicular to the surface is wrong.

20. No. An equipotential line is a line connecting points of equal electric potential. If two equipotential lines crossed, it would indicate that their intersection point has two different values of electric potential simultaneously, which is impossible. As an analogy, imagine contour lines on a topographic map. They also never cross because one point on the surface of the Earth cannot have two different values for elevation above sea level.

21. The Earth's gravitational equipotential lines are roughly circular, so the orbit of the satellite would have to be roughly circular.

22. The potential at point P would be unchanged. Each bit of positive charge will contribute an amount to the potential based on its charge and its distance from point P. Moving charges to different locations on the ring does not change their distance from P, and hence does not change their contributions to the potential at P.
The value of the electric field will change. The electric field is the vector sum of all the contributions to the field from the individual charges. When the charge Q is distributed uniformly about the ring, the y-components of the field contributions cancel, leaving a net field in the x-direction. When the charge is not distributed uniformly, the y-components will not cancel, and the net field will have both x- and y-components, and will be larger than for the case of the uniform charge distribution. There is no discrepancy here, because electric potential is a scalar and electric field is a vector.

23. The charge density and the electric field strength will be greatest at the pointed ends of the football because at the pointed surface there has a smaller radius of curvature than the middle.

24. No. You cannot calculate electric potential knowing only electric field at a point and you cannot calculate electric field knowing only electric potential at a point. As an example, consider the uniform field between two charged, conducting plates. If the potential difference between the plates is known, then the distance between the plates must also be known in order to calculate the field. If the field between the plates is known, then the distance to a point of interest between the plates must also be known in order to calculate the potential there. In general, to find V, you must know E and be able to integrate it. To find E, you must know V and be able to take its derivative. Thus you need E or V in the region around the point, not just at the point, in order to be able to find the other variable.

25. (a) Once the two spheres are placed in contact with each other, they effectively become one larger conductor. They will have the same potential because the potential everywhere on a conducting surface is constant.
(b) Because the spheres are identical in size, an amount of charge $Q/2$ will flow from the initially charged sphere to the initially neutral sphere so that they will have equal charges.
(c) Even if the spheres do not have the same radius, they will still be at the same potential once they are brought into contact because they still create one larger conductor. However, the amount of charge that flows will not be exactly equal to half the total charge. The larger sphere will end up with the larger charge.

26. If the electric field points due north, the change in the po-

tential will be (*a*) greatest in the direction opposite the field, south; (b) least in the direction of the field, north; and (*c*) zero in a direction perpendicular to the field, east and west.

27. In the vicinity of a pointy protrusion, the electric field can be very high. This can lead to a spark inside an electronic device which can make the device to stop functioning.

28. Yes. In regions of space where the equipotential lines are closely spaced, the electric field is stronger than in regions of space where the equipotential lines are farther apart.

29. If the electric field in a region of space is uniform, then you can infer that the electric potential is increasing or decreasing uniformly in that region. For example, if the electric field is 10 V/m in a region of space then you can infer that the potential difference between two points 1 meter apart (measured parallel to the direction of the field) is 10 V. If the electric potential in a region of space is uniform, then you can infer that the electric field there is zero.

30. The electric potential energy of two unlike charges is negative. The electric potential energy of two like charges is positive. In the case of unlike charges, work must be done to separate the charges. In the case of like charges, work must be done to move the charges together.

31. The electric field is the negative of the slope of the *V* vs. *x* graph [Fig. 2.175].

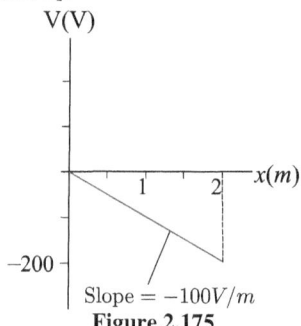

Figure 2.175

32. The electric field is the negative of the slope of the *V* vs. *x* graph. If *V* = constant in some region then *E* = 0 in that region.

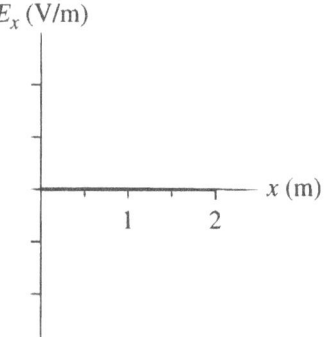

Figure 2.176

33. For contour map shown in Fig.2.91a, electric field at point ①;
$$E_1 = -\left(\frac{10-0}{1}\right)\hat{i} = -10 \text{ V/m } \hat{i}$$
and electric field at point ②;
$$E_2 = -\left(\frac{40-30}{1}\right)\hat{i} = -10 \text{ V/m } \hat{i}$$
here, negative sign shows that electric field is directed along −ve direction of *x* axis.

Now, for contour map shown in Fig.2.91b, electric field at point ①;

$$E_1 = -\left(\frac{10-0}{2}\right)\hat{i} = -5 \text{ V/m } \hat{i}$$
and electric field at point ②;
$$E_2 = -\left(\frac{40-20}{1}\right)\hat{i} = -20 \text{ V/m } \hat{i}$$
here, again negative sign shows that electric field is directed along −ve direction of *x* axis.

34. The electron moves to the right. The electric field is the negative of the slope of the electric potential, so the electric field will be non-zero at $x = 2$ m and points to the left. The electron will therefore experience a net force to the right and accelerate to the right.

35. (a) The electric field vector points in the direction of decreasing potential. Therefore $V_a > V_b$.
(b) $|\Delta V_{ab}| < |\Delta V_{cd}| < |\Delta V_{ef}|$. For a uniform electric field $|\Delta V| = |-E\Delta s| = |E\Delta s|$. The order is determined by the fact that $\Delta s_{ab} < \Delta s_{cd} < \Delta s_{ef}$
(c) Surface 1 is an equipotential surface because it is perpendicular to the electric field vectors. Surface 2 is not perpendicular to the electric field so it is not an equipotential surface.

36. Equal. When a conductor is in electrostatic equilibrium, the entire conductor is at the same potential.

37. (a) $V_1 = V_2$. Both spheres and the wire become one conductor and so are all at the same potential.
(b) Since $V_1 = V_2$, $\frac{1}{4\pi\epsilon_0}\frac{Q_1}{r_1} = \frac{1}{4\pi\epsilon_0}\frac{Q_2}{r_2}$, thus $\frac{Q_1}{r_1} = \frac{Q_2}{r_2}$. Since $r_1 > r_2, Q_1 > Q_2$.
(c) Recall $E_1 = \frac{1}{4\pi\epsilon_0}\frac{Q_1}{r_1^2} = \frac{V_1}{r_1}$ and $E_2 = \frac{1}{4\pi\epsilon_0}\frac{Q_2}{r_2^2} = \frac{V_2}{r_2}$. Since $r_1 > r_2, E_1 < E_2$.

38. **APPROACH**: The potential difference is the negative of the area under the E_x vs. *x* curve.
SOLUTION The potential difference between $x = 1.0$ m and $x = 3.0$ m (Fig.2.96) is-
$$\Delta V = -\frac{1}{2}(200 \text{ V})(3.0 \text{ m} - 1.0 \text{ m}) = -200 \text{ V}$$
The potential difference is negative since the electric field points in the direction of decreasing potential.

39. **APPROACH** The potential difference is the negative of the area under the E_x vs. *x* curve.
SOLUTION The potential difference between the origin and $x = 3.0$ m is
$$\Delta V = V(x = 3.0 \text{ m}) - V(x = 0.0 \text{ m})$$
$$= -\frac{1}{2} \times (3 \text{ m} + 2 \text{ m}) \times 200 \text{ V/m} = -500 \text{ V}$$
Thus, $V(3.0 \text{ m}) = V(0 \text{ m}) - 500 \text{ V}$
$$= -50 \text{ V} - 500 \text{ V} = -550 \text{ V} = -5.50 \text{ kV}.$$
The potential decreases from the origin to $x = 3.0$ m.

40. **APPROACH** The electric field points in the direction of decreasing potential and is perpendicular to the equipotential lines.
SOLUTION Please refer to Figure 2.98. The three equipotential surfaces correspond to potentials of 0 V, 200 V, and 400 V.
The electric field component along a direction of constant potential is $E_s = -dV/ds = 0$ V/m. But, the electric field component perpendicular to the equipotential surface is
$$|\vec{E}| = \frac{\Delta V}{\Delta s} = \frac{400 \text{ V}}{0.02 \text{ m}} = 20 \text{ kV/m}$$
The direction of the electric field vector is "downhill," perpendicular to the equipotential surfaces. That is, the electric field is 20 kV/m downward or $\vec{E} = -20\hat{j}$ kV/m

41. **APPROACH** The electric field is perpendicular to the equipotential lines and points "downhill".

SOLUTION Refer to Figure 2.99. Three equipotential surfaces at potentials of $-200V$, 0 V, and 200 V are shown. The electric field is perpendicular to the equipotential surfaces:
$$E = -\frac{\Delta V}{\Delta s} = -\frac{400\text{ V}}{0.02\text{ m}} = -20\text{ kV/m}$$
The electric field vector is in the third quadrant, 45° below the negative x-axis. That is,
$$\vec{E} = (20\text{ kV/m}, 45°\text{ below } x\text{ axis })$$
In component form, this would be
$$\vec{E} = 20\left[-\left(\frac{\hat{i}+\hat{j}}{\sqrt{2}}\right)\right]\text{ kV/m} = \left(-\frac{20}{\sqrt{2}}\hat{i} - \frac{20}{\sqrt{2}}\hat{j}\right)\text{ kV/m}$$

42. **APPROACH** The electric field is the negative of the slope of the graph of the potential function.
    SOLUTION (a) Please refer to Figure 2.100(a). There are three regions of different slope. For $0\text{ cm} \leq x < 10\text{ cm}$ and $20\text{ cm} \leq x < 30\text{ cm}$,
    $$\frac{\Delta V}{\Delta x} = 0\text{ V/m} \implies E_x = 0\text{ V/m}$$
    For $10\text{ cm} \leq x < 20\text{ cm}$,
    $$\frac{\Delta V}{\Delta x} = \frac{100\text{ V} - (-100\text{ V})}{0.20\text{ m} - 0.10\text{ m}} = 2000\text{ V/m}$$
    $$\implies E_x = -2000\text{ V/m}$$

Figure 2.177

Because $E_s = -dV/ds$, the electric field is zero where the potential is not changing.
(b) Now, refer to Figure 2.100(b). There are three regions of different slope. For $0\text{ cm} \leq x < 2\text{ cm}$,
$$\frac{\Delta V}{\Delta x} = \frac{-50\text{ V} - 0\text{ V}}{0.02\text{ m} - 0\text{ m}} = -2500\text{ V/m} \implies E_x = 2500\text{ V/m}$$
For $x \geq 2\text{ cm}$
$$\frac{\Delta V}{\Delta x} = \frac{50\text{ V} - (-50\text{ V})}{0.03\text{ m} - 0.02\text{ m}} = 10,000\text{ V/m}$$
$$\implies E_x = -10,000\text{ V/m}$$
$\Delta V/\Delta x$ and E_x are the negative of each other.

Figure 2.178

43. The electric potential difference ΔV between two points in a uniform electric field is

Figure 2.179

$$V(x_f) - V(x_i) = -\int E_x dx = -E_x(x_f - x_i)$$

Choosing $x_i = -1.0$ m and $x_f = +1.0$ m,
$$+1000\text{ V} - (-1000\text{ V}) = -E_x[1.0\text{ m} - (-1.0\text{ m})]$$
$$\implies E_x = -1.0\text{ kV/m}$$
Alternatively, $x_i = -1.0$ m and $x_f = -1.0$ m. For this choice,
$$-1000\text{ V} - (+1000\text{ V}) = -E_x[-1.0\text{ m} - (1.0\text{ m})]$$
$$\implies E_x = -1.0\text{ kV/m}$$
The choice of initial and final positions does not change the physical nature of the electric field or the potential difference.

44. **APPROACH** The electric field is the negative of the derivative of the potential function.
    SOLUTION (a) The component of the electric field in the s-direction is $E_s = -dV/ds$. For the given potential,
    $$\frac{dV}{dx} = \frac{d}{dx}\left(100x^2\text{ V}\right) = 200x\text{ V/m} \implies E_x = -200x\text{ V/m}$$
    (b) At $x = 1$ m, $E_x = -200 \times 1$ V/m $= -0.2$ kV/m.
    The potential increases with x, so the electric field must point in the $-x$-direction.

45. **(a)** Since $E_x = -dV/dx$, we have
    $$E_x = -\frac{d}{dx}\left(100\, e^{-2x}\right)\text{ V/m} = 200\, e^{-2x}\text{ V/m}$$
    At $x = 1.0$ m, $E_x = \left(200\, e^{-2(1.0)}\right)$ V/m $= 27$ V/m
    (b) At $x = 2.0$ m, $E_x = \left(200\, e^{-2(2.0)}\right)$ V/m $= 3.7$ V/m

2.16.5 Problems

1. (a) $V_B - V_A = \Delta U/q = -W/(-e)$
   $= -\left(3.94 \times 10^{-19}\text{ J}\right)/\left(-1.60 \times 10^{-19}\text{C}\right)$
   $= 2.46$ V.
   (b) $V_C - V_A = V_B - V_A = 2.46$ V.
   (c) $V_C - V_B = 0$ (since C and B are on the same equipotential line).

2. (a) The change in potential is the negative of the "area" under the curve. Thus, using the area of a triangle formula, we have
   $$V - V_0 = -\int_0^{x=2}\vec{E}\cdot d\vec{s} = -\int_0^{x=2}E\,dx = -\frac{1}{2}(2)(-20)$$
   here, $V_0 = 10$ V is the potential at $x = 0$. Substituting this value and solving for V yields $V = 30$ V.
   (b) For any region within $0 < x < 3$ m, the area under E_x vs x graph is negative, therefore $-\int\vec{E}\cdot d\vec{s}$ is positive, but for any region for which $x > 3$ m, the area is positive therefore $-\int\vec{E}\cdot d\vec{s}$ is negative. Therefore, $V = V_{\max}$ occurs at $x = 3$ m.
   If V_0 is the electric potential at $x = 0$ and $V_{\max}$ at $x = 3$ m, then
   $$V_{\max} - V_0 = -\int_0^{x=3}\vec{E}\cdot d\vec{s} = -\int_0^{x=3}E\,dx = -\frac{1}{2}(3)(-20)$$
   Since, $V_0 = 10$ V, therefore, above equation gives-
   $$V_{\max} - 10\text{ V} = 30\text{ V} \implies V_{\max} = 40\text{ V}$$
   (c) From $x = 0$ to $x = 3$ m, the area under E vs x graph i.e., $\int_{x=0}^{x=3}E\,dx$ is negative, therefore $V - V_0 = -\int_{x=0}^{x=3}E\,dx$ will be positive. Since, $V_0 = 10$ V, therefore V can not be zero anywhere in this region. Furthe from part (b), we have seen that the voltage at $x = 3$ m is 40 V. Now, further the area under the E vs x graph from $x = 3$ to $x = 4$ is only $\frac{1}{2}(4-3)(20) = +10$ V, therefore the change in potential corresponding to this area is -10 V. So from $x = 0$ to $x = 3$ m potential increases from 10 V to 40 V, then from $x = 3$ m to $x = 4$ m there is a decrease of 10 V and voltage reaches to 40 V $- 10$ V, i.e., 30 V. Thus, potential can not

be zero for any value of $x \leq 4$.
Now, suppose at some $x = x' (> 4\text{ m})$, the electric potential is zero, then

$$V - V_0 = -\int_{x=0}^{x=x'} \vec{E} \cdot d\vec{s} = -\int_{x=0}^{x=x'} E\,dx$$

$$\implies V - V_0 = -\left[\int_{x=0}^{x=3} E\,dx + \int_{x=3}^{x=4} E\,dx + \int_{x=4}^{x=x'} E\,dx\right]$$

$$= -\left[\frac{1}{2}(3)(-20) + \frac{1}{2}(4-3)(20) + (x'-4)(20)\right]$$

$$= -[-30 + 10 + 20x' - 80] = 100 - 20x'$$

Given that $V_0 = 10$ V, therefore corresponding to $V = 0$, we have

$$0 - 10 = 100 - 20x' \implies x' = 5.5\text{ m}$$

3. **(a)** The charge on the sphere is

$$q = 4\pi\epsilon_0 VR = \frac{(200\text{ V})(0.15\text{ m})}{8.99 \times 10^9\text{ N}\cdot\text{m}^2/\text{C}^2} = 3.3 \times 10^{-9}\text{C}$$

(b) The (uniform) surface charge density (charge divided by the area of the sphere) is

$$\sigma = \frac{q}{4\pi R^2} = \frac{3.3 \times 10^{-9}\text{C}}{4\pi(0.15\text{ m})^2} = 1.2 \times 10^{-8}\text{C/m}^2$$

4. $E = \dfrac{|\Delta V|}{d} = \dfrac{25.0 \times 10^3\text{ J/C}}{1.50 \times 10^{-2}\text{ m}}$
$= 1.67 \times 10^6\text{ N/C} = 1.67\text{ MN/C}$

5. From Fig.2.180, we can write-

$$V_B - V_A = -\int_A^B \vec{E} \cdot d\vec{s}$$

$$= -\int_A^C \vec{E} \cdot d\vec{s} - \int_C^B \vec{E} \cdot d\vec{s}$$

$$V_B - V_A = (-E\cos 180°)\int_{-0.300}^{0.500} dy$$
$$- (E\cos 90.0°)\int_{-0.200}^{0.400} dx$$

$$V_B - V_A = (325)(0.800) = +260\text{ V}$$

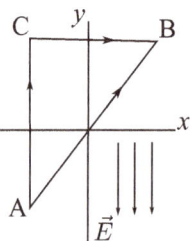

Figure 2.180

6. **(a)** In Fig. 2.181, a charged block with mass m and charge Q is attached to a spring with a spring constant k in an electric field $\vec{E}$ on a smooth surface. Since the block is positively charged, it moves toward a lower electric potential when released from rest, due to the electric force exerted by the electric field.
If at any instant, x represents the extension in spring, then the corresponding decrement in electric potential is Ex. Let's arbitrarily choose $V = 0$ at $x = 0$. For an extension x in the spring, the electric potential decreases to $V = -Ex$ and electric potential energy becomes $U_e = QV = -QEx$.
Between the endpoints of the motion,

$$(K + U_s + U_e)_i = (K + U_s + U_e)_f$$

$$\implies 0 + 0 + 0 = 0 + \tfrac{1}{2}kx_{\max}^2 - QEx_{\max}$$

so $x_{\max} = 2QE/k$

Figure 2.181

(b) At equilibrium,

$$\sum F_x = -F_s + F_e = 0 \implies F_s = F_e \implies kx = QE$$

So the equilibrium position is at $x = QE/k$.

(c) The block's equation of motion is

$$\sum F_x = -kx + QE = m\frac{d^2x}{dt^2}.$$

Let $x' = x - \frac{QE}{k}$ or $x = x' + \frac{QE}{k}$,
so, the equation of motion becomes:

$$-k\left(x' + \frac{QE}{k}\right) + QE = m\frac{d^2(x + QE/k)}{dt^2}$$

$$\implies \frac{d^2x'}{dt^2} = -\left(\frac{k}{m}\right)x'$$

This equation has the form of SHM:

$$\frac{d^2x'}{dt^2} = -\omega^2 x'$$

with angular frequency, $\omega = \sqrt{\dfrac{k}{m}}$
The period of the motion is then

$$T = \frac{2\pi}{\omega} = 2\pi\sqrt{\frac{m}{k}}$$

(d) $(K + U_s + U_e)_i + \Delta E_{\text{mech}} = (K + U_s + U_e)_f$

$$\implies 0 + 0 + 0 - \mu_k mg x_{\max} = 0 + \frac{1}{2}kx_{\max}^2 - QEx_{\max}$$

$$\implies x_{\max} = \frac{2(QE - \mu_k mg)}{k}$$

7. Total charge of the rod of length L: $q = \lambda L$, and total mass of the rod: $m = \mu L$. Arbitrarily take $V = 0$ at the initial point. Then at distance d downfield the electric potential V is given by-
$V = -Ed$ and potential energy $U_e = qV = -\lambda LEd$.

(a) $(K + U)_i = (K + U)_f \implies 0 + 0 = \frac{1}{2}\mu Lv^2 - \lambda LEd$
Solving for v, we get

$$v = \sqrt{\frac{2\lambda Ed}{\mu}} = \sqrt{\frac{2(40.0 \times 10^{-6}\text{C/m})(100\text{ N/C})(2.00\text{ m})}{(0.100\text{ kg/m})}}$$

$$= 0.400\text{ m/s}$$

(b) The same.

8. Arbitrarily take $V = 0$ at point P. Then from Fig.2.182a, the potential at the original position of the charged particle is $-\vec{E} \cdot \vec{s} = Es\cos\theta = -EL\cos\theta$ ($\because |\vec{s}| = L$). From Fig.2.182b, the potential at final position a, $V = -EL$. Given that, the table is frictionless, therefore the forces acting on the particles are–

(a) Normal reaction $\vec{N}$ of table acting perpendicularly outward to the plane of page. It is not shown in diagram.
(b) gravitational force $m\vec{g}$ acting perpendicularly inward to the plane of page. It is also not shown in diagram.
(c) Electric force $q\vec{E}$ in the direction of electric field $\vec{E}$.

Since $\vec{N}$ and $m\vec{g}$ both are perpendicular to the plane of displacement of the particle, therefore, the work done by conservative force $m\vec{g}$ and non-conservative force $\vec{N}$ both are zero. The force which is doing work in this case is the electric force $q\vec{E}$ which is a conservative force in nature, therefore, net mechanical energy of the charged particle in electric field will remain conserved. So, by principle of conservation of mechanical energy between original and final positions of charged particle, we can write

$$(K+U)_i = (K+U)_f \implies 0 - qEL\cos\theta = \frac{1}{2}mv^2 - qEL$$

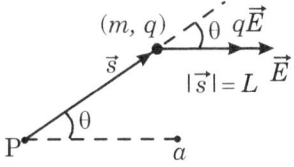

(a) Original position of charged particle

(b) Final position of charged particle

Figure 2.182

Solving for v, we get

$$v = \sqrt{\frac{2qEL(1-\cos\theta)}{m}}$$

$$= \sqrt{\frac{2\,(2.00\times 10^{-6}\text{C})\,(300\text{ N/C})(1.50\text{ m})\,(1-\cos 60.0°)}{0.0100\text{ kg}}}$$

$$= 0.300 \text{ m/s}$$

9. (a) The potential at 1.00 cm is

$$V_1 = k\frac{q}{r} = \frac{(8.99\times 10^9 \text{ N}\cdot\text{m}^2/\text{C}^2)\,(1.60\times 10^{-19}\text{C})}{1.00\times 10^{-2}\text{ m}}$$

$$= 1.44 \times 10^{-7} \text{ V}$$

(b) The potential at 2.00 cm is

$$V_2 = k\frac{q}{r} = \frac{(8.99\times 10^9 \text{ N}\cdot\text{m}^2/\text{C}^2)\,(1.60\times 10^{-19}\text{C})}{2.00\times 10^{-2}\text{ m}}$$

$$= 0.719 \times 10^{-7} \text{ V}.$$

Thus, the difference in potential between the two points is

$$\Delta V = V_2 - V_1 = -7.19\times 10^{-8} \text{ V}.$$

(c) The approach is the same as above except the charge is -1.60×10^{-19}C. This changes the sign of each answer, with its magnitude remaining the same.

That is, the potential at 1.00 cm is -1.44×10^{-7} V.
The potential at 2.00 cm is -0.719×10^{-7} V,
so, $\Delta V = V_2 - V_1 = 7.19\times 10^{-8}$ V.

10. (a) The potential difference is

$$V_A - V_B = \frac{q}{4\pi\epsilon_0 r_A} - \frac{q}{4\pi\epsilon_0 r_B}$$

$$= \left(1.0\times 10^{-6}\text{C}\right)\left(8.99\times 10^9 \text{ N}\cdot\text{m}^2/\text{C}^2\right)$$

$$\left(\frac{1}{2.0\text{ m}} - \frac{1}{1.0\text{ m}}\right)$$

$$= -4.5\times 10^3 \text{ V}$$

(b) Since $V(r)$ depends only on the magnitude of $\vec{r}$, the result is unchanged.

11. **APPROACH** The electric potential for a spherically symmetric charge distribution falls off as $1/r$, where r is the radial distance from the centre of the charge distribution. The electric potential V at the surface of a drop of charge q and radius R is given by $V = q/4\pi\epsilon_0 R$.

SOLUTION (a) With $V = 500$ V and $q = 30\times 10^{-12}$C, we find the radius to be

$$R = \frac{q}{4\pi\epsilon_0 V} = \frac{(8.99\times 10^9 \text{ N}\cdot\text{m}^2/\text{C}^2)\,(30\times 10^{-12}\text{C})}{500 \text{ V}}$$

$$= 5.4\times 10^{-4} \text{ m}$$

(b) After the two drops combine to form one big drop, the total volume is twice the volume of an original drop, so the radius R' of the combined drop is given by $(R')^3 = 2R^3$ and $R' = 2^{1/3}R$. The charge is twice the charge of the original drop: $q' = 2q$. Thus,

$$V' = \frac{1}{4\pi\epsilon_0}\frac{q'}{R'} = \frac{1}{4\pi\epsilon_0}\frac{2q}{2^{1/3}R} = 2^{2/3}V = 2^{2/3}(500 \text{ V})$$

$$\approx 790 \text{ V}$$

☞ A positively charged configuration produces a positive electric potential, and a negatively charged configuration produces a negative electric potential. Adding more charge increases the electric potential.

12. The electric potential is given by-

$$V = \frac{1}{4\pi\epsilon_0}\frac{q}{r} \qquad (1)$$

In applying Eq.1, we are assuming $V\to 0$ as $r\to\infty$. All corner particles are equidistant from the centre, and since their total charge is

$$2q_1 - 3q_1 + 2q_1 - q_1 = 0$$

therefore, their contribution to Eq.1 vanishes. The net potential is due, then, to the two $+4q_2$ particles, each of which is a distance of $a/2$ from the centre:

$$V = \frac{1}{4\pi\epsilon_0}\frac{4q_2}{a/2} + \frac{1}{4\pi\epsilon_0}\frac{4q_2}{a/2} = \frac{16q_2}{4\pi\epsilon_0 a}$$

$$= \frac{16\,(8.99\times 10^9 \text{ N}\cdot\text{m}^2/\text{C}^2)\,(6.00\times 10^{-12}\text{C})}{0.39 \text{ m}}$$

$$= 2.21 \text{ V}$$

13. A charge $-5q$ is at a distance $2d$ from P, a charge $-5q$ is at a distance d from P, and two charges $+5q$ are each at a distance d from P, so the electric potential at P is

$$V = \frac{q}{4\pi\epsilon_0}\left[-\frac{1}{2d} - \frac{1}{d} + \frac{1}{d} + \frac{1}{d}\right]$$

$$= \frac{q}{8\pi\epsilon_0 d} = \frac{(8.99\times 10^9 \text{ N}\cdot\text{m}^2/\text{C}^2)\,(5.00\times 10^{-15}\text{C})}{2\,(4.00\times 10^{-2}\text{ m})}$$

$$= 5.62\times 10^{-4} \text{ V}$$

The zero of the electric potential was taken to be at infinity.

14. First, we observe that $V(x)$ cannot be equal to zero for $x > d$. In fact $V(x)$ is always negative for $x > d$. Now we consider the two remaining regions on the x axis: $x < 0$ and $0 < x < d$.

(a) For $0 < x < d$ we have $d_1 = x$ and $d_2 = d - x$. Now,

$$V(x) = k\left(\frac{q_1}{d_1} + \frac{q_2}{d_2}\right) = \frac{q}{4\pi\epsilon_0}\left(\frac{1}{x} + \frac{-3}{d-x}\right) = 0$$

$$\implies x = d/4.$$

Since, $d = 24.0$ cm, therefore, $x = d/4 = 6.00$ cm.

(b) Similarly, for $x < 0$ the separation between q_1 and a point on the x axis whose coordinate is x is given by $d_1 = -x$; while the corresponding separation for q_2 is $d_2 = d - x$. Now,

$$V(x) = k\left(\frac{q_1}{d_1} + \frac{q_2}{d_2}\right) = \frac{q}{4\pi\epsilon_0}\left(\frac{1}{-x} + \frac{-3}{d-x}\right) = 0$$

$$\implies x = -\frac{d}{2}$$

Since, $d = 24.0$ cm, therefore, $x = -d/2 = -12.0$ cm.

15. Since according to the problem statement there is a point in between the two charges on the x axis where the net electric field is zero, the fields at that point due to q_1 and q_2 must be directed opposite to each other. This means that q_1 and q_2 must have the same sign (i.e., either both are positive or both negative). Thus, the potentials due to either of them must be of the same sign. Therefore, the net electric potential cannot possibly be zero anywhere except at infinity.

16. (a) Since the charges are equal and placed symmetrically, $F = 0$

(b) $F = qE = 0, E = 0$.

(c) $V = 2k\dfrac{q}{r}$

Figure 2.183

$$= 2\left(8.99 \times 10^9 \text{ N} \cdot \text{m}^2/\text{C}^2\right)\left(\frac{2.00 \times 10^{-6}\text{C}}{0.800 \text{ m}}\right)$$
$$= 4.50 \times 10^4 \text{ V} = 45.0 \text{ kV}$$

17. (a) $E_x = \frac{kq_1}{x^2} + \frac{kq_2}{(x-2.00)^2} = 0$

here, $q_1 = q$ and $q_2 = -2q$. Therefore, corresponding to $E_x = 0$ above equation becomes-

$$E_x = k\left(\frac{+q}{x^2} + \frac{-2q}{(x-2.00)^2}\right) = 0.$$

Dividing by k, we get-
$2qx^2 = q(x-2.00)^2 \implies x^2 + 4.00x - 4.00 = 0.$
Therefore $E_x = 0$ when
$$x = \frac{-4.00 \pm \sqrt{16.0 + 16.0}}{2} = -4.83 \text{ m}$$

(Note that the positive root does not correspond to a physically valid situation.)

(b) Since, zero potential points are always closer to that charge which is smaller in magnitude, therefore either zero potential point lies between the charges q and $-2q$ or at the left of q.

Now, corresponding to zero potential, we have
$$V = \frac{kq_1}{|x|} + \frac{kq_2}{|2.00-x|} = 0 \implies k\left(\frac{+q}{|x|} - \frac{2q}{|2.00-x|}\right) = 0.$$
$$\implies \frac{1}{|x|} = \frac{2}{|2.00-x|} \implies 2|x| = |2.00-x|$$
$$\implies 2x = \pm(2.00-x) \quad (\because \text{ if } |x|=|y|, \text{ then } x = \pm y)$$

First considering positive sign, we get
$2x = (2.00 - x) \implies x = 0.667 \text{ m}$
Now, considering negative sign, we get
$2x = -(2.00 - x) \implies x = -2.00 \text{ m}$

Thus, $V = 0$ at $x = \begin{cases} 0.667 \text{ m} & \text{when } 0 \leq x \leq 2.00 \\ -2.00 \text{ m} & \text{when } x < 0 \end{cases}$

18. $V = \sum_i k\frac{q_i}{r_i}$
$$= (8.99 \times 10^9)(7.00 \times 10^{-6})\left[\frac{-1}{0.0100} - \frac{1}{0.0100} + \frac{1}{0.0387}\right]$$
$$= -1.10 \times 10^7 \text{ V} = -11.0 \text{ MV}$$

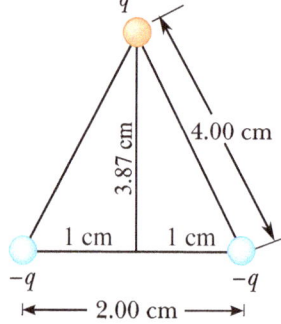

Figure 2.184

19. (a) $U = \frac{qQ}{4\pi\epsilon_0 r}$
$$= \frac{(5.00 \times 10^{-9}\text{C})(-3.00 \times 10^{-9}\text{C})(8.99 \times 10^9 \text{ V} \cdot \text{m/C})}{(0.350 \text{ m})}$$
$$= -3.86 \times 10^{-7} \text{ J}$$
The minus sign means it takes 3.86×10^{-7} J to pull the two charges apart from 35 cm to a much larger separation.

(b) $V = \frac{Q_1}{4\pi\epsilon_0 r_1} + \frac{Q_2}{4\pi\epsilon_0 r_2}$
$$= \frac{(5.00 \times 10^{-9}\text{C})(8.99 \times 10^9 \text{ V} \cdot \text{m/C})}{0.175 \text{ m}}$$
$$+ \frac{(-3.00 \times 10^{-9}\text{C})(8.99 \times 10^9 \text{ V.m/C})}{0.175 \text{ m}} = 103 \text{ V}$$

20. From Fig.2.185, we have-
$U = U_1 + U_2 + U_3 + U_4$
$U = 0 + U_{12} + (U_{13} + U_{23}) + (U_{14} + U_{24} + U_{34})$
$U = 0 + \frac{kQ^2}{s} + \frac{kQ^2}{s}\left(\frac{1}{\sqrt{2}} + 1\right) + \frac{kQ^2}{s}\left(1 + \frac{1}{\sqrt{2}} + 1\right)$
$U = \frac{kQ^2}{s}\left(4 + \frac{2}{\sqrt{2}}\right) = 5.41\frac{kQ^2}{s}$

An alternate way to get the term $\left(4 + \frac{2}{\sqrt{2}}\right)$ is to recognize

Figure 2.185

that there are 4 side pairs and 2 face diagonal pairs.

21. Consider the two spheres as a system.
(a) Conservation of momentum:
$$0 = m_1 v_1 \hat{i} + m_2 v_2 (-\hat{i}) \implies v_2 = \frac{m_1 v_1}{m_2}$$
By conservation of energy,
$$0 = \frac{k(-q_1)q_2}{d} = \frac{1}{2}m_1 v_1^2 + \frac{1}{2}m_2 v_2^2 + \frac{k(-q_1)q_2}{r_1 + r_2}$$
and
$$\frac{kq_1 q_2}{r_1 + r_2} - \frac{kq_1 q_2}{d} = \frac{1}{2}m_1 v_1^2 + \frac{1}{2}\frac{m_1^2 v_1^2}{m_2}$$
Solving for v_1, we get
$$v_1 = \sqrt{\frac{2m_2 k q_1 q_2}{m_1(m_1 + m_2)}\left(\frac{1}{r_1 + r_2} - \frac{1}{d}\right)}$$
Substituting the given value in above expression, we get-
$v_1 = 10.8$ m/s,
and $v_2 = \frac{m_1 v_1}{m_2} = \frac{0.100 \text{ kg}(10.8 \text{ m/s})}{0.700 \text{ kg}} = 1.55$ m/s

(b) If the spheres are conductor, electrons will move around on them with negligible energy loss to place the centers of excess charge on the insides of the spheres. Then just before they touch, the effective distance between charges will be less than $r_1 + r_2$ and the spheres will really be moving faster than calculated in (a).

22. Consider the two spheres as a system.
(a) By conservation of momentum:
$$0 = m_1 v_1 \hat{i} + m_2 v_2(-\hat{i}) \implies v_2 = \frac{m_1 v_1}{m_2}$$
and by conservation of mechanical energy,
$$0 + \frac{k(-q_1)q_2}{d} = \frac{1}{2}m_1 v_1^2 + \frac{1}{2}m_2 v_2^2 + \frac{k(-q_1)q_2}{r_1 + r_2}$$
Substituting the value of v_2 and then rearranging terms, we get
$$\frac{kq_1 q_2}{r_1 + r_2} - \frac{kq_1 q_2}{d} = \frac{1}{2}m_1 v_1^2 + \frac{1}{2}\frac{m_1^2 v_1^2}{m_2}$$
Solving for v_1, we get
$$v_1 = \sqrt{\frac{2m_2 k q_1 q_2}{m_1(m_1 + m_2)}\left(\frac{1}{r_1 + r_2} - \frac{1}{d}\right)}$$
Now, substituting this value of v_1 in expression for v_2, we get

2.16. ANSWER KEYS AND SOLUTIONS

$$v_2 = \left(\frac{m_1}{m_2}\right)v_1 = \sqrt{\frac{2m_1kq_1q_2}{m_2(m_1+m_2)}\left(\frac{1}{r_1+r_2}-\frac{1}{d}\right)}$$

(b) If the spheres are conductor, electrons will move around on them with negligible energy loss to place the centers of excess charge on the insides of the spheres. Then just before they touch, the effective distance between charges will be less than $r_1 + r_2$ and the spheres will really be moving faster than calculated in (a).

23. (a) In an empty universe, the 20 nC charge can be placed at its location with no energy investment. At a distance of 4 cm, it creates a potential

$$V_1 = \frac{kq_1}{r} = \frac{(8.99 \times 10^9 \text{ N} \cdot \text{m}^2/\text{C}^2)(20 \times 10^{-9}\text{C})}{0.04 \text{ m}}$$
$$= 4.50 \text{kV}$$

To place the 10 nC charge there we must put in energy
$$U_{12} = q_2V_1 = \left(10 \times 10^{-9}\text{C}\right)\left(4.5 \times 10^3 \text{ V}\right)$$
$$= 4.50 \times 10^{-5} \text{ J}$$

Next, to bring up the -20 nC charge requires energy
$$U_{23} + U_{13} = q_3V_2 + q_3V_1 = q_3(V_2 + V_1)$$
$$= -20 \times 10^{-9}\text{C}\left(8.99 \times 10^9 \text{ N} \cdot \text{m}^2/\text{C}^2\right)$$
$$\left(\frac{10 \times 10^{-9}\text{C}}{0.04 \text{ m}} + \frac{20 \times 10^{-9}\text{C}}{0.08 \text{ m}}\right)$$
$$= -4.50 \times 10^{-5} \text{ J} - 4.50 \times 10^{-5} \text{ J} = -9.0 \times 10^{-5} \text{ J}$$

The total energy of the three charges is
$$U_{12} + U_{23} + U_{13} = -4.50 \times 10^{-5} \text{ J}$$

(b) The three fixed charges create this potential at the location where the fourth is released:
$$V = V_1 + V_2 + V_3$$
$$= \left(8.99 \times 10^9 \text{ N} \cdot \text{m}^2/\text{C}^2\right)$$
$$\left(\frac{20 \times 10^{-9}}{\sqrt{0.04^2 + 0.03^2}} + \frac{10 \times 10^{-9}}{0.03} - \frac{20 \times 10^{-9}}{0.05}\right)\text{C/m}$$
$$= 3.00 \times 10^3 \text{ V}$$ Energy of the system of four charged objects is conserved as the fourth charge flies away:
$$\left(\frac{1}{2}mv^2 + qV\right)_i = \left(\frac{1}{2}mv^2 + qV\right)_f$$
$$\Rightarrow \quad 0 + \left(40 \times 10^{-9}\text{C}\right)\left(3.00 \times 10^3 \text{ V}\right)$$
$$= \frac{1}{2}\left(2.00 \times 10^{-13} \text{ kg}\right)v^2 + 0$$

Solving for v, we get
$$v = \sqrt{\frac{2(1.20 \times 10^{-4} \text{ J})}{2 \times 10^{-13} \text{ kg}}} = 3.46 \times 10^4 \text{ m/s}$$

24. The original electrical potential energy is
$$U = qV = q\frac{kq}{d} = \frac{kq^2}{d}$$

In the final configuration we have mechanical equilibrium. If k_s is the spring constant, then the spring and electrostatic forces on each charge are $-k_s(2d)$ and $q\frac{kq}{(3d)^2}$ respectively.

In mechanical equilibrium, we have-
$$-k_s(2d) + q\frac{kq}{(3d)^2} = 0 \quad \Rightarrow \quad k_s = \frac{kq^2}{18d^3}$$

In the final configuration the total potential energy is-
$$\frac{1}{2}k_sx^2 + qV = \frac{1}{2}\frac{kq^2}{18d^3}(2d)^2 + q\frac{kq}{3d} = \frac{4}{9}\frac{kq^2}{d}$$

The missing energy must have become internal energy, as the system is isolated:

$$\frac{kq^2}{d} = \frac{4kq^2}{9d} + \Delta E_{int} \quad \Rightarrow \quad \Delta E_{int} = \frac{5}{9}\frac{kq^2}{d}$$

25. (a) Calculation of potential on x–axis
Due to both charges, the net potential $V(x)$ on the x–axis at distance x from the origin is given by
$$V(x) = \frac{kQ_1}{r_1} + \frac{kQ_2}{r_2} = \frac{k(+Q)}{\sqrt{x^2+a^2}} + \frac{k(+Q)}{\sqrt{x^2+(-a)^2}}$$
$$= \frac{2kQ}{\sqrt{x^2+a^2}} = \frac{kQ}{a}\left(\frac{2}{\sqrt{(x/a)^2+1}}\right)$$
$$\Rightarrow \quad \frac{V(x)}{(kQ/a)} = \frac{2}{\sqrt{(x/a)^2+1}}$$

Potential $V(x)$ vs x graph is plotted in Fig.2.186.

(b) Similarly, the net potential $V(y)$ on y–axis at distance

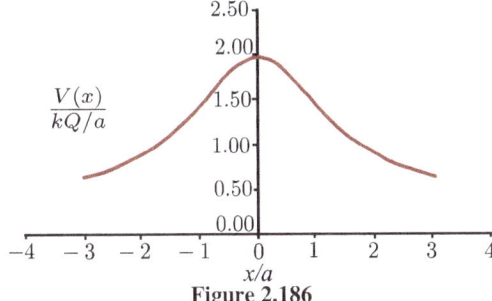

Figure 2.186

y from origin is given by
$$V(y) = \frac{kQ_1}{r_1} + \frac{kQ_2}{r_2} = \frac{k(+Q)}{|y-a|} + \frac{k(-Q)}{|y+a|}$$
$$= \frac{kQ}{a}\left(\frac{1}{|y/a-1|} - \frac{1}{|y/a+1|}\right)$$
$$\Rightarrow \quad \frac{V(y)}{(kQ/a)} = \left(\frac{1}{|y/a-1|} - \frac{1}{|y/a+1|}\right)$$

Potential $V(y)$ vs y graph is plotted in Fig.2.187.

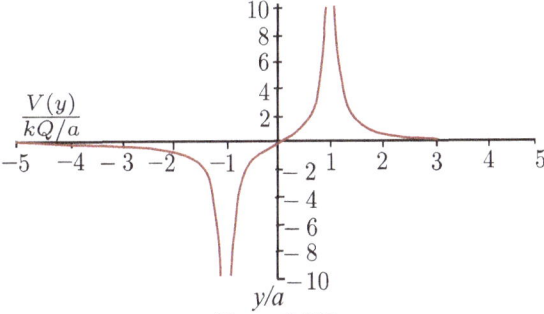

Figure 2.187

26. Using conservation of energy for the alpha particle-nucleus system, we have
$$K_f + U_f = K_i + U_i$$

But $U_i = \frac{kq_\alpha q_{gold}}{r_i}$ and $r_i \approx \infty$.

Thus, $U_i = 0$, $K_f = 0$ (since, at turning point, $v_f = 0$)

Therefore, $U_f = K_i \quad \Rightarrow \quad \frac{kq_\alpha q_{gold}}{r_{min}} = \frac{1}{2}m_\alpha v_\alpha^2$

Solving for r_{min}, we get
$$r_{min} = \frac{2kq_\alpha q_{gold}}{m_\alpha v_\alpha^2}$$
$$= \frac{2(8.99 \times 10^9 \text{ N} \cdot \text{m}^2/\text{C}^2)(2)(79)(1.60 \times 10^{-19}\text{C})^2}{(6.64 \times 10^{-27} \text{ kg})(2.00 \times 10^7 \text{ m/s})^2}$$
$$= 2.74 \times 10^{-14} \text{ m} = 27.4 \text{fm}$$

27. Using conservation of mechanical energy, we have:

$$\frac{keQ}{r_1} = \frac{keQ}{r_2} + \frac{1}{2}mv^2$$

which gives: $v = \sqrt{\frac{2keQ}{m}\left(\frac{1}{r_1} - \frac{1}{r_2}\right)}$

$$= \left[\frac{(2)(8.99 \times 10^9 \text{ N} \cdot \text{m}^2/\text{C}^2)(-1.60 \times 10^{-19}\text{C})(10^{-9}\text{C})}{9.11 \times 10^{-31} \text{ kg}}\right.$$
$$\left.\left(\frac{1}{0.0300 \text{ m}} - \frac{1}{0.0200 \text{ m}}\right)\right]^{1/2}$$

$= 7.26 \times 10^6$ m/s

28. $U = \sum \frac{kq_iq_j}{r_{ij}}$, summed over all pairs of (i, j) where $i \neq j$
Fig.2.188.
$$U = k\left[\frac{q(-2q)}{b} + \frac{(-2q)(3q)}{a} + \frac{(2q)(3q)}{b} + \frac{q(2q)}{a}\right.$$

Figure 2.188

$$\left. + \frac{q(3q)}{\sqrt{a^2+b^2}} + \frac{2q(-2q)}{\sqrt{a^2+b^2}}\right]$$

Substituting the given values, we get
$$U = kq^2\left[\frac{-2}{0.400} - \frac{6}{0.200} + \frac{6}{0.400} + \frac{2}{0.200}\right.$$
$$\left. + \frac{3}{0.447} - \frac{4}{0.447}\right]$$

$$U = (8.99 \times 10^9)(6.00 \times 10^{-6})^2\left[\frac{4}{0.400} - \frac{4}{0.200} - \frac{1}{0.447}\right]$$

$= -3.96$ J

29. Since, the effective force on each charge is diagonally away, therefore each charge moves off on its diagonal line. All charges have equal speeds.
By conservation of mechanical energy, we have
$$\sum(K+U)_i = \sum(K+U)_f$$
$$\Rightarrow 0 + \frac{4kq^2}{L} + \frac{2kq^2}{\sqrt{2}L} = 4\left(\frac{1}{2}mv^2\right) + \frac{4kq^2}{2L} + \frac{2kq^2}{2\sqrt{2}L}$$
$$\Rightarrow \left(2 + \frac{1}{\sqrt{2}}\right)\frac{kq^2}{L} = 2mv^2 \Rightarrow v = \sqrt{\left(1 + \frac{1}{\sqrt{8}}\right)\frac{kq^2}{mL}}$$

30. A cube has 12 edges and 6 faces. Consequently, there are 12 edge pairs separated by s, $2 \times 6 = 12$ face diagonal pairs separated by $\sqrt{2}s$ and 4 interior diagonal pairs separated $\sqrt{3}s$.
$$U = \frac{kq^2}{s}\left[12 + \frac{12}{\sqrt{2}} + \frac{4}{\sqrt{3}}\right] = 22.8\frac{kq^2}{s}$$

31. We can use $W_{q \to \text{final position}} = Q\Delta V_{i \to f}$ to find the work required to move these charges between the given points.
(a) Express the required work in terms of the charge being moved and the potential due to the charge at $x = +a$ and simplify to obtain:
$$W_{+Q \to +a} = Q\Delta V_{\infty \to +a} = Q[V(a) - V(\infty)]$$

$$= QV(a) = Q\left(\frac{kQ}{2a}\right) = \frac{kQ^2}{2a}$$

(b) Express the required work in terms of the charge being moved and the potentials due to the charges at $x = +a$ and $x = -a$ and simplify to obtain:
$$W_{-Q \to 0} = -Q\Delta V_{\infty \to 0} = -Q[V(0) - V(\infty)]$$
$$= -QV(0) = -Q\left[V_{\text{due to charge} \atop \text{at } x = -a} + V_{\text{due to charge} \atop \text{at } x = +a}\right]$$
$$= -Q\left(\frac{kQ}{a} + \frac{kQ}{a}\right) = \frac{-2kQ^2}{a}$$

(c) Express the required work in terms of the charge being moved and the potentials due to the charges at $x = +a$ and $x = -a$ and simplify to obtain:
$$W_{-Q \to 2a} = -Q\Delta V_{0 \to 2a} = -Q[V(2a) - V(0)]$$
$$= -Q\left[V_{\text{due to charge} \atop \text{at } x = -a} + V_{\text{due to charge} \atop \text{at } x = +a} - V(0)\right]$$
$$= -Q\left(\frac{kQ}{3a} + \frac{kQ}{a} - \frac{2kQ}{a}\right) = \frac{2kQ^2}{3a}$$

32. $\Delta K = q|\Delta V| \Rightarrow 7.37 \times 10^{-17} = q(115)$
$\Rightarrow q = 6.41 \times 10^{-19}$ C

33. (a) Energy of the proton-field system is conserved as the proton moves from high to low potential, which can be defined for this problem as moving from 120 V down to 0 V.
$$K_i + U_i + \Delta E_{\text{mech}} = K_f + U_f$$
$$\Rightarrow 0 + qV + 0 = \frac{1}{2}mv_p^2 + 0$$
$$\Rightarrow (1.60 \times 10^{-19}\text{C})(120 \text{ V}) = \frac{1}{2}(1.67 \times 10^{-27} \text{ kg})v_p^2$$
$$\Rightarrow v_p = 1.52 \times 10^5 \text{ m/s}$$

(b) The electron will gain speed in moving the other way, from $V_i = 0$ to $V_f = 120$ V:
$$K_i + U_i + \Delta E_{\text{mech}} = K_f + U_f$$
$$\Rightarrow 0 + 0 + 0 = \frac{1}{2}mv_e^2 + qV$$
$$\Rightarrow 0 = \frac{1}{2}(9.11 \times 10^{-31} \text{ kg})v_e^2$$
$$+ (-1.60 \times 10^{-19}\text{C})(120 \text{ J/C})$$
$$\Rightarrow v_e = 6.49 \times 10^6 \text{ m/s}$$

34. By work-kinetic energy theorem, we have
$$W = \Delta K \Rightarrow -q\Delta V = \frac{1}{2}mv_2^2 - \frac{1}{2}mv_1^2$$
$$\Rightarrow -(-1.60 \times 10^{-19}\text{C})\Delta V$$
$$= 0 - \frac{1}{2}(9.11 \times 10^{-31} \text{ kg})(4.20 \times 10^5 \text{ m/s})^2$$
From which we get, $\Delta V = -0.502$ V.

35. $V = a + bx = 10.0$ V $+ (-7.00$ V/m$)x$
(a) At $x = 0$, $V = 10.0$ V
At $x = 3.00$ m, $V = -11.0$ V
At $x = 6.00$ m, $V = -32.0$ V
(b) $E = -\frac{dV}{dx} = -b = -(-7.00 \text{ V/m}) = 7.00$ N/C in the $+x$ direction

36. (a) For $r < R$, $V = \frac{kQ}{R}$ and $E_r = -\frac{dV}{dr} = 0$
(b) For $r \geq R$, we have $V = \frac{kQ}{r}$
and $E_r = -\frac{dV}{dr} = -\left(-\frac{kQ}{r^2}\right) = \frac{kQ}{r^2}$

37. $V = 5x - 3x^2y + 2yz^2$
Evaluate E at $(1, 0, -2)$
$$E_x = -\frac{\partial V}{\partial x} = -5 + 6xy = -5 + 6(1)(0) = -5$$

2.16. ANSWER KEYS AND SOLUTIONS

$$E_y = -\frac{\partial V}{\partial y} = +3x^2 - 2z^2 = 3(1)^2 - 2(-2)^2 = -5$$

$$E_z = -\frac{\partial V}{\partial z} = -4yz = -4(0)(-2) = 0$$

$$E = \sqrt{E_x^2 + E_y^2 + E_z^2} = \sqrt{(-5)^2 + (-5)^2 + 0^2} = 7.07 \text{ N/C}$$

38. (a) $E_A > E_B$ since $E = \frac{\Delta V}{\Delta s}$

(b) $E_B = -\frac{\Delta V}{\Delta s} = -\frac{(6-2)\text{V}}{2 \text{ cm}} = 200$ N/C down

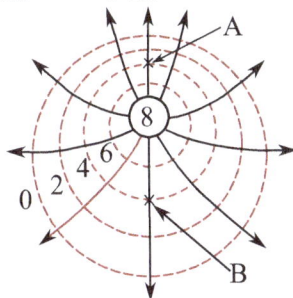

Figure 2.189

(c) Electric field lines are shown in Fig.2.189.

39. $\Delta V = V_{2R} - V_0$

$$= \frac{kQ}{\sqrt{R^2 + (2R)^2}} - \frac{kQ}{R} = \frac{kQ}{R}\left(\frac{1}{\sqrt{5}} - 1\right) = -0.553\frac{kQ}{R}$$

40. The potential is

$$V_P = \frac{1}{4\pi\epsilon_0}\int_{\text{rod}}\frac{dq}{R} = \frac{1}{4\pi\epsilon_0 R}\int_{\text{rod}} dq = \frac{-Q}{4\pi\epsilon_0 R}$$

$$= -\frac{(8.99 \times 10^9 \text{ N}\cdot\text{m}^2/\text{C}^2)(25.6 \times 10^{-12}\text{C})}{3.71 \times 10^{-2} \text{ m}} = -6.20 \text{ V}$$

We note that the result is exactly what one would expect for a point-charge $-Q$ at a distance R. This "coincidence" is due, in part, to the fact that V is a scalar quantity.

41. (a) All the charge is at the same distance R from C, so the electric potential at C is

$$V = \frac{1}{4\pi\epsilon_0}\left(\frac{Q_1}{R} - \frac{6Q_1}{R}\right) = -\frac{5Q_1}{4\pi\epsilon_0 R}$$

$$= -\frac{5(8.99 \times 10^9 \text{ N}\cdot\text{m}^2/\text{C}^2)(4.20 \times 10^{-12}\text{C})}{8.20 \times 10^{-2} \text{ m}} = -2.30 \text{ V}$$

where the zero was taken to be at infinity.

(b) All the charge is at the same distance from P. That distance is $\sqrt{R^2 + D^2}$, so the electric potential at P is

$$V = \frac{1}{4\pi\epsilon_0}\left[\frac{Q_1}{\sqrt{R^2 + D^2}} - \frac{6Q_1}{\sqrt{R^2 + D^2}}\right] = -\frac{5Q_1}{4\pi\epsilon_0\sqrt{R^2 + D^2}}$$

$$= -\frac{5(8.99 \times 10^9 \text{ N}\cdot\text{m}^2/\text{C}^2)(4.20 \times 10^{-12}\text{C})}{\sqrt{(8.20 \times 10^{-2} \text{ m})^2 + (6.71 \times 10^{-2} \text{ m})^2}} = -1.78 \text{ V}$$

42. Letting d denote 0.010 m, we have

$$V = \frac{Q_1}{4\pi\epsilon_0 d} + \frac{3Q_1}{8\pi\epsilon_0 d} - \frac{3Q_1}{16\pi\epsilon_0 d} = \frac{Q_1}{8\pi\epsilon_0 d}$$

$$= \frac{(8.99 \times 10^9 \text{ N}\cdot\text{m}^2/\text{C}^2)(30 \times 10^{-9}\text{C})}{2(0.01 \text{ m})} = 1.3 \times 10^4 \text{ V}$$

43. (a) $[\alpha] = \left[\frac{\lambda}{x}\right] = \frac{C}{m}\cdot\left(\frac{1}{m}\right) = \frac{C}{m^2}$

(b) $V = k\int\frac{dq}{r} = k\int\frac{\lambda dx}{d+x} = k\alpha\int_0^L\frac{xdx}{d+x}$

$$= k\alpha\left[L - d\ln\left(1 + \frac{L}{d}\right)\right]$$

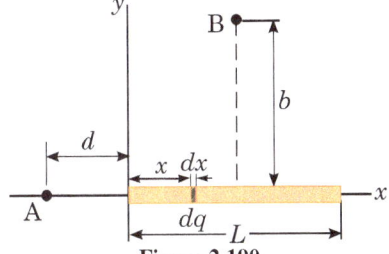

Figure 2.190

44. $V = \int\frac{kdq}{r} = k\int\frac{\alpha x dx}{\sqrt{b^2 + (L/2 - x)^2}}$

Let $z = \frac{L}{2} - x$, then $x = \frac{L}{2} - z$, and $dx = -dz$, therefore-

$$V = k\alpha\int\frac{(L/2 - z)(-dz)}{\sqrt{b^2 + z^2}} = -\frac{k\alpha L}{2}\int\frac{dz}{\sqrt{b^2 + z^2}}$$

$$+ k\alpha\int\frac{zdz}{\sqrt{b^2 + z^2}}$$

$$= -\frac{k\alpha L}{2}\ln\left(z + \sqrt{z^2 + b^2}\right) + k\alpha\sqrt{z^2 + b^2}$$

$$V = -\frac{k\alpha L}{2}\ln\left[\left(\frac{L}{2} - x\right) + \sqrt{\left(\frac{L}{2} - x\right)^2 + b^2}\right]_0^L$$

$$+ k\alpha\sqrt{\left(\frac{L}{2} - x\right)^2 + b^2}\Bigg|_0^L$$

$$V = -\frac{k\alpha L}{2}\ln\left[\frac{L/2 - L + \sqrt{(L/2)^2 + b^2}}{L/2 + \sqrt{(L/2)^2 + b^2}}\right]$$

$$+ k\alpha\left[\sqrt{\left(\frac{L}{2} - L\right)^2 + b^2} - \sqrt{\left(\frac{L}{2}\right)^2 + b^2}\right]$$

$$V = \left[-\frac{k\alpha L}{2}\ln\left[\frac{\sqrt{b^2 + (L^2/4)} - L/2}{\sqrt{b^2 + (L^2/4)} + L/2}\right]\right]$$

45. $dV = \frac{kdq}{\sqrt{r^2 + x^2}}$, where $dq = \sigma dA = \sigma 2\pi r dr$

$$V = 2\pi\sigma k\int_a^b\frac{rdr}{\sqrt{r^2 + x^2}} = 2\pi k\sigma\left[\sqrt{x^2 + b^2} - \sqrt{x^2 + a^2}\right]$$

[see Fig.2.191]

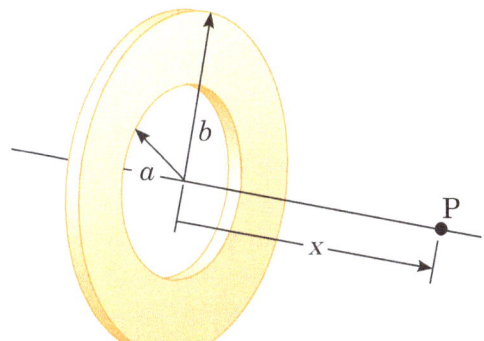

Figure 2.191: Problem 45

46. $V = k\int_{\text{all charge}}\frac{dq}{r} = k\int_{-3R}^{-R}\frac{\lambda dx}{-x} + k\int_{\text{semicircle}}\frac{\lambda ds}{R}$

$$+ k\int_R^{3R}\frac{\lambda dx}{x}$$

$$= -k\lambda\ln(-x)\big|_{-3R}^{-R} + \frac{k\lambda}{R}\pi R + k\lambda\ln x\big|_R^{3R}$$

$\implies \quad V = k \ln \dfrac{3R}{R} + k\lambda\pi + k \ln 3 = k\lambda(\pi + 2\ln 3)$

47. Substituting given values into $V = kq/r$, we get

$7.50 \times 10^3 \text{ V} = \dfrac{(8.99 \times 10^9 \text{ N} \cdot \text{m}^2/\text{C}^2) q}{0.300 \text{ m}}$

$\implies \quad q = 2.50 \times 10^{-7} \text{C}$

Now, from $q = Ne$, number of electrons removed from spherical conductor:

$N = \dfrac{q}{e} = \dfrac{2.50 \times 10^{-7}\text{C}}{1.60 \times 10^{-19}\text{C}} = 1.56 \times 10^{12}$ electrons

48. (a) $E_{max} = 3.00 \times 10^6 \text{ V/m} = \dfrac{kQ}{r^2} = \dfrac{kQ}{r}\left(\dfrac{1}{r}\right) = V_{max}\left(\dfrac{1}{r}\right)$

$V_{max} = E_{max} r = 3.00 \times 10^6 (0.150) = 450$ kV

(b) $\dfrac{kQ_{max}}{r^2} = E_{max} \quad \left\{ \text{or } \dfrac{kQ_{max}}{r} = V_{max} \right\}$

or $\quad Q_{max} = \dfrac{E_{max} r^2}{k} = \dfrac{3.00 \times 10^6 (0.150)^2}{8.99 \times 10^9} = 7.51 \; \mu$C

49. The potential created by the ring at the electron's starting point is

$V_i = \dfrac{kQ}{\sqrt{x_i^2 + a^2}} = \dfrac{k(2\pi\lambda a)}{\sqrt{x_i^2 + a^2}}$

while at the centre, it is $V_f = 2\pi k\lambda$. From conservation of mechanical energy, we have

$0 + (-eV_i) = \dfrac{1}{2} m_e v_f^2 + (-eV_f)$

$\implies \quad v_f^2 = \dfrac{2e}{m_e}(V_f - V_i) = \dfrac{4\pi e k\lambda}{m_e}\left(1 - \dfrac{a}{\sqrt{x_i^2 + a^2}}\right)$

$\implies \quad v_f^2 = \dfrac{4\pi (1.60 \times 10^{-19})(8.99 \times 10^9)(1.00 \times 10^{-7})}{9.11 \times 10^{-31}}$
$\left(1 - \dfrac{0.200}{\sqrt{(0.100)^2 + (0.200)^2}}\right)$

$\implies \quad v_f = \sqrt{1.45 \times 10^7} \text{ m/s}$

50. Take the illustration presented with the problem as an initial picture (Fig.2.124). No external horizontal forces act on the set of four balls, so its centre of mass stays fixed at the location of the centre of the square. As the charged balls 1 and 2 swing out and away from each other, balls 3 and 4 move up with equal y-components of velocity. The maximum-kinetic energy state is shown in Fig.2.192. By conservation of mechanical energy, we have-

$\dfrac{kq^2}{a} = \dfrac{kq^2}{3a} + \dfrac{1}{2}mv^2 + \dfrac{1}{2}mv^2 + \dfrac{1}{2}mv^2 + \dfrac{1}{2}mv^2$

Figure 2.192

$\implies \quad \dfrac{2kq^2}{3a} = 2mv^2 \quad \implies \quad v = \sqrt{\dfrac{kq^2}{3am}}$

51. For an element of area which is a ring of radius r and width dr,

$dV = \dfrac{kdq}{\sqrt{r^2 + x^2}}$

Here, $dq = \sigma dA = Cr(2\pi rdr)$, therefore-

$V = C(2\pi k) \int_0^R \dfrac{r^2 dr}{\sqrt{r^2 + x^2}}$

$= C(\pi k)\left[R\sqrt{R^2 + x^2} + x^2 \ln\left(\dfrac{x}{R + \sqrt{R^2 + x^2}}\right)\right]$

52. **APPROACH** The work done by the electric field of the gold nucleus changes the kinetic energy of the alpha particle-eventually bringing it to rest. We can apply the work-kinetic energy theorem to derive an expression for the distance of closest approach. Because the repulsive Coulomb force $\vec{F}_e$ varies with distance, we'll have to evaluate $\int \vec{F}_e \cdot d\vec{r}$ in order to find the work done on the alpha particles by this force.

SOLUTION Apply the work-kinetic energy theorem to the alpha particle to obtain:

$W_{net} = \int_\infty^{r_{min}} \vec{F}_e \cdot d\vec{r} = \Delta K$

since, $\int_\infty^{r_{min}} \vec{F}_e \cdot d\vec{r} = -\int_\infty^{r_{min}} \dfrac{k(2e)(79e)}{r^2} dr$ and $K_f = 0$

Therefore, $-158 ke^2 \int_\infty^{r_{min}} \dfrac{dr}{r^2} = -K_i$

Evaluating the integral yields:

$-158 ke^2 \left[\dfrac{1}{r}\right]_\infty^{r_{min}} = -\dfrac{158 ke^2}{r_{min}} = -K_i$

Solving for r_{min} we get:

$r_{min} = \dfrac{158 ke^2}{K_i}$

Substituting numerical values in above expression, we get

$r_{min} = \dfrac{158 \left(8.988 \times 10^9 \frac{\text{N} \cdot \text{m}^2}{\text{C}^2}\right)(1.602 \times 10^{-19}\text{C})^2}{5.0 \text{ Mev} \times \frac{1.602 \times 10^{-19} \text{ J}}{\text{eV}}}$

$= 4.6 \times 10^{-14}$ m

2.16.6 Multiple Choice Assignments

2.16.6.1 Level 1

Q.No.	1	2	3	4	5	6	7	8	9
Ans.	A	B	A	C	C	A	B	B	D
Q.No.	10	11	12	13	14	15	16	17	18
Ans.	D	B	A	B	A	C	B	D	A
Q.No.	19	20	21	22	23	24	25	26	27
Ans.	D	D	B	A	A	C	A	B	B
Q.No.	28	29	30	31	32	33	34	35	36
Ans.	A	B	A	B	A	C	B	A	C
Q.No.	37	38	39						
Ans.	B	B	B						

2.16.6.2 Level 2

Q.No.	1	2	3	4	5	6	7	8	9
Ans.	B	B	C	D	B	C	B	D	B
Q.No.	10	11	12	13	14	15	16	17	18
Ans.	B	D	C	B	D	A	C		

2.16.6.3 Level 3

Q.No.	1	2	3	4	5	6	7	8	9
Ans.	A	A	B	B	B	B	A	B	A
Q.No.	10	11	12	13	14	15	16	17	18
Ans.	A	C	A	C	B	B	B	C	B
Q.No.	19	20	21	22	23	24	25	26	27
Ans.	C	A	A	C					

2.16. ANSWER KEYS AND SOLUTIONS

2.16.6.4 Level 4
Section A

Q.No.	1	2	3	4	5	6	7	8	9
Ans.	A	B	D	B	C	A	B	D	D
Q.No.	10	11	12	13	14	15	16	17	18
Ans.	D	C	C	D	A	B	D	D	B
Q.No.	19	20	21	22	23	24	25	26	27
Ans.	C	D	C	B	D	B	B	D	B
Q.No.	28	29	30	31	32	33	34	35	36
Ans.	B	D	C	B	A	C	D	C	D
Q.No.	37	38	39	40	41	42	43	44	45
Ans.	A	B	D	B	D	B	D	C	A
Q.No.	46	47	48						
Ans.	C	5	160						

Section B

Q.No.	1	2	3	4	5	6	7	8
Ans.	B	A	A	D	D	B	B	B
Q.No.	9	10	11	12	13	14	15	16
Ans.	C	C	A	B	ABCD	C	B	C
Q.No.	17	18	19	20	21	22	23	24
Ans.	D	CD	AB	2	6	2	1.73	3.00
Q.No.	25	26	27					
Ans.	ABC	ACD	B					